Power Systems Analysis

PRENTICE-HALL SERIES IN ELECTRICAL
AND COMPUTER ENGINEERING

Leon O. Chua, *Series editor*

Power Systems Analysis

ARTHUR R. BERGEN

*Department of Electrical Engineering
and Computer Sciences*

University of California, Berkeley

PRENTICE-HALL, INC., Englewood Cliffs, New Jersey 07632

Library of Congress Cataloging in Publication Data

Bergen, Arthur R.
  Power systems analysis.

  Bibliography:
  Includes index.
  1. Electric power systems. I. Title
TK1001.B44   1986   621.31   85-9300
ISBN 0-13-687864-4

Editorial/production supervision and interior design: Ellen Denning
Cover design: Photo Plus Art
Manufacturing buyer: Gordon Osbourne
Cover: Courtesy of EPRI, copyright © 1970, EPRI.
On-Line Stability Analysis Study RP90-1. Reprinted
with permission.

Printed in the United States of America

10   9   8   7   6   5   4

ISBN 0-13-687864-4 025

PRENTICE-HALL INTERNATIONAL (UK) LIMITED, *London*
PRENTICE-HALL OF AUSTRALIA PTY. LIMITED, *Sydney*
PRENTICE-HALL CANADA INC., *Toronto*
PRENTICE-HALL HISPANOAMERICANA, S.A., *Mexico*
PRENTICE-HALL OF INDIA PRIVATE LIMITED, *New Delhi*
PRENTICE-HALL OF JAPAN, INC., *Tokyo*
PRENTICE-HALL OF SOUTHEAST ASIA PTE. LTD., *Singapore*
EDITORA PRENTICE-HALL DO BRASIL, LTDA., *Rio de Janeiro*
WHITEHALL BOOKS LIMITED, *Wellington, New Zealand*

To Jane

Contents

APPENDICES   **494**

SELECTED BIBLIOGRAPHY   **521**

INDEX   **523**

Preface

This text was developed from a set of notes used for a two-quarter sequence at the senior level. In the first quarter we covered most of the material through Chapter 9 on stability determination using the equal-area criterion. This part of the course was designed as a "core course" introduction to power systems intended for both power and non-power majors. The arrangement and choice of material in these first nine chapters was influenced by the objective of making this a stand-alone core course for non-majors. We wanted to touch on the following questions: What are the elements constituting a power system, what are the concerns of power system planners and operators, what kinds of mathematical models and methods of analysis are used, and how does the analysis of power systems differ from that of circuits. Although it was necessary to lay the groundwork, we also wanted to cover some topics that would illustrate the challenge, interest, and breadth offered by the power field.

It seemed to us that this was also a good arrangement of material for our power systems majors. For example, covering transient stability first, in Chapter 9, provided some motivation to study more complete generator modeling in Chapter 10. The inclusion of Chapter 8, on generator modeling based on rotating fields, also seemed worthwhile, even for students who had previously taken an energy conversion or machinery course. Students seemed to find the review useful before tackling the more difficult material in Chapter 10. It is recognized, of course, that the two viewpoints complement each other nicely and provide insights unobtainable solely with one viewpoint.

The course has only one prerequisite: a circuits course taken by most of our electrical engineering undergraduates in the junior year. It includes network theorems, two-port network descriptions, and a brief introduction to the Laplace transform, network (transfer) functions, and feedback. We try to use this preparation and tie the course to the notions and notations of circuit and system theory. The students' preparation in feedback theory is usually insufficient for our purposes and we need to cover the root-locus method (Appendix 4) in classroom lectures.

The material in the text has expanded to the point where it seems clear that two semesters rather than two quarters would now be required. If this much time is not available, however, there are alternative pathways through the text. Chapters 10, 11, and 12 might be left out without seriously affecting the coherency of the remaining material; the generator sequence reactances needed in Chapter 13 might be developed without detailed elaboration. For the student with an energy conversion background, Chapters 1, 3, 4, and the first part of Chapter 5, might be excluded. Chapters 3 and 4, on the modeling of transmission lines, could reasonably be left out in a course emphasizing the systems aspects. If pressed for time, much of the material at the ends of Chapters 6, 7, 9, 11, 12, and 13 could be left out without affecting continuity.

There are many whose contributions I would like to acknowledge. Among these, I would particularly like to thank my colleagues Chris De Marco, George Gross, and Felix Wu for valuable advice and comments. To my students, who helped debug much of the material, I offer my appreciation. My sincere thanks to Bettye Fuller and Karen Kerschen for their efforts including, but not limited to, typing the manuscript.

ARTHUR R. BERGEN

Background

chapter 1

1.0 INTRODUCTION

In this chapter we give a simplified description of a power system. The system consists of power sources, called *generating plants* (or *generators*), power end uses, called *loads*, and a transmission and distribution network that connects them. Most commonly the generating plants convert energy from fossil or nuclear fuels, or from falling water, into electrical energy.

1.1 ELECTRIC ENERGY

Electricity is only one of the many forms of energy used in industry, homes, businesses, and transportation. Figure 1.1 is a useful summary of energy sources and their transition into end uses for the United States in 1982. The basic energy sources are shown on the left. About one-third of the resources are used to produce electricity. Of this only about 30% is actually converted into useful power; almost 70% is lost in waste heat. Considering only the useful (net) energy consumption, electricity provides only about 13% of the total; excluding energy used for transportation, the percentage is still only about 20%. Nevertheless, this represents an enormous amount of high-grade energy, whose use is flexible, clean (particularly at the point of use), convenient, and in some cases, irreplaceable by any other energy source.

(Quadrillion Btu)

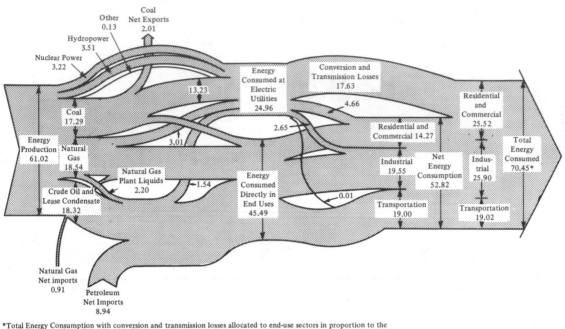

Figure 1.1 Energy flow diagram. [From *Annual Energy Review*, DOE/EIA-3084(82).]

*Total Energy Consumption with conversion and transmission losses allocated to end-use sectors in proportion to the
sectors' use of electricity.
Sum of components does not equal total due to independent rounding; the use of preliminary conversion factors; and the
exclusion of changes in stocks, miscellaneous supply and disposition, and unaccounted for quantities.

Figure 1.2 provides more detail as to the major sources of electrical energy and
some trends in their relative importance. It can be seen that most of the production has
been in so-called conventional steam plants. "Conventional steam" refers to steam
generation by burning coal, petroleum, or gas. In 1982 approximately 2242 billion
kilowatthours of electricity were produced. Of this, coal accounted for approximately
53%, petroleum 6%, natural gas 14% (totaling 73% for conventional steam), hydropower 14%, nuclear power 13%, and others, including gas turbines, about 1%. Note
that nuclear and geothermal power plants also generate steam but not by burning fossil
fuels.

The units used in Fig. 1.1 are quadrillion Btu (10^{15} or "quads"), while those in
Fig. 1.2 are in billion kilowatthours (or 10^9 watthours or gigawatthours). In attempting
to align the figures we can use the conversion factor 1 watt = 3.413 Btu/hr.

Turning to the growth in electricity production, we see in Fig. 1.2 an almost
exponential growth rate until about 1973. Until that time electricity use doubled every
10 years or so. Subsequently the growth rate dropped, reflecting the general slowdown
of worldwide economic growth precipitated in large part by the oil crisis of 1973, and
an increasing awareness of the cost-effectiveness of conservation.

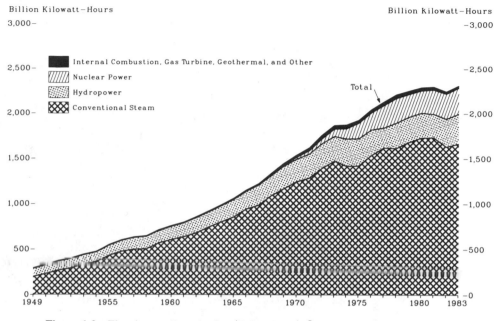

Figure 1.2 Electric energy production (Unites States). [From *Annual Energy Review*, DOE/EIA-0384(82).]

Figure 1.3 shows the growth in installed generating capacity in the United States. In 1982, of the total installed generating capacity of approximately 650 million kilowatts, some 69% was conventional (fossil fuel) steam, 12% was hydropower, 10% was nuclear, 8% was gas turbine, and others about 1%. Comparing these with the production figures given earlier, we see great differences in the rates of utilization of the various sources. Nuclear power has the highest rate, gas turbines the lowest. We will discuss the reasons in a moment.

First it is interesting to calculate an overall utilization factor for 1982. Suppose that it had been possible to utilize the 650 million kilowatt capacity full time. Then the plants would have produced $650 \times 10^9 \times 8760 = 5694 \times 10^{12}$ watthours in 1982. They actually produced 2242×10^{12} watthours. Thus the annual capacity factor or load factor was $2242/5694 = 0.39$ or 39%. We note that before 1973, typical load factors were approximately 60% and we may expect a return to those levels. Why isn't the figure even higher?

There are two main reasons. The first is that generating units are not always available for service. There is "downtime" because of maintenance and other "scheduled outages"; there are also "forced outages" because of equipment failures. The "availability" of fossil-fuel steam turbine units ranges from about 80% to about 92%.

The second reason concerns the load. While there must be enough generating capacity available to meet the requirements of the peak-load demand, the load is

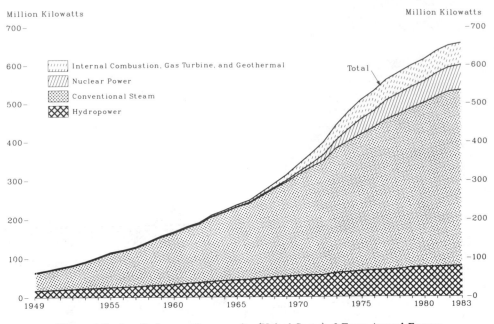

Figure 1.3 Installed generating capacity (United States). [From *Annual Energy Review*, DOE/EIA-0384(82).]

variable, with daily, weekly, and seasonal variations, and thus has a lower average value. The daily variations are roughly cyclic with a minimum value (the *base load*) typically less than one-half of the peak value. A typical daily load curve for a utility is shown in Fig. 1.4. The (weekly) capacity factor for this particular utility is seen to be approximately 60%.

In meeting the varying load requirements, economic considerations make it desirable to utilize fully plants with low (incremental) fuel costs while avoiding the use of plants with high fuel costs. This, in part, explains the use of nuclear plants for base-load service and gas turbines only for peaking-power service; the different rates of utilization of these sources were noted earlier.

Finally, it is interesting to reduce the enormous numbers describing production and capacity to human terms. In 1982 the U.S. population was approximately 230 million. Thus there was an installed capacity of approximately 650/230 = 2.8 kW per

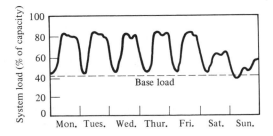

Figure 1.4 Hourly output (typical week).

person. Using the figure 0.40 for the capacity factor, this translates into an average use of energy at the rate 1.1 kW per person. The latter figure is easy to remember and gives an appreciation for the rate of consumption of electricity in this country.

We consider next some typical power plant sources of energy: fossil-fuel steam plants, nuclear plants, and hydroelectric plants.

1.2 FOSSIL-FUEL PLANT

Here coal, oil, or natural gas is burned in a furnace, the combustion products heat water, converting it to steam, and the steam drives a turbine which is mechanically coupled to an electric generator. A schematic diagram showing a typical coal-fired plant is shown in Fig. 1.5. In very brief outline the operation of the plant is as follows. Coal is taken from storage and fed to a pulverizer (or mill), mixed with preheated air, and blown into the furnace, where it is burned. The furnace contains a complex of tubes and drums called a *boiler* through which water is pumped, its temperature rising in the process until it evaporates into steam. The steam passes on to the turbine, while the combustion gases (flue gases) are passed through mechanical and electrostatic precipitators which remove upward of 99% of the solid particles (ash) before being released to the chimney or stack.

The unit just described, with pulverized coal, air, and water as an input and steam as a useful output, is variously called a steam generating unit, or furnace, or boiler. When the combustion process is under consideration the term *furnace* is

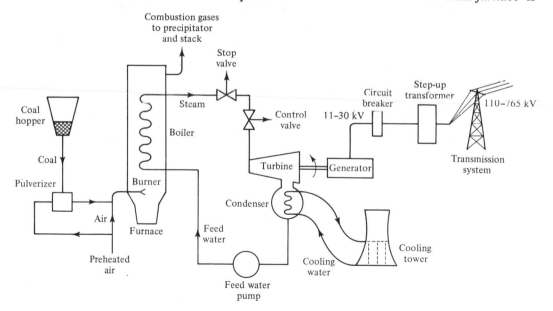

Figure 1.5   Coal-fired power station (schematic).

usually used, while the term *boiler* is more frequently used when the water-steam cycle is under consideration. The steam, at a typical pressure of 3500 psi and a temperature of 1050°F, is supplied through control and stop (shutoff) valves to the steam turbine. The control valve permits the output of the turbine-generator unit (or turbogenerator) to be varied by adjusting steam flow. The stop valve has a protective function; it is normally fully open but can be "tripped" shut to prevent overspeed of the turbine-generator unit if the electrical output drops suddenly (due to circuit-breaker action) and the control valve does not close.

Figure 1.5 suggests a "single-stage" turbine but in practice a more complex "multistage" arrangement is used to achieve relatively high thermal efficiencies. A representative arrangement is shown in Fig. 1.6. Here four turbines are mechanically coupled in tandem and the steam cycle is complex. In rough outline, high-pressure steam from the boiler (superheater) enters the high-pressure (HP) turbine. Upon leaving the HP turbine it is returned to a section of the boiler (reheater) and then directed to the intermediate-pressure (IP) turbine. Leaving the IP turbine, steam (at lower pressure and much expanded) is directed to the two low-pressure (LP) turbines. The exhaust steam from the LP turbines is cooled in a heat exchanger called a *condenser* and, as feedwater, is reheated (with steam extracted from the turbines) and pumped back to the boiler.

Finally we get to the electric generator itself. The turbine turns the rotor of the electric generator in whose stator are embedded three (phase) windings. In the process mechanical power from the turbine drive is converted to three-phase alternating current at voltages in the range 11 to 30 kV line to line at a frequency of 60 Hz in the United States. The voltage is usually "stepped up" by transformers for efficient transmission to remote load centers.

A generator (also called an *alternator* or *synchronous generator*) is shown in longitudinal cross section in Fig. 1.7; the transverse cross section is approximately round. The rotor is called round or cylindrical or smooth. We note that steam-driven turbine-generators are usually two-pole or four-pole, turning at 3600 rpm or 1800 rpm, respectively, corresponding to 60 Hz. The high speeds are needed to achieve high steam turbine efficiencies. At these rotation rates, high centrifugal forces limit rotor diameters to about 3.5 ft for two-pole and 7 ft for four-pole machines.

The average power ratings of the turbine-generator units we have been describing have been increasing, since the 1960s, from about 300 MW to about 600 MW, with maximum sizes up to about 1300 MW. Increased ratings are accompanied by increased rotor and stator size, and with rotor diameters limited by centrifugal forces, the rotor lengths have been increasing. Thus, in the larger sizes, the rotor lengths may be five to six times the diameters. These slender rotors resonate at critical speeds below their rated speeds and care is required in operation to avoid sustained operation at these speeds.

We will consider next the overall efficiency available from a coal-fired power plant. As an example of the best efficiencies available, we can use the unit at the Paradise power plant of the Tennessee Valley Authority, with a capacity of 1150 MW. The net efficiency determined experimentally is about 39% (i.e., 39% of the chemical

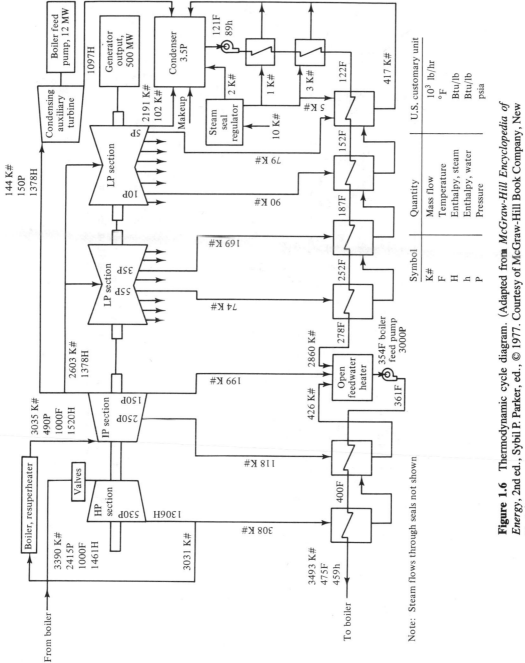

Figure 1.6 Thermodynamic cycle diagram. (Adapted from *McGraw-Hill Encyclopedia of Energy*, 2nd ed., Sybil P. Parker, ed., © 1977. Courtesy of McGraw-Hill Book Company, New York.)

Note: Steam flows through seals not shown

Symbol	Quantity	U.S. customary unit
K#	Mass flow	10^3 lb/hr
F	Temperature	°F
H	Enthalpy, steam	Btu/lb
h	Enthalpy, water	Btu/lb
P	Pressure	psia

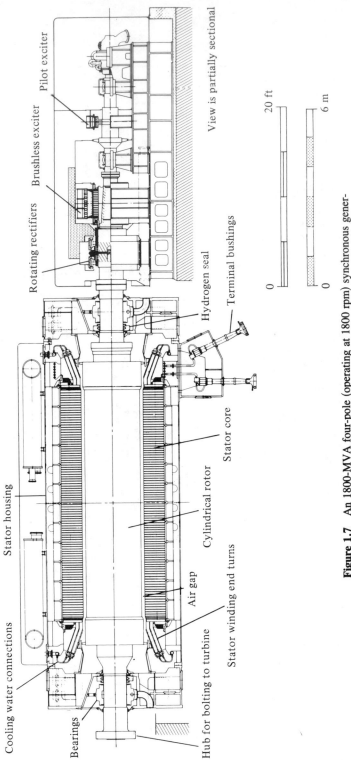

Figure 1.7 An 1800-MVA four-pole (operating at 1800 rpm) synchronous generator; it has a water-cooled stator and rotor windings, and brushless excitation equipment (Courtesy Utility Power Corporation.)

Pilot exciter

Brushless exciter

Rotating rectifiers

View is partially sectional

Hydrogen seal

Terminal bushings

Stator core

Stator housing

Cylindrical rotor

Air gap

Cooling water connections

Stator winding end turns

Bearings

Hub for bolting to turbine

20 ft

6 m

0

0

energy in the coal is converted into electrical energy). This figure compares favorably with the industry average of approximately 30%. Although improved technology will undoubtedly raise the efficiency beyond 39%, there are inherent theoretical limitations (based on thermodynamic considerations) in the conversion process in the turbine. An upper bound is found by considering the maximum efficiency as given by the Carnot cycle efficiency $(T_1-T_2)/T_1$, where T_1 is the temperature in degrees Kelvin of the heat source (superheated steam in our case) available to an ideal heat engine (at constant temperature), and T_2 is the corresponding temperature of the heat sink (the temperature of the cooling water in the condensor in our case). For the Paradise plant these temperatures are approximately 812°K (1003°F) and 311°K (100°F), respectively. Thus the Carnot efficiency is about 62%. In fact, the heat cannot be supplied at 1003°F to all the turbine blades of even just the high-pressure turbine, and for this and other reasons the theoretical efficiency drops to about 53%. Taking practical imperfections into account, the actual turbine efficiency drops to about 89% of the theoretical value, or approximately 47%. Considering, then, typical boiler efficiencies of 88% (for conversion of chemical energy to heat) and generator efficiencies of 99%, we get an overall fractional efficiency of $\eta = 0.47 \times 0.88 \times 0.99 = 0.41$, which, considering the crudeness of the calculations, is a reasonable check of the experimental figure of 0.39. Note the fact that 61% of the fossil-fuel chemical energy is converted to waste heat! The best hope for increased thermodynamic efficiency is to raise the steam temperature, but there are some practical difficulties here.

When we consider the tremendous quantities of waste heat generated in the production of electricity, it is apparent that there is an opportunity to save fuel by the simultaneous generation of electricity and steam (or hot water) for industrial use or space heating. Now called *cogeneration*, such systems have long been common here and abroad. Currently, there is renewed interest in them because the overall energy efficiencies are claimed to be as high as 60 to 65%.

There are different types of cogeneration systems. In a *topping* system electricity is delivered at the "head end" of the cycle and the heat exhaust is a useful by-product. In a *bottoming* system the head end of the cycle is an industrial process in which waste heat is a by-product; this waste heat is then used to drive a turbogenerator. There are variations of these basic schemes as well.

Finally, we note some problems associated with the use of coal-fired power plants. Mining and transportation of coal present safety hazards and other social costs. Coal-fired plants share environmental problems with some other types of fossil-fuel plants: these include "acid rain" and the "greenhouse" effect.

1.3 NUCLEAR POWER PLANT

Controlled nuclear fission is the source of energy in a nuclear power plant. In the process of fission, heat is generated which is transferred to a coolant flowing through the reactor. Water is the most common coolant, but gases, organic compounds, liquid metals, and molten salts have also been used.

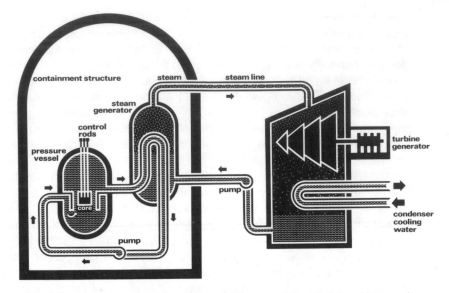

Figure 1.8   Pressurized-water reactor. (Courtesy of Atomic Industrial Forum.)

In this country the two most common types, collectively called *light-water reactors* (to distinguish them from a type using "heavy water"), are the *boiling-water reactor* (*BWR*) and the *pressurized-water reactor* (PWR), and both use water as coolant.

In the BWR the water is allowed to boil in the reactor core; the steam is then directed to the turbine. In the PWR there may be two or more cycles or loops linked by heat exchangers but otherwise isolated from each other. In all but the last stage (the secondary of which carries steam to the turbine) the water is pressurized to prevent the generation of steam. A schematic diagram of a two-loop system with a single heat exchanger (called the steam generator) is shown in Fig. 1.8. The pressurizer is not shown.

Although it might appear that the only difference between a nuclear plant and a fossil-fuel plant is the way the steam is produced (i.e., by nuclear reactor/steam generator rather than furnace/boiler), there are some other differences. For example, nuclear steam generators are presently limited in their temperature output to about 600°F (compared with about 1000°F for a fossil-fuel plant). This has a negative impact on thermal efficiency (30% instead of 40%) and in steam conditions in the turbines. There are of course major differences in the fuel cycle (supply and disposal) and in requirements for plant safety.

1.4 HYDROELECTRIC POWER PLANT

This is an important source of power in the United States, accounting for approximately 13% of the installed generating capacity and energy production. Hydroplants are classified as *high head* (over 100 ft) or *low head*. The "head" refers to the difference of elevation between the upper reservoir above the turbine and the tail race or discharge point just below the turbine. For very high heads (600 to 6000 ft) Pelton wheels or impulse turbines are used; these consist of a bucket wheel rotor with one or more nozzles directed at the periphery. At medium-high heads (120 to 1600 ft) Francis turbines are used. These turn on a vertical axis with the water entering through a spiral casing and then inward through adjustable "gates" (i.e., valves) to a "runner" (i.e., turbine wheel) with fixed blades. The low-head Kaplan type turbine is similar but with adjustable blades in the runner. The efficiency of these turbines is fairly high, in the neighborhood of 88 to 93% when operating at their most efficient points.

Figure 1.9 shows a cross section through a hydroelectric plant with a Kaplan turbine. Some of the generator features should be noted. The rotation axis is vertical. With a speed of 144 rpm we would have 50 poles to generate 60 Hz. These poles are mounted on the periphery of a spoked wheel called a *spider*. There is space between

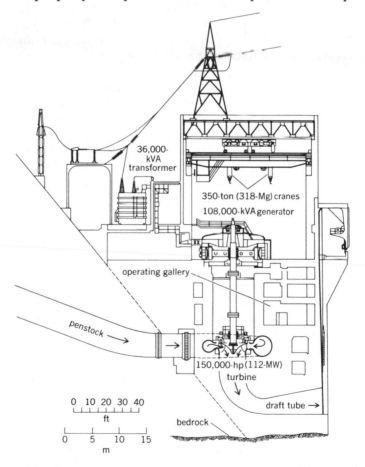

Figure 1.9 Cross section through powerhouse and dam. Grand Coulee plant (Kaplan turbine). (From *Mc-Graw-Hill Encyclopedia of Energy*. 2nd ed., Sybil P. Parker, ed., © 1977. Courtesy of McGraw-Hill Book Company, New York.)

the poles to accommodate the pole windings. The rotor is called by the descriptive title *salient pole*.

We note a highly desirable feature of hydropower plants: the speed with which they may be started up, brought up to speed, connected to the power network, and "loaded" up. This can be done in under 5 minutes, in contrast to many hours in the case of thermal plants; the job is also much simpler and adaptable to remote control. Thus hydropower is well suited for turning on and off at a dispatcher's command to meet changing power needs. We note that when water is in short supply, it is desirable to use the limited available potential energy sparingly, for periods of short duration, to meet the peak-load demands. When water is plentiful with the excess flowing over the spillway of the dam, base-loading use is indicated.

The desirable feature of hydropower in effectively meeting peak demand may be obtained in locations without suitable water flows by using *pumped storage*. In this scheme water is pumped from a lower reservoir to a higher one during off-peak times (generally at night) and the water is allowed to flow downhill in the conventional hydroelectric mode during times of peak demand. Off peak, the generators, operating as motors, drive the turbines in reverse in a pumping mode. The overall efficiency is only about 65 to 70%, but the economics are frequently favorable when one considers the economics of the overall system, including the thermal units. Consider the following. It is not practical to shut down the largest and most efficient thermal units at night, so they are kept "on-line" supplying relatively small (light) loads. When operating in this mode the pumping power may be supplied at low incremental costs. On the other hand, at the time of peak demand the pumped storage scheme provides power that would otherwise have to be supplied by less efficient (older) plants. In a sense, from the point of view of the thermal part of the system, the pumped storage scheme "shaves the peaks" and "fills the troughs" of the daily load-demand curve.

1.5 OTHER ENERGY SOURCES

There are additional energy sources currently used or under development. These include the following: gas turbines, biomass, geothermal, photovoltaic, solar thermal, wind power, wastes as fuel, tidal power, ocean thermal energy conversion (OTEC), magnetohydrodynamic generation, generators driven by diesel engines, fuel cells, wave power, nuclear fusion, and the breeder reactor.

It is in the nature of some of these new sources, for example, photovoltaic and wind power, that they are of small size and widely dispersed geographically. To obtain the benefits of backup power as well as full utilization of the locally generated power, and to eliminate the need for expensive local storage schemes, these small sources may be integrated into the utility network. This poses problems for the utilities. In the past, while the load has been geographically dispersed and somewhat random in its demand, the generation has been in large plants fully under utility control. Now this new generation seems to share a property of the load; it is geographically dispersed and somewhat random in its supply. An additional feature to consider is that these smaller distributed sources may not be owned by the power utility. In this case there is need for special procedures for decentralized operation, maintenance of safety, and pricing of the energy supplies to the utility.

1.6 TRANSMISSION AND DISTRIBUTION SYSTEMS

The sources of central station electric power described in the preceding sections are usually interconnected by a transmission system or network which distributes the power to the various load points or load centers. A small portion of a transmission system which suggests the interconnections is shown on a one-line diagram in Fig. 1.10. Various symbols for generators, transformers, circuit breakers, loads, and the points of connection (nodes) called *buses* are identified in the figure. Figure 1.11 shows a portion of a 300-bus system. The section shown has 69 buses and 15 generators; the circuit breakers and loads are not shown. A different transformer

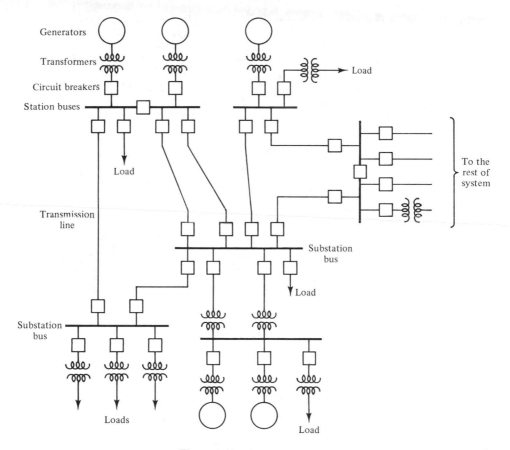

Figure 1.10   One-line diagram.

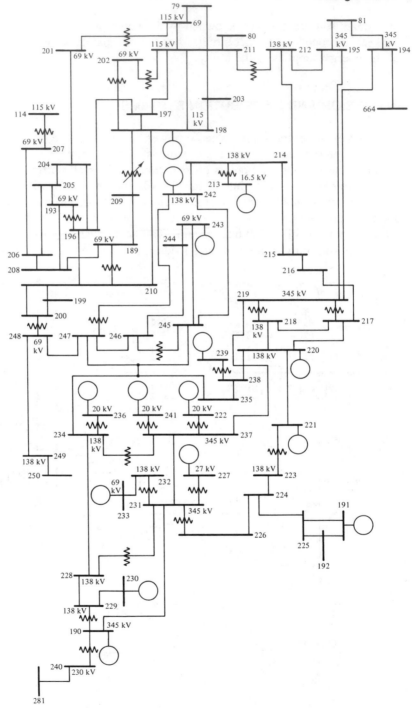

Figure 1.11   One-line diagram of 69-bus system. (Courtesy of EPRI © 1970, *On-Line Stability Analysis Study* RP90-1.)

symbol is used than in Fig. 1.10. On a very different scale Fig. 1.12 suggests the network of major transmission lines spanning the United States.

The generator voltages are in the range 11 to 30 kV; higher generator voltages are difficult to obtain because of insulation problems in the narrow confines of the generator stator. Transformers are then used to "step up" the voltages to the range 110 to 765 kV. In California the backbone of the transmission system is composed mainly of 500-kV, 345-kV, and 230-kV three-phase lines. The voltages refer to voltages from line to line.

One reason for using high transmission-line voltages is to improve energy transmission efficiency. Basically, transmission of a given amount of power (at specified power factor) requires a fixed product of voltage and line current. Thus the higher the voltage, the lower the current can be. Lower line currents are associated with lower resistive losses (I^2R) in the line. Another reason for higher voltages is the enhancement of stability. This will be discussed in later chapters.

A comment is in order about the loads shown in Fig. 1.10. The loads referred to here represent "bulk" loads, such as the distribution system of a town, city, or large industrial plant. Such distribution systems provide power at various voltage levels. Large industrial consumers or railroads might accept power directly at voltage levels of 23 to 138 kV; they would then step down the voltages further. Smaller industrial or commercial consumers typically accept power at voltage levels of 4.16 to 34.5 kV. Residential consumers normally receive single-phase power from pole-mounted distribution transformers at voltage levels of 120/240 V.

Although the transmission-distribution system is actually one interconnected system, it is convenient to separate out the transmission system as we have done in Fig. 1.10. A similar diagram for the distribution system can be drawn with bulk substations replacing the generators as the sources of power and with lower-level loads replacing the bulk power loads shown in Fig. 1.10. Henceforth we will be discussing power systems at the transmission system level; most of the techniques we develop are directly applicable to the distribution system as well.

The transmission systems of most electric utilities in the contiguous United States are interconnected and the systems operate as members of power pools. For example, the utilities in the western United States and Canada belong to the Western Systems Coordinating Council (WSCC). In addition, the power pools are themselves interconnected in a system known as the North American Power Systems Interconnection. The primary objective of the interconnection is improved service reliability; a loss of generation in one area can be made up by utilizing spare capacity in another area. Improved system economy is also achieved along the lines of the earlier discussion concerning advantages of interconnecting hydro with thermal sources.

The end result, then, is a very large system of enormous complexity. The planning of such a system and its efficient and secure operation in real time is a challenging problem.

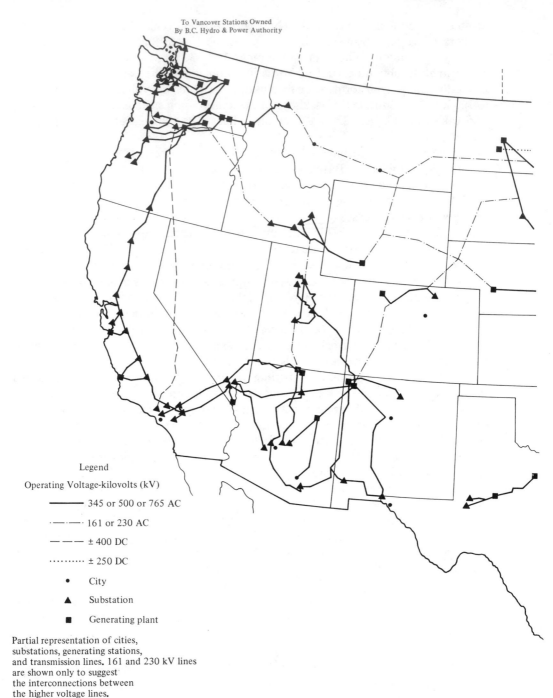

To Vancover Stations Owned
By B.C. Hydro & Power Authority

Legend

Operating Voltage-kilovolts (kV)

———— 345 or 500 or 765 AC

—·—·— 161 or 230 AC

— — — ± 400 DC

·········· ± 250 DC

•   City

▲   Substation

■   Generating plant

Partial representation of cities,
substations, generating stations,
and transmission lines. 161 and 230 kV lines
are shown only to suggest
the interconnections between
the higher voltage lines.
Line locations are approximate.

Figure 1.12 Major transmission in service or authorized. (Courtesy of the Department of Energy.)

To Beauharnois Station
owned by Quebec
Hydroelectric Commission

Adapted from
Department of Energy map

Energy
Information Administration

DOE/EIA-0165(78)

Major extra high voltage
transmission lines

December 31, 1978

0 25 50 100   200   300   400 Miles

0 50 100 200   300   400   500   600   Kilometers

Figure 1.12   (*cont.*)

Basic Principles

chapter 2

2.0 INTRODUCTION

In the steady state, most power system voltages and currents are (at least approximately) sinusoidal functions of time all with the same frequency. We are therefore very interested in sinusoidal steady-state analysis using phasors, impedances, admittances, and complex power. As we will see in later chapters, some of these sinusoidal steady-state relations extend to an important class of transients as well.

We assume that the reader has some familiarity with the basic ideas from circuit theory. In this chapter we first review complex power, then introduce the theorem of conservation of complex power. Next we consider three-phase circuits, balanced three-phase circuits, and per phase analysis. Using per phase analysis we finally consider two cases of the transmission of complex power using a short-transmission-line model.

2.1 COMPLEX POWER SUPPLIED TO A ONE-PORT

We start by reviewing some results from elementary circuit theory. Using associated reference directions, as shown in Fig. 2.1, the instantaneous power supplied to a one-port N is $p(t) = v(t)i(t)$. We now assume that the voltage and current are both sinusoids of angular frequency ω,

$$v(t) = V_{max} \cos (\omega t + \theta_V) \tag{2.1}$$

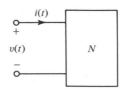

Figure 2.1   Power into a one-port.

$$i(t) = I_{\max} \cos{(\omega t + \theta_I)} \tag{2.2}$$

$V_{\max}$ and $I_{\max}$ are real numbers called the *amplitudes,* and θ_V and θ_I are called the *phases* of the voltage and current, respectively. Sometimes it is convenient to use $\underline{/V}$ for θ_V and $\underline{/I}$ for θ_I. We now calculate $p(t)$ using identities from trigonometry.

$$\begin{aligned} p(t) &= V_{\max}I_{\max} \cos{(\omega t + \theta_V)} \cos{(\omega t + \theta_I)} \\ &= \tfrac{1}{2} V_{\max}I_{\max}[\cos{(\theta_V - \theta_I)} + \cos{(2\omega t + \theta_V + \theta_I)}] \end{aligned} \tag{2.3}$$

We see that $p(t)$ is composed of two parts; there is a constant (average) component and a sinusoidal component of frequency 2ω. An example illustrating the relationship between p, v, and i is shown in Fig. 2.2. We note that $p(t)$ is zero whenever either $v(t)$ or $i(t)$ is zero. Thus, unless v and i are in phase (or 180° out of phase) there will be twice as many zero crossings of $p(t)$ as of $v(t)$ or $i(t)$. The component in (2.3) of frequency 2ω is consistent with this observed behavior. In the particular case shown in Fig. 2.2 the average power is positive, but in general, the average $p(t)$ can be positive, negative, or zero. We next define the power-factor angle:

$$\phi \triangleq \theta_V - \theta_I \tag{2.4}$$

and P, the average power over one period, $T = 2\pi/\omega$. From (2.3) we get

$$\begin{aligned} P &= \frac{1}{T} \int_0^T p(t) \, dt \\ &= \tfrac{1}{2} V_{\max}I_{\max} \cos{\phi} \end{aligned} \tag{2.5}$$

In applications one is normally concerned with P rather than $p(t)$.

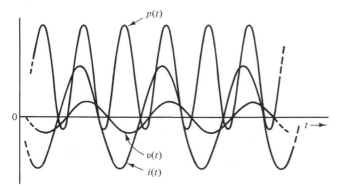

Figure 2.2   Instantaneous power.

It is frequently convenient to calculate P using the phasor representations of $v(t)$ and $i(t)$. At this point we depart from the usual choice of phasors used in circuit theory and define an "effective" phasor V such that

$$v(t) = V_{max} \cos (\omega t + \theta_V) \Leftrightarrow V = \frac{V_{max}}{\sqrt{2}} e^{j\theta_V} \qquad (2.6)$$

Hence

$$v(t) = \text{Re } \sqrt{2} \, V e^{j\omega t} \qquad (2.7)$$

This differs from the circuit theory definition only by the factor $\sqrt{2}$. We note that $|V| = V_{max}/\sqrt{2}$ is the root-mean-square (rms) value of $v(t)$ and hence the value read by the usual ac voltmeter. Suppose that we calculate the average power dissipated in a resistor with resistance R connected to a sinusoidal source of effective voltage V.

$$P = \frac{1}{T} \int_0^T p(t) \, dt = \frac{1}{T} \int_0^T \frac{v^2(t)}{R} \, dt = \frac{|V|^2}{R}$$

This is the same formula we get with dc. Thus if the effective voltage is 120 V, we get the same average heat energy out of the resistor as if the voltage were 120 V dc. This motivates the terminology "effective." A similar discussion holds for effective current through R. Thus we get the simple result

$$P = |I|^2 R = \frac{|V|^2}{R} = |V| \, |I|$$

This simplification extends to more general cases. Thus substituting (effective) phasors in (2.5), we get

$$\begin{aligned} P &= \tfrac{1}{2} V_{max} I_{max} \cos \phi = |V| \, |I| \cos \phi \\ &= \text{Re } |V| \, e^{j\underline{/V}} \, |I| \, e^{-j\underline{/I}} = \text{Re } VI^* \end{aligned} \qquad (2.8)$$

where * designates the complex conjugate. As may be seen, the factor $\tfrac{1}{2}$, which would otherwise appear in (2.8), is eliminated. Since the variable, power, is so extensively used, the simplification is worthwhile.

The quantity $\cos \phi$ that occurs in (2.8) is known as the *power factor* (PF).

$$\text{PF} \triangleq \cos \phi \qquad (2.9)$$

Unlike the case of dc, the product of voltage and current does not give the power. To find the power we need to multiply the product of voltage and current by the power factor.

Power engineers frequently use the terminology *lagging* or *leading* power factor, as in the following description. "A load draws 200 kW at a power factor of 0.707 lagging." The term "lagging" means that I lags V. Thus we deduce that $\phi = 45°$.

In (2.8) we see that $P = \text{Re } VI^*$. This raises the question of the significance of $\text{Im } VI^*$ and of VI^* itself. In fact, these quantities are just as important as P, as we will attempt to show. First, let us define complex power S and reactive power Q:

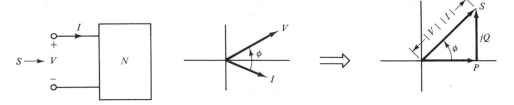

Figure 2.3   Complex power into N.

$$S \triangleq VI^*$$

$$Q \triangleq \text{Im } VI^* \qquad (2.10)$$

Thus

$$S = VI^* = |V|\,|I|\,e^{j\phi} = P + jQ \qquad (2.11)$$

In (2.11) S is described in both polar and rectangular forms and it is seen that $/S = \phi$. It may be helpful to summarize the results as shown in Fig. 2.3. Using associated reference directions for V and I, S is the complex power into N. If V leads I as shown, both P and Q (delivered to N) are positive.

What is the physical significance of Q? We may gain an appreciation for what Q means physically by considering the following cases. We start with the case where N is an inductor.

Example 2.1

For $Z = j\omega L$:

(a) Calculate Q.
(b) Calculate the instantaneous power into L.
(c) Compare.

Solution   (a) Using (2.11), we have

$$S = VI^* = ZII^* = Z\,|I|^2 = j\omega L\,|I|^2$$

$$Q = \text{Im } S = \omega L\,|I|^2$$

(b) Suppose that the current is given by $i(t) = \sqrt{2}\,|I|\,\cos(\omega t + \theta)$; then

$$v(t) = L\frac{di}{dt} = -\sqrt{2}\,\omega L\,|I|\,\sin(\omega t + \theta)$$

$$p(t) = v(t)i(t) = -2\omega L\,|I|^2\,\sin(\omega t + \theta)\cos(\omega t + \theta)$$

$$= -\omega L\,|I|^2\,\sin 2(\omega t + \theta)$$

(c) Comparing the results in parts (a) and (b), we find

$$p(t) = -Q\,\sin 2(\omega t + \theta)$$

Thus Q is the amplitude or maximum value of the instantaneous power into N.

In this example we see that although the average power P supplied to the inductor is zero, there is still an instantaneous power supplied (to sustain the changing energy in the magnetic field) whose maximum value is Q.

Exercise 1.   Repeat Example 2.1 in the case of a capacitor (i.e., $Z = 1/j\omega C$).

We consider next a more general example.

Example 2.2

Consider a network with a driving-point impedance Z.

(a) Find expressions for P and Q.

(b) Find $p(t)$ in terms of P and Q.

(c) Suppose that the network is a series RLC. Interpret the result in part (b).

Solution   (a) $S = VI^* = ZII^* = \text{Re } Z \,|I|^2 + j \text{ Im } Z \,|I|^2 = P + jQ$. Thus

$$P = \text{Re } Z \,|I|^2 = |Z|\,|I|^2 \cos \underline{/Z}$$

$$Q = \text{Im } Z \,|I|^2 = |Z|\,|I|^2 \sin \underline{/Z}$$

(b) It is convenient to choose $i(t) = \sqrt{2}\,|I| \cos \omega t$; then $v(t) = \sqrt{2}\,|Z|\,|I| \cos (\omega t + \underline{/Z})$. Using trigonometric identities, we have

$$p(t) = v(t)i(t) = |Z|\,|I|^2 \cos (\omega t + \underline{/Z}) \cos \omega t$$

$$= |Z|\,|I|^2(\cos \underline{/Z} + \cos (2\omega t + \underline{/Z}))$$

$$= |Z|\,|I|^2(\cos \underline{/Z} + \cos 2\omega t \cos \underline{/Z} - \sin 2\omega t \sin \underline{/Z})$$

$$= P(1 + \cos 2\omega t) - Q \sin 2\omega t$$

(c) In this case $Z = R + j\omega L + 1/j\omega C$. From part (a) we find that $P = R\,|I|^2$ and $Q = Q_L + Q_C$, where $Q_L \triangleq \omega L\,|I|^2$ and $Q \triangleq -(1/\omega C)\,|I|^2$ are the reactive powers into L and C, respectively. Thus we can write

$$p(t) = P(1 + \cos 2\omega t) - Q_L \sin 2\omega t - Q_C \sin 2\omega t$$

The first term is the instantaneous power into R (with average value P). The second and third terms are the instantaneous power into L and C, respectively. It is interesting to note that $Q = Q_L + Q_C = 0$ if $\omega^2 LC = 1$ (i.e., in the case where the inductive and capacitive reactances cancel).

Suppose that we have a more general RLC network. We find, as in Example 2.2(c), that the complex power supplied to the network at its terminals equals the sum of complex powers consumed by the individual elements of the network. An extension of this idea will be discussed in Section 2.2.

We turn next to the importance of Q. Power system planners and operators are interested in Q, as well as P, generation and flow. The reactive-power component of a customer's load as well as the average power must be supplied and there is a cost associated with this supply. From Fig. 2.3 we can see that for a given $|V|$ and P, $|I|$

increases with $|Q|$. Increased $|I|$ means increased heating losses in generators, transformers, and transmission lines, and also increased cost of equipment capable of supplying the higher currents. There is also a cost associated with the maintenance of voltage within acceptable limits at the terminals of transmission lines supplying relatively large $|Q|$. We will return to this matter later in the chapter.

In addition to its importance in system planning and operation, the variable Q is essential in power systems analysis. In circuits we deal primarily with the complex numbers V and I. In power systems, whose function is to deliver power at acceptable voltages, we deal instead with the complex numbers V and S.

Finally, we consider terminology and descriptive units (Table 2.1). It should be noted that, strictly speaking, the units in each case are watts. However, it reduces confusion and is extremely convenient to use the descriptive units to designate the quantity. For example, we speak of supplying a load of 10 kW or 15 MVA. By this we mean that $P = 10$ kW or $|S| = 15$ MVA. Note that the expression "A load draws 200 kW at a power factor of 0.707 lagging" completely specifies S, since we deduce that $P = 200$ kW and $\phi = 45°$; thus $S = 200 + j200$ kVA.

The term *load* is also used to indicate the connected device. For example, we speak of a motor load or a load $Z = 1.0 + j2.0$.

TABLE 2.1   TERMINOLOGY AND DESCRIPTIVE UNITS

Quantity	Terminology	Descriptive units
S	Complex power	Voltamperes: VA, kVA, MVA
$|S|$	Apparent power	Voltamperes: VA, kVA, MVA
P	Average or real or active power	Watts: W, kW, MW
Q	Reactive power	Voltamperes reactive: VAr, kVAr, MVAr

2.2 CONSERVATION OF COMPLEX POWER

In working with complex power we make extensive use of the theorem of conservation of complex power. A statement of the theorem follows.

Theorem of Conservation of Complex Power.   For a network supplied by independent sources all at the same frequency, the sum of the complex power supplied by the independent sources equals the sum of the complex power received by all the other branches of the network.

Implicit in the statement above is the assumption that all the voltages and currents are sinusoids.

For a single source with elements in series the proof of the theorem is immediate, by using Kirchhoff's voltage law (KVL). The same is true for elements in parallel by use of Kirchhoff's current law (KCL). In the general case the most direct proof uses

Tellegen's theorem and may be found in circuit theory textbooks. We emphasize that the theorem states that conservation applies to reactive power as well as active power.

In applying the theorem we frequently find it convenient to replace part of the network by an equivalent independent source. For example, in Fig. 2.4, assuming no coupling between N_1 and N_2 except through the terminals, as shown, we replace N_1 by a source. The source is either the voltage V_1 or the current I_1 at the terminal interface. Then applying the theorem to N_2, we get

$$S_1 + S_2 + S_3 = \sum_i S_i \tag{2.12}$$

where on the right side of (2.12) we sum all the complex powers delivered to the individual branches inside N_2.

This replacement of a network or branch by a source equal to either the voltage or current the network or branch supplies is physically very reasonable (under the conditions assumed). In circuit theory textbooks the replacement is the subject of the so-called *substitution theorem,* and it will be convenient to refer to it by that name.

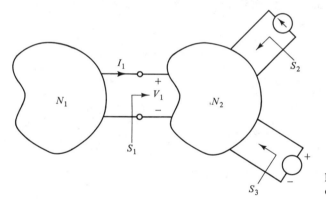

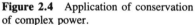

Figure 2.4   Application of conservation of complex power.

Example 2.3

For the circuit shown in Fig. E2.3, find S_2 in terms of S_1, C, and V.

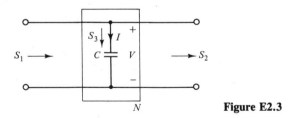

Figure E2.3

Solution   Using the theorem of conservation of complex power, we have

$$S_1 - S_2 = S_3$$

where

$$S_3 = VI^* = VY^*V^* = -j\omega C |V|^2$$

Thus

$$S_2 = S_1 + j\omega C |V|^2$$

$$P_2 = P_1$$

$$Q_2 = Q_1 + \omega C |V|^2$$

We note that $Q_2 > Q_1$; thus it is reasonable and convenient to consider C as a source of reactive power!

Example 2.4

Consider the circuit shown in Fig. E2.4. Assume that $|V_2| = |V_1|$ and show that $S_2 = -S_1^*$.

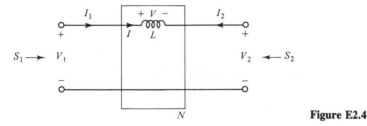

Figure E2.4

Solution

$$S_1 + S_2 = S_3 = VI^* = j\omega L |I|^2$$

Thus

$$P_1 + P_2 = 0$$

$$Q_1 + Q_2 = \omega L |I|^2$$

Now using $|V_2| = |V_1|$, we find

$$\left. \begin{array}{l} S_1 = V_1 I^* \\ S_2 = -V_2 I^* \end{array} \right\} \Rightarrow |S_1| = |S_2| \Rightarrow P_1^2 + Q_1^2 = P_2^2 + Q_2^2$$

But since $|P_2| = |P_1|$ we get $|Q_2| = |Q_1|$. As a consequence, $Q_1 = Q_2 = \frac{1}{2}\omega L |I|^2$. We have finally

$$\left. \begin{array}{l} P_2 = -P_1 \\ Q_2 = Q_1 \end{array} \right\} \Rightarrow S_2 = -S_1^*$$

Exercise 2.   What if the inductor L is replaced by a more general two-terminal reactive network? Is the result in Example 2.4 still true?

In the next example we show one form of one-line diagram showing generators, loads, and transmission lines and the points of interconnection called buses. Some of the complex powers are specified, others are to be determined. We use a double-subscript notation S_{ij} to indicate complex power leaving bus i entering the transmission line connected to bus j. S_{Gi} and S_{Di} are the complex powers from the ith generator and to the ith load, respectively.

Example 2.5

In Fig. E2.5, assume that $S_{ij} = -S_{ji}^*$. This is consistent with modeling the transmission lines as inductors, as in Example 2.4. Find S_{13}, S_{31}, S_{23}, S_{32}, and S_{G3}.

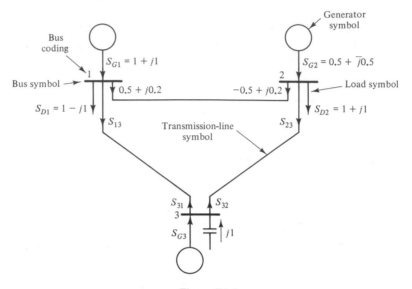

Figure E2.5

Solution   Considering each bus as a network with no internal branches, we get complex power balance for each bus. Thus the complex powers at each bus satisfy KCL. We get

$$S_{13} = (1 + j1) - (1 - j1) - (0.5 + j0.2) = -0.5 + j1.8$$

$$S_{31} = -S_{13}^* = 0.5 + j1.8$$

Similarly,

$$S_{23} = (0.5 + j0.5) - (1 + j1) - (-0.5 + j0.2) = -j0.7$$

$$S_{32} = -j0.7$$

Finally,

$$S_{G3} = (0.5 + j1.8) - j0.7 - j1 = 0.5 + j0.1$$

Example 2.6

Calculate the complex power S transferred from N_1 to N_2 assuming that the only coupling is through the terminals, as shown in Fig. E2.6(a).

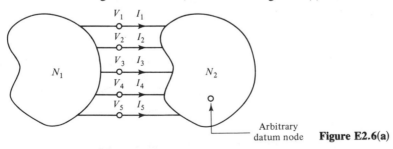

Arbitrary datum node   **Figure E2.6(a)**

Solution   Pick any datum node in N_2 and let the V_i be the voltages with respect to this node. In the spirit of the substitution theorem we replace N_1 by the sources in Fig. E2.6(b). Using the theorem of conservation of complex power, it is clear that

$$S = \sum_{i=1}^{5} V_i I_i^*$$

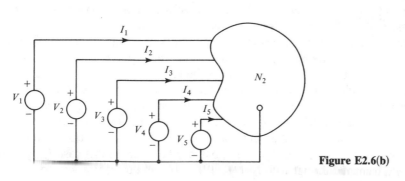

Figure E2.6(b)

Example 2.7

In Fig. E2.7, calculate an expression for the complex power transferred to N_2.

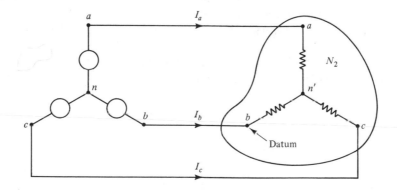

Figure E2.7

Solution   Any node in N_2 can be used as a datum node. If we pick n' as the datum node, then using the result in Example 2.6, we get

$$S = V_{an'}I_a^* + V_{bn'}I_b^* + V_{cn'}I_c^*$$

On the other hand, if we pick b as the datum node, we get

$$S = V_{ab}I_a^* + V_{bb}I_b^* + V_{cb}I_c^* = V_{ab}I_a^* + V_{cb}I_c^*$$

This latter result is the basis for the "two-wattmeter method" for reading three-phase power. The reader may check that the two expressions for S yield identical results.

2.3 BALANCED THREE-PHASE

Power is supplied by three-phase generators. It is then transferred and distributed in the form of three-phase power except at the lowest voltage levels of the distribution system, where single phase is used.

Figure E2.7 illustrates a three-phase (3ϕ) circuit. The circuit is a *balanced* 3ϕ circuit if the impedances are equal and the three voltage source phasors differ only in their angles, with 120° angle differences between any pair.

There are a number of reasons for the popularity of (balanced) three-phase over single-phase supply. Among these we note that three-phase equipment is more efficient and makes more effective use of material (conductors and/or iron), and hence costs less, than single-phase equipment of the same total power-handling capability. As an illustration we can consider the saving in conductor material in a 3ϕ transmission line as compared with an equivalent set of three 1ϕ transmission lines. Three 1ϕ circuits are shown in Fig. 2.5. Six conductors are required to carry the current. Suppose now that $V_{aa'}$, $V_{bb'}$, and $V_{cc'}$ form a balanced 3ϕ set. In this case I_a, I_b, and I_c will also be a balanced 3ϕ set with $I_a + I_b + I_c = 0$, and there is a way to eliminate the return wires. We connect points a', b', and c' together (to a "neutral" point n); the sources are said to be connected in *wye*. We also connect the impedances in wye. The return wires are now in parallel carrying a current $I_a + I_b + I_c = 0$. Since they are carrying no current, they are redundant and may be removed. We now have a 3ϕ transmission line with the same current-handling capacity but with only half as many conductors as in the original configuration of 1ϕ circuits. The saving in I^2R line losses should also be noted. An additional advantage of 3ϕ over 1ϕ supply will be noted in Section 2.5.

Let us consider next a more general balanced three-phase system. For simplicity we will consider a model consisting only of impedances and ideal sources. A balanced

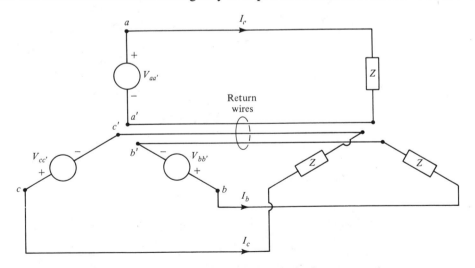

Figure 2.5 Balanced three-phase and single-phase compared.

three-phase system is made up of balanced three-phase sources of the same phase sequence and a network with three-phase symmetry. A three-phase source is balanced if it is composed of three individual sources in a wye or delta configuration and the source voltages (or currents) differ only in their angles with 120° angle differences between any pair. For example in Fig. 2.6, case I, line-neutral voltages $E_{an} = 1 \underline{/0°}$, $E_{bn} = 1 \underline{/-120°}$, and $E_{cn} = 1 \underline{/120°}$ are a balanced three-phase source. In case II, line-line voltages $E_{ab} = 100 \underline{/10°}$, $E_{bc} = 100 \underline{/-110°}$, and $E_{ca} = 100 \underline{/130°}$ are also a balanced three-phase source.

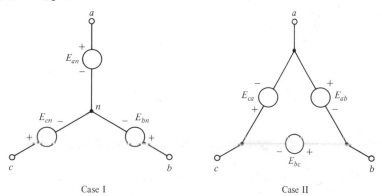

Case I   Case II

Figure 2.6 Wye and delta sources.

Note that if one observes the instantaneous line-neutral voltages associated with the phasors, then for the choice of phasors in case I the voltages reach their maximum values in the order *abc*; in case II the line-line voltages $e_{ab}(t)$, $e_{bc}(t)$, and $e_{ca}(t)$ also reach their maximum values in the order *abc*. In either case, we say the phase sequence is an *abc* or *positive sequence*.

Balanced voltages may also have a phase sequence *acb* or *negative sequence*. For example, in case I, the voltages may be $E_{an} = 1 \underline{/0°}$, $E_{bn} = 1 \underline{/120°}$, and $E_{cn} = 1 \underline{/-120°}$. It is important that all the sources connected to the same network have the *same* phase sequence, and special care is taken when connecting a new source to an existing network to match the phase sequence.

Given an actual system, whether the sequence is *abc* or *acb* depends on nothing more than how one labels the wires. To avoid confusion, unless stated to the contrary, we will always assume a labeling such that under the balanced conditions we are describing the sequence is *abc* or positive sequence.

We point out that the connection of sources shown in Fig. 2.5, case II, is clearly impractical unless, as in the balanced case, $E_{ab} + E_{bc} + E_{ca} = 0$. In the case shown, with $E_{ab} + E_{bc} + E_{ca} = 0$, although the circulating current is indeterminate, we will assume that it is zero.

A symmetric three-phase network is shown in Fig. 2.7. We have labeled the phases *a*, *b*, and *c*, but because of the symmetry the labeling is purely arbitrary. The grouping of elements in phases *a*, *b*, and *c* into triples of equal impedances is the important feature.

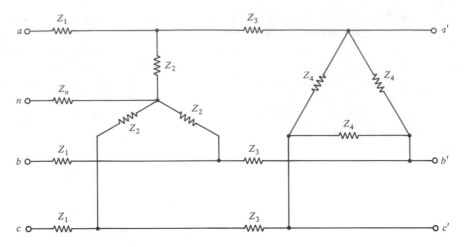

Figure 2.7 Symmetric three-phase network.

From the point of view of analysis, balanced three-phase has a desirable feature; it is possible to carry out the analysis using a much simpler "per phase" circuit with one-third or fewer elements than in the original circuit.

As a preliminary to introducing per phase analysis, consider Example 2.8.

Example 2.8 Calculation of Neutral-Neutral Voltage

Given the balanced circuit shown in Fig. E2.8, calculate $V_{n'n}$.

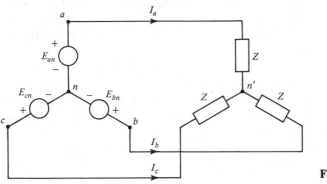

Figure E2.8

Solution Using the admittance $Y = 1/Z$, we have

$$I_a = Y(E_{an} - V_{n'n})$$

$$I_b = Y(E_{bn} - V_{n'n})$$

$$I_c = Y(E_{cn} - V_{n'n})$$

Adding the equations, we have

$$I_a + I_b + I_c = Y(E_{an} + E_{bn} + E_{cn}) - 3YV_{n'n}$$

Now applying KCL to node n or n', we get $I_a + I_b + I_c = 0$. Since E_{an}, E_{bn}, and E_{cn} are a balanced set of sources, they sum to zero. Therefore, $V_{n'n} = 0$.

Note 1: The condition $E_{an} + E_{bn} + E_{cn} = 0$ is less stringent than the assumed conditions (i.e., we do not require E_{an}, E_{bn}, and E_{cn} to be balanced).

Note 2: With $V_{n'n} = 0$ we can calculate $I_a = YE_{an}$, $I_b = YE_{bn}$, and $I_c = YE_{cn}$. Thus the phases are completely decoupled.

Example 2.9

Repeat Example 2.8 when there is an impedance Z_n connected between neutrals n and n'. Assume that $Z_n + Z/3 \neq 0$.

Solution If $Z_n = 0$, then obviously $V_{n'n} = 0$. If $Z_n \neq 0$, we can calculate $Y_n = 1/Z_n$ and we can repeat the calculation in Example 2.8 as follows. We get

$$I_a = Y(E_{an} - V_{n'n})$$

$$I_b = Y(E_{bn} - V_{n'n})$$

$$I_c = Y(E_{cn} - V_{n'n})$$

$$I_{nn'} = -Y_n V_{n'n}$$

where $I_{nn'}$ is the current into the node n'. Adding the four equations and using KCL at node n', we get

$$0 = -(3Y + Y_n)V_{n'n}$$

Thus, since $3Y + Y_n \neq 0$, $V_{n'n} = 0$ and $I_{nn'} = 0$. *Note:* Since the result holds for $Z_n = \infty$ and $Z_n = 0$, we can replace Z_n by an open circuit or short circuit as convenient. Usually, we use a short circuit since it makes explicit the useful condition $V_{nn'} = 0$.

Suppose that we consider a more complicated balanced 3ϕ network: for example, the one in Fig. 2.8. It it still true that the neutrals are at the same potential, with zero neutral currents, and that we may replace Z_n and $Z_{n'}$ by short circuits. We will show this by an alternative to the direct calculation used in Examples 2.8 and 2.9. In Fig. 2.8 assume that

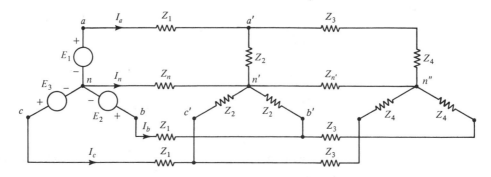

Figure 2.8   Balanced three-phase network.

1. $E_1 + E_2 + E_3 = 0$.
2. The network has three-phase symmetry.

We will now show that $V_{n'n} = V_{n''n} = 0$ (i.e., all the neutral points are at the same voltage).

Proof. The circuit in Fig. 2.8 has $E_{an} = E_1$, $E_{bn} = E_2$, and $E_{cn} = E_3$. Assume that with this 3ϕ input the response $V_{n'n} = V_0$. Next, shift the sources E_1, E_2, and E_3 so that $E_{bn} = E_1$, $E_{cn} = E_2$, and $E_{an} = E_3$. With this 3ϕ input and taking into account the three-phase network symmetry we get the same values of voltages and currents in the network as before but shifted with respect to the phases. However, the neutral voltages do not change. We still get $V_{n'n} = V_0$. Next let $E_{cn} = E_1$, $E_{an} = E_2$, $E_{bn} = E_3$, and once again $V_{n'n} = V_0$. Finally, let the 3ϕ input be the sum of the previous three 3ϕ inputs (i.e., $E_{an} = E_{bn} = E_{cn} = E_1 + E_2 + E_3$); by assumption 1 this input is zero. Considering the previous results, by application of the principle of superposition we get $V_{n'n} = 3V_0$. But since the 3ϕ input is zero, we know that $V_{n'n} = 0$ in this case. Thus $V_0 = 0$ and the response $V_{n'n}$ to the given set of sources is $V_{n'n} = V_0 = 0$. Clearly, the same argument shows that $V_{n''n} = 0$ as well.

We note that with all neutrals at the same potential the solution of the network is greatly simplified. For example, in Fig. 2.8 we can solve for I_a by considering only phase a quantities. This is shown in Fig. 2.9. I_b and I_c may be found by considering similar phase b and phase c circuits. The three circuits are effectively decoupled; the only connection between them is their common neutral, where the phase a, b, and c neutral current contributions sum to zero. We note, finally, that we can be efficient about calculating I_b and I_c once I_a is known. If E_1, E_2, and E_3 are a positive (negative) sequence set, then I_a, I_b, and I_c have the same property.

Note that in Fig. 2.8 the sources and loads are wye connected. If the Z_4 loads were connected in delta, n and n' would still be at the same potential but the decoupling (as in Fig. 2.9) would not occur. The same problem arises if the sources were connected in delta.

Fortunately, we can replace the deltas by equivalent wyes. Consider first the delta-wye transformation of loads, which replaces a given delta by a wye that is equivalent in its terminal behavior.

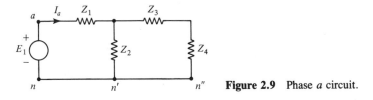

Figure 2.9 Phase a circuit.

Delta-Wye Load Transformation (Symmetrical Case).   Given a delta with impedances Z_Δ, a wye with impedances $Z_\lambda = Z_\Delta/3$ has the same terminal behavior.

Proof. See Fig. 2.10. Because of symmetry it suffices to show the equivalence for I_a in terms of V_{ab} and V_{ac}. For the delta,

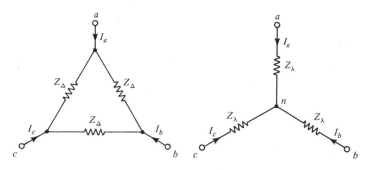

Figure 2.10 Δ-Y transformation.

$$I_a = \frac{V_{ab}}{Z_\Delta} + \frac{V_{ac}}{Z_\Delta} = \frac{V_{ab} + V_{ac}}{Z_\Delta} \qquad (2.13)$$

For the wye,

$$V_{ab} = Z_\lambda(I_a - I_b)$$

$$V_{ac} = Z_\lambda(I_a - I_c)$$

$$V_{ab} + V_{ac} = Z_\lambda(2I_a - (I_b + I_c))$$

But KCL at node n gives $I_a = -(I_b + I_c)$; hence

$$V_{ab} + V_{ac} = 3Z_\lambda I_a \qquad (2.14)$$

Comparing (2.13) and (2.14), we find that $Z_\lambda = Z_\Delta/3$.

Note: The equivalence of the delta and the wye is only external to the terminals a, b, and c. If internal behavior is of interest, one must not forget to go back to the delta.

We next develop the relationship between line-neutral and line-line voltages in balanced systems. This will permit us to replace delta-connected (line-line) sources with equivalent wye-connected (line-neutral) sources. Since the results apply to general voltages (not just source voltages), we will use the V notation rather than the E notation.

Delta-Wye Source Transformation.   Starting with the general relationship between line-neutral and line-line voltages, we have

$$V_{ab} = V_{an} - V_{bn}$$

$$V_{bc} = V_{bn} - V_{cn} \qquad (2.15)$$

$$V_{ca} = V_{cn} - V_{an}$$

If we now introduce the constraint that V_{an}, V_{bn}, and V_{cn} are a positive-sequence set, then, for arbitrary V_{an}, we get $V_{bn} = V_{an}e^{-j2\pi/3}$ and $V_{cn} = V_{an}e^{j2\pi/3}$. Substituting in (2.15), we find that V_{ab}, V_{bc}, and V_{ca} are also a positive-sequence set. We can relate the "leading" terms V_{ab} and V_{an}. We get

$$V_{ab} = (1 - e^{-j2\pi/3})V_{an} = \sqrt{3}\, e^{j\pi/6}V_{an} \qquad (2.16)$$

The transformation also works in the other direction; if V_{ab}, V_{bc}, and V_{ca} are positive sequence, so are V_{an}, V_{bn}, and V_{cn}, with

$$V_{an} = \frac{1}{\sqrt{3}}e^{-j\pi/6}V_{ab} \qquad (2.17)$$

and we can thus find the equivalent wye voltages.

The relationships (2.15) to (2.17) can also be expressed graphically as described below. Draw the line-line voltage phasors V_{ab}, V_{bc}, and V_{ca} in an equilateral triangle, as shown in Fig. 2.11. Label the vertices a, b, and c. Find the center of the triangle and label it n. Draw the inscribed set of line-neutral voltages V_{an}, V_{bn}, and V_{cn}, as shown.

If the voltages are negative sequence, the relationship is found to be

$$V_{ab} = \sqrt{3}\, e^{-j\pi/6}V_{an} \qquad (2.18)$$

and

$$V_{an} = \frac{1}{\sqrt{3}}e^{j\pi/6}V_{ab} \qquad (2.19)$$

Note 1: In either the positive-sequence or negative-sequence case, the line-line voltage magnitudes are $\sqrt{3}$ times the line-neutral voltage magnitudes.

Note 2: The term "line voltage" instead of "line-line voltage" is also used. The term "phase voltage" instead of "line-neutral voltage" is also used.

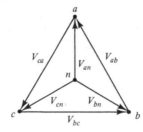

Figure 2.11 Line-neutral voltages in terms of line-line voltages.

Example 2.10

Suppose that there are a set of (balanced) positive-sequence line voltages with $V_{ab} = 1 \underline{/0°}$. Find V_{an}, V_{bn}, and V_{cn}.

Solution The relationship between balanced line and phase voltages is shown in Fig. 2.11. We then find that

$$V_{an} = \frac{1}{\sqrt{3}} \underline{/-30°}$$

$$V_{bn} = \frac{1}{\sqrt{3}} \underline{/-150°}$$

$$V_{cn} = \frac{1}{\sqrt{3}} \underline{/90°}$$

Exercise 3.   Suppose that in (2.15), V_{ab}, V_{bc}, and V_{ca} are a (balanced) positive-sequence set and we try to solve for V_{an}, V_{bn}, and V_{cn} without the constraint that they also are a (balanced) positive-sequence set. What goes wrong mathematically? How do you interpret the difficulty physically?

2.4 PER PHASE ANALYSIS

The stage has now been set to introduce the powerful method of per phase analysis. The justification for the method follows directly from the following theorem.

Balanced Three-Phase Theorem.   Assume that we are given a

1. balanced three-phase (connected) system with
2. all loads and sources wye connected, and
3. in the circuit model there is no mutual inductance between phases.

Then

(a) all the neutrals are at the same potential,
(b) the phases are completely decoupled, and
(c) all corresponding network variables occur in balanced sets of the same sequence as the sources.

Outline of Proof. The fact that the neutrals are at the same potential and the network phases are decoupled follows just as in the case discussed following Fig. 2.8; using superposition, the number of sources can be arbitrary. Without mutual inductance between the phases they are completely decoupled and with balanced (positive-sequence) sources it is clear that the responses in phases *b* and *c* lag the corresponding responses in phase *a* by 120° and 240°, respectively. The net result is that corresponding responses occur in balanced (positive-sequence) sets. The same idea holds in the case of negative-sequence sources.

Method: Per Phase Analysis.   We next turn to the method of per phase analysis. Given a balanced three-phase network with no mutual inductance between phases:

1. Convert all delta-connected sources and loads into equivalent wye connections.

2. Solve for the desired phase a variables using the phase a circuit with all neutrals connected.

3. The phase b and phase c variables can then be determined by inspection; subtract 120° and 240°, respectively, from the phase angles found in step 2, in the usual case of positive-sequence sources. Add 120° and 240° in the case of negative-sequence sources.

4. If necessary, go back to the original circuit to find line-line variables or variables internal to delta connections.

Note: In using per phase analysis we always pick the neutral as datum. In this case it is simpler to use a single-subscript notation for phase voltages. We will use V_a rather than V_{an}, etc.

Example 2.11

Given the balanced three-phase system shown in Fig. E2.11(a), find $v_1(t)$ and $i_2(t)$.

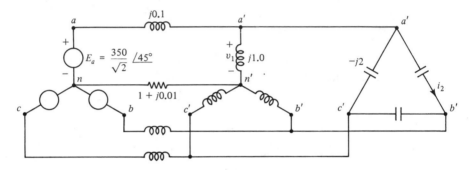

Figure E2.11(a)

Solution   Replace delta by equivalent wye, $Z_\lambda = -j2/3$. Using per phase analysis, we consider the per phase (phase a) circuit shown in Fig. E2.11(b). Note that while V_1 appears in the per phase circuit, I_2 has been suppressed in the process of the $\Delta - Y$ conversion. Carrying out the calculation to find V_1, we note that the equivalent parallel impedance from a' to n' is $-j2$. Using the voltage-divider law,

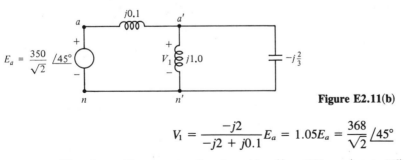

Figure E2.11(b)

$$V_1 = \frac{-j2}{-j2 + j0.1} E_a = 1.05 E_a = \frac{368}{\sqrt{2}} \underline{/45°}$$

The phasor V_1 represents the sinusoid $v_1(t) = 368 \cos(\omega t + 45°)$. From the original circuit we see that to find $i_2(t)$, we need to find $V_{a'b'}$. Thus

$$V_{a'b'} = V_{a'n'} - V_{b'n'} = \sqrt{3}\, e^{j\pi/6} V_{a'n'}$$

$$= \frac{638}{\sqrt{2}} \underline{/75°}$$

Then using the impedance of the capacitor, we find that

$$I_{a'b'} = \frac{319}{\sqrt{2}} \underline{/165°}$$

and it follows that the corresponding sinusoidal function is

$$i_2(t) = 319 \cos (\omega t + 165°)$$

2.5 BALANCED THREE-PHASE POWER

One of the important advantages of balanced three-phase operation is that the instantaneous power delivered to a load is a constant. We will show this in a moment. First, referring to Example 2.7 with the neutral point as the datum, we get, using our single-subscript notation,

$$S_{3\phi} = V_a I_a^* + V_b I_b^* + V_c I_c^* \tag{2.20}$$

Since we are assuming that V_a, V_b, V_c and I_a, I_b, I_c occur in balanced sets, we can write for the positive-sequence case,

$$S_{3\phi} = V_a I_a^* + V_a e^{-j2\pi/3} I_a^* e^{j2\pi/3} + V_a e^{j2\pi/3} I_a^* e^{-j2\pi/3} \tag{2.21}$$

$$S_{3\phi} = 3V_a I_a^* = 3S \tag{2.22}$$

where S is the per phase complex power. The negative-sequence case gives the same result. Thus to calculate the total three-phase power we can find the phase a power (by per phase analysis) and simply multiply by 3. Next, we calculate the instantaneous three-phase power, using (2.3) and (2.4).

$$p_{3\phi}(t) = v_a(t)i_a(t) + v_b(t)i_b(t) + v_c(t)i_c(t) \tag{2.23}$$

$$= |V||I| [\cos \phi + \cos (2\omega t + \underline{/V} + \underline{/I})]$$

$$+ |V||I| \left[\cos \phi + \cos \left(2\omega t + \underline{/V} + \underline{/I} - \frac{4\pi}{3} \right) \right]$$

$$+ |V||I| \left[\cos \phi + \cos \left(2\omega t + \underline{/V} + \underline{/I} + \frac{4\pi}{3} \right) \right]$$

$$= 3|V||I| \cos \phi$$

$$= 3P \tag{2.24}$$

The easiest way to check that the double-frequency terms cancel is to add the phasor representations.

The fact that the instantaneous power delivered to the load is constant should be contrasted with the case of single-phase power illustrated in Fig. 2.2. An advantage of constant power for supplying ac motors is that it is consistent with constant speed and torque. With single-phase motors there are torque pulsations accompanying the power pulsations which can be troublesome in some applications. Thus we have yet another advantage of 3ϕ supply over 1ϕ supply.

2.6 COMPLEX POWER TRANSMISSION (SHORT LINE)

As an application of the method of per phase analysis and an introduction to the problem of power transmission, we consider the simple 3ϕ balanced system shown in Fig. 2.12.

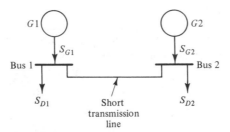

Figure 2.12   One-line diagram.

We wish to consider only the power transferred by the transmission line. Assume that the generators maintain certain 3ϕ voltages at buses 1 and 2, and these are known. In our analysis we can replace these voltages by sources. Anticipating the result of Chapter 4, we model each phase of the short transmission line by a series RL circuit. We are then led to the circuit model in Fig. 2.13. Since the system is balanced, we can use per phase analysis. Figure 2.14 shows the per phase circuit and labels certain quantities of interest. To simplify the notation the phase a subscript has been dropped. S_{12} is the (phase a) complex power from bus 1 to the transmission line connected to bus 2. A similar definition holds for S_{21}. We note that the corresponding three-phase quantities may be obtained by simply multiplying by 3. We next obtain expressions for S_{12} and S_{21} in terms of V_1, V_2, and Z. Assume the following notation.

$$V_1 = |V_1| e^{j\theta_1} \qquad V_2 = |V_2| e^{j\theta_2}$$
$$Z = |Z| e^{j\angle Z} \qquad \theta_{12} \triangleq \theta_1 - \theta_2$$

(2.25)

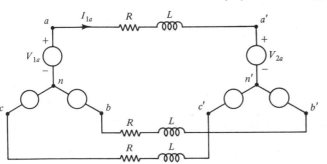

Figure 2.13   Circuit model.

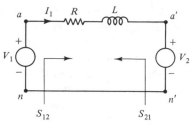

Figure 2.14 Per phase circuit.

We call θ_{12} the power angle. Using (2.11), we have

$$S_{12} = V_1 I_1^* = V_1 \left(\frac{V_1 - V_2}{Z} \right)^* = \frac{|V_1|^2}{Z^*} - \frac{V_1 V_2^*}{Z^*}$$

$$= \frac{|V_1|^2}{|Z|} e^{j\angle Z} - \frac{|V_1||V_2|}{|Z|} e^{j\angle Z} e^{j\theta_{12}} \tag{2.26}$$

and

$$S_{21} = \frac{|V_2|^2}{|Z|} e^{j\angle Z} - \frac{|V_2||V_1|}{|Z|} e^{j\angle Z} e^{-j\theta_{12}} \tag{2.27}$$

Although there is perfect symmetry between (2.26) and (2.27), it is helpful to describe bus 1 as sending power to bus 2. The power sent by V_1 (or bus 1) is given by (2.26). The power received by V_2 (or bus 2) is given by

$$-S_{21} = -\frac{|V_2|^2}{|Z|} e^{j\angle Z} + \frac{|V_2||V_1|}{|Z|} e^{j\angle Z} e^{-j\theta_{12}} \tag{2.28}$$

For a given transmission line (Z fixed) the complex power sent or received depends on $|V_1|$, $|V_2|$, and θ_{12}. From the point of view of control, $|V_1|$ is affected most directly by the field current in generator 1, $|V_2|$ by the field current in generator 2, and θ_{12} by the difference in mechanical power inputs to the two generators. To increase θ_{12} we can increase the mechanical power into generator 1 while decreasing the mechanical power into generator 2.

Under normal operation $|V_1|$ and $|V_2|$ are kept within fairly strict limits at generator buses while θ_{12} varies considerably. It is of interest, therefore, to consider (2.26) and (2.28) as a function of θ_{12} with $|V_1|$, $|V_2|$, and Z fixed as parameters. With this objective in mind we get (2.26) and (2.28) in the form

$$S_{12} = C_1 - Be^{j\theta_{12}} \tag{2.29}$$

$$-S_{21} = C_2 + Be^{-j\theta_{12}} \tag{2.30}$$

where

$$C_1 = \frac{|V_1|^2}{|Z|} e^{j\angle Z} \qquad C_2 = -\frac{|V_2|^2}{|Z|} e^{j\angle Z} \qquad B = \frac{|V_1||V_2|}{|Z|} e^{j\angle Z}$$

It is useful to graph the dependence of S_{12} and $-S_{21}$ on θ_{12}. As θ_{12} is varied,

(2.29) and (2.30) show that S_{12} and $-S_{21}$ trace out circles in the complex plane. These S_{12} and $-S_{21}$ circles are called *sending-end circles* and *receiving-end circles*, respectively. The center of the sending-end circle is C_1 and that of the receiving-end circle is C_2. Both circles have the same radius $|B|$. If $\theta_{12} = 0$, then, in the complex plane, C_1, C_2, and B are collinear. These considerations indicate the construction shown in Fig. 2.15. Concerning the graphs, we make the following comments:

1. The circles do not intersect if $|V_1| \neq |V_2|$.
2. As θ_{12} increases from zero, the active power sent and received increases. (The active power sent is greater than that received by the amount of the losses in the transmission line.) From the geometry of the graphs we note there is an "ultimate" limit to the active power received, which occurs for $\theta_{12} = \underline{/Z}$, and a similar limit for the active power sent, which occurs for $\theta_{12} = 180° - \underline{/Z}$. These ultimate active powers depend on Z, $|V_1|$, and $|V_2|$. In fact, as we will see in Chapter 4, more restrictive limits prevail under normal operating conditions.

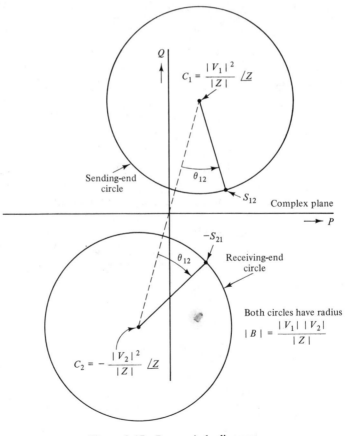

Figure 2.15 Power circle diagram.

3. For most lines the resistance is small compared to the inductance. Assume, then, as an approximation, that $R = 0$ and $Z = jX$. In this case there are no transmission-line losses, and the active power sent equals the active power received. In fact, from (2.26), (2.27), or Fig. 2.15,

$$P_{12} = -P_{21} = \frac{|V_1||V_2|}{X} \sin \theta_{12} \qquad (2.31)$$

$$Q_{12} = \frac{|V_1|^2}{X} - \frac{|V_1||V_2|}{X} \cos \theta_{12} \qquad (2.32)$$

$$Q_{21} = \frac{|V_2|^2}{X} - \frac{|V_2||V_1|}{X} \cos \theta_{12} \qquad (2.33)$$

In this case the ultimate transmission capability is $|V_1||V_2|/X$.

4. If an attempt is made to increase θ_{12} beyond the ultimate transmission capability, "synchronism" between the two generators is lost; we will be discussing this phenomenon in Chapter 9 If the two generators lose synchronism or "fall out of step," they no longer generate voltages of the same frequency. Although the generators are physically connected together by the transmission line, effective exchange of power ceases. We will consider a case in Example 2.12.

5. How can the ultimate transmission capability be increased in an effort to strengthen the tie between buses 1 and 2? The answer may be inferred from (2.31). We can increase the voltage levels (i.e., $|V_1|$ and $|V_2|$) and decrease X. In Chapter 3 we will consider ways to decrease the line inductance by careful line design. Another method is to decrease the series reactance by insertion of series capacitors. This latter process is called *series compensation*.

6. Under normal operating conditions for high voltage lines $|V_1| \approx |V_2|, /\underline{Z} \approx 90°$, and θ_{12} typically is small, less than approximately $10°$. In that case there is reasonably good decoupling between the control of the flow of active versus reactive power; active-power flow couples strongly with θ_{12}, and reactive-power flow couples strongly with $|V_1| - |V_2|$. The easiest way to see this is to do a sensitivity analysis using (2.31) to (2.33). Evaluate the various partial derivatives with respect to θ_{12} and $|V_1|$, $|V_2|$ to establish the result claimed

Exercise 4.   Suppose in Fig. 2.11 that the bus 2 voltages are zero because of a 3ϕ short circuit. Find S_{12} using the power circle diagram. Does the result make sense physically?

Exercise 5.   Prove the statement made in comment 1 above. *Hint:* If $|V_1| \neq |V_2|$, then $(|V_1| - |V_2|)^2 > 0$.

Example 2.12

Suppose that the two 3ϕ generators in Fig. 2.13 are "out of step." Assume that the voltages at buses 1 and 2 are balanced with the phase a voltages given by

$$v_1(t) = \sqrt{2}|V_1| \cos (\omega_1 t + \theta_1^0)$$

$$v_2(t) = \sqrt{2}\,|V_2|\,\cos\,(\omega_0 t + \theta_2^0)$$

Assume that $\omega_1 \approx \omega_0$ and find an expression for the active power transmitted down an approximately lossless line.

Solution With $\omega_1 \approx \omega_0$ it is useful to write

$$v_1(t) = \sqrt{2}\,|V_1|\,\cos\,[\omega_0 t + \theta_1^0 + (\omega_1 - \omega_0)t]$$

and to consider $v_1(t)$ to be a sinusoid of frequency ω_0 and a slowly varying phase $\theta_1^0 + (\omega_1 - \omega_0)t$. Assuming that the phase varies slowly enough, we can still use sinusoidal steady-state analysis as an approximation, treating the phase as a parameter. As the phase varies, we move along a continuum of steady-state solution values. This is an example of so-called pseudo-steady-state analysis. Applied to our problem we use (2.31) with $\theta_{12} = \theta_1 - \theta_2 = \theta_1^0 + (\omega_1 - \omega_0)t - \theta_2^0$ and get a short-term active power which varies with time.

$$P_{12} = -P_{21} = \frac{|V_1||V_2|}{X}\,\sin\,[(\omega_1 - \omega_0)t + \theta_{12}^0]$$

We note that the long-term average of active power is zero. This mode of operation is not only ineffective but is highly undesirable since it is accompanied by very large line currents. In practice, protective devices would remove the out of step generator.

Example 2.13

For a balanced three-phase transmission line with $Z = 1\,\underline{/85°}$, $\theta_{12} = 10°$, find S_{12} and $-S_{21}$ for

(a) $|V_1| = |V_2| = 1.0$
(b) $|V_1| = 1.1$, $|V_2| = 0.9$

We note that the use of voltages with numerical values such as 1.0, 1.1, or 0.9 anticipates a normalization or scaling to be introduced in Chapter 5.

Solution In getting numerical results it is generally easier to use (2.26) and (2.28) rather than the power circle diagram. Using the formulas, we find the following results.

(a) With $|V_1| = |V_2| = 1$,

$$S_{12} = 1\,\underline{/85°} - 1\,\underline{/95°} = 0.1743$$

$$-S_{21} = -1\,\underline{/85°} + 1\,\underline{/75°} = 0.1717 - j0.0303$$

In particular, we note that $Q_{12} = 0$ and $-Q_{21} = -0.0303$.
(b) With $|V_1| = 1.1$, $|V_2| = 0.9$,

$$S_{12} = 1.21\,\underline{/85°} - 0.99\,\underline{/95°} = 0.1917 + j0.2192$$

$$-S_{21} = -0.81\,\underline{/85°} + 0.99\,\underline{/75°} = 0.1856 + j0.1493$$

We note that P_{12} has not changed much but Q_{12} and $-Q_{21}$ have changed considerably from the values in part (a).

Example 2.14

For the system shown in Fig. E2.14(a), all quantities are per phase values.

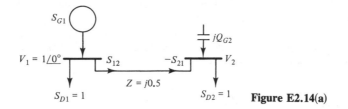

Figure E2.14(a)

(a) Pick Q_{G2} so that $|V_2| = 1$.
(b) In this case, what is $\underline{/V_2}$?
(c) If $Q_{G2} = 0$, can we supply the load S_{D2}?
(d) If yes, what is V_2?

Solution   In parts (a) and (b) we can use (2.31) to (2.33) or the power circle diagram shown in Fig. 2.15. We will use the equations. Since $S_{D2} = 1$ is real while Q_{G2} is purely imaginary, it is clear $P_{12} = -P_{21} = 1$. Then using (2.31),

$$P_{12} = \frac{|V_1||V_2|}{X} \sin \theta_{12} = 2 \sin \theta_{12} = 1$$

Thus $\theta_{12} = 30°$ and $\underline{/V_2} = -30°$. Using (2.33), we obtain

$$Q_{G2} = Q_{21} = \frac{|V_2|^2}{X} - \frac{|V_2||V_1|}{X} \cos \theta_{12} = 2 - 2 \cos 30° = 0.268$$

Note that $Q_{G2} > 0$ is consistent with a capacitor source.

For parts (c) and (d), Fig. 2.15 offers some advantages over using equations. If $Q_{G2} = 0$, then $-S_{21} = S_{D2} = 1$. The question is: Can we find a $|V_2|$ such that the receiving-end circle passes through the point $-S_{21} = 1$ in the complex plane for some θ_{12}? In geometric terms the requirements are as shown in Fig. E2.14(b). It may be seen that a solution can be found if

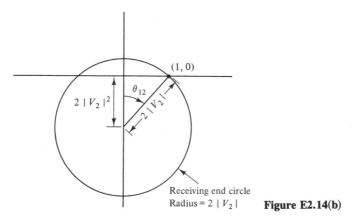

Receiving end circle
Radius = $2|V_2|$ Figure E2.14(b)

$$(2|V_2|^2)^2 + 1^2 = (2|V_2|)^2$$

has a real solution for $|V_2|$. Let $x = |V_2|^2$; then $4x^2 - 4x + 1 = 0$ has a real solution $x = \frac{1}{2}$. Thus $|V_2| = 1/\sqrt{2}$. From the geometry of the figure we also find that $\theta_{12} = 45°$. Hence we can supply the load at a voltage $V_2 = 0.707 \underline{/-45°}$.

We notice the improvement in voltage in (a) versus (c), which can be attributed to the injection of reactive power by the capacitor source at bus 2. In fact, capacitors are frequently used for this purpose.

2.7 COMPLEX POWER TRANSMISSION (RADIAL LINE)

In Section 2.6 we considered the case of a line with generators at both ends. We could assume voltage maintenance or "support" at both ends. Now we will consider the case of a radial line; at the near end there is voltage support but at the far end of the line there is a complex power load without a generator or capacitor bank to maintain the voltage. We wish to study how the voltage at the far end of the line varies with load. The per phase diagram for the case to be considered is shown in Fig. 2.16.

We will assume that the load "draws" complex power at a fixed power factor. In this case it is convenient to express S_D as follows:

$$
\begin{aligned}
S_D = V_2 I^* &= |V_2\|I| \, e^{j\phi} \\
&= |V_2\|I| \, (\cos \phi + j \sin \phi) \\
&= P_D(1 + j\beta)
\end{aligned}
\tag{2.34}
$$

Here $\phi = \underline{/V_2} - \underline{/I}$, $\beta = \tan \phi$, and $PF = \cos \phi$. Thus as we vary the load, P_D varies, with β a parameter.

Next we use (2.31) and (2.33):

$$P_D = P_{12} = \frac{|V_1\|V_2|}{X} \sin \theta_{12} \tag{2.35}$$

$$Q_D = -Q_{21} = -\frac{|V_2|^2}{X} + \frac{|V_2\|V_1|}{X} \cos \theta_{12} \tag{2.36}$$

We next eliminate θ_{12} by using $\cos^2 \theta_{12} + \sin^2 \theta_{12} = 1$ and get

$$\left(\beta P_D + \frac{|V_2|^2}{X}\right)^2 = \left(\frac{|V_2\|V_1|}{X}\right)^2 - P_D^2 \tag{2.37}$$

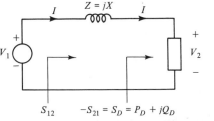

S_{12} $-S_{21} = S_D = P_D + jQ_D$ **Figure 2.16**   Per phase circuit.

A rearrangement gives the quadratic equation in $|V_2|^2$.

$$|V_2|^4 + (2\beta P_D X - |V_1|^2)|V_2|^2 + (1 + \beta^2)P_D^2 X^2 = 0 \tag{2.38}$$

The solutions are given by

$$|V_2|^2 = \frac{|V_1|^2}{2} - \beta P_D X \pm \left[\frac{|V_1|^4}{4} - P_D X(P_D X + \beta|V_1|^2)\right]^{1/2} \tag{2.39}$$

We note the presence of multiple solutions.

Example 2.15

For the system of Fig. 2.16 assume that $|V_1| = 1$ and $X = 0.5$. Use (2.39) to find $|V_2|$ as a function of P_D for load power factors of 1.0, 0.97 leading and 0.97 lagging.

Solution   Using the values provided, we get

$$|V_2|^2 = \frac{1 - \beta P_D \pm [1 - P_D(P_D + 2\beta)]^{1/2}}{2}$$

With PF = 1 we get $\phi = 0$, $\beta = \tan\phi = 0$. With PF = 0.97 leading, $\phi = -14.07°$, $\beta = -0.25$. With PF = 0.97 lagging, $\phi = 14.07°$, $\beta = 0.25$. Plots of $|V_2|$ versus P_D are given in Fig. E2.15 for the three power factors. We note the beneficial effect of a leading power factor load in maintaining the voltage. In all cases, however, there is a point at which voltage "collapse" occurs. For lagging power factor loads this may occur for power levels within normal operating limits of the transmission line.

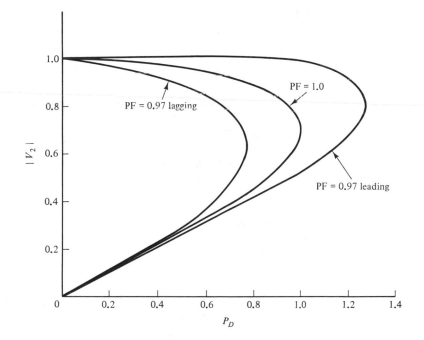

Figure E2.15

The example emphasizes the importance of supplying reactive power at the consumer's location rather than attempt the supply from a great distance.

2.8 SUMMARY

The basic objective of this chapter was to give the reader an appreciation for complex power as a variable and develop some essential analytical tools. The material covered included the definitions of instantaneous power, complex, active, and reactive power and the relations between them. Effective phasors were introduced and used to calculate complex power. Reactive power was related to peak instantaneous power into the reactive elements and thereby a physical interpretation was provided. The theorem of conservation of complex power was given with some applications. The capacitor as a source of reactive power was described. An expression for the complex power delivered to an n-port was derived.

Balanced three-phase power systems were defined and some advantages to their use were indicated. Per phase analysis as a simplifying tool was introduced. Delta-wye transformations of balanced three-phase sources and symmetric loads were derived. As applications, two cases of power transmission were considered. In the first case there was voltage support at both ends of the transmission line; in the second case the (radial) line had no voltage support at its outer terminus. Factors affecting complex power transmission were discussed.

PROBLEMS

2.1. In Fig. 2.1, $v(t) = \sqrt{2} \times 120 \cos (\omega t + 30°)$, $i(t) = \sqrt{2} \times 10 \cos (\omega t - 30°)$.
   (a) Find $p(t)$, S, P, and Q into the network.
   (b) Find a simple (two-element) series circuit consistent with the prescribed terminal behavior as described above.

2.2. With $|V| = 100$ V, the instantaneous power into a network N has a maximum value 1707 W and a minimum value -293 W.
   (a) Find a possible series RL circuit equivalent to N.
   (b) Find $S = P + jQ$ into N.
   (c) Find the maximum instantaneous power into L and compare with Q.

2.3. A certain 1ϕ load draws 5 MW at 0.7 power factor lagging. Determine the reactive power required from a parallel capacitor to bring the power factor of the parallel combination up to 0.9.

2.4. A 3ϕ load draws 200 kW at a PF of 0.707 lagging from a 440-V line. In parallel is a 3ϕ capacitor bank which supplies 50 kVAr. Find the resultant power factor and current (magnitude) into the parallel combination.

2.5. A 1ϕ load draws 10 kW from a 416-V line at a power factor of 0.9 lagging.
   (a) Find $S = P + jQ$.

(b) Find $|I|$.

(c) Assume that $\underline{/I} = 0$ and find $p(t)$.

2.6. The system shown in Fig. P2.6 is balanced (and positive sequence). Assume that

$$Z = 10\,\underline{/-15°}$$

$$V_{ca} = 208\,\underline{/-120°}$$

Find V_{ab}, V_{bc}, V_{an}, V_{bn}, V_{cn}, I_a, I_b, I_c, and $S_{3\phi}$.

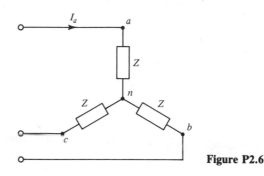

Figure P2.6

2.7. In Fig. P2.7, find the total complex power delivered to the load. Assume that

$$Z_C = -j0.2$$

$$Z_L = j0.1$$

$$R = 10$$

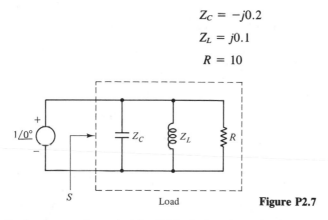

Load   **Figure P2.7**

2.8. In the system shown in Fig. P2.8, find I_a, I_b, and I_c if

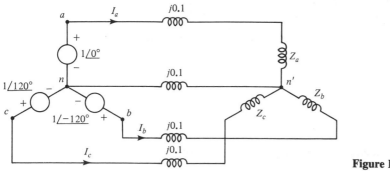

Figure P2.8

(a) $Z_a = Z_b = j1.0$, $Z_c = j0.9$
(b) $Z_a = Z_b = Z_c = j1.0$
Hint: Use per phase analysis in part (b).

2.9.   The system shown in Fig. P2.9 is balanced 4ϕ. Use per phase analysis to find I_a, I_b, I_c, and I_d. *Note:* We need to find a "star" equivalent for the set of voltages.

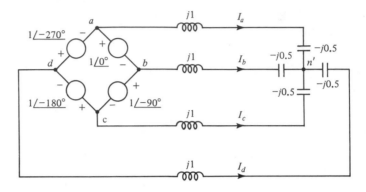

Figure P2.9

2.10.   The system shown in Fig. P2.10 is balanced. Assume that

$$C = 10^{-3} \ F$$

$$R = 1 \ \Omega$$

$$|V_{ab}| = 240 \ V$$

$$\omega = 2\pi \cdot 60$$

Find $|V_{a'b'}|$, $|I_b|$, and $S_{\text{load}}^{3\phi}$. *Hint:* Use per phase analysis.

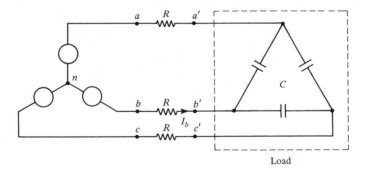

Figure P2.10

2.11.   The system shown in Fig. P2.11 is balanced. Find $V_{a'n}$, $V_{b'n}$, $V_{c'n}$, and $V_{a'b'}$.

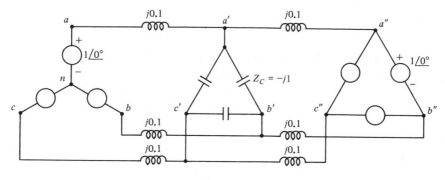

Figure P2.11

2.12. The system shown in Fig. P2.12 is balanced. Assume that

$$\text{load inductors, } Z_L = j10$$

$$\text{load capacitors, } Z_C = -j10$$

Find I_a, I_{cap}, and $S^{3\phi}_{load}$.

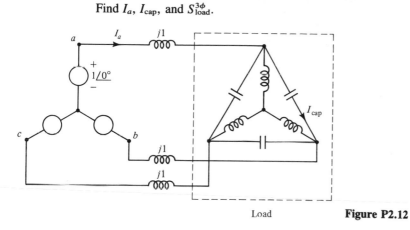

Load   **Figure P2.12**

2.13. Refer to Fig. P2.13 and assume that

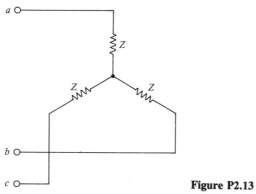

Figure P2.13

$$Z = 100 \underline{/60°}$$

$$V_{ab} = 208 \underline{/0°}$$

(a) If the circuit is balanced positive sequence (abc), find V_{bc}, V_{ca}, I_a, I_b, and I_c.

(b) If the circuit is balanced negative sequence (acb), find V_{bc}, V_{ca}, I_a, I_b, and I_c.

2.14. The system shown in Fig. P2.14 is balanced. Specify Z so that $|V_{a'b'}| > |V_{ab}|$. *Hint:* Use per phase analysis and experiment with the element Z a resistance, an inductance, or a capacitance.

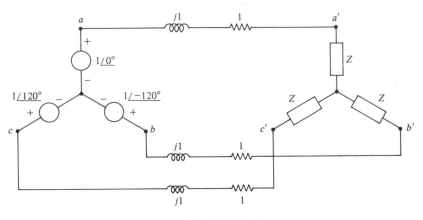

Figure P2.14

2.15. In Fig. P2.15, assume that

$$E_a = \sqrt{2} \underline{/45°}$$

$$E_b = 1 \underline{/-90°}$$

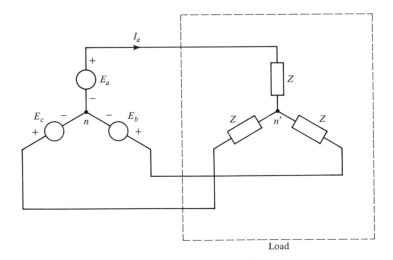

Load

Figure P2.15

$$E_c = 1 \underline{/180°}$$

$$I_a = 1 \underline{/-10°}$$

The load is symmetrical and Z is not given. The sources are not balanced but we note $E_a + E_b + E_c = 0$. Find $S_{load}^{3\phi}$.

2.16. In Fig. P2.16, assume that

$$|V_1| = |V_2| = 1$$

$$Z_{line} = 0.1 \underline{/85°}$$

(a) For what nonzero θ_{12} is S_{12} purely real?

(b) What is the maximum power, $-P_{21}$, that can be received by V_2, and at what θ_{12} does this occur?

(c) When $\theta_{12} = 85°$, what is the active power loss in the line?

(d) For what θ_{12} is $-P_{21} = 1$?

V_1 $\qquad\qquad\qquad\qquad\qquad\qquad\qquad$ V_2

S_{12} $\qquad\qquad\qquad\qquad\qquad\qquad$ $-S_{21}$ $\qquad\qquad$ **Figure P2.16**

2.17. Suppose that we "compensate" the line in Problem 2.16 by inserting series capacitance. The intent is to increase the ultimate power transmission capability. Suppose that $Z_C = -j0.05$. Repeat Problem 2.16(b) and compare results.

2.18. Suppose that we "overcompensate" the line in Problem 2.16 and make the total series reactance negative (capacitive). To be specific, suppose that $Z_C = -j0.2$.

(a) Draw the power circle diagram in this case, labeling it carefully.

(b) For what θ_{12} is the received power $-P_{21} = 1$?

(c) Compare the result with that obtained in Problem 2.16(d). How do you explain the (qualitative) difference?

2.19. Draw the power circle diagram in the case $|V_1| = 1.05$, $|V_2| = 0.95$, $Z_{line} = 0.1 \underline{/85°}$. Find

(a) P_{12max}.

(b) θ_{12} at which we get P_{12max}.

(c) $-P_{21max}$.

(d) θ_{12} at which we get $-P_{21max}$.

(e) Active power loss in the line when $\theta_{12} = 10°$.

2.20. In Fig. P2.20, assume that

$$V_1 = 1 \underline{/0°}$$

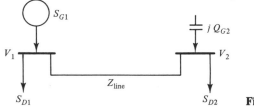

Figure P2.20

$$Z_{\text{line}} = 0.01 + j0.1$$

$$S_{D1} = 0.5 + j0.5$$

$$S_{D2} = 0.5 + j0.5$$

Pick Q_{G2} so that $|V_2| = 1$. In this case what are Q_{G2}, S_{G1}, and $\underline{/V_2}$?

2.21. Repeat Example 2.15 for the case $|V_1| = 1$, load power factor $= 1.0$, and $X = 0.4$. Find $|V_2|$ versus P_D.

2.22. Suppose that we describe the load in Fig. 2.16 as an impedance Z_L of variable magnitude $|Z_L|$ but fixed $\underline{/Z_L}$; in other words, a variable load of fixed power factor. For example, we could be switching on electric lights. We are interested in how $|V_2|$ behaves as we change $|Z_L|$. Assume that $|V_1| = 1$, $X = 0.1$, and that $Z_L = (1/\mu)Z_L^0$ for $0 \leq \mu \leq 1$, and plot $|V_2|$ versus μ for three cases:
(a) $Z_L^0 = j1.0$ (inductive load)
(b) $Z_L^0 = -j1.0$ (capacitive load)
(c) $Z_L^0 = 1.0$ (resistive load)
If we wish to maintain load voltages within narrow limits, are variable reactive loads a problem?

2.23. A 3ϕ transmission line connects buses 1 and 2. Model the line as shown in Fig. 2.13 but with $R = 0$. Measurements at the bus 1 end indicate that the 3ϕ line is carrying 500 MW of active power and 50 MVAr of reactive power. The voltage magnitudes (line-line) are 500 kV at both buses. Assume that at bus 1, $V_{1a} = 500/\sqrt{3}$ kV (i.e., phase angle is zero) and find
(a) The magnitude and phase of I_{1a} (see Fig. 2.13).
(b) The phase difference between V_{1a} and V_{2a}.
(c) The per phase reactance of the line.
(d) The per phase (phase a) instantaneous power from bus 1 into line.
(e) The per phase (phase a) instantaneous power from bus 2 into line.
(f) The total three-phase instantaneous power into line from bus 1.

Transmission-Line Parameters

chapter 3 _____

3.0 INTRODUCTION

In Chapters 3 and 4 we consider the modeling of transmission lines and develop useful lumped-circuit models. In these models a line is characterized by four distributed parameters: series resistance, series inductance, shunt conductance, and shunt capacitance. The series inductance and shunt capacitance represent the effects of the magnetic and electric fields around the conductors. The shunt conductance accounts for leakage currents along insulator strings and ionized pathways in the air; this leakage is usually a small effect and is frequently ignored. The remaining parameters, however, are needed to develop a useful transmission-line model for use in general power system studies.

We are interested in two kinds of questions. The first is descriptive: How are the parameters, particularly inductance and capacitance, related to line geometry (i.e., wire size and configuration)? Systems designers have some choice in the selection of this geometry. One way the selection affects system performance is through the parameter values.

As an example, at the end of Chapter 2 we showed that, at least approximately, for a short line the ultimate power-carrying capability depended inversely on the line inductance. We are therefore particularly interested in how inductance depends on the line geometry and how it may be reduced by careful design.

A second question concerns modeling. In Chapter 2 the advantages of per phase analysis of balanced three-phase circuits were emphasized, the expectation being that this useful tool would be applicable to practical systems, including three-phase transmission lines. This implies that in the circuit model there is no mutual inductance between phases. If one considers the physical three-phase transmission line, there clearly is magnetic coupling between the phases. How, then, does one get rid of the mutual inductance in the circuit model? Similar questions can be asked about the shunt capacitances.

For the reader who has studied transmission lines solely from a field point of view, we note that the circuits approach we will adopt is equivalent for a lossless line. For the practical case of lines with small losses and operating at low frequencies, the use of the circuits approach is completely justifiable.

In the following sections we derive expressions for inductance and capacitance of three-phase lines in terms of line geometry.

3.1 REVIEW OF MAGNETICS

We start by recalling some basic facts from physics.

Ampère's circuital law

$$F = \oint_\Gamma \mathbf{H} \cdot d\mathbf{l} = i_e \qquad (3.1)$$

where F = magnetomotive force (mmf), ampere-turns
        $\mathbf{H}$ = magnetic field intensity, ampere-turns/meter
        $d\mathbf{l}$ = differential path length, meters
        i_e = total instantaneous current linked by the closed path Γ

$\mathbf{H}$ and $d\mathbf{l}$ are vectors in space. In (3.1) we can replace the dot product:

$$\mathbf{H} \cdot d\mathbf{l} = H \, dl \cos \theta \qquad (3.2)$$

where θ is the angle between $\mathbf{H}$ and $d\mathbf{l}$, and we drop the vector notation to indicate scalar quantities. The reference direction for i_e is related to the reference direction for Γ by the familiar right-hand rule.

In words, (3.1) states that the line integral of $\mathbf{H}$ around a closed path Γ is equal to the total current linked. The total current linked may be found unambiguously by considering the algebraic sum (or integral) of the currents crossing any surface, or cap, bounded by the closed path Γ; such a surface is called a *Gaussian surface*.

Even though (3.1) is an implicit relation in $\mathbf{H}$, it is useful for actually calculating H in special cases where we have symmetry. Such a case will be considered in Section 3.2. We note that there is an explicit expression for $\mathbf{H}$, one form, known as the Biot–Savart law, which shows that at any given point $\mathbf{H}$ is a *linear function* of all the differential current elements of the circuit. Thus we can calculate $\mathbf{H}$ using superposition.

Magnetic flux ϕ.   If a linear relationship between flux density and magnetic field intensity is assumed,

$$\mathbf{B} = \mu\mathbf{H} \tag{3.3}$$

where $\mathbf{B}$ = flux density, webers/meter^2
$\quad\quad\mu$ = permeability of medium

The total flux through a surface area A is the surface integral of the normal component of $\mathbf{B}$; that is,

$$\phi = \int_A \mathbf{B} \cdot d\mathbf{a} \tag{3.4}$$

where $d\mathbf{a}$ is a vector with direction normal to the surface element da and with magnitude equal to da. If $\mathbf{B}$ is perpendicular to, and uniform over an area A, (3.4) simplifies to

$$\phi = BA \tag{3.5}$$

where the quantities B and A are scalars.

Flux linkages λ.   The use of flux linkages is very familiar to electrical engineers. Faraday's law gives the relationship between generated electromotive force (emf) in a loop and the time rate of change of flux linkages. The intuitive notion of flux linkages is very simple and can be illustrated by considering Fig. 3.1. Some of the flux links all the turns; ψ_2 is a flux line illustrating this case. Some of the flux, called *leakage flux*, links only some of the turns; ψ_1 and ψ_3 illustrate these components.

If all the flux ϕ links all the N turns of the coil, then

$$\lambda = N\phi \tag{3.6}$$

where λ = flux linkages, in weber turns. If, as in the figure, there is leakage flux, we add the flux linkage contributions turn by turn. In this case

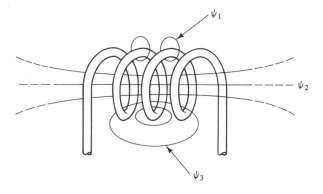

Figure 3.1 Flux linkages.

$$\lambda = \sum_{i=1}^{N} \phi_i \tag{3.7}$$

where ϕ_i is the flux linking the ith turn of the coil.

Inductance.   Assuming (3.3) to be true, the relationship between flux linkages and current is linear and the constant of proportionality is called the *inductance*.

$$\lambda = Li \tag{3.8}$$

Example 3.1

Calculate the inductance of a coil wound closely on a toroidal iron core (Fig. E3.1). Assume

1. Small cross-sectional area.
2. All the flux links all the turns.

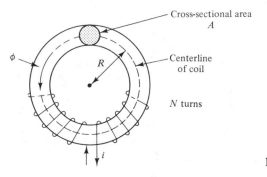

Figure E3.1

Solution   This is a classic example the reader may recall from physics. From physics we can assume that the flux is inside the core and the flux lines form concentric circles. For the choice of reference direction for current shown and using the right-hand rule, we find that the associated reference direction for flux is counterclockwise. Taking a counterclockwise path Γ, aligned with one of the flux lines at a radius r, and applying Ampère's circuital law (3.1), we get

$$F = Hl = Ni$$

where $l = 2\pi r$ in the length of the path Γ. The mmf is N times i because the Gaussian surface bounded by Γ is crossed N times by the multiturn winding. Using (3.3) in the expression above,

$$B = \mu H = \mu \frac{Ni}{l}$$

From this expression and assumption 1, we can see that B does not vary much across the cross-sectional area, and using a value of B corresponding to $l = 2\pi R$ gives a very good approximation. Then

$$\phi = BA = \frac{\mu A}{l} Ni = \frac{Ni}{l/\mu A}$$

We note that ϕ is proportional to $F = Ni$ and inversely proportional to a quantity $l/\mu A$ involving only the geometry and material of the core. This quantity is called *reluctance*. In Appendix 1 the reader will find a short introduction to the more general use of this quantity.

Returning to our calculation of inductance we have, finally, using the assumption that all the flux ϕ links all N turns,

$$L = \frac{\lambda}{i} = \frac{N\phi}{i} = \frac{\mu A N^2}{l} = \frac{\mu A N^2}{2\pi R}$$

3.2 FLUX LINKAGES OF INFINITE STRAIGHT WIRE

We next consider a case of great interest to us in modeling transmission lines. This is the case of the flux linkages of an infinite straight wire. The infinite wire is an approximation of a reasonably long wire and is simpler to consider for the calculation of parameters. We should note at the outset that a single infinite wire is certainly not an approximation of a practical transmission line but its consideration is important to us as a step in using superposition to find inductance parameters for realistic multiconductor lines.

The notion of flux linkages for a single straight wire needs some justification. More commonly, flux linkages are defined for a coil where the interlocking of the turns of the coil and the lines of flux are easy to visualize. The present case is less familiar but the interlocking is also clear; there is no way to visualize how the lines of flux may be detached from the infinite wire. Alternatively, and consistent with this, we may imagine the wire to close at infinity ($-\infty$ and $+\infty$ the same point), thus establishing a kind of "one-turn coil" with the return path at infinity.

Assume then a

1. Straight infinitely long wire of radius r.
2. Uniform current density in the wire. Total current is i.

In the case of dc or low-frequency ac, assumption 2 is justified in practice.

From elementary physics we know that the flux lines form concentric circles (i.e., **H** is tangential). Associated reference directions for flux and current are shown in Fig. 3.2. Because of angular symmetry it suffices to consider $H(x)$.

Case 1: Assume that $x > r$ (i.e., points outside the conductor). Ampère's circuital law applied to the path Γ_1 gives

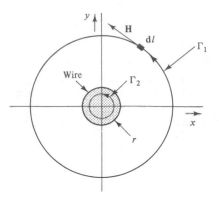

Figure 3.2 Wire carrying current i.

$$\oint_{\Gamma_1} \mathbf{H} \cdot d\mathbf{l} = H \cdot 2\pi x = i$$

$$H = \frac{i}{2\pi x} \tag{3.9a}$$

We note that for current out of the paper $\mathbf{H}$ is directed upward for x positive. *Case 2:* Assume that $x \leq r$ (i.e., points inside the conductor). Ampère's law applied to Γ_2 gives

$$H \cdot 2\pi x = i_e = \frac{\pi x^2}{\pi r^2} i$$

where we have used the assumption of uniform current density. Then

$$H = \frac{x}{2\pi r^2} i \tag{3.9b}$$

To find the flux density we use (3.3) but write $\mu = \mu_r \mu_0$, where in the SI system of units, $\mu_0 = 4\pi \times 10^{-7}$ is the permeability of free space and μ_r is the relative permeability. Thus

$$B = \mu_r \mu_0 H \tag{3.10}$$

Outside the conductor, in air, $\mu_r \approx 1$, and inside $\mu_r \approx 1$, if the conductor is non-magnetic (e.g., copper or aluminum). In most cases, however, steel strands are included for strength and $\mu_r > 1$.

We next consider the flux linkages associated with the infinite wire. It is evident that the flux linkages of the infinite wire are infinite, and we are led to consider, instead, the flux linkages of the wire per meter of length and consider only the flux within a finite radius R. To find these flux linkages, consider Fig. 3.3. The flux per meter surrounding the wire out to a radius R is the same as the flux crossing the 1 meter $\times$ R meter rectangle. For convenience the rectangle is taken to be in the horizontal plane.

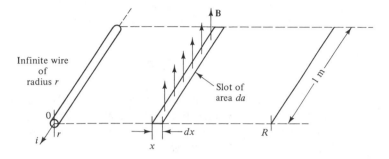

Figure 3.3 Flux crossing rectangle.

There are two components to the flux linkages. They are the linkages due to the flux outside the wire and some additional linkages due to flux inside the wire. We consider the two cases separately and then add the results.

1. *Flux linkages outside the wire:* Because each line of flux interlocks with the wire only once, the total flux linkages are numerically the same as the total flux crossing the rectangle in Fig. 3.3 from the radius r to R. Thus, using the subscript 1 for the external flux contributions,

$$\lambda_1 = \phi_1 = \int_A \mathbf{B} \cdot d\mathbf{a} = \int_r^R B(x)\, dx = \mu_0 \int_r^R \frac{i}{2\pi x}\, dx$$

$$= \frac{\mu_0 i}{2\pi} \ln \frac{R}{r} \tag{3.11}$$

Here we have integrated over the differential areas $da = 1 \cdot dx$ shown in Fig. 3.3, with B found from (3.9a) by using $B = \mu_0 H$.

2. *Flux linkages inside the wire:* This portion of the flux links only part of the wire, and it would not be reasonable to assign it full weight in calculating flux linkages. A careful analysis based on energy justifies the idea that each flux contribution to the flux linkages should be weighted by the fraction of the total current linked. The result, although intuitively reasonable, is by no means obvious. Then using (3.9b), (3.10), and the assumption of uniform current density, an integration over the rectangle inside the wire yields

$$\lambda_2 = \mu_r \mu_0 \int_0^r \frac{x}{2\pi r^2} \frac{\pi x^2}{\pi r^2} i\, dx$$

$$= \frac{\mu_r \mu_0 i}{8\pi} \tag{3.12}$$

The result is independent of wire radius. Adding (3.11) and (31.2), the total flux linkages per meter are

$$\lambda = \lambda_1 + \lambda_2 = \frac{\mu_0 i}{2\pi}\left(\frac{\mu_r}{4} + \ln\frac{R}{r}\right)$$

$$= 2 \times 10^{-7} i\left(\frac{\mu_r}{4} + \ln\frac{R}{r}\right) \tag{3.13}$$

In the second line we have substituted the numerical value of μ_0 to get a form more useful for calculations. We will freely exchange between the two forms but prefer the first form for development of the theory. Regarding (3.13), note that λ increases without bound as R goes to infinity. This is a difficulty we will dispose of when the practical general case is considered.

3.3 FLUX LINKAGES; MANY-CONDUCTOR CASE

We turn now to the practical case. Suppose that instead of one conductor we have n round conductors as suggested by Fig. 3.4. Some of these conductors are the "return" conductors for the rest, but it is equivalent and more convenient to think of each wire as having its return path at infinity (i.e., the case we considered in the preceding section). Strictly speaking, the assumption of uniform current density in each inductor needs to be justified once the fields and currents interact. However, if the distance between conductors is fairly large compared with their radii, the assumption is justified and is convenient because we can then use superposition.

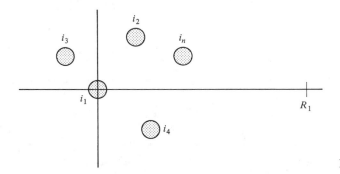

Figure 3.4 Many conductors.

We are still interested in calculating the flux linkages of conductor 1 up to a radius R_1 from the origin, but now we add the contributions due to conductors 2, 3, . . . , n. We use superposition. Consider the contribution of the kth conductor current as shown in Fig. 3.5. Assume that all other currents are zero. The flux lines due to i_k are concentric circles, as shown in Fig. 3.5. The flux line ψ_1 does not link conductor 1. The flux line ψ_3 does. The flux line ψ_5 links conductor 1 but is beyond the radius R_1; it does not qualify. The flux lines ψ_2 and ψ_4 mark the approximate extremes of flux lines that qualify. Thus we are interested in all the flux generated by i_k that passes between the points b and c on the x axis. The calculation is much simpler

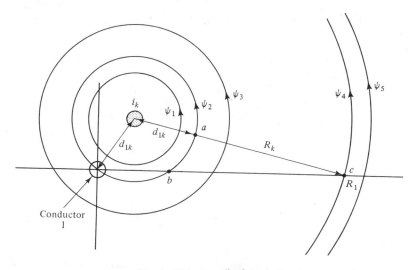

Figure 3.5   Contribution of i_k.

however if we find the flux crossing the surface defined by the points a and c because then the flux and surface are perpendicular. Defining the distance from the center of the kth conductor to the point c as R_k, the flux to be calculated is simply that between a radius d_{1k} and R_k. But this calculation was already done earlier; the result given in (3.11) is directly applicable to the radii presently under consideration. We get as the contribution of i_k to the flux linkages of coil 1,

$$\lambda_{1k} = \frac{\mu_0 i_k}{2\pi} \ln \frac{R_k}{d_{1k}} \tag{3.14}$$

In the calculation the partial flux linkages of conductor 1 are neglected.

We sum the remaining contributions similarly to get the total flux linkage of coil 1 up to a radius R_1 from conductor 1.

$$\lambda_1 = \frac{\mu_0}{2\pi} \left[i_1 \left(\frac{\mu_r}{4} + \ln \frac{R_1}{r_1} \right) + i_1 \ln \frac{R_2}{d_{12}} + \cdots + i_n \ln \frac{R_n}{d_{1n}} \right] \tag{3.15}$$

We note that in general $\lambda_1 \to \infty$ as $R_1 \to \infty$, but in the practical case we can avoid this difficulty by making a key assumption regarding the instantaneous currents in the wires. Assume that

$$i_1 + i_1 + \cdots + i_n = 0 \tag{3.16}$$

in which case we can show that (3.15) does not diverge as $R_1 \to \infty$. The assumption (3.16) is reasonable for a transmission line under normal operating conditions. It is certainly the intention of the system designers that all the currents should flow in the wires of the transmission line. The condition, which may be violated under abnormal (fault) conditions, is reconsidered in Chapter 13.

Rewriting (3.15) in a way that permits the assumption (3.16) to be introduced,

$$\lambda_1 = \frac{\mu_0}{2\pi}\left[i_1\left(\frac{\mu_r}{4} + \ln\frac{1}{r_1}\right) + i_2 \ln\frac{1}{d_{12}} + \cdots + i_n \ln\frac{1}{d_{1n}}\right]$$

$$+ \frac{\mu_0}{2\pi}(i_1 \ln R_1 + i_2 \ln R_2 + \cdots + i_n \ln R_n) \qquad (3.17)$$

To the second part of (3.17) add

$$-\frac{\mu_0}{2\pi}(i_1 \ln R_1 + i_2 \ln R_1 + \cdots + i_n \ln R_1)$$

This is legitimate since, by (3.16), we have added zero. Then the second part of (3.17) becomes

$$\frac{\mu_0}{2\pi}\left(i_1 \ln\frac{R_1}{R_1} + i_2 \ln\frac{R_2}{R_1} + \cdots + i_n \ln\frac{R_n}{R_1}\right)$$

the first term is zero and in the limit as $R_1 \rightarrow \infty$, each other term R_k/R_1 tends to 1, and each logarithmic term then tends to zero. Thus in the limit as $R_1 \rightarrow \infty$, (3.17) reduces to

$$\lambda_1 = \frac{\mu_0}{2\pi}\left[i_1\left(\frac{\mu_r}{4} + \ln\frac{1}{r_1}\right) + \cdots + i_n \ln\frac{1}{d_{1n}}\right] \qquad (3.18)$$

Equation (3.18) can be made more symmetrical by noting the identity

$$\frac{\mu_r}{4} = \ln \epsilon^{\mu_r/4} \qquad (3.19)$$

Thus

$$\frac{\mu_r}{4} + \ln\frac{1}{r_1} = \ln \epsilon^{\mu_r/4} + \ln\frac{1}{r_1} = \ln\frac{1}{r_1}\epsilon^{\mu_r/4} = \ln\frac{1}{r_1'} \qquad (3.20)$$

where

$$r_1' \triangleq r_1\epsilon^{-\mu_r/4}$$

Then (3.18) becomes

$$\lambda_1 = \frac{\mu_0}{2\pi}\left(i_1 \ln\frac{1}{r_1'} + i_2 \ln\frac{1}{d_{12}} + \cdots + i_n \ln\frac{1}{d_{1n}}\right) \qquad (3.21)$$

For a nonmagnetic wire, $\mu_r = 1$, and $r_1' = r_1e^{-1/4} = 0.7788r_1 \approx 0.78r_1$. Interpreted physically, r_1' is the radius of an equivalent hollow conductor with the same flux linkages as the solid conductor of radius r_1. Ampère's circuital law makes it clear that the hollow conductor has no internal flux. However, since r_1' is less than r_1, the external flux is increased to compensate for this lack.

Note that the flux linkages of conductor 1 depend on all the currents $i_1, i_2, \ldots, i_n$. The equation thus is in the form $\lambda_1 = l_{11}i_1 + l_{12}i_2 + \cdots + l_{1n}i_n$, where the inductance parameters l_{1j} depend on geometry. Similarly, with obvious modifications, the flux linkages per meter of the general kth conductor are

$$\lambda_k = \frac{\mu_0}{2\pi}\left(i_1 \ln \frac{1}{d_{k1}} + \cdots + i_k \ln \frac{1}{r_k'} + \cdots + i_n \ln \frac{1}{d_{kn}} \right) \qquad (3.22)$$

in the form $\lambda_k = l_{k1}i_1 + l_{k2}i_1 + \cdots + l_{kn}i_n$. Thus it is clear that there is mutual inductance in the circuit description.

Example 3.2

Calculate the inductance per meter of each phase of a three-phase transmission line (Fig. E3.2). Assume that

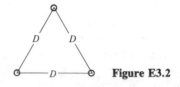

Figure E3.2

1. Conductors are equally spaced, D, and have equal radii r.
2. $i_a + i_b + i_c = 0$.

Solution   Using (3.21) for phase a, wc have

$$\lambda_a = \frac{\mu_0}{2\pi}\left(i_a \ln \frac{1}{r'} + i_b \ln \frac{1}{D} + i_c \ln \frac{1}{D} \right)$$

$$= \frac{\mu_0}{2\pi}\left(i_a \ln \frac{1}{r'} - i_a \ln \frac{1}{D} \right)$$

$$= \frac{\mu_0}{2\pi} i_a \ln \frac{D}{r'}$$

Thus

$$l_a = \frac{\lambda_a}{i_a} = \frac{\mu_0}{2\pi} \ln \frac{D}{r'} = 2 \times 10^{-7} \ln \frac{D}{r'} \qquad (3.23)$$

In the derivation use has been made of the equilateral spacing and the fact that $i_a = -(i_b + i_c)$. Because of symmetry we get the same results for λ_b and λ_c. Thus we get the self-inductances $l_a = l_b = l_c = (\mu_0/2\pi) \ln (D/r')$ henrys per meter. We also note that l_a depends on i_a alone (i.e., there are no mutual inductance terms in this particular description). The same is true for b and c phases.

The example illustrates some important points.

1. Although physically there is magnetic coupling between phases, if $i_a + i_b + i_c = 0$ and with equilateral spacing we can model the magnetic effect using only self-inductances. The self-inductances are equal. This then justifies the use of per phase analysis.

2. To reduce the inductance per meter we can try to reduce the spacing between conductors and increase their radii. Reducing spacing can only go so far because of considerations of voltage flashover. On the other hand, there are cost and weight problems associated with increasing the radii of solid conductors and problems of flexibility and ease of handling as well as cost in the case of hollow conductors. The problem is neatly solved by the practice of conductor "bundling" to be considered next.

3.4 CONDUCTOR BUNDLING

Suppose that instead of one conductor per phase there are b conductors in close proximity as compared with the spacing between phases. Such a composite conductor is said to be made up of bundled conductors. The conductors within a bundle are supported at frequent intervals by a conducting frame, as suggested in Fig. 3.6, and thus these conductors are effectively in parallel. In the figure is shown an equilaterally spaced set of phase conductors each composed of four bundled conductors. Typically, two, three, or four conductors per bundle are used.

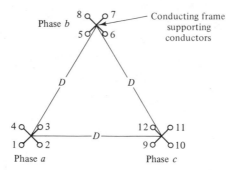

Figure 3.6 Bundled conductors.

We will now consider the configuration of Fig. 3.6 in more detail. Suppose that the spacing between bundle centers is D, where D is large relative to the spacing between the conductors of the same phase. Assume that all the conductors have the same radius r.

Consider the flux linkages of conductor 1 in the phase a bundle. Assume for simplicity that the current in each phase splits equally among the four parallel branches. We can use (3.21) and get the flux linkages of conductor 1,

$$\lambda_1 = \frac{\mu_0}{2\pi}\left[\frac{i_a}{4}\left(\ln\frac{1}{r'} + \ln\frac{1}{d_{12}} + \ln\frac{1}{d_{13}} + \ln\frac{1}{d_{14}}\right)\right.$$
$$+ \frac{i_b}{4}\left(\ln\frac{1}{d_{15}} + \ln\frac{1}{d_{16}} + \ln\frac{1}{d_{17}} + \ln\frac{1}{d_{18}}\right)$$
$$\left. + \frac{i_c}{4}\left(\ln\frac{1}{d_{19}} + \ln\frac{1}{d_{1,10}} + \ln\frac{1}{d_{1,11}} + \frac{1}{d_{1,12}}\right)\right]$$

$$= \frac{\mu_0}{2\pi}\left[i_a \ln \frac{1}{(r'd_{12}d_{13}d_{14})^{1/4}} + i_b \ln \frac{1}{(d_{15}d_{16}d_{17}d_{18})^{1/4}} + i_c \ln \frac{1}{(d_{19}d_{1,10}d_{1,11}d_{1,12})^{1/4}}\right]$$

$$= \frac{\mu_0}{2\pi}\left(i_a \ln \frac{1}{R_b} + i_b \ln \frac{1}{D_{1b}} + i_c \ln \frac{1}{D_{1c}}\right) \tag{3.24}$$

where we have introduced the definitions

$$R_b \triangleq (r'd_{12}d_{13}d_{14})^{1/4}$$

= geometric mean radius (GMR) of bundle

$$D_{1b} \triangleq (d_{15}d_{16}d_{17}d_{18})^{1/4}$$

= geometric mean distance (GMD) from conductor 1 to phase b

$$D_{1c} \triangleq (d_{19}d_{1,10}d_{1,11}d_{1,12})^{1/4}$$

= GMD from conductor 1 to phase c

Noting that $D_{1b} \approx D_{1c} \approx D$, and assuming that $i_a + i_b + i_c = 0$, we get, approximately,

$$\lambda_1 = \frac{\mu_0}{2\pi} i_a \ln \frac{D}{R_b} \tag{3.25}$$

Comparing (3.25) with (3.23), we see that the only difference is that r' is replaced by the (much larger) geometric mean radius of the bundle, R_b.

In calculating the inductance per meter of phase a, we can proceed intuitively. First calculate the inductance per meter of conductor 1. Noting that the current in conductor 1 is $i_a/4$, we are led to

$$l_1 = \frac{\lambda_1}{i_a/4} = 4\left(\frac{\mu_0}{2\pi}\right) \ln \frac{D}{R_b} \tag{3.26}$$

Next, calculating λ_2, we find the same value of R_b as above; the reason is the symmetrical spacing of conductors. In calculating GMDs from conductor 2 to phases b and c, we find that $D_{2b} \approx D_{2c} \approx D$ because of the large spacing between phases. Thus $l_2 \approx l_1$. In this manner we find that $l_1 \approx l_2 \approx l_3 \approx l_4$. Then, since we have four approximately equal inductors (l_1, l_2, l_3, l_4) in parallel,

$$l_a \approx \frac{l_1}{4} = \frac{\mu_0}{2\pi} \ln \frac{D}{R_b} = 2 \times 10^{-7} \ln \frac{D}{R_b} \tag{3.27}$$

Checking the steps in the calculation, we see also that $l_a = l_b = l_c$.

We have been considering the case of four bundled conductors per phase. The reader should check that (3.27) holds also for any other bundled configuration in which the bundled conductors are symmetrically spaced around a circle and the distances between phases are equal and large compared with the spacing of conductors within each phase. Of course, the GMR must be defined appropriately; with b conductors we get

$$R_b \triangleq (r'd_{12}, \ldots, d_{1b})^{1/b} \qquad b \geq 2 \qquad (3.28)$$

Note: We can define $R_b = r'$ if $b = 1$ (i.e., the nonbundled case), and then (3.27) includes (3.23) as a special case.

Example 3.3

Find the GMR of three symmetrically spaced conductors (Fig. E3.3). Assume that $r = 2$ cm and $r' = 2\epsilon^{-1/4} = 1.56$ cm.

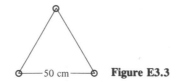

—50 cm—   **Figure E3.3**

Solution

$$R_b = (1.56 \times 50 \times 50)^{1/3} = 15.7 \text{ cm}.$$

Example 3.4

Find the GMR of four symmetrically spaced subconductors (Fig. E3.4). Assume that $r = 2$ cm and $r' = 1.56$ cm.

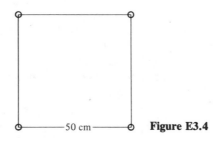

—50 cm—   **Figure E3.4**

Solution

$$R_b = (1.56 \times 50 \times 50 \times 50\sqrt{2})^{1/4} = 22.9 \text{ cm}.$$

In Example 3.4 the inductance of the line has been reduced by a factor 2.69 compared with a line with single conductors of the same radius or by a factor 1.44 compared with single conductors with the same amount of material as in the bundles.

Three additional comments about bundling:

1. If we view the bundle as an approximation of a hollow conductor, the reason for the increased "radius" is intuitively clear.
2. The larger "radius" helps in another respect. At high voltages, above approximately 230 kV, the electric field strength near conductors is sufficiently high to ionize the air nearby. This phenomenon, called *corona,* has an undesirable effect since it is associated with line losses, radio interference, and audible noise. All other things being equal, the larger the conductor radius, the less the electric field

strength at the surface of the conductor. Bundling is beneficial since it effectively increases the conductor radius.

3. Compared with a single conductor of the same cross-sectional area, bundled conductors, having a larger surface area exposed to the air, are better cooled. Thus higher currents may be carried without exceeding thermal limits. More on this point later.

Exercise 1.   In checking the validity of comment 1 above, consider the GMR of a sequence of b symmetrically spaced subconductors all at a fixed radius 1 (Fig. EX1). As $b \to \infty$ the spacing between conductors necessarily tends to zero (as does r') and the configuration looks like a hollow conductor of radius 1. In this case we get $l = (\mu_0/2\pi) \ln (D/1)$ and we expect that $R_b = 1$. Do you believe it? As a check pick $r = 0.1$, pick $b = 31$, obtain the values of d_{ij} graphically, and calculate R_b. Note that for $b = 31$ we approximate a hollow conductor.

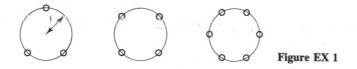

Figure EX 1

The simple result of Example 3.2 has been extended in one direction; with multiple conductors per phase instead of a single conductor we use the GMR, R_b, instead of r'. In Section 3.5 we extend the result in another direction.

3.5 TRANSPOSITION

In practice the equilateral arrangement of phases discussed in the preceding two sections is usually not convenient. It is usually more convenient to arrange the phases in a horizontal or vertical configuration.

In this case symmetry is lost. One way to regain the symmetry and restore balanced conditions is to use the method of transposition of lines. This is illustrated in Fig. 3.7. We can think of this as a top view of three conductors in the same horizontal plane. Equally well, it could be a side view of three conductors in the same vertical plane or even a view of three conductors not in the same plane. In this case, to visualize the transposition, think of a line crew that sets out the insulators on the

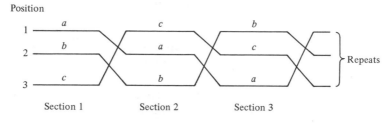

Figure 3.7   Transposed lines.

poles or towers in an arbitrary but fixed configuration labeled 1, 2, 3 on every pole. Another crew then strings the wires, rotating each phase, at intervals, through each insulator position. For example, phase a starts in position 1, gets rotated to position 2, then position 3, and then the cycle repeats.

From the way the line is constructed it seems physically reasonable that the average inductance of each phase will be the same. We now wish to calculate that inductance.

Assume that

1. Each phase occupies each position for the same fraction of the total length of line.
2. Each phase is a single (nonbundled) conductor of radius r.
3. $i_a + i_b + i_c = 0$.

Next we calculate the average flux linkages per meter of phase a. This is

$$\bar{\lambda}_a = \tfrac{1}{3}(\lambda_a^{(1)} + \lambda_a^{(2)} + \lambda_a^{(3)}) \tag{3.29}$$

where $\lambda_a^{(i)}$ is the flux linkage per meter in the ith section, $i = 1, 2, 3$. Substituting in (3.29) from (3.21) and using the transposition cycle in Fig. 3.7, we get

$$\bar{\lambda}_a = \frac{1}{3} \times \frac{\mu_0}{2\pi} \left(i_a \ln \frac{1}{r'} + i_b \ln \frac{1}{d_{12}} + i_c \ln \frac{1}{d_{13}} \right.$$
$$\left. + i_a \ln \frac{1}{r'} + i_b \ln \frac{1}{d_{23}} + i_c \ln \frac{1}{d_{12}} + i_a \ln \frac{1}{r'} + i_b \ln \frac{1}{d_{13}} + i_c \ln \frac{1}{d_{23}} \right) \tag{3.30}$$

Each of the rows in (3.30) is the application of (3.21) to conductor a as it finds itself successively in positions 1, 2, and 3. We can group terms, as follows:

$$\bar{\lambda}_a = \frac{\mu_0}{2\pi} \left(i_a \ln \frac{1}{r'} + i_b \ln \frac{1}{D_m} + i_c \ln \frac{1}{D_m} \right)$$
$$= \frac{\mu_0}{2\pi} i_a \ln \frac{D_m}{r'} \tag{3.31}$$

where $D_m \triangleq (d_{12}d_{23}d_{13})^{1/3}$ = geometric mean of the distances between positions 1, 2, and 3. Thus we get the average inductance per meter:

$$\bar{l}_a = \frac{\mu_0}{2\pi} \ln \frac{D_m}{r'} \tag{3.32}$$

The formula is the same as (3.23) except that D is replaced by D_m. It is clear, because of the transposition, that $\bar{l}_a = \bar{l}_b = \bar{l}_c$. We note that this result is consistent with the use of per phase analysis.

Suppose now that instead of being single round conductors, the conductors are bundled. How does this affect (3.32)? Comparing (3.27) and (3.23), we can guess that the result will be

$$\bar{l}_a = \bar{l}_b = \bar{l}_c = \bar{l} = \frac{\mu_0}{2\pi} \ln \frac{D_m}{R_b} = 2 \times 10^{-7} \ln \frac{D_m}{R_b} \tag{3.33}$$

which, in fact, is correct. This can be shown as follows. If we consider section 1 of the transposed line (using appropriate notational and other modifications of (3.24) to accommodate the unequal spacing of phases), we get the approximate formula for the flux linkages of conductor 1:

$$\lambda_{1a}^{(1)} = \frac{\mu_0}{2\pi}\left(i_a \ln \frac{1}{R_b} + i_b \ln \frac{1}{d_{ab}^{(1)}} + i_c \ln \frac{1}{d_{ac}^{(1)}} \right) \tag{3.34}$$

where $d_{ab}^{(1)}$ and $d_{ac}^{(1)}$ are the section 1 spacings between phases a and b, and a and c, respectively. Equation (3.34) corresponds to (3.24) with the approximations $d_{ab}^{(1)} \approx D_{1b}$, $d_{ac}^{(1)} \approx D_{1c}$. We next calculate

$$\bar{\lambda}_{1a} = \tfrac{1}{3}[\lambda_{1a}^{(1)} + \lambda_{1a}^{(2)} + \lambda_{1a}^{(3)}]$$

and get

$$\bar{\lambda}_{1a} = \frac{1}{3} \times \frac{\mu_0}{2\pi}\left(i_a \ln \frac{1}{R_b} + i_b \ln \frac{1}{d_{ab}^{(1)}} + i_c \ln \frac{1}{d_{ac}^{(1)}} + i_a \ln \frac{1}{R_b} \right.$$
$$\left. + i_b \ln \frac{1}{d_{ab}^{(2)}} + i_c \ln \frac{1}{d_{ac}^{(2)}} + i_a \ln \frac{1}{R_b} + i_b \ln \frac{1}{d_{ab}^{(3)}} + i_c \ln \frac{1}{d_{ac}^{(3)}} \right) \tag{3.35}$$

Equation (3.35) corresponds to (3.30) with appropriate changes in notation. Regrouping terms gives us

$$\bar{\lambda}_{1a} = \frac{\mu_0}{2\pi}\left(i_a \ln \frac{1}{R_b} + i_b \ln \frac{1}{D_m} + i_c \ln \frac{1}{D_m} \right) = \frac{\mu_0}{2\pi} i_a \ln \frac{D_m}{R_b} \tag{3.36}$$

where $D_m \triangleq [d_{ab}^{(1)} d_{ab}^{(2)} d_{ab}^{(3)}]^{1/3} = [d_{ac}^{(1)} d_{ac}^{(2)} d_{ac}^{(3)}]^{1/3}$. The notation obscures the equality of the two bracketed terms; the reader should check that they are equal because of the transposition of phases. By exactly the same technique used in going from (3.25) to (3.27), we go from (3.36) to (3.33). It is again evident that $\bar{l}_a = \bar{l}_b = \bar{l}_c$ in this case also.

We note that in practice, even when the lines are not consistently transposed it is convenient and causes little error to calculate the inductance as if they were transposed. Note also that with the definition $R_b = r'$, for $b = 1$, (3.33) covers all combinations of cases we have considered.

Example 3.5

Find the inductance per meter of the 3ϕ line shown in Fig. E3.5. The conductors are aluminum ($\mu_r = 1$), with radius $r = 0.5$ in.

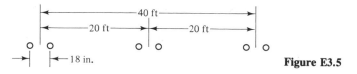

Figure E3.5

Solution   Although for simplicity the problem is stated without precision, it is generally understood that what is required is the *average* inductance of the transposed line. Thus

$$D_m = (20 \text{ ft} \times 20 \text{ ft} \times 40 \text{ ft})^{1/3} = 25.2 \text{ ft}$$

$$R_b = (0.78 \times 0.5 \times 18)^{1/2} = 2.65 \text{ in.} = 0.22 \text{ ft}$$

$$\bar{l}_a = \frac{\mu_0}{2\pi} \ln \frac{D_m}{R_b} = 2 \times 10^{-7} \ln \frac{25.2}{0.22} = 9.47 \times 10^{-7} \text{ H/m}$$

Note: Although line dimensions are in feet the result of the calculation is in henries per meter. We have used $\mu_0 = 4\pi \times 10^{-7}$.

In introducing the calculation of inductance, we have simplified matters by making some reasonable assumptions and approximations. In most cases we can extend the analysis and obtain more accurate results without undue extra labor. A case in point is the calculation of the geometric mean radius (GMR) for stranded conductors. This is an important practical case since almost invariably conductors are stranded rather than solid. The method used to calculate the GMR for bundled conductors extends easily to this application. The results are tabulated in standard references for the different types of conductors. For example, for the aluminum conductor, steel reinforced (ACSR), labeled "Bluejay," with 7 (central) strands of steel and 45 strands of aluminum, the outside diameter is 1.259 in. and the GMR is 0.4980 in. For stranded conductors we get more accurate results by substituting the conductor GMR in place of r' in all our formulas. However, even if we ignore the stranding, our results are usually quite accurate. In the Bluejay example just quoted, if we neglect the stranding (and the presence of iron) and assume a solid round aluminum conductor of the same outside diameter, we get $r' = 0.7788 \times 1.259/2 = 0.4903$ in., compared with the GMR of 0.4980. This very small error is typical for ACSR conductors.

A similar extension permits the use of an exact geometric mean distance (GMD) between phases instead of the approximation used here. For the details of these calculations, the reader is referred to standard references.

Although inductance is the most important transmission-line parameter, capacitance is also important and must be considered in modeling medium-length and long lines. We turn next to the calculation of this parameter.

3.6 REVIEW OF ELECTRIC FIELDS

We start with some basic facts from physics.

Gauss's law.   Gauss's law states that

$$\int_A \mathbf{D} \cdot d\mathbf{a} = q_e \tag{3.37}$$

where $\mathbf{D}$ = electric flux density vector, coulombs/meter^2
          $d\mathbf{a}$ = differential area da with direction normal to the surface, meters^2
          A = total closed surface area, meters^2
          q_e = algebraic sum of all charges enclosed by A, coulombs

As in the case of Ampère's circuital law, Gauss's law is implicit in $\mathbf{D}$, the variable of interest, but can be usefully applied when symmetries are present, as in the following example.

Example 3.6

Find the field of an infinite uniformly charged straight round wire.

Solution Draw a cylindrical Gaussian surface concentric with the wire and h meters long (Fig. E3.6, where the charge on the wire is q c/m of length). Considerations of symmetry indicate that $\mathbf{D}$ is radial and constant in magnitude over the curved portion of the cylinder (it is zero on the end caps). Thus using Gauss's law,

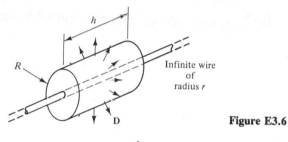

Figure E3.6

$$\int_A \mathbf{D} \cdot d\mathbf{a} = D\,2\pi R h = qh$$

Then

$$D = \frac{q}{2\pi R} \qquad R \geq r \tag{3.38}$$

where D is the scalar version of $\mathbf{D}$. Knowing that $\mathbf{D}$ is directed radially, we also have

$$\mathbf{D} = \mathbf{a}_r \frac{q}{2\pi R} \qquad R \geq r \tag{3.39}$$

where $\mathbf{a}_r$ is a radially directed unit vector.

Electric field E.   In a homogeneous medium the electric field intensity $\mathbf{E}$ is related to $\mathbf{D}$ by

$$\mathbf{D} = \epsilon\mathbf{E} \tag{3.40}$$

where the units of $\mathbf{E}$ are volts per meter, and in free space, in the SI system of units, $\epsilon = \epsilon_0 = 8.854 \times 10^{-12}$ farads per meter. In other media we may write $\epsilon = \epsilon_r\epsilon_0$, where ϵ_r is called the relative permittivity. The relative permittivity for dry air may be taken to be 1.0 with negligible error.

Voltage difference $v_{\beta\alpha}$. We can next find the voltage difference between any two points P_α and P_β by integrating $\mathbf{E}$ (in volts per meter) along any path joining the two points. From physics,

$$v_{\beta\alpha} \triangleq v_{P_\beta} - v_{P_\alpha} = -\int_{P_\alpha}^{P_\beta} \mathbf{E} \cdot d\mathbf{l} \qquad (3.41)$$

We will now apply the result in Example 3.6 to find an expression for line capacitance.

3.7 LINE CAPACITANCE

Roughly speaking, capacitance relates charge to voltage and hence we consider next the voltage differences associated with the infinite charged line of Example 3.6. Since the electric field is radial, we can think of a path in which we get from P_α to P_β by first moving parallel to the wire and then at a fixed radius until we can reach P_β by a purely radial path. The integrand in (3.41) will be zero for the first two segments, and for the third segment we get

$$v_{\beta\alpha} = v_{P_\beta} - v_{P_\alpha} = -\int_{R_\alpha}^{R_\beta} \frac{q}{2\pi\epsilon R}\, dR = \frac{q}{2\pi\epsilon} \ln \frac{R_\alpha}{R_\beta} \qquad (3.42)$$

where R_α is the (radial) distance of P_α from the wire and R_β is the (radial) distance of P_β from the wire. Note that if q is positive and P_β is closer to the wire than P_α, its potential will be higher than that of P_α. Note that we have not yet considered the "voltage" of the line. To do this we could let P_β be a point on the line and let P_α be the reference point (i.e., the point whose voltage is the datum or reference voltage). An attractive reference point would be the point at infinity. Unfortunately, in this case $R_\alpha \to \infty$ and from (3.42) we see the voltage diverges. This difficulty, similar to the one in Section 3.2 in connection with flux linkages, has the same resolution; in the many-conductor case, under reasonable assumptions, the line-to-datum voltages are well defined, even if the datum point is at infinity.

Consider next the multiconductor case illustrated in Fig. 3.8. Shown here are cross sections of an infinite line with n conductors, the charges per meter of these

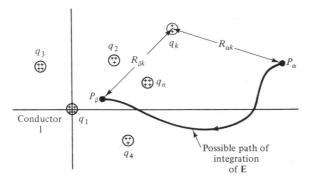

Figure 3.8 Conductor cross sections.

conductors, and a possible path of integration to evaluate the potential of P_β relative to P_α. With conductor radii small compared with the spacing between conductors, we can assume that the effect of the interaction between the charges on different conductors is negligible. In this case we can evaluate (3.41) using superposition. We are thus led to the multiconductor version of (3.42):

$$v_{\beta\alpha} = v_{P_\beta} - v_{P_\alpha} = \frac{1}{2\pi\epsilon} \sum_{i=1}^{n} q_i \ln \frac{R_{\alpha i}}{R_{\beta i}} \tag{3.43}$$

We next make a key assumption regarding the instantaneous charge densities in any cross section of the line. Assume that

$$q_1 + q_2 + \cdots + q_n = 0 \tag{3.44}$$

which is analogous to (3.16). Proceeding as in the development following (3.16), we then rewrite (3.43) in a way that facilitates the introduction of (3.44):

$$v_{\beta\alpha} = \frac{1}{2\pi\epsilon} \sum_{i=1}^{n} q_i \ln \frac{1}{R_{\beta i}} + \frac{1}{2\pi\epsilon} \sum_{i=1}^{n} q_i \ln R_{\alpha i} \tag{3.45}$$

To the second part of (3.45) add

$$-\frac{1}{2\pi\epsilon} \sum_{i=1}^{n} q_i \ln R_{\alpha 1} \tag{3.46}$$

Because of (3.44) we have added zero. On the other hand, the second part of (3.45) now becomes

$$\frac{1}{2\pi\epsilon} \sum_{i=1}^{n} q_i \ln \frac{R_{\alpha i}}{R_{\alpha 1}} \tag{3.47}$$

Now we are in a position to pick point P_α as reference and move it out toward infinity. As we do this, all the ratios $R_{\alpha i}/R_{\alpha 1} \to 1$ and each logarithmic term individually tends toward zero. In the limit (3.47) becomes zero. We can use the notation v_β to indicate voltage at any point P_β with respect to the reference.

$$v_\beta = \frac{1}{2\pi\epsilon} \sum_{i=1}^{n} q_i \ln \frac{1}{R_{\beta i}} \tag{3.48}$$

We note that the formula holds for an arbitrary point P_β as long as it is not inside a wire. $R_{\beta i}$ is the distance between the (otherwise) arbitrary point P_β and the center of the ith conductor. If the point β is on the ith conductor surface, then $R_{\beta i}$, the distance between the surface of the ith conductor and its center, reduces to the radius.

The main application of (3.48) is to relate transmission-line voltages and charges. For example, we find the voltage at a point on the surface of conductor 1 in Fig. 3.8.

$$v_1 = \frac{1}{2\pi\epsilon} \left(q_1 \ln \frac{1}{R_{11}} + q_2 \ln \frac{1}{R_{12}} + \cdots + q_n \ln \frac{1}{R_{1n}} \right) \tag{3.49}$$

where R_{11} is the radius of conductor 1, R_{12} the distance from the point to the center of conductor 2, and so on. Now while physical considerations indicate that the surface of conductor 1 is an equipotential surface, application of (3.49) gives slightly different results for different choices of points on the surface. This discrepancy is expected in the light of the approximations made in the derivation. A more detailed investigation of the potential fields indicates that R_{ij} should be replaced by d_{ij}, the spacing between centers of conductors. With small radii compared with interconductor spacings, the differences are insignificant.

Thus we have

$$v_1 = \frac{1}{2\pi\epsilon}\left(q_1 \ln \frac{1}{r_1} + q_2 \ln \frac{1}{d_{12}} + \cdots + q_n \ln \frac{1}{d_{1n}}\right) \tag{3.50}$$

With obvious modifications the equation for the kth conductor voltage is

$$v_k = \frac{1}{2\pi\epsilon}\left(q_1 \ln \frac{1}{d_{k1}} + \cdots + q_k \ln \frac{1}{r_k} + \cdots + q_n \ln \frac{1}{d_{kn}}\right) \tag{3.51}$$

In matrix notation

$$\mathbf{v} = \mathbf{Fq} \tag{3.52}$$

where $\mathbf{v}$ is an n-vector with components $v_1, v_2, \ldots v_n$, $\mathbf{q}$ is an n-vector with components $q_1, q_2, \ldots, q_n$, and $\mathbf{F}$ is an $n \times n$ matrix with typical element $f_{ij} = (1/2\pi\epsilon) \ln (1/d_{ij})$. To find the capacitance parameters, we need the inverse relationship,

$$\mathbf{q} = \mathbf{Cv} \tag{3.53}$$

where $\mathbf{C} = \mathbf{F}^{-1}$.

Note that, in general, there is mutual capacitance between conductors; that is, the charges depend on all the voltages, and vice versa. However, in an important special case we can model with only self-capacitances.

Example 3.7

Calculate an expression for the capacitance per meter of a three-phase transmission line [Fig. E3.7(a)]. Assume that

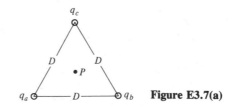

Figure E3.7(a)

1. Conductors are equally spaced, D, and have equal radii r.
2. $q_a + q_b + q_c = 0$.

Solution   Using a, b, c notation instead of 1, 2, 3, we find v_a using (3.50):

$$v_a = \frac{1}{2\pi\epsilon}\left(q_a \ln\frac{1}{r} + q_b \ln\frac{1}{D} + q_c \ln\frac{1}{D}\right)$$

$$= \frac{q_a}{2\pi\epsilon}\ln\frac{D}{r} \tag{3.54}$$

where we have used the assumption $q_a + q_b + q_c = 0$. In the same way we find identical relationships between v_b and q_b, and between v_c and q_c. We are now ready to calculate the three capacitances. To interpret the result physically, however, we digress briefly. Consider the potential of a point p equidistant from the conductors a, b, and c. Using (3.48) gives us

$$v_p = \frac{1}{2\pi\epsilon}\left(q_a \ln\frac{1}{R_{pa}} + q_b \ln\frac{1}{R_{pb}} + q_c \ln\frac{1}{R_{pc}}\right) = 0$$

Thus the point p is at the same potential as the point at infinity and may be taken as the datum point. In fact, construct an imaginary geometric line parallel to the conductors passing through the point P. Every point on this line is at the datum potential. Picking this also to be the neutral potential, we get the relationships between phase-to-neutral voltages and phase charges via three equal phase-neutral (distributed) capacitances:

$$c_a = c_b = c_c = c = \frac{2\pi\epsilon}{\ln(D/r)} \qquad \text{F/m to neutral} \tag{3.55}$$

arranged as shown in Fig. E3.7(b). This symmetric circuit may also be taken to be a representation of the physical cross section of the transmission line.

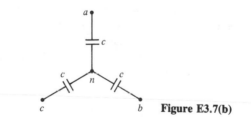

Figure E3.7(b)

In the more general case of unequal interphase spacings and bundled subconductors as described in Section 3.4, we assume that the lines are transposed and get

$$\bar{c}_a = \bar{c}_b = \bar{c}_c = \bar{c} = \frac{2\pi\epsilon}{\ln(D_m/R_b^c)} \qquad \text{F/m to neutral} \tag{3.56}$$

where D_m is the GMD between phases and

$$R_b^c \triangleq (rd_{12}\cdots d_{1b})^{1/b} \qquad b \geq 2 \tag{3.57}$$

We note that the only difference between (3.57) and (3.28) is that r is used instead of r'. If $b = 1$, it is convenient to define $R_b^c = r$ so that we can use (3.56) to cover the nonbundled case.

Example 3.8

Find phase-neutral capacitance and capacitative reactance per mile for a three-phase line with $D_m = 35.3$ ft. Conductor diameter = 1.25 in.

Solution   In air, $\epsilon = \epsilon_r \epsilon_0 = 1.0 \epsilon_0 = 8.854 \times 10^{-12}$. Using this value in (3.56), we have

$$\bar{c} = \frac{2\pi\epsilon}{\ln (D_m/r)} = \frac{2\pi \times 8.854 \times 10^{-12}}{\ln [(35.3 \times 12)/(1.25/2)]} = 8.53 \times 10^{-12} \text{ F/m}$$

We next calculate the susceptance in mhos per meter.

$$B_c = \omega\bar{c} = 2\pi \times 60 \times 8.53 \times 10^{-12} = 3.216 \times 10^{-9} \text{ mho/m}$$

$$= 3.216 \times 10^{-9} \times 1609.34 = 5.175 \times 10^{-6} \text{ mho/mile}$$

Finally, we have the phase-neutral reactance per mile:

$$|X_c| = \frac{1}{B_c} = \frac{1}{5.175 \times 10^{-6}} = 0.193 \text{ M}\Omega\text{/mile}$$

Exercise 2.   Comparing $\bar{l}$ in (3.33) and $\bar{c}$ in (3.56), we see that there is an approximately inverse relationship. Assume that $b = 2$, $\mu_r = 1$, $D_m/R_b = 15$, and show that $\bar{l}\bar{c} \approx \mu_0\epsilon_0$. We note that $(\mu_0\epsilon_0)^{-1/2} = 2.998 \times 10^8$ m/sec is the velocity of light in a vacuum (i.e., a universal constant).

Before leaving the subject of line capacitance it should be noted that we are neglecting the effect of the (conducting) earth under the transmission line. Charges are induced in the earth, and these have some effect on the calculated values of capacitance. The effect is usually quite small for lines of reasonable height operating under normal nonfault conditions.

3.8 TYPICAL PARAMETER VALUES

We conclude with some parameter values for high-voltage lines. Consider the three lines in Fig. 3.9. Some typical data concerning the lines are given in Table 3.1. A few comments on the tabulated values follow.

1. The spacing between the phases increases with the voltage rating. However, by increasing the number of conductors per bundle, the inductance actually decreases with voltage rating.

2. Line resistance increases with temperature and with frequency. In Table 3.1 are shown the line resistance for dc and 60 Hz. At dc the distribution of current across each conductor cross section is uniform. Because of inductive effects, the current distribution changes as the frequency is increased, with higher current densities at the surface. Although the current averaged over the cross section

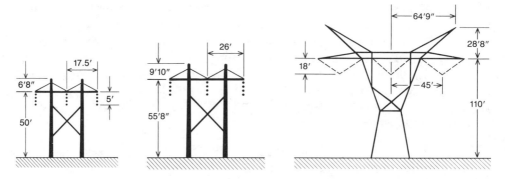

Figure 3.9 Typical high-voltage transmission lines. Copyright © 1977, Electric Power Research Institute. EPRI-EM-285, Synthetic Electric Utility Systems for Evaluating Advances Technologies. Reprinted with permission.

remains the same, the I^2R losses increase and this may be viewed as the reason for the increased R.

Checking the table, we see that the resistance at 60 Hz is only 5% higher than at dc. This is evidence that there is only a slight departure from the uniform current distribution assumed in our calculation of inductance.

3. The resistance is relatively small compared with the inductive reactance. In calculating power flows, voltages, and currents, the resistances have a minor effect compared with the inductive reactances. In some calculations they are neglected. However, in other calculations they are of vital importance. Examples are the calculation of line thermal limits considered in Chapter 4, and in problems of economic operation considered in Chapter 7.

TABLE 3.1 LINE DATA

	Line voltage (kV)		
	138	345	765
Conductors per phase (18-in. spacing)	1	2	4
Number of strands aluminum/steel	54/7	45/7	54/19
Diameter (in.)	0.977	1.165	1.424
Conductor GMR (ft)	0.0329	0.0386	0.0479
Current-carrying capacity per conductor (A)	770	1010	1250
Bundle GMR-R_b (ft)	0.0329	0.2406	0.6916
Flat phase spacing (ft)	17.5	26.0	45.0
GMD phase spacing (ft)	22.05	32.76	56.70
Inductance (H/m $\times$ 10^{-7})	13.02	9.83	8.81
X_L (Ω/mi)	0.789	0.596	0.535
Capacitance (F/m $\times$ 10^{-12})	8.84	11.59	12.78
$|X_c|$ (MΩ/mi)	0.186	0.142	0.129
Resistance (Ω/mi), dc, 50°C	0.1618	0.0539	0.0190
Resistance (Ω), 60 Hz, 50°C	0.1688	0.0564	0.0201
Surge impedance loading (MVA)	50	415	2268

4. For future reference we list surge impedance loading for the three lines. This is a measure of line power-handling capability considered in Chapter 4.

5. On each transmission tower in Fig. 3.9 the reader may have noticed two triangular structures above the horizontal cross member. These structures support "ground" wires which are electrically connected to each tower and thereby to ground. The purpose of these wires is to shield the phase conductors in the event of lightning strokes, and to provide a low-impedance path in the event of a phase-to-ground fault.

In Chapter 4 we consider the modeling of transmission lines using the distributed parameters described in this chapter.

3.9 SUMMARY

For a three-phase line with transposition and bundling, the average per phase inductance is given by

$$l = \frac{\mu_0}{2\pi} \ln \frac{D_m}{R_b} = 2 \times 10^{-7} \ln \frac{D_m}{R_b} \qquad \text{H/m} \tag{3.33}$$

where for convenience we have dropped the $\bar{l}$ notation. D_m is the GMD between phases [i.e., $D_m = (d_{ab}d_{bc}d_{ca})^{1/3}$]. R_b is the GMR of each phase bundle given by $R_b = (r'd_{12} \cdots d_{1b})^{1/b}$, where $r' = r \exp(-\mu_r/4)$ and d_{ij} is the distance between conductors i and j. We are assuming that each phase has the same symmetrical arrangement of bundled conductors all of the same radius r. If there is no bundling, $R_b = r'$. If the GMR of the (stranded) conductors is given, it should be used in the formulas in place of r'.

The formula for average capacitance to neutral is

$$c = \frac{2\pi\epsilon}{\ln(D_m/R_b^c)} \qquad \text{F/m to neutral} \tag{3.56}$$

where $\epsilon = \epsilon_r\epsilon_0 = 1.0\epsilon_0 = 8.854 \times 10^{-12}$ in air, D_m has already been defined, and $R_b^c = (rd_{12} \cdots d_{1b})^{1/b}$ is the same as R_b except r is used instead of r'. We are again assuming a symmetrical arrangement of bundled subconductors all of the same radius r. Without bundling $R_b^c = r$.

Equations (3.33) and (3.56) make the dependence of l and c on geometry quite clear. In this connection the approximately inverse relationship of l and c should be noted.

PROBLEMS

3.1. Suppose that a 60-Hz single-phase power line and an open-wire telephone line are parallel to each other and in the same horizontal plane. The power line spacing is 5 ft; the telephone wire spacing is 12 in. The nearest conductors of the two lines are 20 ft apart. The power line current is 100 A. Find the magnitude of the induced "loop" or "round-trip" voltage per mile in the telephone line. *Hint:* Following the approach in (3.11), find the flux linkages of the telephone wire pair due to the power line currents. Use Faraday's law to find the loop voltage. Assume that the telephone wire radius is negligible.

3.2. Repeat Problem 3.1 but assume that the telephone line has been displaced vertically by 10 ft [i.e., the nearest conductors of the two lines are now $(20^2 + 10^2)^{1/2}$ ft apart].

3.3. Repeat Problem 3.1 but now assume a three-phase power line with a flat horizontal spacing of 5 ft between phases. The power line currents are balanced three-phase each of magnitude 100A. *Hint:* Use phasors to add the flux contributions.

3.4. Suppose that in Problem 3.1 the telephone wires are transposed every 1000 ft. Calculate the induced loop voltage in a 1-mile length of telephone wire.

3.5. Justify the formula $l = 4 \times 10^{-7} \ln (D/r')$ henrys per meter for the round-trip inductance of a single-phase line made up of round wires of radius r separated by a distance D between centers. For simplicity, assume that $D \gg r$ and make reasonable simplifications. Note that the inductance "per conductor" is the same as the per phase inductance of a three-phase line, but the inductance is doubled because there are two conductors.

3.6. Reconsider Example 3.2 but assume that each of the conductors is hollow. Find an expression for the inductance l in this case.

3.7. Reconsider Example 3.2 but assume that each of the conductors is stranded with a central wire of radius r surrounded by six wires each of the same radius. Assume that $D \gg r$ and justify the approximate formula

$$l = 2 \times 10^{-7} \ln \frac{D}{R_s}$$

where

$$R_s = [(d_{11}d_{12} \cdots d_{17})^{1/7}(d_{21}d_{22} \cdots d_{27})^{1/7} \cdots (d_{71}d_{12} \cdots d_{77})^{1/7}]^{1/7}$$

and where $d_{ii} \triangleq r'$ and d_{ij} is the distance between centers of the *i*th and *j*th strands. R_s is called the geometric mean radius (GMR) of the stranded conductor. *Hint:* Notice the similarity with Fig. 3.6 and proceed by analogy with the derivation of (3.27). The only difference is that since the seven inductances $l_1, l_2, \ldots, l_7$ are not all equal, we need to use the more general rule for combining parallel inductors: namely, $l^{-1} = l_1^{-1} + l_2^{-1} + \cdots + l_7^{-1}$.

3.8. Given an aluminum 52,620-circular mil conductor composed of seven strands, each strand with a diameter of 0.0867 in. and an outside diameter of 0.2601 in. Find R_s, the GMR, using the formula in Problem 3.7 and compare with the manufacturer's figure of 0.00787 ft.

3.9. Calculate the per phase inductance (per meter) of the 765-kV line described in Section 3.8. Note the flat horizontal spacing of 45 ft between phases. Assume that four conductors per bundle are placed at the corners of a square (with 18 in. on a side). Use the values of GMR specified in place of r'.

3.10. Repeat Problem 3.9 neglecting the stranding and using the specified (outside) diameter of 1.424 in. as if each individual conductor were solid aluminum. What is the percent error compared with the result in Problem 3.9?

3.11. Calculate the per phase inductive reactance of the line in Example 3.5 in ohms per mile.

3.12. Repeat Problem 3.9 for the 345-kV line in Section 3.8.

3.13. Repeat Problem 3.12 neglecting the stranding (i.e., using r' instead of the GMR).

3.14. Calculate the per phase inductive reactance of the line in Problem 3.12 in ohms per mile.

3.15. Calculate the phase-neutral capacitance per meter of the 765-kV line described in Section 3.8.

3.16. Calculate the capacitive reactance of the line in Problem 3.15 in megohms per mile. *Hint:* First calculate the susceptance per mile and then take the reciprocal to find $|X_c|$.

3.17. Calculate the phase-neutral capacitance per meter of the 345-kV line described in Section 3.8.

3.18. Calculate the capacitive reactance of the line in Problem 3.17 in megohms per mile.

3.19. Calculate the product of inductance and capacitance values found for the 765-kV line in Problems 3.9 and 3.15 and compare with the value of $\mu_0 \epsilon_0$.

3.20. Repeat Problem 3.19 for the 345-kV line using the results of Problems 3.12 and 3.17.

Transmission-Line Modeling

chapter 4 _____

4.0 INTRODUCTION

In Chapter 3 the distributed per phase inductance and capacitance of a transmission line was related to line geometry. Some typical values of distributed series reactance, series resistance, and shunt (capacitive) reactance were given. These elements may be used in the per phase equivalent circuit of a three-phase line operating under balanced conditions. We now use this per phase model to derive the relationship between voltages and currents at the terminals of the transmission line.

4.1 DERIVATION OF TERMINAL *V, I* RELATIONS

We consider the transmission line in the sinusoidal steady state. Thus we may use phasors and impedances. Assume that

$$z = r + j\omega\iota = \text{series impedance per meter}$$

$$y = g + j\omega c = \text{shunt admittance per meter to neutral}$$

The lowercase letters are used for the distributed parameters; we will reserve uppercase letters for lumped (total) impedances and admittances. We also note the use of the symbol ι for (distributed) inductance and l for total length of line. The per phase circuit is shown in Fig. 4.1. The per phase terminal voltages and currents are V_1 and

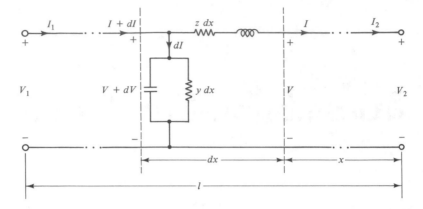

Figure 4.1 Transmission line.

I_1 at the left end and V_2 and I_2 at the right end. This is similar to the setup in Fig. 2.14 except that, because of the shunting elements, it can no longer be assumed that $I_2 = I_1$. With the given reference directions for current it is useful to think of the left side (side 1) as the sending end of the line and the right side (side 2) as the receiving end.

A typical differential section of line length dx is shown in Fig. 4.1. The series impedance of the differential section is then $z\,dx$. The shunt admittance is $y\,dx$; its location within the differential section is of no consequence. Note that the receiving end is located at $x = 0$, the sending end is at $x = l$.

Applying Kirchhoff's voltage law (KVL) and Kirchhoff's current law (KCL) to the section, we get

$$dV = Iz\,dx$$
$$dI = (V + dV)y\,dx \approx Vy\,dx \tag{4.1}$$

where we neglect the products of differential quantities. Thus we get two first-order linear differential equations,

$$\frac{dV}{dx} = zI$$
$$\frac{dI}{dx} = yV \tag{4.2}$$

or one second-order linear equation, either

$$\frac{d^2V}{dx^2} = yzV = \gamma^2V \tag{4.3}$$

or

$$\frac{d^2I}{dx^2} = yzI = \gamma^2I \tag{4.4}$$

where $\gamma \triangleq \sqrt{yz}$ is called the *propagation constant*. Without resistance (i.e., if $z = j\omega\iota$ and $y = j\omega c$), $\gamma = \sqrt{-\omega^2 \iota c} = j\omega\sqrt{\iota c}$, which is purely imaginary. If resistance is included, γ is complex, in the form $\gamma = \alpha + j\beta$. In either case with γ complex we do not expect real solutions for (4.3), (4.4).

Using the standard method of solving linear ordinary differential equations, we determine that the characteristic equation is $s^2 - \gamma^2 = 0$ and the characteristic roots are thus s_1, $s_2 = \pm\gamma$. The general solution for V is then

$$V = k_1 e^{\gamma x} + k_2 e^{-\gamma x} \tag{4.5}$$

$$= (k_1 + k_2)\frac{e^{\gamma x} + e^{-\gamma x}}{2} + (k_1 - k_2)\frac{e^{\gamma x} - e^{-\gamma x}}{2}$$

$$= K_1 \cosh \gamma x + K_2 \sinh \gamma x \tag{4.6}$$

where $K_1 = k_1 + k_2$, $K_2 = k_1 - k_2$. Equation (4.6) is convenient for introducing the boundary conditions. We note that the solution of (4.4) for I also is in the form (4.5) or (4.6).

From Fig. 4.1 we see when $x = 0$, $V = V_2$, which implies that $K_1 = V_2$. Also, when $x = 0$, $I = I_2$. Thus from (4.2),

$$\frac{dV(0)}{dx} = zI_2 \tag{4.7}$$

The notation $dV(0)/dx$ means $dV(x)/dx$ evaluated at $x = 0$. From (4.6)

$$\frac{dV}{dx} = -K_1\gamma \sinh \gamma x + K_2\gamma \cosh \gamma x \tag{4.8}$$

Using (4.7) in (4.8), we get, using the definition of γ,

$$K_2 = \frac{z}{\gamma}I_2 = \frac{z}{\sqrt{zy}}I_2 = \sqrt{\frac{z}{y}}I_2 = Z_c I_2$$

where $Z_c \triangleq \sqrt{z/y}$ is called the *characteristic impedance* of the line. We can now eliminate K_1 and K_2 in (4.6). A similar development for the solution of I is left for the reader. Then we get the pair of equations

$$V = V_2 \cosh \gamma x + Z_c I_2 \sinh \gamma x$$
$$I = I_2 \cosh \gamma x + \frac{V_2}{Z_c} \sinh \gamma x \tag{4.9}$$

We are particularly interested in the terminal conditions (i.e., V and I when $x = l$). In this case

$$V_1 = V_2 \cosh \gamma l + Z_c I_2 \sinh \gamma l$$
$$I_1 = I_2 \cosh \gamma l + \frac{V_2}{Z_c} \sinh \gamma l \tag{4.10}$$

Equation (4.10) is the desired relationship between the per phase voltages and currents at the two ends of the transmission line. The relationship is specified by the series impedance z and shunt admittance y in combination in their effect on the propagation constant $\gamma = \sqrt{zy}$ and the characteristic impedance $Z_c = \sqrt{z/y}$.

The reader may wonder why the problem was set up so that the result in (4.10) gives V_1 and I_1 explicitly in terms of V_2 and I_2, rather than the other way around. The reason can be stated clearly in the case of a radial transmission line. Ordinarily, it is the load (its voltage and complex power) which is specified and we wish to calculate the supply-side variables needed to satisfy the load requirements. This is another manifestation of the notion that under normal operating conditions a power system is "driven" by the power demanded by the loads. However, the inverse relation is easily found for those occasions when we seek V_2 and I_2 in terms of V_1 and I_1.

Example 4.1

A 60-Hz 138-kV 3ϕ transmission line is 225 mi long. The distributed line parameters are

$$r = 0.169 \ \Omega s/mi \qquad \iota = 2.093 \ mH/mi \qquad c = 0.01427 \ \mu F/mi \qquad g = 0$$

The transmission line delivers 40 MW at 132 kV with 95% power factor lagging. Find the sending-end voltage and current. Find the transmission-line efficiency.

Solution   We first find z and y for $\omega = 2\pi \cdot 60$.

$$z = 0.169 + j0.789 = 0.807 \ \underline{/77.9°} \ \Omega/mi$$

$$y = j5.38 \times 10^{-6} = 5.38 \times 10^{-6} \ \underline{/90°} \ mhos/mi$$

We then calculate Z_c and γl.

$$Z_c = \sqrt{\frac{z}{y}} = 387.3 \ \underline{/-6.05°} \ \Omega$$

$$\gamma l = 225\sqrt{zy} = 0.4688 \ \underline{/83.95°} = 0.0494 + j0.466$$

In (4.10) we need $\sinh \gamma l$ and $\cosh \gamma l$. We calculate these quantities next.

$$2 \sinh \gamma l \ \triangleq \ e^{\gamma l} - e^{-\gamma l} = e^{0.0494}e^{j0.466} - e^{-0.0494}e^{-j0.466}$$

$$= 1.051 \ \underline{/0.466} \ rad \ -0.952 \ \underline{/-0.466} \ rad$$

Thus

$$\sinh \gamma l = 0.452 \ \underline{/84.4°}$$

In the calculation above it is important to note that in the formula $e^{j\theta} = \cos \theta + j \sin \theta = 1 \ \underline{/\theta}$, θ is in radians. Thus $e^{j0.466} = 1 \ \underline{/0.466}$ rad. Continuing the calculation gives us

$$2 \cosh \gamma l = e^{\gamma l} + e^{-\gamma l} = 1.790 \ \underline{/1.42°}$$

$$\cosh \gamma l = 0.8950 \ \underline{/1.42°}$$

We now have all the transmission-line quantities needed in (4.10). It remains to find V_2 and I_2. The quantities in the statement of the problem refer to 3ϕ and line-line quantities. Thus

$$|V_2| = 132 \times 10^3/\sqrt{3} = 76.2 \text{ kV}$$

It is convenient to pick $\underline{/V_2} = 0$, in which case we get $V_2 = 76.2 \underline{/0°}$ kV. We next find the per phase power supplied to the load.

$$P_{\text{load}} = \frac{40 \times 10^6}{3} = 13.33 \text{ MW}$$

Since we are also given the power factor (0.95 lagging) we can find I_2. We have

$$P_{\text{load}} = 0.95|V_2||I_2|$$

Thus $|I_2| = 184.1$. Since I_2 lags V_2 by $\cos^{-1} 0.95 = 18.195°$,

$$I_2 = 184.1 \underline{/-18.195°}$$

Finally, plugging all these values into (4.10) gives us

$$V_1 = V_2 \cosh \gamma l + Z_c I_2 \sinh \gamma l$$

$$= 76.2 \times 10^3 \times 0.8950 \underline{/1.42°} + 387.3 \underline{/-6.05°}$$

$$\times 184.1 \underline{/-18.195°} \times 0.452 \underline{/84.4°}$$

$$= 68.20 \times 10^3 \underline{/1.42°} + 32.23 \times 10^3 \underline{/60.155°}$$

$$= 89.28 \underline{/19.39°} \text{ kV}$$

Multiplying by $\sqrt{3}$ we find that the sending-end line-line voltage has magnitude 154.64 kV. We next use (4.10) to find I_1.

$$I_1 = I_2 \cosh \gamma l + \frac{V_2}{Z_c} \sinh \gamma l$$

$$= 162.42 \underline{/14.76°}$$

Thus the magnitude of the sending-end current is 162.42 A. We now can calculate the efficiency of transmission. Since the per phase output power is 13.33 MW while the corresponding input power is

$$P_{12} = \text{Re } V_1 I_1^* = 89.28 \times 10^3 \times 162.42 \cos (19.39° - 14.76°)$$

$$= 14.45 \text{ MW}$$

we have

$$\eta = \frac{13.33}{14.45} = 0.92$$

That is, the efficiency of transmission is 92%.

Example 4.2

Suppose that a radial line is terminated in its characteristic impedance Z_c. Find the driving-point impedance V_1/I_1, the voltage "gain" $|V_2|/|V_1|$, the current gain $|I_2|/|I_1|$, the complex power gain $-S_{21}/S_{12}$, and the real power efficiency—P_{21}/P_{12}.

Solution   If the line is terminated in Z_c, then $V_2 = Z_c I_2$ and (4.10) becomes

$$V_1 = V_2(\cosh \gamma l + \sinh \gamma l) = V_2 e^{\gamma l} = V_2 e^{\alpha l} e^{j\beta l}$$

$$I_1 = I_2(\cosh \gamma l + \sinh \gamma l) = I_2 e^{\gamma l} = I_2 e^{\alpha l} e^{j\beta l}$$

Thus

$$\frac{V_1}{I_1} = \frac{V_2}{I_2} = Z_c$$

and we see that the driving-point impedance is Z_c. From the first equation we can calculate

$$|V_1| = |V_2| e^{\alpha l} \Rightarrow \frac{|V_2|}{|V_1|} = e^{-\alpha l}$$

From the second equation we find that

$$\frac{|I_2|}{|I_1|} = e^{-\alpha l}$$

Noting the reference direction for I_2, the complex power gain may be calculated as follows:

$$-S_{21} = V_2 I_2^* = V_1 e^{-\alpha l} e^{-j\beta l} I_1^* e^{-\alpha l} e^{j\beta l}$$

$$= S_{12} e^{-2\alpha l}$$

Thus

$$\frac{-S_{21}}{S_{12}} = e^{-2\alpha l}$$

Alternatively, we may observe that

$$V_1 = Z_c I_1 \Rightarrow V_1 I_1^* = Z_c |I_1|^2$$

while

$$V_2 I_2^* = Z_c |I_2|^2$$

Thus

$$\frac{-S_{21}}{S_{12}} = \frac{|I_2|^2}{|I_1|^2} = e^{-2\alpha l}$$

Finally, since α is real, we have

$$\eta = \frac{-P_{21}}{P_{12}} = e^{-2\alpha l}$$

Example 4.3

Repeat Example 4.2 for the case of a lossless line. In addition, find Z_c, γ, and P_{12}.

Solution In a lossless line $r = g = 0$. Thus we find that

$$Z_c = \sqrt{\frac{z}{y}} = \sqrt{\frac{j\omega l}{j\omega c}} = \sqrt{\frac{l}{c}} = \sqrt{\frac{L}{C}}$$

where $L = l\iota$ = total inductance of the line
$\quad\quad\quad$ $C = lc$ = total capacitance of the line

Z_c is seen to be a real quantity. Next we find that

$$\gamma = \sqrt{zy} = \sqrt{j\omega\iota j\omega c} = j\omega\sqrt{\iota c}$$

This is a purely imaginary number. We see that in the description $\gamma = \alpha + j\beta$, $\alpha = 0$, and $\beta = \omega\sqrt{\iota c}$. Since $\alpha = 0$, all the ratios calculated in Example 4.2 are unity. Thus

$$\frac{|V_2|}{|V_1|} = \frac{|I_2|}{|I_1|} = \frac{-S_{21}}{S_{12}} = \frac{-P_{21}}{P_{12}} = \eta = 1$$

Finally, $P_{12} = \text{Re } V_1 I_1^* = \text{Re } Z_c |I_1|^2 = Z_c |I_1|^2$ since Z_c in this case is purely real. An alternate expression based on $I_1 = V_1/Z_c$ is frequently more useful:

$$P_{12} = \frac{|V_1|^2}{Z_c}$$

A comment on this last result in Example 4.3: Frequently, in the case of a lossless line, the terminology "surge impedance" is used for Z_c instead of characteristic impedance. A lossless line operating at its nominal voltage, terminated in its surge impedance Z_c, is said to be surge impedance loaded (SIL). The transmitted per phase power in this case is designated P_{SIL}. Using this notation we see that the per phase power transmitted under surge impedance loading is

$$P_{SIL} = \frac{|V_1|^2}{Z_c} \tag{4.11}$$

Multiplying by 3, the corresponding three-phase power is given by a similar expression $P_{SIL}^{3\phi} = |V_1^{ll}|^2/Z_c$, where $|V^{ll}|$ is the line-line voltage magnitude.

4.2 WAVES ON TRANSMISSION LINES

The terms "propagation constant" and "characteristic impedance" are part of a very useful description of the transmission line in terms of "incident" and "reflected waves." We can introduce the idea by looking at $k_1 e^{\gamma x}$ and $k_2 e^{-\gamma x}$, the two voltage components in (4.5), whose sum is the phasor V evaluated x meters from the right terminal of the line. These voltage components have the following interpretation. The first term describes a voltage wave traveling to the right (incident wave); the second term describes a voltage wave traveling to the left (reflected wave). We may see this by writing $\gamma \triangleq \alpha + j\beta$, where $\alpha \geq 0$ is called the *attenuation constant* and β is called the *phase constant*. Substituting for γ in (4.5) and finding the instantaneous voltage as a function of t and x gives us

$$v(t, x) = \sqrt{2}\ \text{Re }k_1 e^{\alpha x} e^{j(\omega t + \beta x)} + \sqrt{2}\ \text{Re }k_2 e^{-\alpha x} e^{j(\omega t - \beta x)}$$
$$= v_1(t, x) + v_2(t, x) \tag{4.12}$$

Neglecting α for the moment (small in any case) and considering $v_2(t, x)$, we see that for fixed x, v_2 is a sinusoidal function of t; for fixed t, v_2 is a sinusoidal function of x.

Suppose that as t increases we look at (scan) the voltage v_2 at points x which also increase in accordance with the formula

$$\omega t - \beta x = \text{constant}$$

In this case v_2 remains constant! We are thus looking at a fixed point on a voltage wave which is traveling to the left with a velocity

$$\frac{dx}{dt} = \frac{\omega}{\beta} = \frac{\omega}{\text{Im } \sqrt{zy}}$$

The reason the wave is traveling to the left is that, from Fig. 4.1, increasing x means moving from right to left. The effect of the neglected term $e^{-\alpha x}$ is to attenuate the wave as it moves to the left. This wave is called the *reflected wave*.

A similar interpretation holds for $v_1(x)$ as an incident wave moving to the right. From (4.12) we can deduce that if the line is of infinite length, and $\alpha > 0$, there is no reflected wave. For an infinite line, from (4.10) we can verify that the driving-point impedance of the line is Z_c. We can deduce that a line of finite length terminated in Z_c also has a driving-point impedance Z_c and no reflected wave.

The importance of the interpretation of (4.5) in terms of incident and reflected waves lies in the application to electromagnetic transients arising from lightning strokes and switching. Although this case is more complicated than the steady-state case (we must deal with partial rather than ordinary differential equations), the results can be interpreted once again in terms of incident and reflected waves.

We mention one result of such a treatment. If an incident lightning voltage pulse (or surge) traveling down a line hits the open-circuited end of the line (or an open switch) there will be a reflected wave generated such that the voltage at the line termination will approximately double. In designing the insulation for transmission lines and equipment connected to it (such as transformers), it is vitally important to take this doubling into account!

4.3 TRANSMISSION MATRIX

Section 4.2 was a digression from our main interest in the relationships between steady-state terminal voltages and currents. Considering (4.10), we note that it is in the form

$$V_1 = AV_2 + BI_2$$
$$I_1 = CV_2 + DI_2 \tag{4.13}$$

where

$$A = \cosh \gamma l \qquad B = Z_c \sinh \gamma l$$
$$C = \frac{1}{Z_c} \sinh \gamma l \qquad D = \cosh \gamma l \tag{4.14}$$

Note: If γ is complex, A, B, C, and D also are complex.

The A, B, C, D parameters are called *transmission parameters*. The matrix

$$\mathbf{T} = \begin{bmatrix} A & B \\ C & D \end{bmatrix} \tag{4.15}$$

is called a *transmission matrix* or *chain matrix*. By direct calculation det $\mathbf{T} = AD - BC = \cosh^2 \gamma l - \sinh^2 \gamma l = 1$. Thus the inverse exists and in fact is

$$\mathbf{T}^{-1} = \begin{bmatrix} D & -B \\ -C & A \end{bmatrix} \tag{4.16}$$

The advantage of the transmission-matrix description is that the $\mathbf{T}$ matrix for a cascade of two-ports is the product of individual $\mathbf{T}$ matrices. For example, in Fig. 4.2 we find that

$$\begin{bmatrix} V_1 \\ I_1 \end{bmatrix} = \mathbf{T}_1 \begin{bmatrix} V_2 \\ I_2 \end{bmatrix} = \mathbf{T}_1 \mathbf{T}_2 \begin{bmatrix} V_3 \\ I_3 \end{bmatrix}$$

which indicates that the correct transmission matrix for the cascade is

$$\mathbf{T} = \mathbf{T}_1 \mathbf{T}_2 \tag{4.17}$$

The result, that det $T = 1$, holds in general for two-port networks composed of (linear time-invariant) resistors, capacitors, inductors, coupled inductors, and transformers. This provides a useful check of analytical or numerical work.

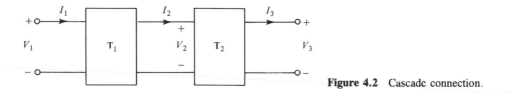

Figure 4.2 Cascade connection.

4.4 LUMPED-CIRCUIT EQUIVALENT

We next wish to derive a lumped-circuit equivalent for the transmission line. We will find a Π-equivalent circuit which has the same A, B, C, D parameters as the transmission line [i.e., the A, B, C, D parameters given in (4.14)]. We note that a T-equivalent circuit may also be derived. In either case, for electrical engineers there are many advantages to the circuit representation. Among these is a better sense of the physical behavior of the line.

We want to pick Z' and Y' so that the circuit in Fig. 4.3 has the same $\mathbf{T}$ matrix as in (4.14). To find the $\mathbf{T}$ matrix for the circuit we can write KVL and KCL equations attempting ultimately to relate V_1 and I_1 to V_2 and I_2. This is illustrated in the following calculation.

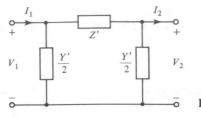

Figure 4.3 Π-equivalent circuit.

$$V_1 = V_2 + Z'\left(I_2 + \frac{Y'}{2}V_2\right)$$

$$= \left(1 + \frac{Z'Y'}{2}\right)V_2 + Z'I_2 \tag{4.18}$$

$$I_1 = \frac{Y'}{2}V_1 + \frac{Y'}{2}V_2 + I_2$$

$$= Y'\left(1 + \frac{Z'Y'}{4}\right)V_2 + \left(1 + \frac{Z'Y'}{2}\right)I_2 \tag{4.19}$$

In (4.19) we used (4.18) to eliminate V_1. From (4.18) and (4.19) we see that the A, B, C, and D parameters for the circuit are

$$A = 1 + \frac{Z'Y'}{2} \qquad\qquad B = Z'$$

$$C = Y'\left(1 + \frac{Z'Y'}{4}\right) \qquad D = 1 + \frac{Z'Y'}{2} \tag{4.20}$$

Equating the B parameters in (4.20) and (4.14) and making some substitutions, we get

$$Z' = Z_c \sinh \gamma l = \sqrt{\frac{z}{y}} \sinh \gamma l = zl\frac{\sinh \gamma l}{\gamma l} = Z\frac{\sinh \gamma l}{\gamma l} \tag{4.21}$$

where $Z \triangleq zl$ is the total series impedance of line. We note that for $|\gamma l| \ll 1$, which is often the case for power lines, $(\sinh \gamma l)/\gamma l \approx 1$, and in this case it is useful to view $(\sinh \gamma l)/\gamma l$ as a correction factor which multiplies the total series impedance of the line to give Z' exactly. If the exact value is not needed, we may use $Z' \approx Z$.

Equating the A parameters in (4.20) and (4.14), we get

$$1 + \frac{Z'Y'}{2} = \cosh \gamma l \tag{4.22}$$

Using (4.21) and solving for $Y'/2$, we get

$$\frac{Y'}{2} = \frac{\cosh \gamma l - 1}{Z_c \sinh \gamma l} = \frac{1}{Z_c}\tanh\frac{\gamma l}{2} \tag{4.23}$$

The trigonometric identity in (4.23) can be checked as follows:

$$\frac{\cosh \gamma l - 1}{\sinh \gamma l} = \frac{e^{\gamma l} + e^{-\gamma l} - 2}{e^{\gamma l} - e^{-\gamma l}} = \frac{(e^{\gamma l/2} - e^{-\gamma l/2})^2}{(e^{\gamma l/2} + e^{-\gamma l/2})(e^{\gamma l/2} - e^{-\gamma l/2})}$$

$$= \frac{e^{\gamma l/2} - e^{-\gamma l/2}}{e^{\gamma l/2} + e^{-\gamma l/2}} = \tanh \frac{\gamma l}{2}$$

We can derive an alternative expression for (4.23) as follows:

$$\frac{1}{Z_c} = \frac{1}{\sqrt{z/y}} = \frac{y}{\sqrt{zy}} = \frac{yl}{\gamma l} = \frac{Y}{\gamma l} \tag{4.24}$$

where $Y \triangleq yl$ is the total line-neutral admittance of line. Then

$$\frac{Y'}{2} = \frac{Y}{2} \frac{\tanh (\gamma l/2)}{\gamma l/2} \tag{4.25}$$

Note that for $|\gamma l| \ll 1$, $(\tanh \gamma l/2)/(\gamma l/2) \approx 1 \Rightarrow Y'/2 \approx Y/2$. As in the case of (4.21), it is useful to think of $(\tanh \gamma l/2)/(\gamma l/2)$ as a correction factor (close to 1) which multiplies the total line-neutral admittance of the line to give Y' exactly. And $Y' \approx Y$.

Now that Z and Y have been defined, it is useful to note the following alternative expressions for Z_c and γl.

$$Z_c = \sqrt{\frac{z}{y}} = \sqrt{\frac{zl}{yl}} = \sqrt{\frac{Z}{Y}} \tag{4.26}$$

and

$$\gamma l = \sqrt{zyl} = \sqrt{zl \, yl} = \sqrt{ZY} \tag{4.27}$$

Example 4.4

Find the Π-equivalent circuit for the transmission line described in Example 4.1.

Solution   In Example 4.1 we have already calculated Z_c and $\sinh \gamma l$; thus from (4.21),

$$Z' = Z_c \sinh \gamma l = 387.3 \; \underline{/-6.04°} \times 0.452 \; \underline{/84.4°} = 175.06 \; \underline{/78.35°}$$

To emphasize the point concerning correction factors, we can also use an alternative calculation:

$$Z' = Z \frac{\sinh \gamma l}{\gamma l} = 181.57 \; \underline{/77.9°} \times 0.9642 \; \underline{/0.45°} = 175.07 \; \underline{/78.35°}$$

Here the correction factor is $0.9642 \; \underline{/0.45°}$, which is reasonably close to 1.0. Similarly, using (4.23) we get

$$\frac{Y'}{2} = 614.57 \times 10^{-6} \; \underline{/89.8°} \text{ mho}$$

For comparison we note that

$$\frac{Y}{2} = 605.25 \times 10^{-6} \; \underline{/90°} \text{ mho}$$

is a reasonably good approximation to $Y'/2$. Now that we have calculated Z' and $Y'/2$, the Π-equivalent circuit is specified.

The example indicates the possibility of simplifying the transmission-line model by using Z instead of Z' and $Y/2$ instead of $Y'/2$. We consider this question in the next section.

4.5 SIMPLIFIED MODELS

The circuit in Fig. 4.3 is equivalent to the equations given in (4.10). Sometimes it is more convenient to use one representation over the other, but either may be used to give the exact relationships between the terminal voltages and currents of a transmission line. For a *long* line the use of the exact circuit and/or equations is indicated. However, for *medium*-length lines, it turns out that the circuit and equations may be greatly simplified. Thus, from (4.21) and (4.25) we can see that if $|\gamma l| \ll 1$, we can replace Z' by Z, and Y' by Y. In this case the circuit elements in Fig. 4.3 may be found without the tedious calculation of the correction factors. For so-called *short* lines, with Y small, we can even leave out the shunting elements; thus we obtain the very simple model used in Section 2.6.

Experience indicates the following classification of lines to be reasonable.

Long line ($l > 150$ mi, approximately): Use the Π-equivalent circuit model with Z' and $Y'/2$ given by (4.21) and (4.23). Of course, instead of the circuit model we can use (4.10).

Medium-length line ($50 < l < 150$ mi, approximately): Use the circuit model with Z and $Y/2$ instead of Z' and $Y'/2$, where $Z = zl$ and $Y = yl$. This is called the *nominal* Π-equivalent circuit.

Short line ($l < 50$ mi, approximately): Same as the medium-length line except that we neglect $Y/2$.

Example 4.5

Consider the receiving-end voltage of a lossless open-circuited line and compare the results by use of the three models. V_1 is a fixed voltage.

Solution By "open circuited" we mean $I_2 = 0$. By "lossless" we mean $\alpha = 0$, or equivalently $\gamma = j\beta$.

Model 1: Long-line model. It is easiest to use (4.10) rather than the circuit model and we find that

$$V_1 = V_2 \cosh \gamma l = V_2 \cos \beta l$$

Model 2: Medium-length line model. Using the nominal Π-equivalent circuit we get

$$V_1 = \left[1 + \frac{ZY}{2}\right]V_2 = \left[1 + \frac{(\gamma l)^2}{2}\right]V_2 = \left[1 - \frac{(\beta l)^2}{2}\right]V_2$$

The terms in parentheses are seen to be the first two terms in the series expansion of cos βl.

Model 3: Short-line model. We get

$$V_1 = V_2$$

Thus it appears we have retained only the first term in the series expansion of cos βl and have completely lost the property observed in the first two models—that the voltage at the (open) receiving end is higher than at the sending end, for small βl.

Let us compare these models as they predict V_2 for different length lines. A figure of $\beta = 0.002$ rad/mi is representative for a 60-Hz open-wire line. For a 50-mi line, $\beta l \approx 0.1$ rad and thus

Model 1 $\Rightarrow V_1 = 0.995004 V_2$
Model 2 $\Rightarrow V_1 = 0.995000 V_2$
Model 3 $\Rightarrow V_1 = V_2$

The errors in calculating V_2 are certainly negligible in using the simpler models.
But consider a 200-mi line, $\beta l \approx 0.4$.

Model 1 $\Rightarrow V_1 = 0.921 V_2$
Model 2 $\Rightarrow V_2 = 0.920 V_2$
Model 3 $\Rightarrow V_3 = V_2$

The differences between models 1 and 2 are still negligible but there is about a 8% error in using the short-line model.
Finally, consider a 600-mi line, $\beta l \approx 1.2$.

Model 1 $\Rightarrow V_1 = 0.362 V_2$
Model 2 $\Rightarrow V_1 = 0.280 V_2$
Model 3 $\Rightarrow V_1 = V_2$

The difference between the exact (long-line) model and the nominal Π-equivalent model is significant. The short-line model is totally inaccurate.

4.6 COMPLEX POWER TRANSMISSION (LONG OR MEDIUM LINE)

In Section 2.6 we considered the problem of complex power transmission using a short-line model. Let us see what modifications are necessary in the analysis using the long or medium-length-line models of Section 4.5. We consider the Π-equivalent circuit in Fig. 4.4.

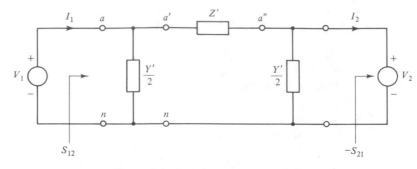

Figure 4.4   Complex power transmission.

The easiest way to find S_{12} is to observe the following fact based on conservation of power. The sending-end power, S_{12}, is equal to the power consumed in $Y'/2$ plus the power supplied to the rest of the network through the terminals $a'n$. This latter power depends only on the current and voltage at the terminals $a'n$, and therefore on V_1, V_2, and Z'. In fact, we have already considered this case in Section 2.6 and can use (2.26) with only the substitution of Z' for Z. Thus adding the contributions,

$$S_{12} = \frac{Y'^*}{2}|V_1|^2 + \frac{|V_1|^2}{Z'^*} - \frac{|V_1||V_2|}{Z'^*}e^{j\theta_{12}} \tag{4.28}$$

where the first term is the power consumed in $Y'/2$ and the last two terms are from (2.26). By the same technique, the received power, $-S_{21}$, is equal to the power received from the network through the terminals $a''n$ minus the power consumed in (the right hand) $Y'/2$. Using (2.28), we get for the received power

$$-S_{21} = -\frac{Y'^*}{2}|V_2|^2 - \frac{|V_2|^2}{Z'^*} + \frac{|V_1||V_2|}{Z'^*}e^{-j\theta_{12}} \tag{4.29}$$

We see that, except for the additional constant terms, (4.28) has the same form as (2.26), and (4.29) has the same form as (2.28). The additional constant terms are relatively small in the usual case.

Thus the appearance of the power circle diagram usually changes only a little. The new power circle diagram is shown in Fig. 4.5. The main change is that the centers of the circles are shifted as shown. *Note:* With Y' almost purely reactive, the shift is approximately vertical; thus the active power transfer is unaffected. This conclusion is consistent with the physical setup as well. It appears that the general conclusions based on the simpler model are still valid, although we can expect some numerical discrepancies.

Exercise 1.   Using the short-line approximation, we showed that for $|V_2| \neq |V_1|$ the power circles do not intersect. Is this also true in the present case?

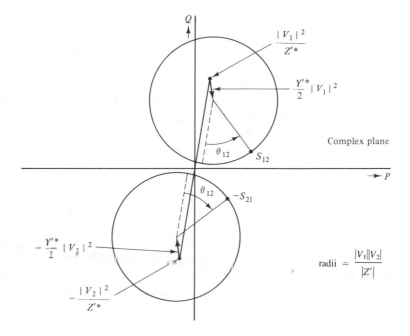

Figure 4.5   Power circle diagram.

4.7 POWER-HANDLING CAPABILITY OF LINES

Power lines are limited in their ability to deliver power. The two most important limits can be understood by considering thermal effects and stability.

When current flows in conductors there are I^2R losses and heat is generated. This loss of power reduces transmission efficiency, a factor we consider in Chapter 7. Of more consequence to the present discussion, the heat generated by line losses also causes a temperature rise. Since line temperature of overhead lines must be kept within safe limits to prevent excessive line sag between transmission towers (minimum ground clearances must be maintained) and to prevent irreversible stretching (at around 100°C), this imposes limitations on the maximum safe current a line can carry. Clearly, this limit depends on the line design (conductor size and geometry, spacing between towers, etc.) and operating conditions (ambient temperature, wind velocity, etc.). The technique of bundling that we discussed in Chapter 3 as a means of reducing line inductance and electrostatic stress also has a beneficial effect here. The greater spacing between subconductors and the increased surface area help in heat dissipation with an accompanying improvement in allowable safe current levels.

We note that for cables there are even more stringent thermal limits because of the more limited possibilities for heat transfer. If the cable gets too hot, the insulation will begin to deteriorate and may fail in time.

In either case thermal considerations impose a current-handling limitation for a given line. A line has a current "rating" which provides for safe operation. Operation above the line rating (particularly for an extended period of time) is not recommended.

We also find that lines are designed for operation at a given voltage level. The choice of conductor size and geometry, spacing between phases, and selection of insulations all are appropriate for the intended voltage level. In addition, the transmission system is operated so as to keep transmission-line voltages reasonably close to the nominal values; the reason for this is the desirability (and statutory requirement) that the voltage at the consumer's location be held within reasonable bounds.

With limitations on maximum current and voltage there is a corresponding limitation on the MVA which can safely be transmitted, and hence on the MW as well. Thus we might have a 345-kV transmission line (345-kV line-line) with a thermal rating of 1600 MVA (three-phase). At 100% power factor it could transmit 1600 MW as well; since at other power factors its MW capability is reduced, the 1600-MW figure stands as a uniform limitation on active power transmission. Finally, it should be pointed out that the thermal limitations on terminating equipment (such as transformers) may be more restrictive than those pertaining to the line itself.

The reader may recall that we discussed another limitation on power-handling capability in Chapter 2. In the case of a short lossless line with voltage support at both ends, we found an ultimate transmission capability corresponding to a power angle $\theta_{12} = 90°$. In fact, to have reasonable expectations of maintaining synchronism (i.e., stability), θ_{12} should be more strictly limited to maximum values in the neighborhood of 40 to 50°. In this case the "stability" limit is perhaps 65 to 75% of the ultimate transmission capability. It turns out that for short lines the power-handling capability is set by thermal limits rather than stability limit.

For long lines the reverse is true. We will now consider the stability limit for long lines. For simplicity assume a lossless line with equal voltage magnitudes at each end. In the lossless case, as was shown in Example 4.3, $Z_c = \sqrt{L/C}$ is real and $\gamma = \alpha + j\beta = j\beta$ is a purely imaginary number. We also calculate Y' and Z' for this case.

$$Y' = Y\frac{\tanh (\gamma l/2)}{\gamma l/2} = j\omega C\frac{\tan \beta l/2}{\beta l/2} \tag{4.30}$$

$$Z' = Z_c \sinh \gamma l = jZ_c \sin \beta l \tag{4.31}$$

Thus we see that Y' is the admittance of a pure capacitance and Z' is the impedance of a pure inductance. It makes sense since the line is assumed lossless!

Next we calculate the active power transmitted. From (4.28) and (4.29), we find that $P_{12} = -P_{21}$ and

$$P_{12} = \frac{|V_1|^2}{Z_c}\frac{\sin \theta_{12}}{\sin \beta l} \tag{4.32}$$

where we have used the fact that Y' and Z' are purely imaginary.

Using the definition of P_{SIL} from (4.11) in (4.32), we get

$$P_{12} = P_{\text{SIL}}\frac{\sin \theta_{12}}{\sin \beta l} \tag{4.33}$$

We note that, for fixed θ_{12}, as line length is increased, βl increases and P_{12} decreases. A particularly severe restriction occurs for very long lines with $\beta l = \pi/2$. In this case even if we maximize active power flow by picking $\theta_{12} = \pi/2$, we cannot exceed $P_{12} = P_{SIL}$. In practice we can only realize a fraction of the P_{SIL} value if θ_{12} is kept within safe limits. The reader should now be in a position to interpret the $P_{SIL}^{3\phi}$ figures given for the three transmission lines in Section 3.9.

Finally, we note that in the lossless case, as shown in Example 4.3, $\beta = \omega\sqrt{lc}$. As discussed in Section 3.8, the product lc is virtually constant for the (open-wire) transmission lines we have been considering. Thus in (4.33), for fixed θ_{12}, the "drop off" of P_{12}/P_{SIL} with length l is virtually the same for all (open-wire) transmission lines.

Example 4.6

Assume that $\beta = 0.002$ rad/mi and $\theta_{12} = 45°$. Find P_{12}/P_{SIL} as a function of line length.

Solution   In (4.33), βl is in radians. We will convert to degrees. Thus $\beta l = 0.002l$ rad $= 0.1146l$ deg.

$$\frac{P_{12}}{P_{SIL}} = 0.707 \frac{1}{\sin 0.1146l}$$

The stability limit is shown in Fig. E4.6. For comparison a typical thermal limit is also shown. We see that for short lines the thermal limit governs, whereas for long lines the stability limit prevails.

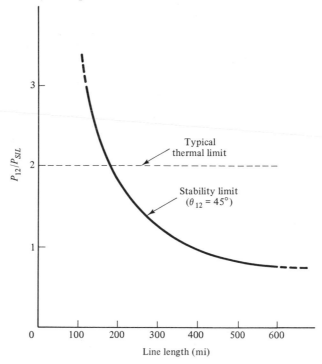

Figure E4.6

4.8 SUMMARY

We may summarize the results of the analysis by discussing the three Π-equivalent circuits. The most accurate circuit involving the use of the elements Z' and $Y'/2$ is recommended when dealing with (long) transmission lines of lengths greater than approximately 150 mi; Z' is specified in (4.21) and $Y'/2$ is specified in (4.25). For medium-length lines in the range of approximately 50 to 150 mi the simpler nominal Π-equivalent circuit may be used instead. Here we replace Z' and $Y'/2$ by Z and $Y/2$, where Z is the total series impedance ($Z = zl$) and Y is the total shunt admittance ($Y = yl$). For a short line of length under about 50 mi, the shunt elements may be neglected, making the circuit even simpler. We would like to draw attention to the very simple final results of the analysis. Although we started with a complicated circuit with distributed parameters (the 3ϕ line), we were able to model this for per phase analysis by one of three very simple Π-equivalent lumped circuits!

Alternatively, equations may be used. In particular, the long-line equation (4.10) is frequently used. In working with networks in cascade, matrix operations involving the *ABCD* parameters (or transmission matrices) are very useful.

Finally, we note that thermal effects limit the power-handling capability of short and medium-length lines, whereas stability requirements impose the limitations on long lines.

PROBLEMS

4.1. Given a 138-kV three-phase line with series impedance $z = 0.17 + j0.79\ \Omega/\text{mi}$ and shunt admittance $y = j5.4 \times 10^{-6}$ mho/mi, find the characteristic impedance Z_c, the propagation constant γ, the attenuation constant α, and the phase constant β.

4.2. Repeat Problem 4.1 for a 765-kV 3ϕ line with series impedance $0.02 + j0.54\ \Omega/\text{mi}$ and shunt admittance $y = j7.8 \times 10^{-6}$ mho/mi. Compare with the values found in Problem 4.1.

4.3. Given a transmission line described by (4.10), we perform two tests and obtain the following results.

    1. Open-circuit test ($I_2 = 0$):

$$Z_{oc} = \frac{V_1}{I_1} = 800\ \underline{/-89^\circ}$$

    2. Short-circuit test ($V_2 = 0$):

$$Z_{sc} = \frac{V_1}{I_1} = 200\ \underline{/77^\circ}$$

Find the characteristic impedance Z_c and find γl.

4.4. Jack and Jill measure Z_{oc} and Z_{sc} as in Problem 4.3. Jill says: "With this information we can now calculate V_1 and I_1 in terms of V_2 and I_2 for *any* termination." Agree or disagree with Jill and give your reasons.

4.5. Given a transmission line described by a total series impedance $Z = zl = 20 + j80$ and a total shunt admittance $Y = yl = j5 \times 10^{-4}$. Find Z_c, γl, $e^{\gamma l}$, $\sinh \gamma l$, and $\cosh \gamma l$.

4.6. Suppose that the line in Problem 4.5 is terminated in its characteristic impedance (i.e., $V_2/I_2 = Z_c$). Find the efficiency of transmission in this case (i.e., find $\eta = -P_{21}/P_{12}$).

4.7. Show that the driving-point impedance of an infinite transmission line is Z_c, independent of termination Z_L. Assume that r, l, g, $c > 0$. *Note:* The result also holds under the weaker conditions r, l, $c > 0$, $g = 0$.

4.8. The 138-kV 3ϕ line described in Problem 4.1 is 150 mi long and is delivering 15 MW at 132 kV at a 100% power factor. Find the sending-end voltage and current, the power angle θ_{12} (i.e., $/\underline{V_{a1}} - /\underline{V_{2a}}$), and the transmission efficiency. Use the long-line model.

4.9. Specify the Π-equivalent circuit corresponding to the line in Problem 4.8.

4.10. Repeat Problem 4.8 when the power factor is 90% lagging.

4.11. Suppose that in Problem 4.8 the load is removed while the sending-end voltage magnitude is held constant at the value determined in the problem. Find the new receiving-end voltage. Find the new sending-end current.

4.12. Repeat Problem 4.8 using the nominal Π-equivalent circuit (i.e., the medium-length line model).

4.13. Repeat Problem 4.8 using the short-line model.

4.14. Given a 200-mi transmission line with $r = 0.1 \ \Omega/\text{mi}$, $l = 2.0 \ \text{mH/mi}$, $c = 0.01 \ \mu\text{F/mi}$, and $g = 0$, find Π-equivalent circuits using (a) the long-line model, (b) the medium-line (nominal Π) model, and (c) the short-line model.

4.15. The 765-kV 3ϕ line described in Problem 4.2 is 400 mi long and delivers 100 MW at 750 kV at 95% power factor lagging. Find the sending-end voltage and current and the transmission efficiency. Use the long-line model.

4.16. Suppose that the load in Problem 4.15 is removed while the sending-end voltage magnitude is held constant. Find the receiving-end voltage and the sending-end current and complex power into the transmission line.

4.17. Repeat Problem 4.15 using the nominal Π-equivalent circuit.

4.18. Find the *ABCD* constants of the 150-mi line in Problem 4.8.

4.19. Find the *ABCD* constants of the 150-mi line in cascade with another identical 150-mi line by using the transmission matrices of the 150-mi lines.

4.20. Using the nominal Π-equivalent circuit, find an expression for the per phase complex power consumed by the line in terms of V_1, V_2, Y, and Z.

4.21. To maintain a safe "margin" of stability, system designers have decided that the power angle θ_{12} for a particular transmission line can be no greater than 45°. We wish to transmit 500 MW down a 300-mi line and need to pick a transmission-line voltage level. Consider 138-, 345-, and 765-kV lines. Which voltage level(s) would be suitable? As a first approximation assume that $|V_1| = |V_2|$ and the lines are lossless. You can assume that $\beta = 0.002 \ \text{rad/mi}$ in all three cases.

4.22. Neglect g and show that for $r \ll \omega l$ the phase constant $\beta \approx \omega \sqrt{lc}$. This helps explain the remarkable constancy of β for lines of varying geometry.

Transformer Modeling and the Per Unit System

5.0 INTRODUCTION

The generation of power in large generating stations is at line voltages typically in the range of 11 to 30 kV. Efficient and effective transmission over long distances requires much higher voltages, typically line voltages of 138, 230, or 345 kV, with a trend toward higher voltages (765 kV) for really long distance transmission. Distribution of power takes place at much lower voltages, typically at 2400 or 4160 V, which is usually "stepped down" further to voltages of 440 V for typical industrial use or 240/120 V for commercial or residential use.

Power transformers provide a trouble-free and efficient means of shifting from one voltage level to another. For this application they normally have a fixed voltage ratio. Transformers can also be used for controlling voltage and/or power flow under varying operating conditions. For this application they have a voltage ratio that can be changed in small increments, on command. These transformers are called *regulating transformers*.

In this chapter we consider the modeling of transformers for use in system studies. We then consider a normalization of system variables which greatly simplifies the consideration of large systems in which there are many voltage levels. This normalization is called per unit (p.u.) normalization, or conversion to the per unit system. We start by considering a linear model of a single-phase transformer.

5.1 SINGLE-PHASE TRANSFORMER MODEL

A two-winding transformer is shown schematically in Fig. 5.1. The diagram is intended to illustrate the working principle of a transformer. In the actual physical layout certain steps are taken to increase the magnetic coupling between the primary and secondary coils; the windings are either wound concentrically or are sectionalized and interleaved.

We assume that the magnetic fluxes can be split into three distinct components. There is a mutual flux Φ_m contained within the magnetic core which thus links all the turns of the primary and secondary windings. There is a leakage flux Φ_{l1} which is assumed to link only the primary circuit and, similarly, a secondary leakage flux Φ_{l2}. These components of flux are shown schematically in Fig. 5.1.

We pick the reference direction for i_1 arbitrarily, then assign the reference direction for i_2' so that with i_1 and i_2' both positive the resultant mutual fluxes tend to add. In the usual way, the reference directions for the mutual flux Φ_m and the leakage fluxes Φ_{l1} and Φ_{l2} are found by the right-hand rule. The reference directions for voltages are associated reference directions as shown in Fig. 5.1. We also (redundantly in this case) mark the terminals using the dot notation, and also show the standard marking of power transformers using X_1 and X_2 for low-voltage winding and H_1 and H_2 for high-voltage winding.

We remind the reader that the dot notation indicates that the mutual flux components, due to the currents i_1 and i_2', tend to add when these currents both enter (or

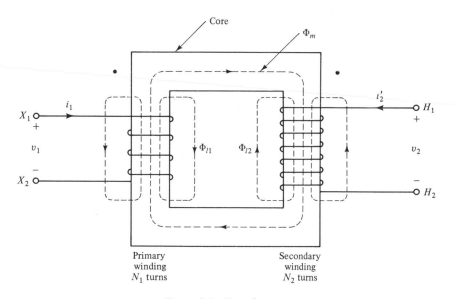

Figure 5.1   Transformer.

leave) the dotted terminals. The reader can check that the dots are correctly placed in Fig. 5.1 by using the right-hand rule.

It is convenient to start the mathematical modeling by describing that version of Fig. 5.1 called an *ideal transformer*. Later using the ideal transformer as the key element we will add a few series and parallel elements to obtain a more realistic model of a physical transformer.

An ideal transformer has the following physical properties:

1. No losses.
2. No leakage fluxes.
3. Magnetic core has infinite permeability.

From a modeling point of view, a physical power transformer is reasonably close to an ideal transformer, with losses in the order of 0.5% (of the transformer power rating), leakage fluxes in the order of 5% of the mutual flux, and the high permeability of special alloy steels.

The assumption of no losses implies that there is no resistance in the circuit model corresponding to Fig. 5.1. The assumption of no leakage fluxes Φ_{l1} and Φ_{l2} means that the same flux Φ_m links both primary and secondary windings. Thus we have $\lambda_1 = N_1 \Phi_m$ and $\lambda_2 = N_2 \Phi_m$ as the flux linkages of the primary and secondary circuits, respectively. The terminal voltages are then given by

$$v_1 = \frac{d\lambda_1}{dt} = N_1 \frac{d\Phi_m}{dt}$$

$$v_2 = \frac{d\lambda_2}{dt} = N_2 \frac{d\Phi_m}{dt} \tag{5.1}$$

and the voltage gain is

$$\frac{v_2}{v_1} = \frac{N_2}{N_1} = n \tag{5.2}$$

where n is the ratio of secondary turns to primary turns. Frequently, it is also convenient to work with the reciprocal of n. This is called the *transformer turns ratio* and is designated $a = N_1/N_2 = 1/n$.

Continuing with our discussion of the ideal transformer, we note that with the assumed reference directions for current (see Fig. 5.1), the primary and secondary magnetomotive forces (mmfs) tend to add and the total mmf is

$$F = N_1 i_1 + N_2 i_2' = R\Phi_m \tag{5.3}$$

where R is the reluctance of the core. Now since the core is assumed to have infinite permeability, the reluctance is zero. Thus

$$F = N_1 i_1 + N i_2' = 0 \tag{5.4}$$

and as a consequence, from (5.3),

$$\frac{i_2'}{i_1} = -\frac{N_1}{N_2} = -\frac{1}{n} = -a \qquad (5.5)$$

To eliminate the minus sign it is convenient to define $i_2 = -i_2'$, and we then get the circuit model with reference directions as shown in Fig. 5.2. Correspondingly, instead of (5.5) we get

$$\frac{i_2}{i_1} = \frac{N_1}{N_2} = \frac{1}{n} = a \qquad (5.6)$$

with the advantage that we get a positive current gain.

We consider next an example illustrating the impedance transformation properties of ideal transformers.

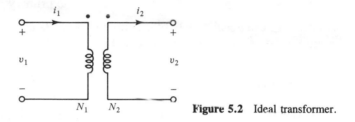

Figure 5.2 Ideal transformer.

Example 5.1

Consider the circuits shown in Fig. E5.1. We wish to pick Z_1 (in terms of Z_2) so that the terminal behavior of the two circuits is identical. The transformer turns ratios, $a = N_1/N_2$, are the same in both circuits.

Solution The terminal behavior will be identical if the two circuits have the same two-port parameters. The transmission matrices used in Chapter 4 are convenient to use. Using the *ABCD* parameters defined in (4.13), we have

$$\mathbf{T}_{xfmr} = \begin{bmatrix} 1 & 0 \\ 0 & \frac{1}{a} \end{bmatrix} \qquad \mathbf{T}_{N_1} = \begin{bmatrix} 1 & Z_1 \\ 0 & 1 \end{bmatrix} \qquad \mathbf{T}_{N_2} = \begin{bmatrix} 1 & Z_2 \\ 0 & 1 \end{bmatrix}$$

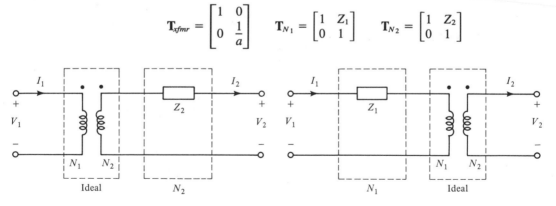

Figure E5.1

for the transmission matrices of ideal transformer, network N_1 and network N_2, respectively. For equivalent terminal behavior, using (4.17), we require that

$$
\mathbf{T} = \begin{bmatrix} a & 0 \\ 0 & \dfrac{1}{a} \end{bmatrix} \begin{bmatrix} 1 & Z_2 \\ 0 & 1 \end{bmatrix} = \begin{bmatrix} 1 & Z_1 \\ 0 & 1 \end{bmatrix} \begin{bmatrix} a & 0 \\ 0 & \dfrac{1}{a} \end{bmatrix}
$$

Carrying out the matrix multiplications, we get

$$
\mathbf{T} = \begin{bmatrix} a & aZ_2 \\ 0 & \dfrac{1}{a} \end{bmatrix} = \begin{bmatrix} a\dfrac{Z_1}{a} \\ 0\dfrac{1}{a} \end{bmatrix}
$$

Thus if $Z_1 = a^2 Z_2$, the terminal behavior of the two circuits will be identical.

Note 1: Replacing Z_2 in the secondary of the ideal transformer by $Z_1 = a^2 Z_2$ in the primary is sometimes called "referring" Z_2 to the primary.

Note 2: We can equally well refer to primary impedance into the secondary; we must multiply the primary impedance value by $n^2 = 1/a^2$.

Exercise 1

(a) Show that the result of Example 5.1 is still true if instead of a series impedance there is a shunt impedance (i.e., we replace the secondary shunt impedance Z_2 with an equivalent primary shunt impedance $Z_1 = a^2 Z_2$).

(b) Now show that we can refer a ladder network (in the secondary side) to the primary simply by multiplying every impedance by a^2. *Hint:* Do it step by step, shifting one element at a time.

(c) Do you believe it is true that any two-port (not just a ladder) can be referred to the primary by multiplying each impedance by a^2?

We next consider a more realistic model for a physical transformer. With leakage fluxes present instead of $\lambda_1 = N_1 \Phi_m$, $\lambda_2 = N_2 \Phi_m$, we have

$$
\begin{aligned}
\lambda_1 &= \lambda_{l1} + N_1 \Phi_m \\
\lambda_2 &= \lambda_{l2} + N_2 \Phi_m
\end{aligned}
\tag{5.7}
$$

Φ_m is the mutual flux which links all the turns of the primary and secondary coils. λ_{l1} and λ_{l2} are the flux linkages (including partial flux linkages) of the fluxes Φ_{l1} and Φ_{l2} which link the primary and secondary coils individually. As shown schematically in Fig. 5.1, the paths for the leakage flux, largely in air, have high reluctance and hence the leakage fluxes (and flux linkages) are relatively small in magnitude.

Assuming linearity of the magnetic core, $\lambda_{l1} = L_{l1} i_1$ and $\lambda_{l2} = L_{l2} i_2'$. Including the effects of series resistance, we then get

$$v_1 = r_1 i_1 + \frac{d\lambda_1}{dt} = r_1 i_1 + L_{l1} \frac{di_1}{dt} + N_1 \frac{d\Phi_m}{dt}$$

$$v_2 = r_2 i_2' + \frac{d\lambda_2}{dt} = r_2 i_2' + L_{l2} \frac{di_2'}{dt} + N_2 \frac{d\Phi_m}{dt}$$

(5.8)

The reader should compare (5.8) and (5.1) and note that if we add appropriate series resistance and leakage inductance to the primary and secondary circuits of an ideal transformer, the voltage relations in (5.8) will be satisfied.

We next consider the effect of finite permeability. With finite permeability the core mmf is no longer zero and thus we cannot use (5.4) to relate i_1 and i_2'. Neglecting the small leakage fluxes, however, we can still use (5.3). First, consider the following application of (5.3). Suppose that $i_2' = 0$ (i.e., the transformer secondary is open circuited). In the case of an ideal transformer the primary current would then be zero, but in the more practical case now under consideration some current would still flow to sustain the magnetic field. This current, i_m, is called *primary magnetization current* and its value, from (5.3), is

$$i_m = \frac{R\Phi_m}{N_1}$$

(5.9)

Thus $N_1 i_m = R\Phi_m$ and we can introduce this in (5.3). Dividing by N_1, we get

$$i_1 = i_m - \frac{N_2}{N_1} i_2' = i_m + \frac{N_2}{N_1} i_2$$

(5.10)

Finally, using (5.9), we can replace the voltage $e_1 \triangleq N_1(d\Phi_m/dt)$ by

$$e_1 \triangleq N_1 \frac{d\Phi_m}{dt} = L_m \frac{di_m}{dt}$$

(5.11)

where $L_m \triangleq N_1^2/R$. Using (5.8), (5.10), and (5.11), we then get the circuit model in Fig. 5.3. In the model we have replaced i_2' (with reference direction inward) by its negative, i_2 (with reference direction outward).

The circuit model in Fig. 5.3 has the desirable feature that its elements closely match those of the physical model. This helps in understanding and improving the model. For example, L_m represents the (assumed) linear relationship between the mutual flux Φ_m and the core mmf; if a more accurate nonlinear model is desired, a nonlinear inductor may be substituted. Also, in the model, conductor (copper) losses have been accounted for (by r_1 and r_2), but core losses have not. The reader may recall from physics the phenomena of hysteresis and eddy currents, which in the environment of the cyclic flux variation in the iron core are responsible for these core losses. These losses may be accounted for in the circuit model by paralleling L_m with a resistor. We will not, however, undertake the improvements just indicated. Certainly, a transformer designer needs an improved model, but for our uses, considering the larger system, the simple model in Fig. 5.3 is usually adequate.

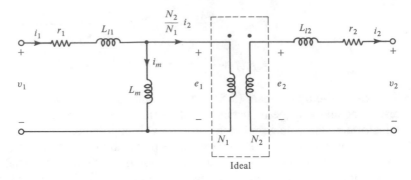

Figure 5.3 Transformer equivalent circuit.

In fact we will now consider a simplification. We can refer the secondary elements to the primary side of the ideal transformer as described in Example 5.1. We then have a so-called T-equivalent circuit on the primary side. In fact, in practice, the element values are such (low series, high shunt impedances) that the magnetizing reactance and referred secondary impedances may be interchanged, resulting in the circuit model shown in Fig. 5.4.

For most systems studies the model may be even further simplified. The series resistance r is much smaller than the leakage reactance $X_l = \omega L_l$, and may usually be neglected (i.e., shorted out); the shunt magnetizing inductance L_m is much greater than L_l and may frequently be neglected (i.e., open circuited).

The model in Fig. 5.4 makes it quite simple to discuss the behavior of the transformer in terms of the physical parameters. For simplicity suppose that we neglect r in the following discussion. The fact that L_m is large and L_l is small makes the operation of the power transformer quite understandable in terms of Fig. 5.4. For example, suppose that the transformer is supplying a lagging load (current lags voltage). Suppose that the voltage gain $n = N_2/N_1 = 1$. Then from Fig. 5.4 we get

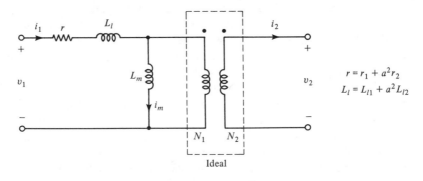

$$r = r_1 + a^2 r_2$$
$$L_l = L_{l1} + a^2 L_{l2}$$

Figure 5.4 Simplified equivalent circuit.

a relationship between phasors similar to that shown in Fig. 5.5. In this figure we arbitrarily pick V_2 and I_2 and "work back" to V_1 and I_1. More specifically, starting with V_2 and I_2, we can find $I_m = V_2/jX_m$, then $I_1 = I_m + I_2$, then $V_1 = V_2 + jX_lI_1$. We note that if I_m is small (X_m large), then $I_1 \approx I_2$. If X_l is small, then $V_2 \approx V_1$. We note that if the supply voltage $|V_1|$ is fixed, the effect of increasing $|I_2|$ (while maintaining the lagging load PF) is to cause $|V_2|$ to drop. If we neglect the parallel magnetizing inductance (open circuit it), then $I_1 = I_2$. If we neglect the series leakage inductance (short circuit it), then $V_1 = V_2$.

An additional advantage of the model is that L_l, L_m, and N_2/N_1 may be directly determined by test. This is illustrated in the following example.

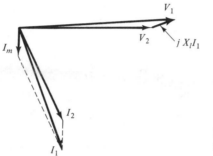
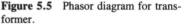

Figure 5.5 Phasor diagram for transformer.

Example 5.2

A single-phase transformer has the following ratings: 200 MVA, 200/400 kV. Assume that the primary voltage is 200 kV (and the secondary voltage is 400 kV). Two tests are performed: an open-circuit test and a short-circuit test. In the open-circuit test the secondary is left open and the rated primary voltage is applied. This causes 10 A to flow in the primary. In the short-circuit test the secondary terminals are connected together (short circuited) and a reduced voltage applied to the primary until rated primary current flows. The required voltage is found to be 21.0 kV. Find an equivalent circuit for the transformer neglecting resistance.

Solution Rated primary current $= 200 \times 10^6/(200 \times 10^3) = 1000$ A. From Fig. 5.4 we note that in the short-circuit test L_m is effectively short circuited, leaving only L_l in the circuit. Thus

$$X_l = \frac{21 \times 10^3}{1000} = 21 \ \Omega$$

In the open-circuit test the total reactance seen at the primary terminals is

$$X_l + X_m = \frac{200 \times 10^3}{10} = 20,000 \ \Omega$$

Thus

$$X_m = 20,000 - 21 = 19,979 \approx 20,000 \ \Omega$$

We next pick N_2/N_1 so that the open-circuit voltages are in the ratio 400:200. Strictly speaking, we should use the formula (based on the voltage-division law)

$$\frac{V_2}{V_1} = \frac{N_2}{N_1} \frac{X_m}{X_m + X_l} \tag{5.12}$$

But since $X_m \gg X_l$, as a practical matter we can take

$$\frac{N_2}{N_1} = \frac{V_2}{V_1} = \frac{400}{200}$$

Thus we get the circuit shown in Fig. E5.2.

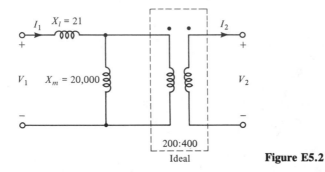

Figure E5.2

Note: If the transformer is operating under rated conditions ($|I_1| = 1000$ A, $|V_1| = 200$ kV), the voltage drop in X_l is 21 kV, which is about 10% of the rated voltage. Thus we get an idea of the significance of X_l. Similarly, the current through X_m is approximately 200 kV/20 kΩ = 10 A, which is only about 1% of the rated current. This is thus seen to be a fairly negligible current; for this reason in most systems applications we would neglect X_m (i.e., we replace it by an open circuit).

It is interesting to compare the model in Fig. 5.4 with the L_1, L_2, M model familiar from circuit theory. We consider this briefly in the next example.

Example 5.3

For the transformer of Example 5.2, find $X_1 = \omega L_1$, $X_2 = \omega L_2$, and $X_m = \omega M$ of the L_1, L_2, M model.

Solution   Using the equations familiar from circuit theory,

$$V_1 = jX_1 I_1 + jX_m I_2'$$
$$V_2 = jX_m I_1 + jX_2 I_2' \tag{5.13}$$

it is easy to find the values of X_1, X_2, and X_m using the test data. For example, in the open-circuit test we get $|V_1| = X_1 |I_1| \Rightarrow X_1 = 200 \times 10^3/10 = 20,000$. Other values are $X_2 \approx 80,000$ and $X_m = 40,000$. Although certainly useful for calculations, the parameters do not give nearly as good a grasp of the physical behavior of the transformer as the model in Fig. 5.4.

5.2 THREE-PHASE TRANSFORMER CONNECTIONS

There are basically four different ways in which single-phase transformers may be connected into so-called three-phase banks. They may be connected wye-wye, delta-delta, wye-delta, or delta-wye, as shown in Fig. 5.6. All the connections are shown with primary (low voltage) on the left and secondary (high voltage) on the right.

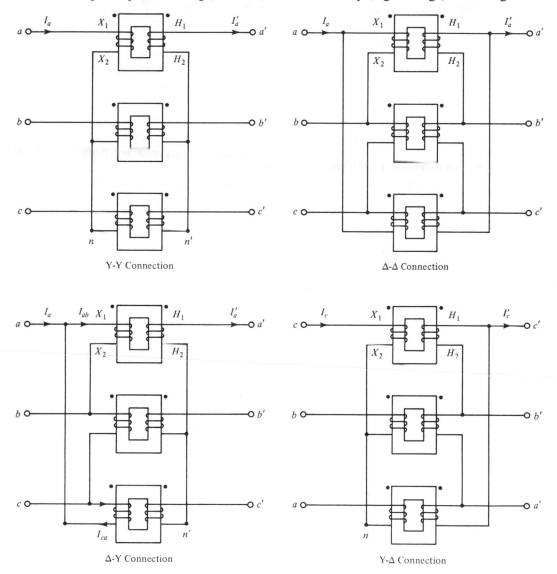

Figure 5.6 Standard transformer connections for three phase.

The favored connection is the delta-wye with the wye on the high voltage side. The advantages of this connection include the following. There is a neutral on the high-voltage side which may be grounded; the effect of this ground will be considered in Chapter 13 when short circuits are considered. There is a voltage gain of $\sqrt{3}$ just by virtue of the connection, in addition to the voltage gain due to the turns ratio of the individual single-phase transformers. The delta connection on the primary side serves a useful function as well, with respect to unbalanced operation and/or in the presence of nonsinusoidal current or voltage waveforms.

Considering now some of the other connections, the delta-delta connection offers the possibility of operating, at least on an emergency basis, with one (single-phase) transformer removed. This is called the *open-delta connection*. The wye-wye connection is seldom used because of problems with unbalanced operation and harmonics. The advantages and disadvantages of the different connections are fully discussed in standard references on transformers. We will simply recognize the presence of all four types of connections and consider the modeling thereof. We note that there are also three-phase transformers in which the three sets of phase coils share the core material. Two types are shown in Fig. 5.7.

The advantage of a three-phase transformer is that it is possible to save core material and get a cheaper and more efficient transformer. It is exactly analogous to the saving of wire and energy in a three-phase transmission line versus three single-phase transmission lines. There are, however, some practical disadvantages as well. A fault in any phase of the transformer would normally require the entire unit to be removed from service, whereas a similar fault in a single-phase transformer would only require that unit to be replaced. Also, there may be some difficulties in shipping the larger units. In any case the theory of the three-phase transformers is virtually the same as for three-phase banks made up of single-phase units and we will not consider them separately.

Rather than draw the connections in detail as in Fig. 5.6, it is more convenient to use a schematic as illustrated in Fig. 5.8 for the case of the wye-wye connection.

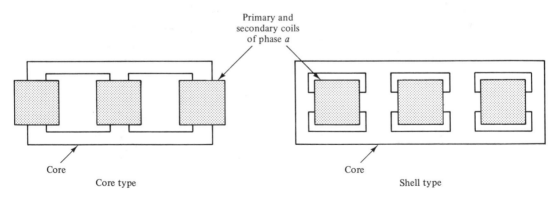

Figure 5.7 Three-phase transformers.

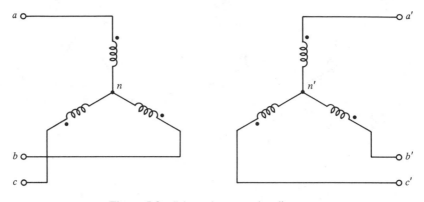

Figure 5.8 Schematic connection diagram.

Windings drawn in parallel directions are primary and secondary windings of the same single-phase transformer. Thus the dots indicate that $V_{a'n'}$ and V_{an} are (approximately) in phase. We will occasionally use the even simpler schematic diagram of Fig. 5.9 drawn for the case of a delta-wye connection. In this figure the transformer windings are indicated by heavy lines. Parallel lines indicate correspondingly primary and secondary windings. Although the polarity dots are not shown explicitly, they are assumed to be at the same ends of the corresponding lines. Thus $V_{a'n'}$ and V_{ab} are approximately in phase. Using this kind of representation, we can replace Fig. 5.6 by Fig. 5.10.

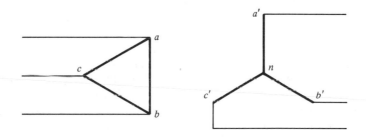

Figure 5.9   Schematic connection diagram.

We will also frequently use a one-line diagram to indicate the transformer connection. Figure 5.11 is a one-line-diagram representation of the transformer connection in Fig. 5.9. Of course, the wye-wye or delta-delta connection may be indicated similarly.

In the next section we are going to simplify the descriptions of these three-phase transformer banks, for the case of a balanced system, by introducing their per phase representations. As a preliminary we would like to consider the approximate behavior of the three-phase banks by assuming that each single-phase transformer is ideal.

In the case of the wye-wye and delta-delta banks, we find that only the voltage

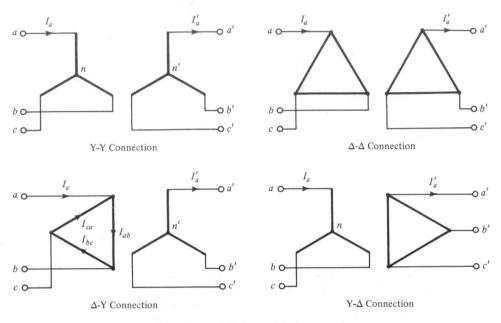

Figure 5.10   Standard transformer connections.

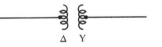

Figure 5.11   One-line diagram.

and current levels are changed. For example, in Fig. 5.10, $V_{a'b'} = nV_{ab}$ and $I_a' = I_a/n$, where n is the voltage gain of the single-phase transformer. In the case of the wye-delta connection, if we compare either line-line or phase-neutral voltages, there is a phase shift as well as a voltage- and current-level change. For example, for the delta-wye connection shown in Fig. 5.10, assuming positive-sequence voltages for V_{an}, V_{bn}, and V_{cn}, we get $V_{a'n'} = nV_{ab} = n(V_{an} - V_{bn}) = \sqrt{3}ne^{j\pi/6}V_{an}$. Thus $V_{a'n'}$ is related to V_{an} by the complex gain $K_1 \triangleq \sqrt{3}ne^{j\pi/6}$. We get identical relationships for the remaining phases. Thus

$$V_{a'n'} = K_1 V_{an} \qquad V_{b'n'} = K_1 V_{bn} \qquad V_{c'n'} = K_1 V_{cn} \qquad (5.14)$$

and identical relationships hold for the line voltages as well. Thus there is a voltage gain of $\sqrt{3}n$ and a phase advance or lead of 30° of the secondary voltages relative to the corresponding primary voltages. Note the $\sqrt{3}$ step-up ratio beyond the step-up ratio of the single-phase transformers. If for the same connection (see Fig. 5.10) we now consider the current relationships, we get, again assuming positive-sequence conditions,

$$I_a = I_{ab} - I_{ca} = n(I'_a - I'_c)$$
$$= \sqrt{3}ne^{-j\pi/6}I'_a \qquad (5.15)$$
$$= K_1^* I'_a$$

where K_1^* is the complex conjugate of K_1. A similar result holds for phases b and c. Thus

$$I'_a = \frac{I_a}{K_1^*} \qquad I'_b = \frac{I_b}{K_1^*} \qquad I'_c = \frac{I_c}{K_1^*} \qquad (5.16)$$

Note that the per phase complex power gain of the transformer bank is unity. We can see this as follows. Let S' and S be the per phase complex power output and input, respectively. Using (5.14) and (5.16), we get

$$S' = V_{a'n'}I'^*_a = K_1 V_{an}\left(\frac{I_a}{K_1^*}\right)^* = V_{an}I^*_a = S \qquad (5.17)$$

This is entirely consistent with the model of the ideal transformer as a device that neither stores nor dissipates any energy. In fact, the relationship between currents in (5.16) could have been derived, alternatively, by utilizing this physical property.

Exercise 2.   If we assume that V_{an}, V_{bn}, and V_{cn}, are a negative-sequence set, show that (5.14) and (5.16) still hold, with a gain change $K = K_2 \triangleq \sqrt{3}ne^{-j\pi/6}$. Note that $K_2 = K_1^*$.

Exercise 3.   Show that if we assume positive-sequence voltages and consider the wye-delta connection in Fig. 5.10 (again with ideal transformers), then (5.14) and (5.16) still holds, with $K = K_3 = (n/\sqrt{3})e^{j\pi/6}$. What is the gain if V_{an}, V_{bn} and V_{cn} are a negative-sequence set?

Exercise 4.   Show that, in general, if the positive-sequence voltage gain is K, the negative-sequence gain will be its complex conjugate K^*. *Hint:* Note that taking the complex conjugate transforms a positive-sequence set into a negative-sequence set.

In all the cases we have been considering the indicated relationships between primary and secondary voltages and primary and secondary currents are very similar to those of an ideal transformer. It will therefore be convenient to extend the notion of an ideal transformer to cover this case. We will define a *complex ideal transformer* with complex voltage gain K and complex current gain $1/K^*$. The complex power gain is $K(1/K^*)^* = 1$. The impedance transformation properties are the following: When terminated in a symmetric wye-connected load, the per phase driving-point impedance on the primary side is

$$\frac{V_{an}}{I_a} = \frac{V_{a'n'}/K}{K^*I_a'} = \frac{1}{|K|^2}Z_L \qquad (5.18)$$

where Z_L is the load impedance in the secondary.

These properties are summarized in Table 5.1 and compared with those of an ideal transformer. We see that the two types of ideal transformers have many properties in common. In particular we note that the complex power gain, the impedance transformation properties, and the ratios of secondary to primary voltage magnitudes and current magnitudes are independent of the phase angle $\underline{/K}$ and the dependence on $|K|$ and n are identical. Note that in cases where $\underline{/Z}$ is involved, it shifts the phase angle of the voltage and current the same amount in going from primary to secondary.

TABLE 5.1 TRANSFORMER PROPERTIES

	Complex ideal transformer	Ideal transformer
Voltage gain	$K = |K|\ \underline{/K}$	n
Current gain	$\dfrac{1}{K^*} = \dfrac{1}{|K|}\ \underline{/K}$	$\dfrac{1}{n}$
Complex power gain	1	1
Secondary impedance Z_L referred to primary	$\dfrac{1}{|K|^2}Z_L$	$\dfrac{1}{n^2}Z_L$

While we have been considering standard three-phase transformer connections, many other connections are possible. In the delta-wye, for example, we could change the labeling of the conductors a', b', c' to c', a', b'. We would then find the (positive-sequence) voltage gain to be $K = -j\sqrt{3}n$. Or, in the wye-wye case, we could change the labeling of the conductors a', b', c' to b', c', a' and find the (positive-sequence) voltage gain $K = ne^{j2\pi/3}$. In these nonstandard cases all the properties in Table 5.1 still hold.

We will follow the industry practice and assume the use of the standard connections shown in Fig. 5.10 unless stated to the contrary. For convenience the various gains of the standard connections are tabulated in Table 5.2.

TABLE 5.2 VOLTAGE GAIN, K, OF COMPLEX IDEAL TRANSFORMER

Standard connection	Positive-sequence gain[a]	Negative-sequence gain
Delta-wye	$\sqrt{3}ne^{j\pi/6}$	$\sqrt{3}ne^{-j\pi/6}$
Wye-delta	$\dfrac{n}{\sqrt{3}}e^{j\pi/6}$	$\dfrac{n}{\sqrt{3}}e^{-j\pi/6}$

[a]n is the voltage gain of the ideal transformer in the single-phase transformer model.

Finally, we note that while the complex ideal transformer may be viewed as a generalization of the ideal transformer, it must be remembered that its use is limited to the case of balanced three-phase sinusoids, while the gain of an ideal transformer is valid for any waveform.

This concludes our discussion of three-phase ideal transformer interconnections. We turn next to practical cases in which the leakage and magnetizing reactances appear in the model of the single-phase transformers.

5.3 PER PHASE ANALYSIS

With transformers in the network there is the possibility that a simplified model of the power system network is not (conductively) connected. In this case there does not seem to be any reason to assume that the neutrals of the separate parts are at the same potential. In fact, when more complete modeling is considered, including the grounding of neutrals and/or capacitance to ground, it appears that with balanced inputs and a symmetric network, all the neutrals are at the same potential and all the responses are also balanced. Under these conditions, in Fig. 5.12, for example, we can assume that n' and n are at the same potential and that V_{an}, V_{bn}, and V_{cn} are also a balanced set of voltages.

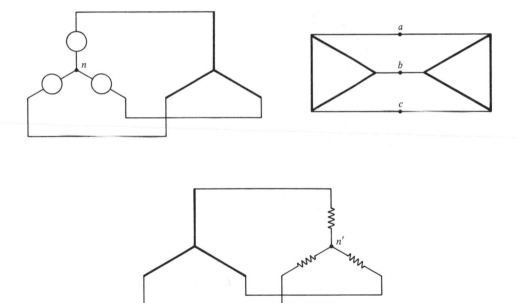

Figure 5.12   Illustrating separate parts.

In the following we will frequently show the assumption that the neutrals are at the same potential (under balanced conditions) by showing a connection between neutrals in the per phase diagram.

We now consider the per phase modeling of the four types of transformer connections given in Section 5.2. For simplicity, we will be neglecting resistance. It can always be reintroduced later in series with the leakage reactance L_l.

Per Phase Diagram (Wye-Wye)

Using the equivalent circuit of each single-phase transformer shown in Fig. 5.4, we get the following three-phase connection. In Fig. 5.13 we observe a symmetric network composed of L_l's and L_m's in cascade with a Y-Y interconnection of ideal transformers. Assuming balanced conditions, we then get the per phase circuit shown in Fig. 5.14, where we explicitly show our assumption that n' and n are at the same potential. It is instructive to compare Figs. 5.14 and 5.4.

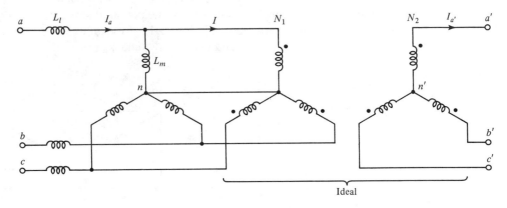

Figure 5.13   Three-phase wye-wye connection.

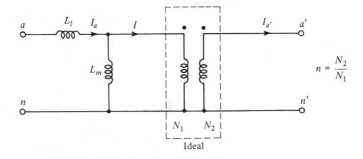

Figure 5.14   Per phase equivalent circuit (wye-wye).

Per Phase Diagram (Delta-Delta)

The three-phase circuit is shown in Fig. 5.15. The primaries and secondaries of the ideal transformers are the windings identified with the dots with N_1 turns in the primary and N_2 turns in the secondary.

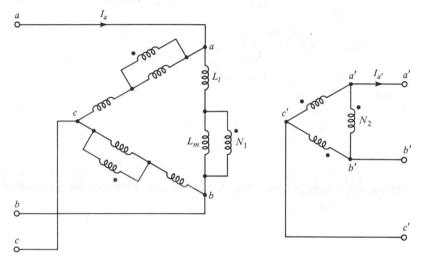

Figure 5.15   Three-phase delta-delta connection.

We are now going to show that for balanced conditions the circuit in Fig. 5.15 is equivalent to that in Fig. 5.16. We can show the equivalence by writing an equation for each of the circuits, but it is less tedious to compare behavior under open-circuit and short-circuit conditions. We will effectively be demonstrating that the two circuits have the same *ABCD* parameters (for the per phase two-port) and hence are equivalent.

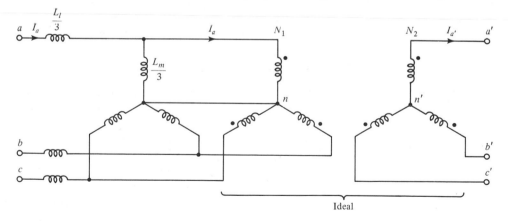

Figure 5.16   Circuit equivalent to Fig. 5.15.

Secondaries open circuited. In this case there is zero current in the ideal transformers. From the primary side of Fig. 5.15 we see a delta with $L_l + L_m$ in each branch. In Fig. 5.16 we see an equivalent wye with $(L_l + L_m)/3$ in each branch. Thus for the (balanced) input voltages we get the same input currents in both cases. Comparing output voltages, the same voltage-divider ratio assures that they will also be the same.

Secondaries short circuited. In this case the secondary voltages of the ideal transformers are zero and thus the primary voltages are zero as well. In effect, L_m and $L_m/3$ are shorted out and the currents through them are therefore zero. From the primary sides we see a delta of L_l and a wye of $L_l/3$. Therefore, for a given set of (balanced) input voltages, the input currents are equal. Checking the output currents, in the case of Fig. 5.15 we get a current gain of $1/n$ in the ideal transformers, which translates into a line current gain of $1/n$, while in the case of Fig. 5.16 we get the same result.

On the basis of the equivalence of Figs. 5.15 and 5.16, we can now use the per phase equivalent circuit in Fig. 5.17. This circuit should be compared with Fig. 5.14. The only difference is that L_l and L_m of the single-phase transformers are divided by 3.

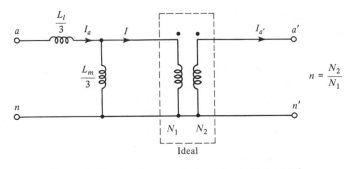

Figure 5.17 Per phase equivalent circuit (delta-delta).

Per Phase Diagram (Delta-Wye)

If we repeat the calculations just concluded for the case where the secondary in Fig. 5.15 is connected in a wye (as shown in Fig. 5.10), we find that the primary-side driving-point impedances are the same for the open-circuited and short-circuited secondaries. However, there are changes in the open-circuit voltage gain and short-circuit current gain. In the open circuit, for positive-sequence inputs, the voltage gain is $K_1 L_m/(L_l + L_m)$, where K_1 is the complex voltage gain of the ideal-transformer delta-wye connection. This is to be contrasted with $nL_m(L_l + L_m)$ as the gain in the delta-delta connection. In the short-circuit case the line current gain is $1/K_1^*$ rather than $1/n$. Thus it appears that in the equivalent circuit corresponding to Fig. 5.17, the complex gain K_1 should be used instead of n. We are then led to the per phase equivalent circuit shown in Fig. 5.18. Comparing with Fig. 5.17, we see the change

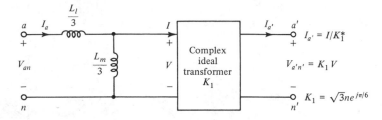

Figure 5.18   Per phase equivalent circuit (delta-wye).

to a complex ideal transformer. Note the block diagram representation of the complex ideal transformer. Using the standard block diagram convention, the arrow directed toward the block designates the input side. Thus V and I are inputs and $V_{a'n'} = K_1 V$ and $I_a' = I/K_1^*$ are outputs.

Sometimes it is convenient to represent the complex ideal transformer more explicitly. We can then replace Fig. 5.18 with Fig. 5.19. In Fig. 5.19, n is the voltage gain of each 1ϕ transformer in the 3ϕ bank. Note that in Fig. 5.19 we have the cascade of a conventional ideal transformer (with voltage gain $\sqrt{3}n$) and a phase shifter (which advances the phase by 30°). It is convenient to refer to the voltage gain as a "connection-induced voltage gain" and to the phase shift as a "connection-induced phase shift." The phase shifter is represented in block diagram form with the arrow indicating the input side.

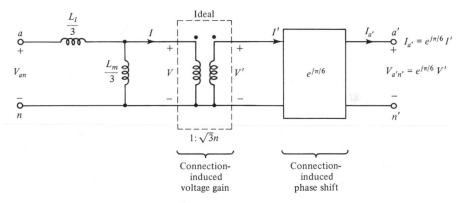

Figure 5.19   Alternate circuit (delta-wye).

Per Phase Diagram (Wye-Delta)

Finally, we consider the wye-delta connection shown in Fig. 5.10. Using the equivalent circuit for each single-phase transformer, we get Fig. 5.20, in which a symmetric network of L_l's and L_m's is in cascade with a Y-Δ interconnection of ideal transformers. As discussed in Section 5.2, with positive-sequence variables, the gain of the bank of ideal transformers is $K_3 = (n/\sqrt{3})e^{j\pi/6}$. The per phase equivalent circuit is shown in Fig. 5.21. The complex ideal transformer may also be replaced by a more

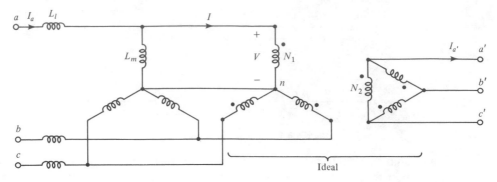

Figure 5.20   Three-phase wye-delta connection.

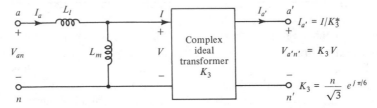

Figure 5.21   Per phase equivalent circuit (wye-delta).

explicit circuit similar to Fig. 5.19, but with the connection-induced voltage gain equal to $n/\sqrt{3}$.

All the per phase equivalent circuits are similar, being the cascade of a (voltage-divider) network and of an ideal or complex ideal transformer. This similarity will be enhanced further by developments to be introduced in Section 5.4.

Finally, we note that in all these models the series resistance r may be included. With Y-connected primaries we add r in series with L_l. With Δ-connected primaries we add $r/3$ in series with $L_l/3$. We conclude this section with two examples.

Example 5.4

We are given a system involving a step-up transformer with the one-line diagram shown in Fig. E5.4(a). The transformer bank is made up of identical 1ϕ transformers each specified by $X_l = 0.21\ \Omega$, $n = N_2/N_1 = 10$. Each generator phase is modeled by a Thévenin equivalent circuit. The transformer bank is delivering 100 MW at 0.9 PF lagging to a substation bus whose voltage is 230 kV.

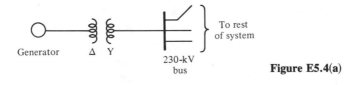

Figure E5.4(a)

(a) Find the primary current, primary voltage (line-line), and 3ϕ complex power supplied by the generator.

(b) Find the phase shift between the primary and secondary voltages.

Solution Assuming positive-sequence operation and the standard Δ-Y connection, we can use Fig. 5.19 [see Fig. E5.4(b)]. Since resistance and magnetizing reactance are not given, we assume that they may be neglected.

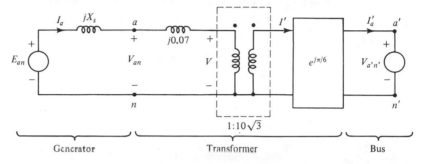

Figure E5.4(b)

(a) From the data given, we have $|V_{a'n'}| = 230/\sqrt{3} = 132.8$ kV. The phase angle is not given. For convenience in calculations we can choose $\underline{/V_{a'n'}} = 0$. Then $V_{a'n'} = 132.8 \underline{/0°}$. The 3ϕ power delivered to the bus is specified. Thus we can calculate the per phase complex power into the bus:

$$S' = \frac{100 \times 10^6}{0.9 \times 3} \underline{/25.84°} = 37.04 \underline{/25.84°} \text{ MVA}$$

Next we calculate I_a' from

$$I_a'^* = \frac{S'}{V_{a'n'}} = \frac{37.04 \times 10^6 \underline{/25.84°}}{132.8 \times 10^3} = 278.9 \underline{/25.84°} \text{ A}$$

Thus $I_a' = 278.9 \underline{/-25.84°}$ A. We next find I_a:

$$I_a = 10\sqrt{3} \, e^{-j\pi/6} I_a' = 4830.6 \underline{/-55.84°} \text{ A}$$

Recalling that the phase angle of $V_{a'n'}$ was not given (an angle of zero was arbitrarily chosen), only the primary current *magnitude* can be determined. In this case we take the assignment "find the primary current" to mean "find the primary current magnitude," and this magnitude is 4830.6 A.

We are now in a position to calculate V_{an}.

$$V_{an} = V + j0.07 I_a$$

$$= \frac{1}{10\sqrt{3}} 132.8 \times 10^3 \underline{/-30°} + j0.07 \times 4830.6 \underline{/-55.84°}$$

$$= 7667.2 \underline{/-30°} + 338.1 \underline{/34.16°}$$

$$= 7820.5 \underline{/-27.77°} \text{ V}$$

Just as in the case of the current I_a, only the voltage magnitude can be determined. Thus we have the primary voltage (line-line), $\sqrt{3} \times 7820.5 = 13.55$ kV.

Next we can calculate the per phase complex power into the transformer:

$$S = V_{an}I_a^* = 7820.5 \; \underline{/-27.77°} \times 4830.6 \; \underline{/55.84°}$$

$$= 37.778 \; \underline{/28.07°} \text{ MVA}$$

To find the 3ϕ power delivered by the generator, we multiply by 3.

$$S^{3\phi} = 3S = 113.33 \; \underline{/28.07°} \text{ MVA}$$

Since the complex power depends on the *difference* in the phase angles between V_{an} and I_a (and not on their individual values) the arbitrary choice of phase angle for $V_{a'n'}$ is of no consequence. This then completes part (a).

(b) The phase shift between primary and secondary voltages (either phase-neutral or line-line) may now be determined since we have V_{an} for the choice of $V_{a'n'}$. Thus the secondary phase leads the primary by $0° - (-27.77°) = 27.77°$.

Some comments on the example follow.

1. If the transformers were ideal, the angle difference calculated in part (b) would be $30°$ rather than $27.77°$. Also, the transformer input and output complex powers would be identical.

2. In part (b) it is essential to model the phase shift in the complex ideal transformer. In part (a), however, the phase shift could have been left out without affecting the calculation of voltage magnitude, current magnitude, and complex power.

Example 5.5

Suppose that we are given a system with the one-line diagram shown in Fig. E5.5(a). We have in cascade, the following: a source, a step-up transformer bank (primary on the left), a transmission line, a step-down transformer bank (primary on the right), and an impedance load. The transformer banks are the same as in Example 5.4. Assume that the generator terminal voltage (line-line) is 13.8 kV and find the generator current, the transmission-line current, the load current, the load voltage, and the complex power delivered to the load.

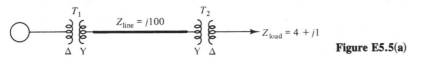

T_1 $\quad\quad Z_{\text{line}} = j100 \quad\quad$ T_2

Δ Y $\quad\quad\quad\quad\quad\quad$ Y Δ $\quad Z_{\text{load}} = 4 + j1$

Figure E5.5(a)

Solution   Using Fig. 5.19, we have the circuit diagram shown in Fig. E5.5(b). From the problem specification we have $|V_1| = 13.8/\sqrt{3} = 7.97$ kV. For convenience we may choose $V_1 = 7.97 \; \underline{/0°}$ kV. The easiest way to calculate the generator current I_1 is to find the driving-point impedance to the right of the terminals identified by the voltage V_1. Starting from the load, we have $Z_3 \triangleq Z_{\text{load}} + j0.07 = 4 + j1.07$ Ω. Referring this impedance to the secondary side of the ideal complex transformer, we get

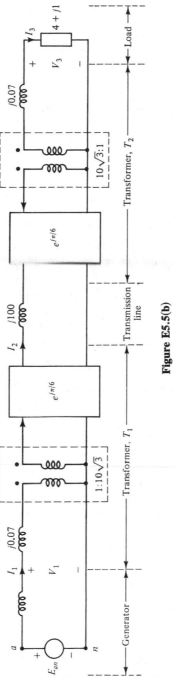

Figure E5.5(b)

$(10\sqrt{3})^2 Z_3 = 1200 + j321 \ \Omega$. Then adding Z_line, we get $Z_2 \triangleq Z_\text{line} + 1200 + j321 = 1200 + j421 \ \Omega$. Referred to the primary side of T_1, we get $Z_2/(10\sqrt{3})^2 = 4.00 + j1.4033$. Finally, we get the driving-point impedance $Z_1 \triangleq 4.00 + j1.4733 = 4.263 \ \underline{/20.22°}$. Thus we get

$$I_1 = \frac{V_1}{Z_1} = 1869.1 \ \underline{/-20.22°} \ \text{A}$$

We next calculate I_2, I_3, V_3, and S_load:

$$I_2 = \frac{1}{10\sqrt{3}} e^{j\pi/6} I_1 = 107.9 \ \underline{/9.78°} \ \text{A}$$

$$I_3 = 10\sqrt{3} e^{-j\pi/6} I_2 = 1869.1 \ \underline{/-20.22°} \ \text{A}$$

$$V_3 = Z_\text{load} I_3 = 7.71 \ \underline{/-6.18°} \ \text{kV}$$

$$S_\text{load} = V_3 I_3^* = Z_\text{load} |I_3|^2 = 14.4 \ \underline{/14.0°} \ \text{MVA}$$

The *relative* phase angles of the voltages and currents are significant, but as in Example 5.4, the *individual* values depend completely on our (arbitrary) selection of $\underline{/V_1}$. Thus we can only specify voltage and current magnitudes. Thus we have

generator current $= |I_1| = 1869.1$ A

transmission-line current $= |I_2| = 107.9$ A

load current $= |I_3| = 1869.1$ A

load voltage (line-line) $= \sqrt{3} \ |V_3| = 13.35$ kV

3ϕ complex power $= 3S_\text{load} = 43.21 \ \underline{/14.0°}$ MVA

Exercise 5. Convince yourself that the 30° phase shifts can be ignored and all the results above concerning voltage and current magnitudes and complex power will be obtained.

Note: In Examples 5.4 and 5.5 we have so-called *radial* systems, in which none of the transmission lines and/or transformers are in parallel. In these and similar examples, for the usual power system problem, we can completely ignore the connection-induced phase shifts of the 3ϕ Δ-Y or Y-Δ transformer banks. In more complicated cases this may not be true. We consider these cases next.

5.4 NORMAL SYSTEMS

Typical power transmission systems are richly interconnected, with many loops and parallel paths in the one-line diagrams. There are many advantages to this arrangement in terms of reliability and operating flexibility. Aside from this rationale, there are also historical factors; transmission-line voltages and power-handling capabilities have increased over the years, with the modern additions overlaying the existing network.

Where these loops or parallel paths include transformers, special care must be exercised in system design to avoid large "circulating" currents. The following example illustrates the nature of the problem.

Example 5.6

We are given a one-line diagram for two transformer banks in parallel [Fig. E5.6(a)]. The 1ϕ transformers in the two banks are specified as follows:

Y-Y bank: $n = 10, X_l = 0.05$

Y-Δ bank: $n = \sqrt{3} \times 10, X_l = 0.05$

The line-neutral voltage at bus 1 has magnitude 8 kV. The load at bus 2 is Y connected; each impedance, $Z_{\text{load}} = 100 \underline{/0°}$. Find $|I_1'|$, $|I_2'|$, and $|I_{\text{load}}|$.

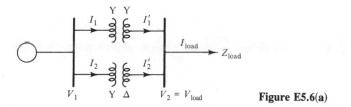

Figure E5.6(a)

Solution   Using Figs. 5.14 and 5.21, and assuming the phase of V_1 is zero, we get the per phase equivalent circuit shown in Fig. E5.6(b). Writing KVL for the I_1 loop, we get

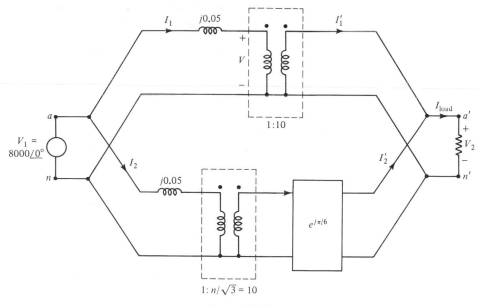

Figure E5.6(b)

$$V_1 = j0.05I_1 + V$$

$$= j0.05I_1 + \frac{1}{10} \times 100(I_1' + I_2')$$

Replacing I_1 by $10I_1'$, we get

$$V_1 = (10 + j0.5)I_1' + 10I_2' = 8000 \tag{1}$$

Similarly, writing KVL for the I_2 loop, we get

$$V_1 = j0.05I_2 + \tfrac{1}{10}e^{-j\pi/6} \times 100(I_1' + I_2')$$

Replacing I_2 by $10I_2'e^{-j\pi/6}$, we get

$$V_1 = 10I_1'e^{-j\pi/6} + (10 + j0.5)I_2'e^{-j\pi/6}$$

Multiplying by $e^{j\pi/6}$, we can simplify and we get

$$10I_1' + (10 + j0.5)I_2' = 8000 \underline{/30°} \tag{2}$$

Solving (1) and (2) for I_1' and I_2', we get

$$I_1' = 3755.0 \underline{/-164.85°} \text{ A}$$

$$I_2' = 4527.2 \underline{/14.88°} \text{ A}$$

$$I_{\text{load}} = I_1' + I_2' = 772.5 \underline{/13.57°} \text{ A}$$

Notice that $|I_1'|$ and $|I_2'|$ are much larger than $|I_{\text{load}}|$. The physical interpretation is that there is a large component of the current which circulates from one transformer bank to the other without entering the load. This circulating current serves no useful purpose. In fact, it is harmful, wasting energy and possibly overheating the transformers. *Note:* Even if Z_{load} is removed and $I_{\text{load}} = 0$, we still get a large circulating current. The basic problem is that the connection-induced phase shifts in the two parallel transformer banks are different. In practice this connection would not be used.

Example 5.7

Repeat Example 5.6 but replace the Y-Δ transformer bank with a Y-Y transformer bank with $n = 20$.

Solution   In this case both transformer banks have the same connection-induced phase shifts (zero), but the open-circuit secondary voltages are different. Paralleling the steps in Example 5.6, we get

$$(10 + j0.5)I_1' + 10I_2' = 8000 \tag{1}$$

$$5I_1' + (5 + j1)I_2' = 8000 \tag{2}$$

Solving for I_1' and I_2', we get

$$I_1' = 3,260.76 \underline{/76.40°}$$

$$I_2' = 3,213.39 \underline{/-86.58°}$$

$$I_{\text{load}} = I_1' + I_2' = 959.23 \underline{/-2.29°}$$

Once again, as in Example 5.6, we get a very large circulating current. The problem this time is the mismatch in the turns ratios of the transformers, and in normal practice we would not parallel these transformers.

Example 5.8

Repeat Example 5.7 but with $n = 10$ for both banks of transformers.

Solution   By symmetry $I_2' = I_1'$ and we get

$$(20 + j0.5)I_1' = 8000$$

Then

$$I_1' = I_2' = 399.9 \underline{/-1.43°}$$

$$I_{load} = I_1' + I_2' = 799.8 \underline{/-1.43°}$$

We see an equal load division in this case without circulating current.

The examples illustrate the need for care in paralleling transformers. The normal arrangement is to parallel transformer banks that have the same (connection-induced) phase shift and the same open-circuit step-up or step-down voltage ratio. We want to avoid *inadvertent* mismatches in (connection-induced) phase shifts and voltage ratios. This same idea extends to more general parallel paths. It is to be noted, however, that occasionally we *deliberately* introduce phase shift and/or turns ratio anomalies for complex power flow control; we will consider some examples later. Considering these cases to be exceptional, it is useful to define a "normal" system as follows.

Definition.   A system is *normal* if the product of the complex ideal transformer gains in every pair of parallel paths is the same.

Alternatively, we have, for a normal system, for every pair of parallel paths:

1. The same product of connection-induced voltage gains
2. The same sum of connection-induced phase shifts

We note that the open-circuit line voltage magnitudes for three-phase transformer banks are usually specified in the transformer ratings. From these we can calculate the voltage magnitude ratios, which are, ordinarily, numerically indistinguishable from the connection-induced voltage gains referred to in the above. This then gives a convenient way to check condition 1.

Concerning condition 2, for simplicity we will assume a more stringent condition. We will assume (unless otherwise stated) that delta-wye or wye-delta transformer banks, if present at the terminals of transmission lines, have been wired in a standard, uniform pattern so that the connection-induced phase shifts cancel, line by line.

We consider an example of a normal system in Fig. 5.22. The three-phase transformer bank (open-circuit) line-line voltage magnitude ratings are given. Thus in the top path we step up by a factor of 10 and down by the same factor; the product of voltage gains is $10 \times 0.1 = 1$. Similarly, for the bottom path we get

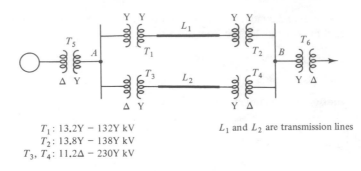

T_1: 13.2Y − 132Y kV L_1 and L_2 are transmission lines
T_2: 13.8Y − 138Y kV
T_3, T_4: 11.2Δ − 230Y kV

Figure 5.22   Normal system.

$(230/11.2) \times (11.2/230) = 1$. We assume also that T_3 and T_4 have been wired with the same phase-shift advance (say 30°) from primary to secondary. Thus the connection-induced phase shifts cancel along the bottom path. The rest of the system does not have parallel paths (i.e., is "radial") and need not be considered in this connection.

Example 5.9

Suppose that in Fig. 5.22 the wiring of the Δ-Y banks is uniform, with θ the phase advance from primary to secondary. Show that in the per phase equivalent circuit of the transmission link consisting of T_3, L_2, and T_4, the θ phase shifts cancel (and may be left out) if we are considering terminal behavior.

Solution   Representing the Δ-Y banks by Fig. 5.19 and the transmission line by *ABCD* parameters, we get the circuit shown in Fig. E5.9. Using the *ABCD* parameters, we get

$$V_1' e^{j\theta} = AV_2' e^{j\theta} - BI_2' e^{j\theta}$$

$$I_1' e^{j\theta} = CV_2' e^{j\theta} - DI_2' e^{j\theta}$$

Multiplying by $e^{-j\theta}$, we get

$$V_1' = AV_2' - BI_2'$$

$$I_1' = CV_2' - DI_1'$$

Thus, as long as the connection-induced phase shifts are equal (from primary to secondary), as we have assumed, they do not enter in the relation between V_1', I_1' and V_2', I_2'. It follows that if we are relating V_1, I_1 and V_2, I_2, we can leave out the phase-shifting blocks.

Note: After finding V_1', I_1', V_2', I_2', we can then find the transmission-line currents and voltages by advancing the phases by θ.

Leaving out the phase-shifting blocks in the per phase diagram is a worthwhile simplification. It is natural to designate this diagram as follows:

Terminology.   A *simplified per phase diagram* is a per phase diagram in which all the connection-induced phase shifts (i.e., the phase shifters) have been left out.

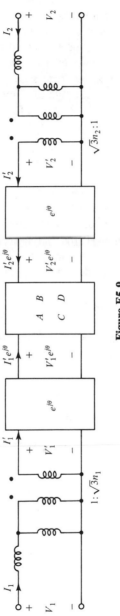

Figure E5.9

Suppose we have a normal system. For the usual power system problem (where the phases of the voltage sources are not specified), we can find the complex power flows, and the voltage and current magnitudes by using the simplified per phase diagram. Although we will not offer a proof, we draw attention to the results in Examples 5.4, 5.5, 5.7, and 5.9, and to related comments following the examples. Thus we can say that one advantage of a normal system, from an analysis point of view, is that we can use a simplified per phase diagram.

We turn now to a second advantage of a normal system, from an analysis point of view. It permits the unambiguous introduction of a useful normalization called per unit normalization.

5.5 PER UNIT NORMALIZATION

In power system calculations a normalization of variables called per unit normalization is almost always used. It is especially convenient if many transformers and voltage levels are involved.

The idea is to pick base values for quantities such as voltages, currents, impedances, power, and so on, and to define the quantity in per unit as follows.

$$\text{quantity in per unit} = \frac{\text{actual quantity}}{\text{base value of quantity}}$$

A vital point is that the base variables are picked to satisfy the same kind of relationship as the actual variables. For example, corresponding to the equation between actual variables (complex numbers),

$$V = ZI \tag{5.19}$$

we have the equation for base quantities (real numbers),

$$V_B = Z_B I_B \tag{5.20}$$

Dividing (5.19) by (5.20), we get

$$\frac{V}{V_B} = \frac{Z}{Z_B} \frac{I}{I_B} \tag{5.21}$$

or

$$V_{\text{p.u.}} = Z_{\text{p.u.}} I_{\text{p.u.}}$$

Equation (5.21) has the same form as (5.19), which implies that we can do circuit analysis using (5.21) exactly as with (5.19). The p.u. subscript indicates per unit and is read "per unit."

Example 5.10

Given the circuit shown in Fig. E5.10(a), pick $V_B = 100$ and $Z_B = 0.01$, and find I_B, $V_{\text{p.u.}}$, $Z_{\text{p.u.}}$, $I_{\text{p.u.}}$, and I.

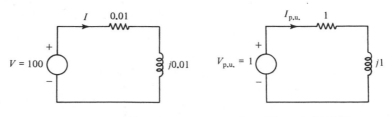

Figure E5.10(a)   Figure E5.10(b)

Solution $I_B = V_B/Z_B = 10^4$. Then

$$V_{p.u.} = \frac{100}{100} = 1$$

$$Z_{p.u.} = \frac{0.01 + j0.01}{0.01} = 1 + j1$$

$$I_{p.u.} = \frac{V_{p.u.}}{Z_{p.u.}} = \frac{1}{1 + j1} = 0.707 \underline{/-45°}$$

$$I = I_{p.u.} \times I_B = (0.707 \underline{/-45°}) \times 10^4 = 7070 \underline{/-45°}$$

Note: In solving for $I_{p.u.}$ we can also use the original circuit with all quantities replaced by per unit quantities. This is shown in Fig. E5.10(b).

What we have done for Ohm's law can also be done in the case of power calculations. For example, corresponding to

$$S = VI^* \tag{5.22}$$

we have

$$S_B = V_B I_B \tag{5.23}$$

and

$$S_{p.u.} = V_{p.u.} I^*_{p.u.} \tag{5.24}$$

Note that since the two equations (5.20) and (5.23) involve four base quantities V_B, I_B, Z_B, and S_B, specifying any two base quantities determines the remaining two base quantities. For example, if we pick V_B, S_B, we can solve for $I_B = S_B/V_B$ and $Z_B = V_B/I_B = V_B^2/S_B$.

The extension to other related variables is routine. For example,

$$S_{p.u.} = P_{p.u.} + jQ_{p.u.}$$

where

$$P_{p.u.} = \frac{P}{S_B} \qquad Q_{p.u.} = \frac{Q}{S_B}$$

In the case of impedance we have

$$Z_{\text{p.u.}} = R_{\text{p.u.}} + jX_{\text{p.u.}}$$

where

$$R_{\text{p.u.}} = \frac{R}{Z_B} \qquad X_{\text{p.u.}} = \frac{X}{Z_B}$$

We note that

$$Y_B \triangleq \frac{I_B}{V_B} = \frac{1}{Z_B}$$

If the network includes transformers, there is an advantage to picking different bases for the two sides of the transformer. Suppose that we make the ratio of voltage bases equal to the connection-induced voltage ratio in the per phase equivalent circuit, or what we will take to be equivalent, the ratio of secondary and primary (open-circuit) voltage ratings of the three-phase banks. Let S_B be the same on both sides. For example, consider a transformer bank connected wye-wye. Using Fig. 5.14 with a slight change in notation, we have Fig. 5.23. Pick convenient bases V_{1B}, S_B, I_{1B}, and Z_{1B}, for side 1. Then we get for the corresponding quantities on side 2,

$$V_{2B} = nV_{1B}, \quad S_B, \quad I_{2B} = \frac{1}{n}I_{1B}, \quad Z_{2B} = n^2Z_{1B}$$

All quantities on the left side are now normalized with respect to the bases sub 1; all on the right side with respect to the bases sub 2. In particular, we get

$$V_{2\text{p.u.}} = \frac{V_2}{V_{2B}} = \frac{nV}{nV_{1B}} = V_{\text{p.u.}} \tag{5.25}$$

and

$$I_{2\text{p.u.}} = \frac{I_2}{I_{2B}} = \frac{(1/n)I}{(1/n)I_{1B}} = I_{\text{p.u.}} \tag{5.26}$$

In consideration of (5.25) and (5.26), we can redraw Fig. 5.23 without an ideal transformer.

Consider next the case of a wye-delta transformer bank. Assume that the system is normal and we are able to use the simplified per phase diagram. In this case in Fig.

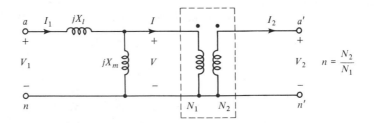

Figure 5.23 Per phase equivalent circuit (wye-wye).

5.21 we discard the 30° phase shift. Relating the bases as described above (by $|K_3| = n/\sqrt{3}$), we again get Fig. 5.24.

Consider the delta-delta case given in Fig. 5.17. Again, relating bases by n, we get Fig. 5.24 except that $X_{lp.u.} = \frac{1}{3}(X_l/Z_{1B})$ and $X_{mp.u.} = \frac{1}{3}(X_m/Z_{1B})$. For the delta-wye case, using the simplified per phase diagram and relating the bases by $|K_1| = \sqrt{3}n$, we get Fig. 5.24, with the same values of $X_{lp.u.}$ and $X_{mp.u.}$ as in the delta-delta case.

The very interesting result is that for a normal per phase system, using the simplified per phase diagram and per unit normalization, we can dispense with all the ideal transformers and all the phase shifters. We call the simplified per phase per unit diagram an *impedance diagram*.

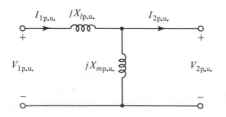

Figure 5.24   Transformer impedance diagram.

5.6 PER UNIT THREE-PHASE QUANTITIES

Three-phase quantities may also be normalized by picking appropriate three-phase bases. In a very natural way we define

$$S_B^{3\phi} \triangleq 3S_B \tag{5.27}$$

$$V_{iB}^{ll} \triangleq \sqrt{3}\, V_{iB} \tag{5.28}$$

where S_B and V_{iB} are the per phase quantities discussed earlier, and the V_{iB}^{ll} refer to the line-line voltages. Since

$$S_{p.u.}^{3\phi} = \frac{S^{3\phi}}{S_B^{3\phi}} = \frac{3S}{3S_B} = S_{p.u.} \tag{5.29}$$

and

$$|V_{ip.u.}^{ll}| = \frac{|V_i^{ll}|}{V_{iB}^{ll}} = \frac{\sqrt{3}\,|V_i|}{\sqrt{3}\,V_{iB}} = |V_{ip.u.}| \tag{5.30}$$

we see that numerically, the distinction between per phase and three-phase quantities expressed in per unit may be unimportant. Power engineers sometimes need not specify whether a per unit voltage is line-line or line-neutral. For example, if we say that the voltage (magnitude) is 1 p.u., this means that the line-line voltage (magnitude) is 1 p.u. (i.e., equal to its base value) and that the line-neutral voltage (magnitude) is also 1 p.u. (i.e., equal to its base value). A similar vagueness is permissible relative to three-phase and single-phase power. Consider also the formula for Z_B. We can calculate Z_B using either single-phase or three-phase quantities.

$$Z_{iB} = \frac{V_{iB}^2}{S_B} = \frac{(\sqrt{3}\ V_{iB})^2}{3S_B} = \frac{(V_{iB}^{ll})^2}{S_B^{3\phi}} \qquad (5.31)$$

Example 5.11

Given three single-phase transformers with the following nameplate ratings, find the impedance diagram for the wye-wye, wye-delta, delta-delta, and delta-wye connections, picking the voltage and power bases for the three-phase bank "induced" by the nameplate ratings. What is meant by "induced" will be clearer from the example.

Nameplate ratings (1 ϕ transformers) 1000 kVA

13.2–66 kV

$X_l = 0.1$ p.u.

$X_m = 100$ p.u.

Solution   The significance of the per unit reactances specified in the nameplate ratings is the following. The manufacturer has picked impedance bases in accordance with the nameplate volt-ampere and voltage ratings. Thus for the 1ϕ transformers, we have

$$\tilde{Z}_{1B} = \frac{\tilde{V}_{1B}^2}{\tilde{S}_B} = \frac{(13.2 \times 10^3)^2}{1000 \times 10^3} = 174\ \Omega$$

$$\tilde{Z}_{2B} = \frac{\tilde{V}_{2B}^2}{\tilde{S}_B} = \frac{(66 \times 10^3)^2}{1000 \times 10^3} = 4356\ \Omega$$

where the tilda indicates the nameplate ratings. We have been using models in which the reactances are in the primary. Thus the "actual" reactances (referred to the primary) are

$$X_l = 0.1 \times 174 = 17.4\ \Omega\ \text{(actual)}$$

$$X_m = 100 \times 174 = 17,400\ \Omega\ \text{(actual)}$$

Now consider the three-phase interconnections of these single-phase transformers. If we connect the primaries in wye (the secondaries can be wye or delta) and pick $S_B^{3\phi}$ and V_{1B}^{ll} induced by the nameplate ratings, we get $S_B^{3\phi} = 3 \times 1000$ kVA and $V_{1B}^{ll} = \sqrt{3}\ 13.2$ kV (so that 1 p.u. of line voltage corresponds to rated voltage on the transformer primary). Then using (5.31), we find that $Z_{1B} = 174\ \Omega$. Then

$$X_{l\ \text{p.u.}} = \frac{17.4}{174} = 0.1\ \Omega$$

$$X_{m\ \text{p.u.}} = \frac{17,400}{174} = 100\ \Omega$$

in the impedance diagram. Of course, these are just the per unit values supplied by the manufacturer.

     If the primaries of the single-phase transformers are connected in delta and we pick $S_B^{3\phi} = 3 \times 1000$ kVA and $V_{1B}^{ll} = 13.2$ kV (so that 1 p.u. of the line voltage corresponds to rated voltages on the transformer primary), then $Z_{1B} = (13.2 \times 10^3)^2/(3 \times 1000 \times 10^3) = 174/3 = \tilde{Z}_B/3$. In the simplified circuit diagram for the delta-delta and delta-wye connections, we find that the per unit reactances are $X_l/3Z_{1B}$ and $X_m/3Z_{1B}$ and thus

$$X_{l\,\text{p.u.}} = \frac{X_l}{3\tilde{Z}_B/3} = \frac{X_l}{Z_B} = 0.1 \; \Omega$$

$$X_{m\,\text{p.u.}} = 100 \; \Omega$$

Thus we get, in every case, the same per unit values for X_l and X_m as in the case of the single-phase transformer. In all four cases we have the impedance diagram shown in Fig. E5.11. Since an impedance diagram is understood to be a per unit diagram, we have abandoned the cumbersome p.u. notation.

The interesting result of the example is that, provided that we use the bases induced by the nameplate ratings, the per unit reactances have the same numerical values in the wye-wye, wye-delta, delta-wye, and delta-delta connections, as in the single phase case.

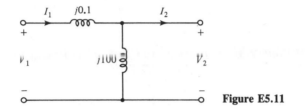

Figure E5.11

Example 5.12

A three-phase transformer bank rated at 5000 kVA, $13.8\Delta - 138Y$ kV, has $X_l = 0.1$ p.u.. Find X_l (actual) referred to the low voltage, i.e., delta, side.

Solution Power and voltage ratings of three-phase equipment are always given in three-phase, line-line terms. Using (5.31), we have

$$Z_{1B} = \frac{(13.8 \times 10^3)^2}{5000 \times 10^3} = 38.09 \; \Omega$$

Then referred to the low voltage side

$$X_l = X_{l\text{p.u.}} \times Z_{1B} = 3.809 \; \Omega \; \text{(actual)}$$

5.7 CHANGE OF BASE

With several items of equipment, with different ratings, it is not usually possible to pick base values so that they are always the same as the nameplate ratings. It is then necessary to recalculate the per unit values on the new basis. The key idea is that $Z_{\text{p.u.}}$ depends on Z_B, but, of course, Z_{actual} does not. We note the relationship between old and new values:

$$Z_{\text{actual}} = Z_{\text{p.u.}}^{\text{old}} Z_B^{\text{old}} = Z_{\text{p.u.}}^{\text{new}} Z_B^{\text{new}} \tag{5.32}$$

Then

$$Z_{\text{p.u.}}^{\text{new}} = Z_{\text{p.u.}}^{\text{old}} \, \frac{Z_B^{\text{old}}}{Z_B^{\text{new}}}$$

$$= Z_{\text{p.u.}}^{\text{old}} \left[\frac{V_B^{\text{old}}}{V_B^{\text{new}}} \right]^2 \frac{S_B^{\text{new}}}{S_B^{\text{old}}} \tag{5.33}$$

Note: In applying (5.33), we can substitute three-phase and/or line-line values.

Example 5.13

A three-phase generator has Thévenin output reactance $X = 0.2$ p.u. based on a generator nameplate rating of 13.2 kV, 30,000 kVA. The new base is 13.8 kV, 50,000 kVA. Find new X in p.u.

Solution   Using (5.33) with three-phase quantities,

$$X_{\text{p.u.}}^{\text{new}} = 0.2 \left(\frac{13.2}{13.8} \right)^2 \frac{50,000}{30,000} = 0.305 \text{ p.u.}$$

5.8 PER UNIT ANALYSIS OF NORMAL SYSTEM

If we have a normal system, we can greatly simplify the solution of the usual power system problem by using an impedance diagram. The rationale was described in Section 5.5 and the procedure is summarized as follows.

Procedure for a per unit analysis

1. Pick a volt-ampere base for the whole system.
2. Pick one base voltage arbitrarily. Relate all the others by the ratio of the magnitudes of the open-circuit line voltages of each transformer bank.
3. Find the impedance bases in the different sections and express all impedances in consistent per unit terms.
4. Draw the impedance diagram for the entire system, and solve for desired per unit quantities.
5. Convert back to actual quantities if desired.

In step 1 it is convenient to pick the three-phase volt-ampere rating of one of the generators or transformer banks. In step 2 it is convenient to pick as a base voltage the rated voltage of the unit picked in step 1. In this case in step 3, the manufacturer's specification of impedance in per unit may be used directly. If not, in step 3 use (5.33). In step 4 we note that the impedance diagram will not have phase shifters or ideal transformers. Finally, we note that for a normal system the procedure described in step 2 is feasible; we get a consistent set of base voltages.

The procedure is illustrated in the next two examples.

Example 5.14

Consider a system with the one-line diagram shown in Fig. E5.14(a). The three-phase transformer nameplate ratings are listed. The transformer reactances are given in percent; 10% = 0.1 p.u. The transmission line and load impedances are in actual ohms. The generator terminal voltage (magnitude) is 13.2 kV (line-line). Find the generator current, the transmission-line current, the load current, the load voltage, and the power delivered to the load.

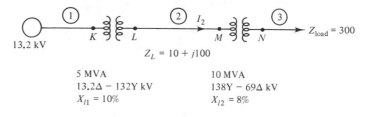

Figure E5.14(a)

Solution   The system is normal, and we can ignore transformation-induced phase shifts. We wish to derive an impedance diagram. In Fig. E5.14(a), three sections (1, 2, and 3) are identified. We will need to pick appropriate bases for these three sections.

Step 1: Pick a common $S_B^{3\phi}$ for the entire system; for example, pick $S_B^{3\phi} = 10$ MVA.

Step 2: Pick one voltage base: for example, $V_{2B}^{ll} = 138$ kV. Relate the other voltage bases by ratios of transformer (line-line) voltage ratings: $V_{1B}^{ll} = 13.8$ kV, $V_{3B}^{ll} = 69$ kV. Here the subscripts refer to the labeled sections, 1, 2, and 3.

Step 3: Find the impedance bases for the three sections using (5.31), and calculate the per unit line and load impedance values.

$$Z_{3B} = \frac{(69 \times 10^3)^2}{10 \times 10^6} = 476 \quad \Rightarrow \quad Z_{\text{load}} = \frac{300}{476} = 0.63 \text{ p.u.}$$

$$Z_{2B} = \frac{(138 \times 10^3)^2}{10 \times 10^6} = 1904 \quad \Rightarrow \quad Z_{\text{line}} = 5.25 \times 10^{-3}(1 + j10) \text{ p.u.}$$

Using (5.33), we can express X_{t1} relative to the new base values.

$$X_{t1}^{\text{new}} = 0.1 \left(\frac{13.2}{13.8}\right)^2 \left(\frac{10}{5}\right) = 0.183 \text{ p.u.}$$

We do not have to recalculate X_{t2}, because the base values are the same as the nameplate values. Thus

$$X_{t2} = 0.08 \text{ p.u.}$$

Finally, we express the source (line-line) voltage in per unit.

$$|E_s| = \frac{13.2}{13.8} = 0.96 \text{ p.u.}$$

To do the circuit analysis it is convenient to pick $E_s = 0.96 \underline{/0°}$ p.u., but since this choice is arbitrary, there is no significance to the absolute phases of the quantities resulting from the analysis.

Step 4: We are now able to draw the impedance diagram [Fig. E5.14(b)]. The points labeled *KLMN* correspond to the points similarly labeled on the one-line diagram. The element values shown are expressed in per unit. We next find $I_{\text{p.u.}}$ by circuit analysis.

$$I_{\text{p.u.}} = \frac{0.96}{Z_{\text{total}}} = \frac{0.96}{0.709 \,\underline{/26.4^\circ}} = 1.35 \,\underline{/-26.4^\circ}$$

Note that the same $I_{\text{p.u.}}$ represents different currents in sections 1, 2, and 3 because the base values are different. We can also calculate the load voltage,

$$V_{3\text{p.u.}} = 0.63 I_{\text{p.u.}} = 0.8505 \,\underline{/-26.4^\circ}$$

and the load power,

$$S_{L \text{ p.u.}} = V_{3\text{p.u.}} I^*_{\text{p.u.}} = Z_{L \text{ p.u.}} |I_{\text{p.u.}}|^2 = 0.63 \times 1.35^2 = 1.148$$

Step 5: We next find I_{1B}, I_{2B}, I_{3B} and then the actual current magnitudes.

$$I_{1B} = \frac{S_B}{V_{1B}} = \frac{10 \times 10^6}{3} \times \frac{\sqrt{3}}{13.8 \times 10^3} = 418.4$$

$$I_{2B} = \frac{13.2}{132} \times 418.4 = 41.84$$

$$I_{3B} = \frac{138}{69} \times 41.84 = 83.67$$

Then

generator current $= |I_1| = 1.35 \times 418.4 = 584.8$ A

transmission-line current $= |I_2| = 1.35 \times 41.84 = 56.48$ A

load current $= |I_3| = 1.35 \times 83.67 = 112.95$ A

We can calculate the actual load voltage using $V_{3\text{p.u.}}$ and V^{ll}_{3B}:

load voltage $= |V^{\text{ll}}_3| = 0.8505 \times 69$ kV $= 58.68$ kV

The actual load power can be found using $S_{L \text{ p.u.}}$ and $S^{3\phi}_B$:

load power $= S_{L \text{ p.u.}} S^{3\phi}_B = 1.148 \times 10$ MVA $= 11.48$ MVA

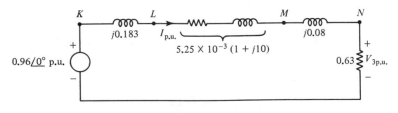

Figure E5.14(b)

Example 5.15

Consider the same basic system as in Example 5.14 but with these changes. The generator voltage and load impedance are not specified. Instead, we are given the following information. The voltage (magnitude) at the load is 63 kV and the three-phase load power is 5.0 MW at a PF of 0.9 lagging. Find the load current, the generator voltage, and the generator power.

Solution   Using the same choice of bases as in Example 5.14, we have

$$|V_{3\text{p.u.}}| = \frac{63}{69} = 0.913$$

We can pick $V_{3\text{p.u.}} = 0.913 \,\underline{/0°}$. We also have

$$P_{D3\text{p.u.}} = \frac{5}{10} = 0.5$$

Taking the p.f. into account, we have

$$P_{D3\text{p.u.}} = |V_{3\text{p.u.}}||I_{\text{p.u.}}| \times 0.9$$

Thus we can solve for $I_{\text{p.u.}}$. We get $|I_{\text{p.u.}}| = 0.608$. Since PF = 0.9 lagging implies that the current lags the voltage by 25.84°, we have, in fact, $I_{\text{p.u.}} = 0.608 \,\underline{/-25.84°}$. Altering the impedance diagram in Example 5.14 to fit the present case, we have the circuit shown in Fig. E5.15. It is now easy to find $E_{sp.u.}$.

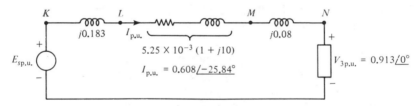

Figure E5.15

$$E_{sp.u.} = V_{3p.u.} + Z_{\text{total}}I_{\text{p.u.}}$$

$$= 0.913 + 0.709 \,\underline{/26.4°} \times 0.608 \,\underline{/-25.84°}$$

$$= 1.34 \,\underline{/0.18°}$$

Also, the generator power is

$$S_{G\text{p.u.}} = E_{sp.u.}I_{\text{p.u.}}^* = 1.34 \,\underline{/0.18°} \times 0.608 \,\underline{/25.84°}$$

$$= 0.8147 \,\underline{/26.02°}$$

We now can calculate actual quantities by multiplying the per unit values by the appropriate bases. We get

load current = $|I_3| = 0.608 \times 83.67 = 50.87$ A

generator voltage = $|E_s^{\text{II}}| = 1.34 \times 13.8$ kV = 18.49 kV

generator power = $S_G^{3\phi} = 0.8147 \,\underline{/26.02°} \times 10$ MVA = 8.147 $\,\underline{/26.02°}$ MVA

The simplicity in solving the problems in Examples 5.14 and 5.15 may be contrasted with the more complicated solution of a similar problem in Example 5.5.

The following is a summary of some of the advantages of the per unit system for analysis.

1. The system is simplified by elimination of transformers in the per phase diagram.
2. Constants are more uniform. Actual ohmic values differ widely for equipment of different sizes (ratings) but are fairly constant when expressed in per unit. This makes it possible to estimate unknown per unit impedances and/or spot obvious data errors.
3. The numbers coming out of the analysis are more easily interpreted physically. For example, suppose that there is a line drop of 50 V (actual). The number in itself does not have much significance. Relative to a 50-kV base it is negligible, but on a 500-V base it is excessive. Expressed in per unit we see its relative size.

Finally, we note that for simplicity, we have described the procedure for per unit normalization of *normal* systems; however, the normalization may be applied more generally. In this case, although it is not possible to eliminate all the ideal transformers and/or phase shifters, the per unit normalization still offers many advantages.

5.9 REGULATING TRANSFORMERS FOR VOLTAGE AND PHASE ANGLE CONTROL

We wish to consider briefly the use of so-called regulating transformers to adjust voltage magnitude and/or phase. These transformers add a small component of voltage, typically less than 0.1 p.u., to the line or phase voltages. In the case of voltage magnitude control a setup similar to Fig. 5.25 is employed. An adjustable portion of the voltage V_{an} is fed into the primary of the series transformer whose secondary is in series with phase a. Assuming ideal transformers, it is easy to grasp the basic voltage-boosting mechanism. An application of this device is to maintain voltage automatically on radial feeders under varying load conditions. In loop systems reactive power flow may also be controlled by this device.

Voltage phase angle may be adjusted by the system of Fig. 5.26. Windings shown in parallel are the primary and secondary of the same single-phase transformer. We can best understand the operation by assuming ideal transformers. Then $V_{cc'} = \rho V_{ab}$, $V_{aa'} = \rho V_{bc}$, $V_{bb'} = \rho V_{ca}$, where ρ is a small positive number. Noting that

$$V_{a'b'} = V_{a'a} + V_{ab} + V_{bb'}$$

$$= V_{ab} + \rho(V_{ca} - V_{bc})$$

and assuming a positive-sequence set of voltages V_{ab}, V_{bc}, and V_{ca}, we find

$$V_{a'b'} = V_{ab} + \rho(e^{j2\pi/3} - e^{-j2\pi/3})V_{ab}$$

$$= V_{ab}(1 + j\rho\sqrt{3})$$

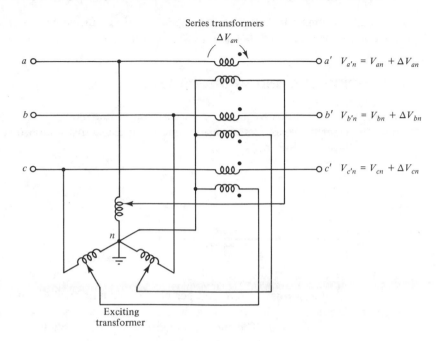

Figure 5.25   Regulating transformer for voltage control.

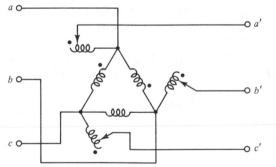

Figure 5.26   Regulating transformer for phase angle control.

Thus $V_{a'b'}$ leads V_{ab}. The phases of the remaining voltages are similarly advanced. By adjusting ρ (i.e., the taps on the secondaries), the phase advance is adjustable. For small ρ the voltage magnitude is relatively unaffected.

We next consider an application. Suppose that there are two transmission links in parallel and that their reactances are the same. Then they will share the transmitted power equally since they both have the same power angle. On the other hand, it may be that their thermal limits are different and we would like to load them independently. This is where the phase-angle control we have been describing can be used effectively, to increase the power angle of the link with the higher thermal limit.

Finally, we note that if regulating transformers are used, the system in general is no longer normal. In that case the impedance diagram has real or complex ideal transformers even after the per unit normalization.

5.10 TRANSMISSION LINE AND TRANSFORMERS

We conclude this chapter by joining the transmission-line model developed in Chapter 4 with the transformer model. Consider one particular transmission link joining two buses, as shown in Fig. 5.27.

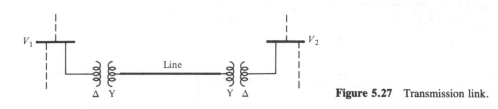

Figure 5.27   Transmission link.

Assuming a normal system and per phase, per unit analysis, we may use the simplified circuit diagram in Fig. 5.24 for the transformers and the circuit in Fig. 4.3 for the transmission line. The most detailed description, including transformer resistances, gives us the circuit shown in Fig. 5.28.

In almost all cases we can simplify this circuit. The transformer resistances and magnetizing reactances are usually neglected; for short lines the other shunt elements are also neglected. In this case we end up with a very simple series *RL* circuit joining buses 1 and 2.

In any case we are usually only interested in terminal (two-port) behavior and can calculate which ever two-port parameters are desired from Fig. 5.28. In Chapter 6 we will be interested in the two-port admittance parameters. These may be calculated by doing circuit analysis on Fig. 5.28. The following example illustrates one among several techniques for accomplishing this.

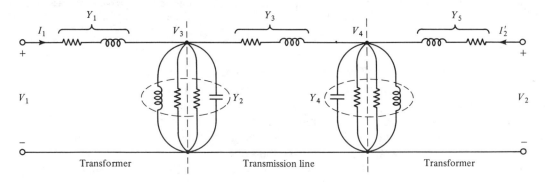

Figure 5.28   Impedance diagram for line section.

Example 5.16

Given the circuit of Fig. 5.28, find an expression for the two-port admittance parameters.

Solution   Using nodal analysis, we find that

$$
\begin{bmatrix} I_1 \\ I_2' \\ \text{---} \\ 0 \\ 0 \end{bmatrix} = \begin{bmatrix} Y_1 & 0 & -Y_1 & 0 \\ 0 & Y_5 & 0 & -Y_5 \\ \text{---} & \text{---} & \text{---} & \text{---} \\ -Y_1 & 0 & Y_1 + Y_2 + Y_3 & -Y_3 \\ 0 & -Y_5 & -Y_3 & Y_3 + Y_4 + Y_5 \end{bmatrix} \begin{bmatrix} V_1 \\ V_2 \\ \text{---} \\ V_3 \\ V_4 \end{bmatrix}
$$

We next suppress V_3 and V_4. This is conveniently done by matrix algebra. The above is in the form

$$
\begin{bmatrix} \mathbf{I} \\ \mathbf{0} \end{bmatrix} = \begin{bmatrix} \mathbf{A}_{11} & \mathbf{A}_{12} \\ \mathbf{A}_{21} & \mathbf{A}_{22} \end{bmatrix} \begin{bmatrix} \mathbf{V} \\ \mathbf{V}' \end{bmatrix}
$$

where $\mathbf{I}$ and $\mathbf{V}$ are the terminal quantities of interest and we seek to eliminate $\mathbf{V}$. Thus we get two matrix equations:

$$
\mathbf{I} = \mathbf{A}_{11}\mathbf{V} + \mathbf{A}_{12}\mathbf{V}'
$$

$$
\mathbf{0} = \mathbf{A}_{21}\mathbf{V} + \mathbf{A}_{22}\mathbf{V}'
$$

Using the second equation, we solve for $\mathbf{V}'$ in terms of $\mathbf{V}$ and then substitute in the first equation. The result is

$$
\mathbf{I} = [\mathbf{A}_{11} - \mathbf{A}_{12}\mathbf{A}_{22}^{-1}\mathbf{A}_{21}]\mathbf{V} = \mathbf{YV}
$$

where $\mathbf{Y}$ is the desired 2×2 matrix of two-port admittance parameters. We assume that $\mathbf{A}_{22}$ has the required inverse.

Example 5.17

Suppose that we neglect resistances in the circuit in Fig. 5.28 and assume that $Z_1 = Z_3 = Z_5 = j0.1$ and $Y_2 = Y_4 = j0.01$. Thus we are assuming inductive series elements and (net) capacitive shunt elements.

(a) Find the two-port admittance parameters.
(b) Find a Π-equivalent circuit having the same two-port admittance parameters.

Solution   (a) Using the results of Example 5.16, we have $\mathbf{A}_{11} = -j10 \times \mathbf{1}$ and $\mathbf{A}_{12} = \mathbf{A}_{21} = j10 \times \mathbf{1}$, where $\mathbf{1}$ is the 2×2 identity matrix. Also,

$$
\mathbf{A}_{22} = j \begin{bmatrix} -19.99 & 10 \\ 10 & -19.99 \end{bmatrix}
$$

with inverse

$$
\mathbf{A}_{22}^{-1} = \frac{1}{j299.6} \begin{bmatrix} -19.99 & -10 \\ -10 & -19.99 \end{bmatrix}
$$

Performing the indicated matrix operations, we get the two-port admittance matrix

$$\mathbf{Y} = \mathbf{A}_{11} - \mathbf{A}_{12}\mathbf{A}_{22}^{-1}\mathbf{A}_{21} = \begin{bmatrix} -j3.328 & j3.338 \\ j3.338 & -j3.328 \end{bmatrix}$$

This completes part (a).

(b) The reader should check that the Π-equivalent circuit with the same two-port admittance matrix is as shown in Fig. E5.17. We note that $Y \approx Y_2 = Y_4 = j0.01$ and $Z \approx Z_1 + Z_3 + Z_5 = j0.3000$. This suggests some simplifications in deriving the Π-equivalent circuit.

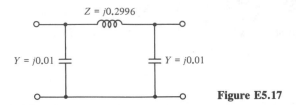

Figure E5.17

In any case, for a normal system we can always find a Π-equivalent circuit for a transmission link, given the two-port admittance parameters. In this way we model a transmission link by a very simple circuit composed of three elements.

Our system model is now complete enough to consider a very basic and important problem called the problem of power (or load) flow analysis. We will do this in Chapter 6.

5.11 SUMMARY

In this chapter we have derived per phase models for the four most common types of interconnections of 1ϕ transformers to make up 3ϕ banks. The model consists of the cascade of a network of reactances and either a conventional ideal transformer (in the case of the Y-Y or Δ-Δ connection) or a complex ideal transformer (in the case of the Δ-Y or Y-Δ connection). For the conventional ideal transformer, in going from primary to secondary the voltage gain is n, the current gain is $1/n$, the (complex) power gain is 1, and any secondary impedance referred to the primary is multiplied by $1/n^2$. In the case of the complex ideal transformer, we get, corresponding to the above, K, $1/K^*$, 1, and $1/|K|^2$, where the voltage gain K is a complex number that depends on the turns ratio of the 1ϕ transformers, the type of interconnection, and on whether we are operating in a positive- or negative-sequence mode.

There are some simplifications and other advantages to introducing the per unit normalization. In this case, in the usual type of analysis problem involving a normal system, all ideal and complex ideal transformers may be eliminated from the circuit diagram.

The impedance diagram for a transmission link composed of a transmission line terminated at both ends by 3ϕ transformer banks may be further simplified. Using circuit analysis, we obtain a Π-equivalent circuit with only three impedances.

PROBLEMS

5.1. An ideal 1ϕ transformer has a voltage gain of 10. The secondary is terminated in a load impedance $Z_L = 30 + j40$.
  (a) Find the (primary) driving-point impedance.
  (b) If the primary voltage is 120 V, find the primary current, secondary current, complex power into the load, and complex power into the transformer primary.

5.2. In Fig. P5.2, the transformer is ideal. Find the (primary) driving-point impedance (at the terminals a–n) for the circuit.
  (a) As shown.
  (b) With n and n' solidly connected.
  Hint: In part (b) assume that $V_{an} = 1$ V and find the current I_a.

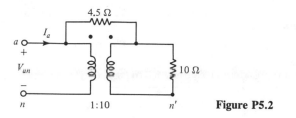

Figure P5.2

5.3. A 120/240-V 1ϕ transformer is connected as an "autotransformer" (Fig. P5.3). Suppose that $V_1 = 120 \underline{/0°}$ and the transformer is ideal. Find V_2, I_2, and I_1.

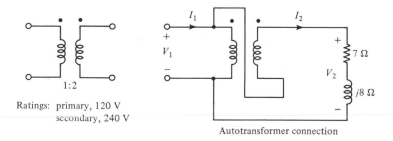

Autotransformer connection

Figure P5.3

5.4. A 1ϕ transformer has the following ratings: 10 kVA, 240/2400 V. In the open-circuit test with rated primary voltage (240 V), the primary current is 0.85 A and the secondary voltage is 2400 V. A short-circuit test is performed as follows: The primary is short circuited and a reduced secondary voltage is applied. Rated secondary circuit is achieved with a secondary voltage of 121 V. Neglect resistance and find X_l, X_m, and $n = N_2/N_1$ (i.e., the circuit elements in Fig. 5.4).

5.5. A 24-kV feeder (line) supplies a 1ϕ load through a step-down transformer. The feeder impedance is $50 + j400$ Ω, the transformer voltage rating is 24/2.4 kV, and the (series) impedance is $0.2 + j1.0$ Ω referred to the low-voltage side. The feeder sending-end voltage, $|V_s|$, is adjusted so that at full load, 200 kW at 0.9 PF lagging is supplied to the load at a voltage $|V_L|_{\text{full load}} = 2300$ V.

   (a)　Find $|V_s|$.
   (b)　Find $|V_R|$, the feeder receiving-end voltage.
   (c)　Find the overall transmission efficiency [i.e., $\eta = P_{\text{load}}/(P_{\text{load}} + P_{\text{losses}})$].
   (d)　Suppose that the load is now removed while $|V_s|$ remains unchanged at its value found in part (a). We then find $|V_L|_{\text{no load}} = 0.1 |V_s|$. Find the "percent voltage regulation," where

$$\text{percent voltage regulation} = \frac{|V_L|_{\text{no load}} - |V_L|_{\text{full load}}}{|V_L|_{\text{full load}}} \times 100$$

5.6. To see the effect of variations in power factor on efficiency and voltage regulation, repeat parts (a), (c), and (d) in Problem 5.5 with PF = 0.7 lagging. Is the regulation better or worse? Better means less variation between no-load and full-load voltage magnitude. Is the efficiency better or worse?

5.7. In Fig. P5.7, the voltages at a, b, and c are balanced 3ϕ. Transformers are ideal and have the same turns ratio. Find the voltages $V_{a'b'}$, $V_{b'c'}$, and $V_{c'a'}$ in terms of V_{ab}, V_{bc}, and V_{ca}.
   (a)　Are they balanced 3ϕ?
   (b)　If not, how would you reconnect the transformers to obtain balanced 3ϕ?

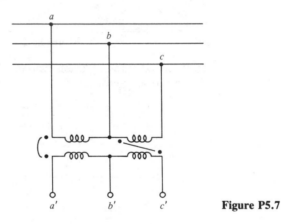

Figure P5.7

5.8. The transformers shown in Fig. P5.8 are labeled in a nonstandard manner. Assume that the 1ϕ transformers are ideal (each with voltage gain n) and find the positive-sequence per phase equivalent circuit relating $V_{a'n'}$ to V_{an}.

5.9. Given the strange connection of 1ϕ transformers shown in Fig. P5.9, assume $E_a = E_b = E_c = 1 \underline{/0°}$ and find I_a.

5.10. A 10 MVA 3ϕ load is to be served by a 3ϕ transformer bank at a (line-line) voltage of 13.8 kV. The supply-side (line-line) voltage is 138 kV. Various 1ϕ transformers are available in the warehouse but all have windings with voltage ratings under 100 kV and current ratings under 250 A.
   (a) Is it possible to serve the load using only three 1ϕ transformers?
   (b) If so, specify the 3ϕ connection and the low- and high-voltage ratings of the 1ϕ transformers.

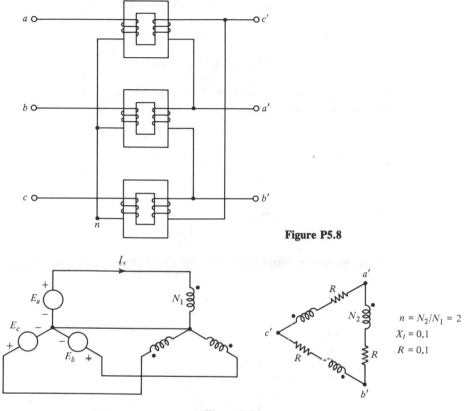

Figure P5.8

Figure P5.9

5.11. We are given the connection of 1ϕ transformers shown in Fig. P5.11. The primaries are on the left. For each 1ϕ transformer, $n = 1$, $X_l = 150$ Ω. The voltage sources are positive-sequence with $V_{an} = V_{a'n'} = 8.0$ kV $\underline{/0°}$. Use the per phase circuit in Fig. 5.18 and find I_a and I_a'.

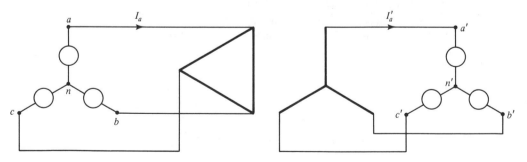

Figure P5.11

5.12. We are given the standard step-down connection shown in Fig. P5.12. The primary is on the right side. Assume that the voltage sources are positive-sequence with $V_{ab} = 13.8$ kV $\underline{/0°}$. The transformer banks are made up of 1ϕ transformers: $X_l = 1.0$ Ω, $n = N_2/N_1 = 10$.

   (a) Find I_a, I'_a, $V_{a'b'}$, and $S^{3\phi}_{\text{load}}$ using the per phase equivalent circuit of Fig. 5.18. *Note:* The diagram needs to be reversed.

   (b) Repeat part (a) but neglect the connection-induced phase shifts in Fig. 5.18.

   (c) Compare the results in parts (a) and (b).

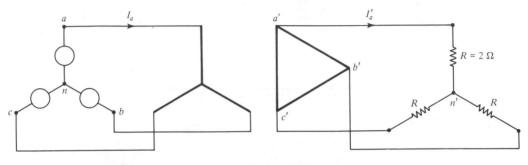

Figure P5.12

5.13. (a) Repeat Example 5.6 with Z_{load} removed. Compare the circulating current with the result in the text.

   (b) Repeat part (a) with the leakage reactances reduced by a factor of 2. How does this affect the size of the circulating current?

5.14. A 1ϕ generator is represented by a Thévenin equivalent circuit: 1320 V in series with $Z_s = 2 \underline{/84°}$ Ω. A load, $Z_L = 50 \underline{/60°}$ Ω is connected across the terminals. Draw per unit diagrams for the following choices of bases.

   (a) $V_B = 1000$ V, $S_B = 100$ kVA

   (b) $V_B = 1320$ V, $S_B = 50$ kVA

   Using the per unit diagrams, do circuit analysis to find the per unit load voltage, current, and complex power in each case. Convert per unit quantities to actual quantities and show they are the same in both cases.

5.15. Draw an impedance diagram for the system whose one-line diagram is shown in Fig. P5.15. The 3ϕ and line-line ratings are given below.

> Generator: 30 MVA, 13.8 kV, $X_s = 0.10$ p.u.
> Motor: 20 MVA, 13.8 kV, $X_s = 0.08$ p.u.
> T_1: 20 MVA, 13.2–132 kV, $X_l = 0.10$ p.u.
> T_2: 15 MVA, 138–13.8 kV, $X_l = 0.12$ p.u.
> Line: $20 + j100$ Ω (actual)

Pick the generator ratings for the bases in the generator section.

Figure P5.15

5.16. Using the impedance diagram in Problem 5.15, assume that the motor voltage is 13.2 kV when the motor draws 15 MW at a power factor of 0.85 leading.

    (a) Find the following quantities in per unit: motor current, transmission-line current, generator current, generator terminal voltage, sending-end transmission-line voltage, and complex power supplied by generator.

    (b) Convert the per unit quantities found in part (b) into actual units (i.e., amperes, volts, and volt-amperes).

5.17. Using the impedance diagram in Problem 5.15, assume that the motor is replaced by a wye-connected load impedance with $Z_L = 20 \,\underline{/45°}\, \Omega$ in each leg. The generator terminal voltage is 13.2 kV. Find the voltage and current at the load in per unit and actual units.

Power Flow Analysis

chapter 6

6.0 INTRODUCTION

We consider next the important problem of power flow, or *load flow,* as it is frequently called. In this analysis the transmission system is modeled by a set of buses or nodes interconnected by transmission links. Generators and loads, connected to various nodes of the system, inject and remove power from the transmission system.

The model is appropriate for solving for the steady-state powers and voltages of the transmission system. The calculation is analogous to the familiar problem of solving for the steady-state voltages and currents in a circuit, and is just as fundamental. It is an integral part of most studies in system planning and operation and is, in fact, the most common of power system computer calculations.

To suggest the variety of possible studies, consider the system with the one-line diagram shown in Fig. 6.1. The systems considered by power engineers would undoubtedly be larger, with as many as thousands of buses and thousands of transmission links.

In the figure, the S_{Gi} are the injected (complex) generator powers and the S_{Di} are the (complex) load powers. The V_1 are the complex (phasor) bus voltages. Transformers are assumed to have been absorbed into the generator, load, or transmission-line models and are not shown explicitly. It should be understood that we are restricting our attention to the transmission system, which transmits the bulk power from the generators to the bulk power substations. Thus the load powers shown in Fig. 6.1 represent the bulk power loads supplied to large industrial consumers and/or to a

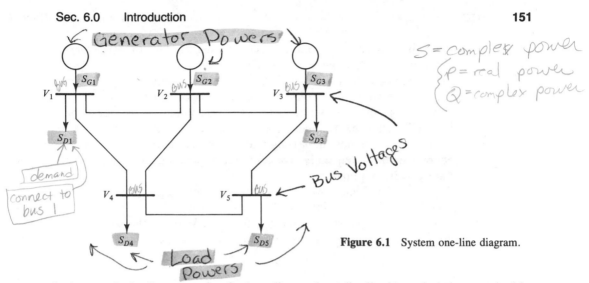

Generator Powers

$S = $ complex power
$\{P = $ real power
$Q = $ complex power

demand
connect to bus 1

Bus Voltages

Load Powers

Figure 6.1 System one-line diagram.

"subtransmission" system for further dispersal to distribution substations and ultimately a network of distribution feeders. While we are considering only the top layer of a multilayer system (the backbone of the overall system), it should be noted that the techniques developed in this chapter are applicable to the different layers of the system. A figure similar to Fig. 6.1 might represent a subtransmission system with power injections from the transmission substations, and possibly some of the older (lower-voltage) generators as well. The reason we concentrate on the transmission system is its basic importance and the fact that there are some interesting and vital problems, such as stability, unique to it.

The purpose of a power system is to deliver the power the customers require in real time, on demand, within acceptable voltage and frequency limits, and in a reliable and economic manner. We are concerned here only with the implications of this objective on the operation and design of the system at the transmission level.

In the analysis we assume that the load powers S_{Di} are known constants. This assumption conforms to the driving nature of the customers' demand, wherein we may take it to be the input, and to the (usually) slowly varying nature of it, wherein we may take it to be constant. The effect of the actual variations in S_{Di} with time can be studied by considering a number of different cases; for each we assume steady-state conditions. Frequently the cases treated are the ones for which some difficulties in meeting system requirements may be expected. What are some of these difficulties? It may be that the voltage magnitudes are not within acceptable limits, or one or more lines are (thermally) overloaded, or that the stability margin for a transmission link is too small (i.e., the power angle across a transmission link is too great), or that a particular generator is overloaded. Other studies relate to contingencies such as the emergency shutdown of a generator, or the loss of one or more transmission links due to equipment failure. With a given loading, the system may be functioning normally, but upon a "single (or multiple) contingency outage" the system is overloaded in some sense. In system operation it is desirable to operate the system in such a way that it is not overloaded in any way nor will it become so in the event of a likely emergency; in system planning there is a need to consider alternative plans to assure that these same objectives are met when the addition goes "on line."

In system operation and planning it is also extremely important to consider the economy of operation. For example, we wish to consider among all the possible allocations of generation assignments what is optimal in the sense of minimum "production" costs (i.e., the fuel cost per hour to generate all the power needed to supply the loads). We note in passing that the objectives (i.e., economy of operation and secure operation) frequently give conflicting operating requirements, and compromises are usually required.

This list of problems, which is far from complete, gives an idea of the range of problems in which we are interested. In considering all these problems we need to know the relationships between the S_{Di}, S_{Gi}, and V_i. The relationships are given by equations called power (or load) flow equations. These are derived in the next section.

6.1 POWER FLOW EQUATIONS

We restrict attention to three-phase balanced system operation, so that per phase analysis may be used. We have already considered a special case of the power flow equations in Section 2.6. Just as in the two-bus case, it is convenient to work with the power at each bus *injected* into the transmission system. Thus we define the complex per phase "bus power," S_i, as follows:

$$S_i \triangleq S_{Gi} - S_{Di}$$

S_i is what is left of S_{Gi} after stripping away the "local" load S_{Di}. We can visualize S_i by "splitting" a bus. For example, in Fig. 6.1 we may split bus 3 as shown in Fig. 6.2. S_3 is the net bus power injected into bus 3. Using conservation of complex power, we also have for the ith bus,

$$S_i = \sum_{k=1}^{n} S_{ik} \qquad i = 1, 2, \ldots, n \tag{6.1}$$

where we sum S_{ik} over all the transmission links connected to the ith bus.

We also define the "bus current" I_i:

$$I_i = I_{Gi} - I_{Di} = \sum_{k=1}^{n} I_{ik} \qquad i = 1, 2, \ldots, n$$

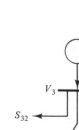

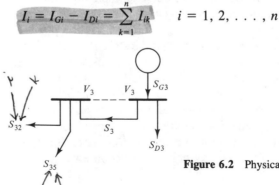

Figure 6.2 Physical significance of S_3.

I_i is the total phase a current entering the transmission system. For bus 3 in Fig. 6.1 we may visualize I_3 as shown in Fig. 6.3, where all the currents shown are phase a currents.

Next we introduce the (per phase) model of the transmission links. Assuming a normal system, as discussed in Section 5.10, we assume that we have replaced each transmission link by a per phase Π-equivalent circuit. Then, taking the neutral as the reference node, we can relate all the (phase a) bus currents I_i to the (phase-neutral) bus voltages V_i by using nodal analysis. The technique is clarified by considering an example.

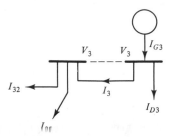

Figure 6.3 Physical significance of I_3.

Example 6.1

Given the simplified circuit diagram shown in Fig. E6.1, write the admittance matrix relating the bus currents I_i to the bus voltages V_i. We note that the transmission links are represented by their Π-equivalent circuits, whose elements are identified by an arbitrary numbering scheme.

Solution   We wish to find the I_i in terms of the V_i. For clarity, elements that are connected together, and hence are incident to the same node, have been shown spread out in the drawing; their identification with the same node has been preserved by duplicating the node numbering as shown in the figure. Applying KCL at nodes 1, 2, and 3, we find that

$$I_1 = (Y_1 + Y_2 + Y_4 + Y_5)V_1 - Y_2V_2 - Y_5V_3$$
$$I_2 = -Y_2V_1 + (Y_2 + Y_3 + Y_8 + Y_9)V_2 - Y_8V_3$$
$$I_3 = -Y_5V_1 - Y_8V_2 + (Y_5 + Y_6 + Y_7 + Y_8)V_3$$

In matrix notation

$$\mathbf{I} = \mathbf{Y_{bus}V}$$

where

$$\mathbf{I} \triangleq \begin{bmatrix} I_1 \\ I_2 \\ I_3 \end{bmatrix}, \quad \mathbf{V} \triangleq \begin{bmatrix} V_1 \\ V_2 \\ V_3 \end{bmatrix}$$

$$\mathbf{Y_{bus}} \triangleq \begin{bmatrix} Y_1 + Y_2 + Y_4 + Y_5 & -Y_2 & -Y_5 \\ -Y_2 & Y_2 + Y_3 + Y_8 + Y_9 & -Y_8 \\ -Y_5 & -Y_8 & Y_5 + Y_6 + Y_7 + Y_8 \end{bmatrix}$$

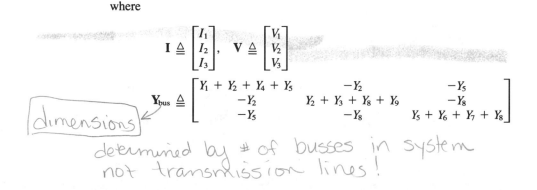

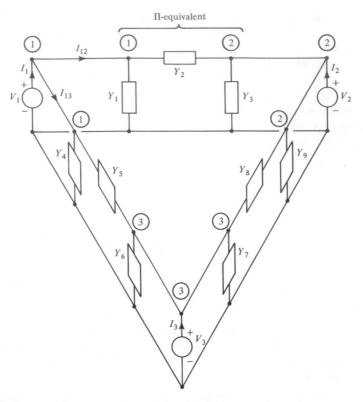

Figure E6.1

Concerning $\mathbf{Y}_{bus}$, the matrix called the *bus admittance matrix*, we note the following well-known properties.

1. It is symmetric.
2. y_{ii}, the ith self-admittance (on the diagonal), is equal to the sum of those admittances of the Π-equivalent circuits which are connected to the ith node.
3. y_{ik}, the ikth element of $\mathbf{Y}_{bus}$ (off the diagonal), is equal to the negative of the admittance of the Π-equivalent circuit element which connects nodes i and k.

On the basis of these rules we can write $\mathbf{Y}_{bus}$ by inspection.

The result found in the example extends to an arbitrary number of nodes and we can state it concisely as follows. In the relation $\mathbf{I} = \mathbf{Y}_{bus}\ \mathbf{V}$, $\mathbf{I}$ and $\mathbf{V}$ are n-vectors and $\mathbf{Y}_{bus}$ is an $n \times n$ matrix with elements.

$y_{ii} = \sum$ admittances of Π-equivalent circuit elements incident to the ith bus

$y_{ik} = -$ admittance of Π-equivalent circuit element bridging the ith and kth buses

Note the following points.

1. The reader may recognize the result from circuit theory; it is merely an application of the general way of writing admittance matrices by inspection in the absence of dependent sources and mutual inductance.
2. If there is no direct link between nodes i and k, the admittance of the bridging element is zero (i.e., $y_{ik} = 0$). The structure of realistic power systems is such that many zeros are present in $\mathbf{Y}_{bus}$.
3. Later, in Chapter 13, we will have use of $\mathbf{Z}_{bus} \triangleq \mathbf{Y}_{bus}^{-1}$. We might note that, unlike $\mathbf{Y}_{bus}$, $\mathbf{Z}_{bus}$ has mainly nonzero entries.

Exercise 1. Consider the structure of $\mathbf{Y}_{bus}$ for the system in Fig. 6.1. What fraction of the entries in $\mathbf{Y}_{bus}$ are zero?

We are now in a position to specify the bus currents. Using $\mathbf{I} = \mathbf{Y}_{bus} \mathbf{V}$, we get for the ith component,

$$I_i = \sum_{k=1}^{n} y_{ik} V_k \qquad i = 1, 2, \ldots, n \tag{6.2}$$

We next calculate the ith bus power. Using (6.2) gives us

$$S_i = V_i I_i^*$$

$$= V_i \left(\sum_{k=1}^{n} y_{ik} V_k \right)^* \tag{6.3}$$

$$= V_i \sum_{k=1}^{n} y_{ik}^* V_k^* \qquad i = 1, 2, \ldots, n$$

Suppose that we let

$$V_i \triangleq |V_1| e^{j\underline{/V_i}} = |V_i| e^{j\theta_i}$$

$$\theta_{ik} \triangleq \theta_i - \theta_k$$

$$y_{ik} \triangleq g_{ik} + jb_{ik}$$

Note that we use a *polar* representation for (complex) voltage, but a *rectangular* representation for (complex) admittance. The g_{ik} are called *conductances*, the b_{ik} are called *susceptances*. Then (6.3) becomes

$$S_i = \sum_{k=1}^{n} |V_i| |V_k| e^{j\theta_{ik}} (g_{ik} - jb_{ik})$$

$$= \sum_{k=1}^{n} |V_i| |V_k| (\cos \theta_{ik} + j \sin \theta_{ik})(g_{ik} - jb_{ik}) \qquad i = 1, 2, \ldots, n \tag{6.4}$$

Equations (6.3) and (6.4) are two equivalent forms of the (complex) power flow equations.

Exercise 2.   Show that (6.4) reduces to (2.26) and (2.27) for the case considered in Section 2.6.

Example 6.2

Suppose that in Example 6.1 all the shunt elements are capacitors with an admittance $Y_C = j0.01$, while all the series elements are inductors with an impedance $Z_L = j0.1$. Find the power flow equations in the form (6.3).

Solution $Z_L = j0.1 \Rightarrow Y_L = -j10$. Substituting Y_C and Y_L in $\mathbf{Y}_{bus}$, we get

$$\mathbf{Y}_{bus} = \begin{bmatrix} -j19.98 & j10 & j10 \\ j10 & -j19.98 & j10 \\ j10 & j10 & -j19.98 \end{bmatrix}$$

Using (6.3), we get

$$S_1 = j19.98|V_1|^2 - j10V_1V_2^* - j10V_1V_3^*$$

$$S_2 = -j10V_2V_1^* + j19.98|V_2|^2 - j10V_2V_3^*$$

$$S_3 = -j10V_3V_1^* - j10V_3V_2^* + j19.98|V_3|^2$$

Note that the relationship between bus powers and bus voltages is quadratic and hence nonlinear.

6.2 THE POWER FLOW PROBLEM

The power flow problem may now be stated with some precision. The formulation is based on operational considerations of the power industry as well as mathematical considerations. We will discuss some of these considerations and introduce some commonly used terminology.

Some buses are supplied by generators. We can call these *generator buses*. Other buses without generators are called *load buses*. In Fig. 6.1, buses 1, 2, and 3 are generator buses and buses 4 and 5 are load buses.

Operational consideratons indicate that at a generator bus the active power P_{Gi} and the voltage magnitude $|V_i|$ may be specified (by varying turbine power and generator field current). At all buses we assume that the S_{Di} are specified. In terms of bus power, then, we see that at a generator bus $P_i = P_{Gi} - P_{Di}$ may be specified, while at a load bus $S_i = -S_{Di}$ is specified. Thus at some buses P_i and $|V_i|$ may be specified, at others P_i and Q_i. One important point must be noted. In general, we cannot specify all the P_i's independently. There is a constraint imposed by the need to balance active power. With a lossless transmission system, the sum of the P_i's over all the buses equals zero. Thus one of the P_i's is determined by specification of the rest. On the other hand, with a lossy system the sum of the P_i's must equal the $|I|^2R$ losses in the transmission system. A problem arises because these losses are not known accurately in advance of the power flow calculation. The resolution of this problem is simple and effective and takes care of both cases. For a calculation of steady-state values, this choice is arbitrary; for convenience we specify P_i at all buses but one. The injected power at this bus is left open "to take up the slack" and balance the active powers. It is conventional, but completely arbitrary, to number the buses so that the generator assigned this function is connected to bus 1. For this generator we do not specify P_1, or equivalently P_{G1}, but rather specify $V_1 = |V_1| \underline{/V_1}$. Choosing $\underline{/V_1}$ amounts to no more than picking a time reference. For a calculation of steady-state values, this choice is arbitrary; for convenience we ordinarily pick $\theta_1 = \underline{/V_1} = 0$.

In summary, there are three types of "sources" at the different buses:

1. A voltage source. Assume at bus 1.
2. $P, |V|$ sources. At the other generator buses.
3. P, Q sources. At the load buses.

We also note the following terminology. Bus 1 is called a *slack bus* or *swing bus* or voltage reference bus. We prefer not to use the terminology "voltage reference bus" to avoid confusion with the designation of the neutral as voltage reference (or datum) node. Buses with $P, |V|$ sources are called $P, |V|$, or *voltage control buses*. Buses with only P, Q sources are called P, Q, or *load buses*.

Finally, it should be noted that while ordinarily a bus may be clearly identified as either generator or load bus, in the case of a load bus with capacitors the bus may be identified as a P, Q bus if the capacitors supply a fixed reactive power, or it may be a $P, |V|$ bus if the capacitors are utilized to maintain a specified $P (= 0)$ and $|V|$. Sometimes Q rather than $|V|$ is specified at a generator bus. In this case we include it with the load buses. Unless otherwise indicated, we assume voltage control at generator buses.

We now state two versions of the power flow problem. In both cases we assume that bus 1 is the slack (or swing) bus. In case I we assume that all the remaining buses are P, Q buses.

Case I: Given $V_1, S_2, S_3, \ldots, S_n$,
         find $S_1, V_2, V_3, \ldots, V_n$.

In case II we assume both $P, |V|$ and P, Q buses. We number the buses so that buses $2, 3, \ldots, m$ are $P, |V|$ buses and $m + 1, \ldots, n$ are P, Q buses.

Case II: Given $V_1, (P_2, |V_2|), \ldots, (P_m, |V_m|), S_{m+1}, \ldots, S_n$,
         find $S_1, (Q_2, \underline{/V_2}), \ldots, (Q_m, \underline{/V_m}), V_{m+1}, \ldots, V_n$.

We now discuss the formulation by noting the following points.

1. The formulations above concern *bus* powers (and voltages). However, the S_{Di} and S_{Gi} are involved since $S_i \triangleq S_{Gi} - S_{Di}$. Thus at the P, Q buses $S_i = -S_{Di}$, while at the $P, |V|$ buses, $P_i = P_{Gi} - P_{Di}$.
2. Case I corresponds to the one-generator case (at bus 1). Case II is the more typical case.

3. In both cases we assume that two out of four variables at each bus are given (complex V_i specifies $|V_i|$ and $\underline{/V_i}$, while complex S_i specifies P_i and Q_i) and are asked to find the remaining two variables.

4. From (6.4) we can see that the $\underline{/V_i}$ appear only in the differences θ_{ij}. Thus from these equations we can only solve for the differences. However, since $\underline{/V_1}$ is assumed specified ($= 0$), all the $\underline{/V_i}$ can then be calculated.

5. Equation (6.3) or (6.4) is implicit in the V_i, and it turns out that we need to solve implicit (and nonlinear) equations. Only in the case of solving for S_1 do we have an explicit equation.

6. Once the stated problem has been solved, we know all the V_i and can then solve for power flows or currents on individual transmission links. In particular, we can check if stability margins and line or transformer thermal ratings are satisfied.

7. In the practical case we may pose the problem differently to take into account some practical limits on the dependent variables. Thus in case II we may specify $Q_i^{\min} \leq Q_i \leq Q_i^{\max}$, $i = 2, 3, \ldots, m$. Then in order not to overspecify the problem, it is the practice to relax the specification on a particular $|V_i|$, in the event that a limit on Q_i is reached.

8. Suppose that a particular load is specified by its impedance Z_{Di} instead of its complex power S_{Di}. This fits into the general scheme with the following modification: Set $S_{Di} = 0$, and add $Y_{Di} = 1/Z_{Di}$ to the ith term on the diagonal of $\mathbf{Y}_{\text{bus}}$.

9. The reader may wish to check that, for a normal system, the problem formulation is consistent with the use of Π-equivalent circuits to model the transmission links, i.e., we do not need to model the connection-induced phase shifts of Δ-Y transformer banks.

Before considering the next topic, on solution methods, it is appropriate to ask about existence and uniqueness of the solutions to the problems we posed in cases I and II.

We can easily pose problems that have no solutions or multiple solutions. For example, consider the following example in the case I category. This example is similar to parts of Example 2.13 considered earlier.

Example 6.3

In Fig. E6.3(a) we are given V_1 and S_2 and are asked to find S_1 and V_2. We consider the solution as a function of P_{D2} for $P_{D2} \geq 0$.

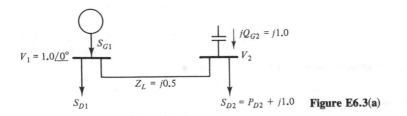

Figure E6.3(a)

Solution   The capacitor in this case injects a specified power, while the voltage is uncontrolled. Thus bus 2 is a P, Q bus. In fact,

$$S_2 = S_{G2} - S_{D2} = -P_{D2}$$

The easiest way to proceed is to use the same approach as in Example 2.13. Noting that $S_2 = S_{21}$, we draw a receiving end circle [Fig. E6.3(b)]. From the geometry of the figure we can see that if there is a solution, $|V_2|$ must satisfy

$$4 |V_2|^4 - 4 |V_2|^2 + P_{D2}^2 = 0$$

which implies that

$$|V_2|^2 = \tfrac{1}{2}(1 \pm \sqrt{1 - P_{D2}^2})$$

Thus if $P_{D2} > 1$, there are no real solutions, and since $|V_2|$ is necessarily real, there are no solutions. If $P_{D2} = 1$, $|V_2| = 1/\sqrt{2} = 0.707$. If $0 \le P_{D2} < 1$, there are *two* real solutions. In the cases where there *are* solutions, we may determine $\underline{/V_2} = -\theta_{12}$ from the figure and $S_1 = S_{12}$ from (6.4) or (2.26).

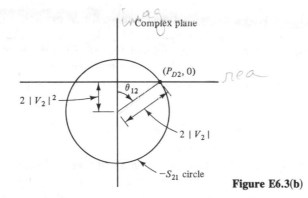

Figure E6.3(b)

In conclusion, we emphasize the probability that there may be no solution or at least no unique solution to the mathematical problem posed. Nevertheless, we expect that most problems which arise in engineering practice have solutions, and even if nonunique, we can identify the practical solution. Usually, the practical solution is the one with voltages (in p.u.) closest to 1.0. This more desirable solution corresponds to the one achieved in practice in the actual power system.

For example, suppose that in Example 6.3, $P_{D2} = 0.5$. We find two solutions for V_2 (i.e., $V_2 = 0.97 \underline{/-15°}$ and $V_2 = 0.26 \underline{/-75°}$). From an operational point of view, the first solution is satisfactory, whereas the second is not. Actually, the first is the one to be expected on the actual power system. The reason has to do with the way a power system is normally operated; the system operating point evolves slowly and continuously. If we consider the past history by which the present loading was achieved, we can expect that initially the received power was zero (thus $V_2 = 1 \underline{/0°}$); then as the received power was slowly increased from zero to 0.5, V_2 varied slowly and continuously from the value $1 \underline{/0°}$ to the nearby value of $0.97 \underline{/-15°}$.

In some cases the power flow equations can be solved analytically. Example 6.3

is an illustration of that. In most cases, however, the solution cannot be found analytically, and the use of iterative methods implemented by digital computer is indicated. We consider next a simple iterative method called the Gauss iterative method and a variant called the Gauss–Seidel iterative method.

6.3 SOLUTION BY GAUSS ITERATION

Consider case I of the power flow problem: given $V_1, S_2, S_3, \ldots, S_n$, find $S_1, V_2, V_3, \ldots, V_n$. We use a form of (6.3) repeated here for convenience,

$$S_1 = V_1 \sum_{k=1}^{n} y_{1k}^* V_k^* \tag{6.3a}$$

$$S_i = V_i \sum_{k=1}^{n} y_{ik}^* V_k^* \qquad i = 2, \ldots, n \tag{6.3b}$$

Note: If we know $V_1, V_2, \ldots, V_n$ we can solve for S_1 explicitly using (6.3a). Since we already do know V_1, it only remains to find $V_2, V_3, \ldots, V_n$. These $n - 1$ unknowns may be found from the $n - 1$ equations of (6.3b). Thus the heart of the problem is the solution of $n - 1$ implicit equations in the unknown $V_2, V_3, \ldots, V_n$, where V_1 and $S_2, S_3, \ldots, S_n$ are known. Equivalently, taking complex conjugates, we can replace (6.3b) by

$$S_i^* = V_i^* \sum_{k=1}^{n} y_{ik} V_k \qquad i = 2, 3, \ldots, n \tag{6.5}$$

We now rearrange (6.5) in a form in which a solution by iteration may be attempted. It should be noted that there are alternative ways of setting up the problem.

Dividing (6.5) by V_i^* and separating out the y_{ii} term, we can rewrite (6.5)

$$\frac{S_i^*}{V_i^*} = y_{ii} V_i + \sum_{\substack{k=1 \\ k \neq i}}^{n} y_{ik} V_k \qquad i = 2, 3, \ldots, n \tag{6.6}$$

or, rearranging,

$$V_i = \frac{1}{y_{ii}} \left[\frac{S_i^*}{V_i^*} - \sum_{\substack{k=1 \\ k \neq i}}^{n} y_{ik} V_k \right] \qquad i = 2, 3, \ldots, n \tag{6.7}$$

Thus we get $n - 1$ implicit nonlinear algebraic equations in the unknown complex V_i in the form

$$V_2 = \tilde{h}_2(V_2, V_3, \ldots, V_n)$$
$$V_3 = \tilde{h}_3(V_2, V_3, \ldots, V_n) \tag{6.8}$$
$$V_n = \tilde{h}_n(V_2, V_3, \ldots, V_n)$$

where the $\bar{h}_i$ are given by (6.7) [e.g., $\bar{h}_2$ is the right-hand side of the first equation (with $i = 2$), etc.]. The numbering in (6.8) is awkward, and we therefore renumber. Define a complex vector $\mathbf{x}$ with components $x_1 = V_2, x_2 = V_3, \ldots, x_N = V_n$. Also renumber the equations in a similar fashion. We then get $N \triangleq n - 1$ equations in the $n - 1$ unknown variables. In vector notation (6.8) is then in the form

$$\mathbf{x} = \mathbf{h}(\mathbf{x}) \tag{6.9}$$

We solve (6.9) by iteration. In the simplest case we use the formula

$$\mathbf{x}^{\nu+1} = \mathbf{h}(\mathbf{x}^\nu) \qquad \nu = 0, 1, \ldots \tag{6.10}$$

where the superscript indicates the iteration number. Thus, starting with an initial value $\mathbf{x}^0$ [which we pick by guessing the solution of (6.9)], we generate the sequence

$$\mathbf{x}^0, \mathbf{x}^1, \mathbf{x}^2, \ldots$$

If the sequence converges (i.e., $\mathbf{x}^\nu \to \mathbf{x}^*$), then

$$\mathbf{x}^* = \mathbf{h}(\mathbf{x}^*)$$

Thus $\mathbf{x}^*$ is a solution of (6.9). The procedure can be visualized in a two-dimension real space as shown in Fig. 6.4. The solution $\mathbf{x}^*$ is also called a *fixed point* of $\mathbf{h}(\cdot)$, which is a good terminology because while other values of $\mathbf{x}$ cause $\mathbf{h}(\mathbf{x})$ to be different than $\mathbf{x}$ (i.e., $\mathbf{h}$ applied to $\mathbf{x}$ moves it), $\mathbf{h}$ applied to $\mathbf{x}^*$ leaves it *fixed*.

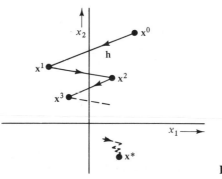

Figure 6.4   Steps in the iteration.

In practice we stop the iterations when the changes in $\mathbf{x}^\nu$ become very small. Thus defining $\Delta\mathbf{x}^\nu \triangleq \mathbf{x}^{\nu+1} - \mathbf{x}^\nu$, we stop when

$$\|\Delta\mathbf{x}^\nu\| \leq \epsilon$$

where ϵ is a small positive number (typically on the order of 0.0001 p.u.) and where $\|\cdot\|$ indicates a "norm." Specific examples of norms are

$$\|\Delta\mathbf{x}\| = \max_i |(\Delta x)_i|$$

called the *sup norm,* and

$$\|\Delta\mathbf{x}\| = \left[\sum_{i=1}^{N}|(\Delta x)_i|^2\right]^{1/2}$$

called the *Euclidean norm*.

We next consider a scalar version of (6.10).

Gauss iteration, $\nu = 0, 1, 2, \ldots$:

$$x_1^{\nu+1} = h_1(x_1^{\nu}, x_2^{\nu}, \ldots, x_N^{\nu})$$

$$x_2^{\nu+1} = h_2(x_1^{\nu}, x_2^{\nu}, \ldots, x_N^{\nu})$$

$$\vdots$$

$$x_N^{\nu+1} = h_N(x_1^{\nu}, x_2^{\nu}, \ldots, x_N^{\nu})$$

(6.11)

In carrying out the computation (normally by digital computer) we process the equations from top to bottom. We now observe that when we solve for $x_2^{\nu+1}$ we already know $x_1^{\nu+1}$. Since $x_1^{\nu+1}$ is (presumably) a better estimate than x_1^{ν}, it seems reasonable to use the "updated" value. Similarly, when we solve for $x_3^{\nu+1}$ we can use the values of $x_1^{\nu+1}$ and $x_2^{\nu+1}$. This line of reasoning leads to the modification called the Gauss–Seidel iteration.

Gauss–Seidel iteration, $\nu = 0, 1, 2, \ldots$:

$$x_1^{\nu+1} = h_1(x_1^{\nu}, x_2^{\nu}, \ldots, x_N^{\nu})$$

$$x_2^{\nu+1} = h_2(x_1^{\nu+1}, x_2^{\nu}, \ldots, x_N^{\nu})$$

$$x_3^{\nu+1} = h_3(x_1^{\nu+1}, x_2^{\nu+1}, x_3^{\nu}, \ldots, x_N^{\nu})$$

$$\vdots$$

$$x_N^{\nu+1} = h_N(x_1^{\nu+1}, x_2^{\nu+1}, \ldots, x_{N-1}^{\nu+1}, x_N^{\nu})$$

(6.12)

Note: Gauss–Seidel is actually easier to program than Gauss and it converges faster.

Example 6.4

In Example 6.3 we found an explicit solution for the power flow equations. Now to illustrate the technique we would like to use Gauss iteration to solve the same problem. Suppose that $P_{D2} = 0.5$. We show this in Fig. E6.4. The problem can be stated as follows: given $V_1 = 1 \underline{/0°}$ and $S_2 = S_{G2} - S_{D2} = -0.5$, find S_1 and V_2, using Gauss iteration.

Solution   We iterate on V_2 using (6.7). With $n = 2$ there is only one equation.

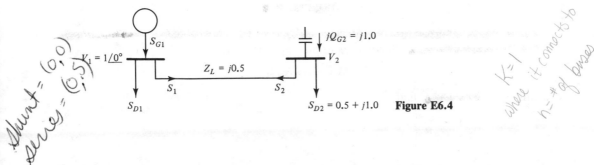

Figure E6.4

$$V_2^{\nu+1} = \frac{1}{y_{22}}\left[\frac{S_2^*}{(V_2^\nu)^*} - y_{21}V_1\right]$$

We next calculate the elements of $\mathbf{Y}_{bus}$. $Z_L = j0.5$ implies that

$$\mathbf{Y}_{bus} = \begin{bmatrix} y_{11} & y_{12} \\ y_{21} & y_{22} \end{bmatrix} = \begin{bmatrix} -j2 & j2 \\ j2 & -j2 \end{bmatrix}$$

Thus, putting in the values for S_2, V_1, y_{11}, and y_{21}, we get

$$V_2^{\nu+1} = -j\frac{0.25}{(V_2^\nu)^*} + 1.0$$

Starting with a guess, $V_2^0 = 1 \underline{/0°}$, we get convergence in about six steps. Even with a very poor guess we get convergence in about eight steps. The results are listed in Table E6.4. The exact solution from Example 6.3 is

$$V_2 = 0.965926 \underline{/-15.000000°}$$

Note: As we showed in Example 6.3, there are two solutions to the equation. The second is $V_2 = 0.258819 \underline{/-75.000000°}$. If we try to converge to this solution by Gauss iteration, we fail. Even with an initial choice as close as $V_2^0 = 0.25 \underline{/-75°}$, we reach the larger of the two solutions.

TABLE E6.4

Iteration number	V_2	V_2
0	$1 \underline{/0°}$	$0.1 \underline{/0°}$
1	$1.030776 \underline{/-14.036243}$	$2.692582 \underline{/-68.198591}$
2	$0.970143 \underline{/-14.036249}$	$0.914443 \underline{/-2.161079}$
3	$0.970261 \underline{/-14.931409}$	$1.026705 \underline{/-15.431731}$
4	$0.966235 \underline{/-14.931416}$	$0.964213 \underline{/-14.089103}$
5	$0.966236 \underline{/-14.995078}$	$0.970048 \underline{/-15.025221}$
6	$0.965948 \underline{/-14.995072}$	$0.965813 \underline{/-14.934752}$
7		$0.966221 \underline{/-15.001783}$
8		$0.965918 \underline{/-14.995310}$

We conclude this section by considering case II of the power flow problem: given V_1, $(P_2, |V_2|)$, . . . , $(P_m, |V_m|)$, S_{m+1}, . . . , S_n, find S_1, $(Q_2, \underline{/V_2})$, . . . , $(Q_m, \underline{/V_m})$, V_{m+1}, . . . , V_n. There is a simple modification of the procedures developed to handle case I which works for case II, which will now be described. Equation (6.7) is the basis for the iteration and we show it here in a form suitable for Gauss iteration in case II:

$$\bar{V}_i^{\nu+1} = \frac{1}{y_{ii}}\left[\frac{P_i - jQ_i^\nu}{(V_i^\nu)^*} - \sum_{\substack{k=1 \\ k \neq i}}^{n} y_{ik}V_k^\nu\right] \qquad i = 2, 3, \ldots, n \qquad (6.13)$$

For the load buses (i.e., $i = m + 1, \ldots, n$), P_i and Q_i are known and the iteration proceeds just as in case I [i.e., in (6.13) the superscript on Q_i and the tilde on $V_i^{\nu+1}$ are ignored]. For the generator buses (i.e., $i = 2, \ldots, m$), Q_i is not specified, but we can do a side calculation to estimate it on the basis of the νth step voltages, already calculated. Thus using (6.3b) yields

$$Q_i^\nu = \operatorname{Im}\left[V_i^\nu \sum_{k=1}^{n} y_{ik}^*(V_k^\nu)^* \right] \qquad i = 2, 3, \ldots, m \qquad (6.14)$$

and this is the value we use in (6.13). We then use (6.13) to calculate $\tilde{V}_i^{\nu+1}$, a preliminary version of $V_i^{\nu+1}$. Since $|V_i|$ is specified for the generator buses we then replace $|\tilde{V}_i^{\nu+1}|$ by $|V_i|_{\text{spec}}$ and obtain $V_i^{\nu+1}$.

We have been describing Gauss iteration. The extension to Gauss–Seidel iteration in case II is exactly the same as in case I. We simply use the most up-to-date values of V_i at each stage of the iteration.

Example 6.5

We consider a simple example illustrating Gauss iteration for case II (see Fig. E6.5). The problem can be stated as follows: given $V_1 = 1$ and $P_2 = -0.75$, $|V_2| = 1$, find S_1, Q_2, $\underline{/V_2}$, using Gauss iteration.

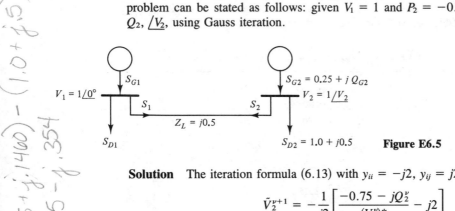

Figure E6.5

Solution   The iteration formula (6.13) with $y_{ii} = -j2$, $y_{ij} = j2$ gives

$$\tilde{V}_2^{\nu+1} = -\frac{1}{j2}\left[\frac{-0.75 - jQ_2^\nu}{(V_2^\nu)^*} - j2 \right]$$

$$= 1 + \frac{0.75 + jQ_2^\nu}{j2(V_2^\nu)^*}$$

where we estimate Q_2^ν using (6.14), *that is*,

$$Q_2^\nu = \operatorname{Im}\{V_2^\nu[y_{21}^*V_1^* + y_{22}^*(V_2^\nu)^*]\}$$

$$= \operatorname{Im}(-j2V_2^\nu + j2|V_2^\nu|^2)$$

$$= 2(1 - \operatorname{Re} V_2^\nu)$$

Starting with a guess, $V_2^0 = 1 \underline{/0^\circ}$, we obtain the iterations shown in Table E6.5. Convergence is obtained after four iterations. The exact solution may easily be found without iteration in this simple case using (2.31) and (2.32). The result is

$$\theta_2 = \underline{/V_2} = -22.0243^\circ \qquad Q_2 = Q_{21} = 0.1460$$

(handwritten marginal notes:)

$S_2 = (.25 + j.1460) - (1.0 + j.5)$

$= -.75 - j.354$

TABLE E6.5

Iteration number	V_2^ν	Q_2^ν	$\tilde{V}_2^{\nu+1}$
0	1 $\underline{/0°}$	0	1.0680 $\underline{/-20.5560°}$
1	1 $\underline{/-20.5560°}$	0.1273	1.0003 $\underline{/-21.9229°}$
2	1 $\underline{/-21.9229°}$	0.1446	1.0000 $\underline{/-22.0169°}$
3	1 $\underline{/-22.0169°}$	0.1459	1.0000 $\underline{/-22.0238°}$
4	1 $\underline{/-22.0238°}$	0.1459	

To complete the problem, we can solve for S_1 using (6.3a):

$$S_1 = V_1[y_{11}^* V_1^* + y_{12}^* V_2^*]$$

$$= j2 - j2 \underline{/22.0238°}$$

$$= 0.7641 \underline{/11.0119°} = 0.7500 + j0.1459$$

which conforms to the expected result.

6.4 MORE GENERAL ITERATION SCHEME

If we use Gauss or Gauss–Seidel, sometimes we get convergence, sometimes not. It is known that convergence (and existence and uniqueness of solutions) is assured if the map $\mathbf{x} \to \mathbf{h}(\mathbf{x})$ is a so-called *contraction mapping*. In general, the conditions required are hard to check and in practice we just try the iterative scheme and hope for convergence. Still, we would like some control over convergence. This is not available using the basic Gauss or Gauss–Seidel scheme. We need and will now derive a more general formula. We will also formulate the problem in a slightly different way. For the general discussion we will use the notation $\mathbf{f}(\mathbf{x})$, reserving $\mathbf{h}(\mathbf{x})$ for the equations defined in (6.9).

Problem.   Solve $\mathbf{f}(\mathbf{x}) = \mathbf{0}$.

Method.   Use the iteration formula $\mathbf{x}^{\nu+1} = \mathbf{\Phi}(\mathbf{x}^\nu)$ with the function $\mathbf{\Phi}$ still to be determined. Starting with an initial value $\mathbf{x}^0$ and assuming convergence, we generate the sequence $\mathbf{x}^0, \mathbf{x}^1, \mathbf{x}^2, \ldots, \mathbf{x}^*$. We assume that $\mathbf{\Phi}(\cdot)$ has the property

$$\mathbf{x}^* = \mathbf{\Phi}(\mathbf{x}^*) \Leftrightarrow \mathbf{f}(\mathbf{x}^*) = \mathbf{0} \tag{6.15}$$

Suppose we now pick $\mathbf{\Phi}$ so that, for any $\mathbf{x}$, it satisfies

$$\mathbf{A}(\mathbf{x})[\mathbf{x} - \mathbf{\Phi}(\mathbf{x})] = \mathbf{f}(\mathbf{x}) \tag{6.16}$$

where $\mathbf{A}(\mathbf{x})$ is a nonsingular matrix. In this case, noting the nonsingularity of $\mathbf{A}(\mathbf{x})$, the reader can check that (6.15) is satisfied for $\mathbf{x}^*$. This gives us a prescription for $\mathbf{\Phi}$. Solving for (6.16) for $\mathbf{\Phi}(\mathbf{x})$, we get

$$\mathbf{\Phi}(\mathbf{x}) = \mathbf{x} - \mathbf{A}(\mathbf{x})^{-1}\mathbf{f}(\mathbf{x})$$

Then the iteration formula is

$$\mathbf{x}^{\nu+1} = \mathbf{x}^{\nu} - [\mathbf{A}(\mathbf{x}^{\nu})]^{-1}\mathbf{f}(\mathbf{x}^{\nu}) \tag{6.17}$$

This is the more general iterative scheme we seek. Note that the flexibility resides in the choice of **A**. The nature of the scheme is such that the accuracy of the final result is independent of the choice of **A**; we continue to iterate until the *mismatch* between $\mathbf{f}(\mathbf{x}^{\nu})$ and **0** is sufficiently small. Of course, we are assuming that the choice of **A** is consistent with convergence of the iterative scheme.

As a special case we consider the solution of $\mathbf{f}(\mathbf{x}) = \mathbf{0}$ with $\mathbf{A}(\mathbf{x})^{-1} = \alpha\mathbf{1}$, where α is a real scalar and **1** is the identity matrix. In this case (6.17) reduces to

$$\mathbf{x}^{\nu+1} = \mathbf{x}^{\nu} - \alpha\mathbf{f}(\mathbf{x}^{\nu}) \tag{6.18}$$

Noting that the solution of (6.9) is equivalent to

$$\mathbf{f}(\mathbf{x}) \triangleq \mathbf{x} - \mathbf{h}(\mathbf{x}) = \mathbf{0}$$

we can substitute in (6.18) and solve (6.9) by the iteration formula

$$\mathbf{x}^{\nu+1} = \mathbf{x}^{\nu} - \alpha(\mathbf{x}^{\nu} - \mathbf{h}(\mathbf{x}^{\nu})) \tag{6.19}$$

If $\alpha = 1$, we get (6.10) [i.e., the Gauss (or Gauss–Seidel) scheme we discussed in Section 6.3]. Choosing a different α we get so-called *accelerated* Gauss (or Gauss–Seidel) where α, the "acceleration" factor, can be chosen to improve convergence properties. We note that α can be either positive or negative.

Example 6.6

To get a feeling for the effect of α on convergence, consider a general scalar example unrelated to power flow. Suppose that we want to solve $f(x) = 0$, where $f(x)$ is plotted in Fig. E6.6(a). Of course, once we plot the graph the zeros of $f(x)$ can be read off directly; however, to observe the effect of α on the behavior of the iteration formula, we

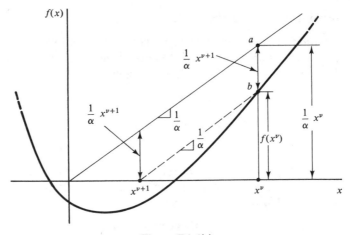

Figure E6.6(a)

proceed as if we did not know the solution. We will specify the iterations by a graphical construction. Let us solve $f(x) = 0$ by a scalar version of (6.18), namely,

$$x^{\nu+1} = x^\nu - \alpha f(x^\nu)$$

or, equivalently,

$$\frac{1}{\alpha} x^{\nu+1} = \frac{1}{\alpha} x^\nu - f(x^\nu)$$

On the graph of $f(x)$ we show an arbitrary abscissa x^ν and the corresponding ordinate $f(x^\nu)$; this identifies the point b. Also shown are two lines of slope $1/\alpha$; one shown dashed passes through the point b, the other passes through the origin. Note point a directly above point b. From the figure it may be seen that the line segment ab, available graphically, is $(1/\alpha) x^{\nu+1} = (1/\alpha) x^\nu - f(x^\nu)$. This line segment is translated to the left until it touches the x axis at a point which must be $x^{\nu+1}$.

The construction justifies the following graphical procedure for obtaining the iterations.

1. Given x^ν, move vertically to $f(x^\nu)$.
2. Return to horizontal axis along a line of slope $1/\alpha$.
3. The horizontal-axis intercept of the line is $x^{\nu+1}$.

Typically, we get the kinds of behavior demonstrated in Fig. E6.6(b). In addition, we can get oscillatory behavior. The sluggish behavior for $\alpha \ll 1$ can also be inferred from (6.18); in the limit if $\alpha = 0$, $\mathbf{x}^\nu = \mathbf{x}^0$, for all ν! Clearly, we would like to avoid a choice of α that leads to the extreme modes of behavior shown in Fig. E6.6(b).

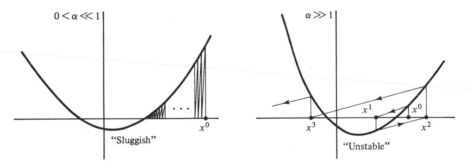

Figure E6.6(b)

Exercise 3. Suppose that in Example 6.6, $f(x) = x^2 - x - 2$.

(a) Find a good value of α and a range of values for x^0 such that the sequence of iterations converges to the zero at $x = 2$.
(b) Repeat part (a) to converge to the zero at $x = -1$.

Exercise 4. Suppose that in Example 6.6, $f(x) = x - 1$. Find a value of α such that the sequence of iterations

(a) Is monotonically convergent.

(b) Converges in one step.

(c) Is oscillatory and convergent.

(d) Is oscillatory and divergent.

(e) Is monotonically divergent.

Exercises 3 and 4 should convince the reader of the advantages of making $1/\alpha$ depend on the slope of $f(x)$ in the neighborhood of the zero of interest. What, then, would be a good general choice of α? From the vantage point of knowing the graph, we can see the advantage of picking $1/\alpha = f'(x^\nu)$, where f' is the derivative of f. Iterations under this scheme are shown (for two different initial conditions) in Fig. 6.5. This graph describes the behavior of the following formula for solving $f(x) = 0$:

$$x^{\nu+1} = x^\nu - [f'(x^\nu)]^{-1}f(x^\nu) \tag{6.20}$$

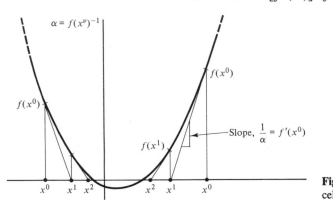

Figure 6.5 Regarding the choice of accelerating factor.

6.5 NEWTON–RAPHSON ITERATION

The iterative scheme specified in (6.20) is the well-known Newton–Raphson (N-R) iteration formula in the one-dimensional case. By analogy we can also easily obtain the n-dimensional Newton–Raphson iteration formula. We replace the scalars x and $f(x)$ by n-vectors $\mathbf{x}$ and $\mathbf{f}(\mathbf{x})$. It is also reasonable that the scalar derivative operator $f'(x)$ must be generalized into an $n \times n$ matrix operator.

One way to obtain the generalization is to write appropriate Taylor series expansions for $f(x + \Delta x)$ and $\mathbf{f}(\mathbf{x} + \Delta \mathbf{x})$ and to compare corresponding terms. In the case of the scalar Taylor series, we get

$$f(x + \Delta x) = f(x) + f'(x)\Delta x + \text{h.o.t.} \tag{6.21}$$

where "h.o.t." stands for "higher-order terms." In the case of the vector Taylor series, instead of one equation we get n scalar equations.

$$f_1(\mathbf{x} + \Delta\mathbf{x}) = f_1(\mathbf{x}) + \frac{\partial f_1(\mathbf{x})}{\partial x_1}\Delta x_1 + \cdots + \frac{\partial f_1(\mathbf{x})}{\partial x_n}\Delta x_n + \text{h.o.t.}$$

$$f_2(\mathbf{x} + \Delta\mathbf{x}) = f_2(\mathbf{x}) + \frac{\partial f_2(\mathbf{x})}{\partial x_1}\Delta x_1 + \cdots + \frac{\partial f_2(\mathbf{x})}{\partial x_n}\Delta x_n + \text{h.o.t.}$$

$$\vdots \tag{6.22}$$

$$f_n(\mathbf{x} + \Delta\mathbf{x}) = f_n(\mathbf{x}) + \frac{\partial f_n(\mathbf{x})}{\partial x_1}\Delta x_1 + \cdots + \frac{\partial f_n(\mathbf{x})}{\partial x_n}\Delta x_n + \text{h.o.t.}$$

where the notation $\partial f_i(\mathbf{x})/\partial x_j$ means the partial derivative of f_i with respect to x_j evaluated at $\mathbf{x}$. Using matrix notation, we get

$$\mathbf{f}(\mathbf{x} + \Delta\mathbf{x}) = \mathbf{f}(\mathbf{x}) + \mathbf{J}(\mathbf{x})\Delta\mathbf{x} + \text{h.o.t.} \tag{6.23}$$

where

$$\mathbf{J}(\mathbf{x}) \triangleq \begin{bmatrix} \dfrac{\partial f_1(\mathbf{x})}{\partial x_1} & \cdots & \dfrac{\partial f_1(\mathbf{x})}{\partial x_n} \\ \vdots & & \vdots \\ \dfrac{\partial f_n(\mathbf{x})}{\partial x_1} & \cdots & \dfrac{\partial f_n(\mathbf{x})}{\partial x_n} \end{bmatrix} \qquad \Delta\mathbf{x} \triangleq \begin{bmatrix} \Delta x_1 \\ \Delta x_2 \\ \vdots \\ \Delta x_n \end{bmatrix} \tag{6.24}$$

$\mathbf{J}(\mathbf{x})$ is called the *Jacobian matrix* of $\mathbf{f}$ evaluated at $\mathbf{x}$. Comparing (6.23) and (6.21), it is reasonably clear that $\mathbf{J}(\mathbf{x})$ is the generalization of the scalar derivative $f'(x)$. In this case, by analogy with the scalar case, (6.20), we get the more general N-R iteration formula,

$$\mathbf{x}^{\nu+1} = \mathbf{x}^{\nu} - [\mathbf{J}(\mathbf{x}^{\nu})]^{-1}\mathbf{f}(\mathbf{x}^{\nu}) \tag{6.25}$$

We can comment on this equation as follows:

1. Usually rather than $\mathbf{J}(\mathbf{x}^{\nu})$ we will use the simpler notation $\mathbf{J}^{\nu}$. Occasionally, to leave room for the inverse symbol, we will use the notation $\mathbf{J}_{\nu}$ instead.
2. Equation (6.25) fits into our general scheme (6.17) with $\mathbf{A}(\mathbf{x}^{\nu}) = \mathbf{J}(\mathbf{x}^{\nu})$.
3. We do not need to use analogy to derive (6.25). We can use the following reasoning. Suppose that we want to solve $\mathbf{f}(\mathbf{x}) = \mathbf{0}$. Try $\mathbf{x}^{\nu}$. Suppose that $\mathbf{f}(\mathbf{x})^{\nu} \neq \mathbf{0}$ but is small (i.e., $\mathbf{x}^{\nu}$ is pretty close to the exact solution). How should we pick the next approximation $\mathbf{x}^{\nu+1}$? One way: Let $\mathbf{x}^{\nu+1} = \mathbf{x}^{\nu} + \Delta\mathbf{x}^{\nu}$ with $\Delta\mathbf{x}^{\nu}$ to be determined; we expect $\Delta\mathbf{x}^{\nu}$ to be small. Then using Taylor series, $\mathbf{f}(\mathbf{x}^{\nu+1}) = \mathbf{f}(\mathbf{x}^{\nu} + \Delta\mathbf{x}^{\nu}) = \mathbf{f}(\mathbf{x}^{\nu}) + \mathbf{J}^{\nu}\Delta\mathbf{x}^{\nu} + \text{h.o.t.}$ Neglecting the h.o.t. we can pick $\Delta\mathbf{x}^{\nu}$ so that $\mathbf{f}(\mathbf{x}^{\nu+1}) = \mathbf{0}$. We get $\Delta\mathbf{x}^{\nu} = \mathbf{x}^{\nu+1} - \mathbf{x}^{\nu} = -[\mathbf{J}^{\nu}]^{-1}\mathbf{f}(\mathbf{x}^{\nu})$, which is the same as (6.25). If the h.o.t. *are* negligible, we get very fast convergence!
4. The h.o.t. are usually negligible as $\Delta\mathbf{x} \to \mathbf{0}$. At the beginning, to improve the initial guess $\mathbf{x}^0$, a few steps of Gauss–Seidel iteration may be used before the N-R iteration is started.

5. A disadvantage of N-R is the need to update **J** every iteration. Sometimes we can update less often and still get good results.

6. In practice we do not evaluate the inverse matrix. Taking inverses is computationally expensive and not really needed. Instead, using

$$\Delta \mathbf{x}^{\nu} \overset{\Delta}{=} \mathbf{x}^{\nu+1} - \mathbf{x}^{\nu} \tag{6.26}$$

we can write (6.25) as

$$\mathbf{J}^{\nu} \, \Delta \mathbf{x}^{\nu} = -\mathbf{f}(\mathbf{x}^{\nu}) \tag{6.27}$$

and solve for $\Delta \mathbf{x}^{\nu}$ by Gaussian elimination; that is, by elementary row operations we convert the Jacobian matrix into upper triangular form and then find $\Delta \mathbf{x}^{\nu}$ by back substitution. Of course, once we know $\Delta \mathbf{x}^{\nu}$ we can find $\mathbf{x}^{\nu+1} = \mathbf{x}^{\nu} + \Delta \mathbf{x}^{\nu}$ and proceed to the next iteration. We emphasize that (6.27) is just a rearrangement of (6.25) to suggest the method of solution for the unknown $\mathbf{x}^{\nu+1}$.

Example 6.7

Given the direct-current (dc) system shown in Fig. E6.7, use the Newton–Raphson method to find the (dc) bus voltages V_1 and V_2 and find P_{G1}.

Solution   The reader should convince himself or herself that the considerations and techniques developed for ac systems hold as well for dc systems. For simplicity we are using the same notation, although all the variables are real quantities. Using the technique in Example 6.1 for forming $\mathbf{Y}_{bus}$ by inspection, we have

$$\mathbf{Y}_{bus} = 100 \begin{bmatrix} 2 & -1 & -1 \\ -1 & 2 & -1 \\ -1 & -1 & 2 \end{bmatrix}$$

In what corresponds to (6.3), we write the power flow equations

$$P_1 = 200V_1^2 - 100V_1V_2 - 100V_1V_3$$

$$P_2 = -100V_2V_1 + 200V_2^2 - 100V_2V_3 \tag{6.28}$$

$$P_3 = -100V_3V_1 - 100V_3V_2 + 200V_3^2$$

Bus 1 is the slack bus, with V_1 known and P_1 unknown. We (temporarily) strip away the first equation and solve the remaining two for the unknown V_2 and V_3.

In using Newton–Raphson the next step is to put the equations in the form

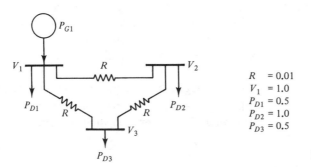

R	= 0.01
V_1	= 1.0
P_{D1}	= 0.5
P_{D2}	= 1.0
P_{D3}	= 0.5

Figure E6.7

$f(x) = 0$. The simplest way is to subtract the left sides from the right sides, or vice versa. It does not matter which we do; the solution algorithm will be identical in either case.

Subtracting left from right and putting in the known values of $P_2 = -P_{D2} = -1.0$, $P_3 = -P_{D3} = -0.5$, $V_1 = 1.0$, we get

$$f_1(x) \triangleq 1.0 - 100V_2 + 200V_2^2 - 100V_2V_3 = 0$$

$$f_2(x) \triangleq 0.5 - 100V_3 - 100V_3V_2 + 200V_3^2 = 0$$

The first component of x is V_2, the second is V_3; with that understanding we will not always use the x_1, x_2 notation.

Next we find the Jacobian matrix,

$$J(x) = \begin{bmatrix} \dfrac{\partial f_1}{\partial x_1} & \dfrac{\partial f_1}{\partial x_2} \\ \dfrac{\partial f_2}{\partial x_1} & \dfrac{\partial f_2}{\partial x_2} \end{bmatrix} = 100\begin{bmatrix} -1 + 4V_2 - V_3 & -V_2 \\ -V_3 & -1 - V_2 + 4V_3 \end{bmatrix}$$

In the 2×2 case it is so easy to find the inverse that we will solve using (6.25); in higher-dimensional cases we would almost certainly use (6.27).

Starting with a "flat profile" (i.e., $V_2 = 1$, $V_3 = 1$), we get

$$J^0 = 100\begin{bmatrix} 2 & -1 \\ -1 & 2 \end{bmatrix} \qquad (J^0)^{-1} = \frac{1}{300}\begin{bmatrix} 2 & 1 \\ 1 & 2 \end{bmatrix} \qquad f(x^0) = \begin{bmatrix} 1.0 \\ 0.5 \end{bmatrix}$$

Then using (6.25) yields

$$x^1 = \begin{bmatrix} 1 \\ 1 \end{bmatrix} - \frac{1}{300}\begin{bmatrix} 2 & 1 \\ 1 & 2 \end{bmatrix}\begin{bmatrix} 1.0 \\ 0.5 \end{bmatrix} = \begin{bmatrix} 0.991667 \\ 0.993333 \end{bmatrix}$$

Continuing to the next iteration gives us

$$J^1 = 100\begin{bmatrix} 1.973333 & -0.991667 \\ -0.993333 & 1.981667 \end{bmatrix} \qquad (J^1)^{-1} = \frac{1}{292.54}\begin{bmatrix} 1.981667 & 0.991667 \\ 0.993333 & 1.973333 \end{bmatrix}$$

$$f(x^1) = \begin{bmatrix} 0.00843 \\ 0.00323 \end{bmatrix}$$

Note that x^1 is really a very good estimate. We want $f(x) = 0$ and already $f(x^1)$ is almost zero! Note the great improvement from $f(x^0)$ to $f(x^1)$ in just one step. Continuing the iteration, we have

$$x^2 = \begin{bmatrix} 0.991667 \\ 0.993333 \end{bmatrix} - \frac{1}{294.54}\begin{bmatrix} 1.981667 & 0.991667 \\ -0.993333 & 1.973333 \end{bmatrix}\begin{bmatrix} 0.00843 \\ 0.00323 \end{bmatrix} = \begin{bmatrix} 0.991599 \\ 0.993283 \end{bmatrix}$$

Noting that the change between x^1 and x^2 is less than 0.0005, we stop the iteration. Alternatively, we can compute $f(x^2)$. We get

$$f(x^2) = \begin{bmatrix} 0.000053 \\ 0.000040 \end{bmatrix}$$

which is suitably small.

Thus the objective of finding the x for which $f(x) = 0$ seems very well met. An interpretation using the power flow equations (6.28) is meaningful. Using the values $V_2 = 0.991599$ and $V_3 = 0.993283$ found after two iterations, the right side of the second equation and the given P_2 on the left side match very well; the mismatch is only 0.000053. The mismatch in the third equation is only 0.000040.

Finally, to complete the problem, we find P_1 by using the first equation in (6.28). We find $P_1 = 1.511800$. Note that since $P_{D1} + P_{D2} = 1.5$, the I^2R loss in the transmission system is 0.011800.

We turn next to a general formulation of the solution to the (ac) power flow equations using the N-R method.

6.6 APPLICATION TO POWER FLOW EQUATIONS

We will start by considering the use of (6.27) to solve the power flow equations case I. For N-R calculations the real form of the power flow equations is used. These may be derived from (6.4) by taking the real and imaginary parts. We get

$$P_i = \sum_{k=1}^{n} |V_i| |V_k| [g_{ik} \cos(\theta_i - \theta_k) + b_{ik} \sin(\theta_i - \theta_k)] \qquad i = 1, 2, 3, \ldots, n$$

$$Q_i = \sum_{k=1}^{n} |V_i| |V_k| [g_{ik} \sin(\theta_i - \theta_k) - b_{ik} \cos(\theta_i - \theta_k) \qquad i = 1, 2, 3, \ldots, n$$

$$(6.29)$$

where $\theta_i \triangleq \underline{/V_i}$.

Just as in the solution by Gauss iteration discussed in Section 6.3, we strip away the first equations (involving P_1 and Q_1). In the remaining equations, this being a case I power flow problem, the P_i and Q_i on the left sides of the equations are specified numbers. The right sides are functions of $|V_i|$ and θ_i. We assume that $|V_1|$ and $\theta_1 (= 0)$ are known. It remains to find the $n - 1$ unknown $|V_i|$ and $n - 1$ unknown θ_i in the equations on the right. It is convenient to define the $(n - 1)$-vectors $\boldsymbol{\theta}$, $|\mathbf{V}|$, and their composite vector $\mathbf{x}$ as follows:

$$\boldsymbol{\theta} = \begin{bmatrix} \theta_2 \\ \vdots \\ \theta_n \end{bmatrix} \qquad |\mathbf{V}| = \begin{bmatrix} |V_2| \\ \vdots \\ |V_n| \end{bmatrix} \qquad \mathbf{x} = \begin{bmatrix} \boldsymbol{\theta} \\ |\mathbf{V}| \end{bmatrix} \qquad (6.30)$$

With this definition the right sides of (6.29) are functions of the unknown $\mathbf{x}$, and we wish to introduce notation which makes that dependence explicit. Thus define the functions $P_i(\mathbf{x})$ and $Q_i(\mathbf{x})$ by

$$P_i(\mathbf{x}) \triangleq \sum_{k=1}^{n} |V_i| |V_k| [g_{ik} \cos(\theta_i - \theta_k) + b_{ik} \sin(\theta_i - \theta_k)] \qquad i = 1, 2, 3, \ldots, n$$

$$Q_i(\mathbf{x}) \triangleq \sum_{k=1}^{n} |V_i| |V_k| [g_{ik} \sin(\theta_i - \theta_k) - b_{ik} \cos(\theta_i - \theta_k)] \qquad i = 1, 2, 3, \ldots, n$$

$$(6.31)$$

The notation is a natural one since for any given $\mathbf{x}$ the right sides are the active and reactive components of the bus power. We can then replace (6.29) with equivalent but notationally simpler power flow equations; at the same time we will strip away the first (active and reactive) equations. We get

$$P_i = P_i(\mathbf{x}) \qquad i = 2, 3, \ldots, n$$
$$Q_i = Q_i(\mathbf{x}) \qquad i = 2, 3, \ldots, n \tag{6.32}$$

In these equations the P_i and Q_i are specified constants, while the $P_i(\mathbf{x})$ and $Q_i(\mathbf{x})$ are specified functions of the unknown $\mathbf{x}$. In the course of the iterations we will be picking a sequence of values $\mathbf{x}^\nu$ in an effort to make the right sides match the given left sides (i.e., to drive the mismatches to zero). We now need to set up the equations in the form $\mathbf{f}(\mathbf{x}) = \mathbf{0}$. As in Example 6.7, we can subtract the left sides of (6.32) from the right sides to get

$$P_i(\mathbf{x}) - P_i = 0 \qquad i = 2, 3, \ldots, n$$
$$Q_i(\mathbf{x}) - Q_i = 0 \qquad i = 2, 3, \ldots, n \tag{6.33}$$

Equation (6.33) identifies the $2n - 2$ components of $\mathbf{f}(\mathbf{x})$. Thus $\mathbf{f}_1(\mathbf{x}) = P_2(\mathbf{x}) - P_2$, $\mathbf{f}_2(\mathbf{x}) = P_3(\mathbf{x}) - P_3, \ldots, \mathbf{f}_{2n-2}(\mathbf{x}) = Q_n(\mathbf{x}) - Q_n$. In matrix notation (6.33) becomes

$$\mathbf{f}(\mathbf{x}) \triangleq \begin{bmatrix} P_2(\mathbf{x}) - P_2 \\ \vdots \\ P_n(\mathbf{x}) - P_n \\ \hline Q_2(\mathbf{x}) - Q_2 \\ \vdots \\ Q_n(\mathbf{x}) - Q_n \end{bmatrix} = \mathbf{0} \tag{6.34}$$

We next consider $\mathbf{J}$, the Jacobian of $\mathbf{f}$. It is convenient to partition as follows:

$$\mathbf{J} = \begin{bmatrix} \mathbf{J}_{11} & \mathbf{J}_{12} \\ \mathbf{J}_{21} & \mathbf{J}_{22} \end{bmatrix} \tag{6.35}$$

Each partition of the matrix $\mathbf{J}$ is $(n - 1) \times (n - 1)$. $\mathbf{J}_{11}$ is made up of terms $\partial P_i(\mathbf{x})/\partial \theta_k$. $\mathbf{J}_{12}$ has the terms $\partial P_i(\mathbf{x})/\partial |V_k|$, $\mathbf{J}_{21}$ the terms $\partial Q_i(\mathbf{x})/\partial \theta_k$, and $\mathbf{J}_{22}$ the terms $\partial Q_i(\mathbf{x})/\partial |V_k|$. These terms may be evaluated more explicitly using (6.31) and we will give some examples quite soon. But first let us find the form of the N-R iteration. For convenience we repeat (6.27):

$$\mathbf{J}^\nu \Delta \mathbf{x}^\nu = -\mathbf{f}(\mathbf{x}^\nu) \tag{6.27}$$

$\mathbf{J}$ has been partitioned; $\mathbf{x}$, and by extension, $\Delta \mathbf{x}$, have partitioned forms. It remains to partition $\mathbf{f}(\mathbf{x})$ and at the same time we would like to get rid of the minus sign in (6.27). Noting (6.34), we define so-called *mismatch vectors:*

$$\Delta \mathbf{P}(\mathbf{x}) = \begin{bmatrix} P_2 - P_2(\mathbf{x}) \\ \vdots \\ P_n - P_n(\mathbf{x}) \end{bmatrix} \qquad \Delta \mathbf{Q}(\mathbf{x}) = \begin{bmatrix} Q_2 - Q_2(\mathbf{x}) \\ \vdots \\ Q_n - Q_n(\mathbf{x}) \end{bmatrix} \tag{6.36}$$

and can express $\mathbf{f}(\mathbf{x})$ as follows:

$$\mathbf{f}(\mathbf{x}) = -\begin{bmatrix} \Delta\mathbf{P}(\mathbf{x}) \\ \Delta\mathbf{Q}(\mathbf{x}) \end{bmatrix} \tag{6.37}$$

Using (6.30), (6.35), and (6.37) in (6.27), we finally get

$$\begin{bmatrix} \mathbf{J}^\nu_{11} & \mathbf{J}^\nu_{12} \\ \mathbf{J}^\nu_{21} & \mathbf{J}^\nu_{22} \end{bmatrix} \begin{bmatrix} \Delta\boldsymbol{\theta}^\nu \\ \Delta|\mathbf{V}|^\nu \end{bmatrix} = \begin{bmatrix} \Delta\mathbf{P}(\mathbf{x}^\nu) \\ \Delta\mathbf{Q}(\mathbf{x}^\nu) \end{bmatrix} \tag{6.38}$$

as a form in which the power flow equations may be solved by N-R iteration. Two comments on (6.38).

1. The right sides are defined by (6.36) and represent the mismatch between the specified values of P and Q and the corresponding values obtained with the trial value $\mathbf{x}^\nu$. As the iteration proceeds, we expect these mismatched terms to go to zero.
2. The components of $\mathbf{x}^\nu$ are specified in (6.30). To find $\mathbf{x}^{\nu+1}$, we solve (6.38) for $\Delta\mathbf{x}^\nu$ and use $\mathbf{x}^{\nu+1} = \mathbf{x}^\nu + \Delta\mathbf{x}^\nu$. At that point we can update the mismatch vector and the Jacobian matrix and continue iterating.

We next consider the calculation of some typical elements of the Jacobian matrices. Using (6.31), we get, for example,

$$\begin{aligned}
\frac{\partial P_2}{\partial \theta_2} &= \frac{\partial}{\partial \theta_2} \sum_{k=1}^n |V_2||V_k| \left[g_{2k} \cos(\theta_2 - \theta_k) + b_{2k} \sin(\theta_2 - \theta_k) \right] \\
&= \sum_{\substack{k=1 \\ k\neq 2}}^n |V_2||V_k| \left[-g_{2k} \sin(\theta_2 - \theta_k) + b_{2k} \cos(\theta_2 - \theta_k) \right]
\end{aligned} \tag{6.39}$$

The $k = 2$ term was removed because for $k = 2$, the right-hand side of the top line is not a function of θ_2. Other typical elements are shown below.

$$\frac{\partial P_2}{\partial \theta_3} = |V_2||V_3| \left[g_{23} \sin(\theta_2 - \theta_3) - b_{23} \cos(\theta_2 - \theta_3) \right]$$

$$\frac{\partial P_2}{\partial |V_2|} = \sum_{k=1}^n |V_k| \left[g_{2k} \cos(\theta_2 - \theta_k) + b_{2k} \sin(\theta_2 - \theta_k) \right] + |V_2| g_{22}$$

$$\frac{\partial P_2}{\partial |V_3|} = |V_2| \left[g_{23} \cos(\theta_2 - \theta_k) + b_{23} \sin(\theta_2 - \theta_3) \right]$$

$$\frac{\partial Q_2}{\partial |V_2|} = \sum_{k=1}^n |V_k| \left[g_{2k} \sin(\theta_2 - \theta_k) - b_{2k} \cos(\theta_2 - \theta_k) \right] - |V_2| b_{22} \tag{6.40}$$

$$\frac{\partial Q_2}{\partial |V_3|} = |V_2| \left[g_{23} \sin(\theta_2 - \theta_k) - b_{23} \cos(\theta_2 - \theta_k) \right]$$

$$\frac{\partial Q_2}{\partial \theta_2} = \sum_{\substack{k=1 \\ k \neq 2}}^{n} |V_2| \, |V_k| \, [g_{2k} \cos (\theta_2 - \theta_k) + b_{2k} \sin (\theta_2 - \theta_k)]$$

$$\frac{\partial Q_2}{\partial \theta_3} = -|V_2| \, |V_3| \, [g_{23} \cos (\theta_2 - \theta_3) + b_{23} \sin (\theta_2 - \theta_3)]$$

The general terms may be inferred from (6.39) and (6.40).

From these equations we can deduce an important property of the Jacobian matrix. The off-diagonal terms in each submatrix are seen to involve the bridging elements of the Π-equivalent circuit of the transmission links. If there is no (direct) connection between two buses, the corresponding bridging admittance is zero and we get a 0 in the Jacobian matrix. In the five-bus system in Fig. 6.1, in $\mathbf{Y}_{bus}$, y_{13}, y_{15}, y_{25}, y_{31}, y_{34}, y_{43}, y_{51}, and y_{52} are all zero. This is 8 out of the 25 total number of y_{ik}. For large systems the sparsity of the nonzero entries is usually much more pronounced. (Typically, at least 50% of the entries are zero; sometimes as many as 95% of the entries are zero.) Thus $\mathbf{J}$ typically is a so-called *sparse matrix*. There are computational techniques available to take advantage of this sparsity in the computer solution of (6.38). For a large system (on the order of 1000 buses) it is essential to do this!

Another saving in computation occurs if some of the buses are generator buses (i.e., in the case II problem introduced earlier). At these buses, we do not have to solve for the $|V_i|$, because they are specified, and the dimensionality of the problem is reduced. To consider what happens in more detail, we go back to the power flow equations (6.29). Suppose for simplicity that the only voltage control bus is bus 2. Then $|V_2|$ is known but Q_2 is not. Consider the equation for Q_2. Once we know $\boldsymbol{\theta}$ and $|V_3|, |V_4|, \ldots, |V_n|$ (we already know $|V_1|$ and $|V_2|$), we can solve for Q_2 explicitly. The situation is just like that encountered when we treated the slack bus, where P_1 and Q_1 could be found explicitly once the remaining equations were solved (implicitly) for the unknown $|\mathbf{V}|$ and $\boldsymbol{\theta}$. The corresponding changes in the analysis are straightforward. The first row of the $|\mathbf{V}|$ vector is stripped away, as is the nth row of $\mathbf{f}(\mathbf{x})$ and the mismatch vector. Correspondingly, the nth column and row of the Jacobian matrix are removed. More generally, by exactly the same arguments, corresponding to buses, 2, 3, . . . , m being voltage control (or $P, |V|$) buses, we strip away the first $m - 1$ rows of $|\mathbf{V}|$, and rows $n, n + 1, \ldots, n + m - 2$ in $\mathbf{f}(\mathbf{x})$ and the mismatch vector. Correspondingly, the following rows and columns of $\mathbf{J}$ are removed: $n, n + 1$, . . . , $n + m - 2$. Note, therefore, that the presence of $P, |V|$ buses reduces the dimensionality of the problem and thereby simplifies matters. By contrast, using Gauss–Seidel iteration the presence of $P, |V|$ buses adds (slighty) to the amount of computation.

Example 6.8

Find θ_2, $|V_3|$, θ_3, S_{G1}, and Q_{G2} for the system shown in Fig. E6.8. The transmission system is assumed to be the same as in Example 6.2.

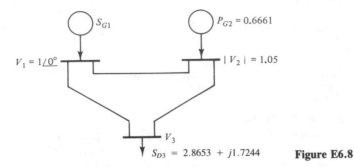

$S_{D3} = 2.8653 + j1.7244$ **Figure E6.8**

Solution   Bus 1 is the slack bus. Bus 2 is a P, $|V|$ bus. Bus 3 is a P, Q bus. We will use Newton–Raphson in the solution, noting that $|V_2|$ is given and therefore we can simplify the computation. The only unknown (implicit) variables are θ_2, θ_3, and $|V_3|$. Thus $\mathbf{J}$ will only be a 3×3 matrix.

From Example 6.2 we know that $b_{ii} = -19.98$ and $b_{ij} = 10$, $i \neq j$. We first write (6.31) for the case at hand, putting in the known values of $|V_1|$, $|V_2|$, θ_1, and the b_{ij}. Thus

$$P_2(\mathbf{x}) = |V_2|\,|V_1|\,b_{21} \sin(\theta_2 - \theta_1) + |V_2|\,|V_3|\,b_{23} \sin(\theta_2 - \theta_3)$$
$$= 10.5 \sin \theta_2 + 10.5|V_3| \sin(\theta_2 - \theta_3) \tag{6.41a}$$

$$P_3(\mathbf{x}) = |V_3|\,|V_1|\,b_{31} \sin(\theta_3 - \theta_1) + |V_3|\,|V_2|\,b_{32} \sin(\theta_3 - \theta_2)$$
$$= 10.0|V_3| \sin \theta_3 + 10.5|V_3| \sin(\theta_3 - \theta_2) \tag{6.41b}$$

Because we know $|V_2|$, we can eliminate the equation involving $Q_2(\mathbf{x})$. We need the equation involving $Q_3(\mathbf{x})$ to solve for $|V_3|$:

$$Q_3(\mathbf{x}) = -[\,|V_3|\,|V_1|\,b_{31} \cos(\theta_3 - \theta_1) + |V_3|\,|V_2|\,b_{32} \cos(\theta_3 - \theta_2) + |V_3|^2 b_{33}\,]$$
$$= -[10|V_3| \cos \theta_3 + 10.5|V_3| \cos(\theta_3 - \theta_2) - 19.98|V_3|^2] \tag{6.41c}$$

We describe the unknown vector and Jacobian matrix in the present case

$$\mathbf{x} = \begin{bmatrix} \theta_2 \\ \theta_3 \\ |V_3| \end{bmatrix}$$

$$\mathbf{J}(\mathbf{x}) = \begin{bmatrix} \dfrac{\partial P_2}{\partial \theta_2} & \dfrac{\partial P_2}{\partial \theta_3} & \dfrac{\partial P_2}{\partial |V_3|} \\[2ex] \dfrac{\partial P_3}{\partial \theta_2} & \dfrac{\partial P_3}{\partial \theta_3} & \dfrac{\partial P_3}{\partial |V_3|} \\[2ex] \dfrac{\partial Q_3}{\partial \theta_2} & \dfrac{\partial Q_3}{\partial \theta_3} & \dfrac{\partial Q_3}{\partial |V_3|} \end{bmatrix}$$

We can find the various partial derivatives from (6.41):

$$\frac{\partial P_2}{\partial \theta_2} = |V_2|\,|V_1|\,b_{21} \cos(\theta_2 - \theta_1) + |V_2|\,|V_3|\,b_{23} \cos(\theta_2 - \theta_3)$$
$$= 10.5 \cos \theta_2 + 10.5|V_3| \cos(\theta_2 - \theta_3)$$

$$\frac{\partial P_2}{\partial \theta_3} = -|V_2||V_3|b_{23} \cos(\theta_2 - \theta_3)$$

$$= -10.5|V_3| \cos(\theta_2 - \theta_3)$$

$$\frac{\partial P_2}{\partial |V_3|} = |V_2|b_{23} \sin(\theta_2 - \theta_3)$$

$$= 10.5 \sin(\theta_2 - \theta_3)$$

$$\frac{\partial P_3}{\partial \theta_2} = -10.5|V_3| \cos(\theta_3 - \theta_2)$$

$$\frac{\partial P_3}{\partial \theta_3} = 10.0|V_3| \cos\theta_3 + 10.5|V_3| \cos(\theta_3 - \theta_2)$$

$$\frac{\partial P_3}{\partial |V_3|} = 10 \sin\theta_3 + 10.5 \sin(\theta_3 - \theta_2)$$

$$\frac{\partial Q_3}{\partial \theta_2} = -10|V_3||V_2| \sin(\theta_3 - \theta_2) = -10.5|V_3| \sin(\theta_3 - \theta_2)$$

$$\frac{\partial Q_3}{\partial \theta_3} = 10|V_3| \sin\theta_3 + 10|V_3||V_2| \sin(\theta_3 - \theta_2)$$

$$= 10|V_3| \sin\theta_3 + 10.5|V_3| \sin(\theta_3 - \theta_2)$$

$$\frac{\partial Q_3}{\partial |V_3|} = -[10 \cos\theta_3 + 10.5 \cos(\theta_3 - \theta_2) - 39.96|V_3|]$$

We are ready to start iterating using (6.38). We note that $P_2 = P_{G2} = 0.6661$, $P_3 = -P_{D3} = -2.8653$, and $Q_3 = -Q_{D3} = -1.7244$; of course, these quantities remain constant through the entire iterative process. With an initial guess $\theta_2^0 = \theta_3^0 = 0$, $|V_3| = 1.0$, using (6.36) and (6.41) we get

$$\begin{bmatrix} \Delta P_2 \\ \Delta P_3 \\ \Delta Q_3 \end{bmatrix}^0 = \begin{bmatrix} P_2 \\ P_3 \\ Q_3 \end{bmatrix} - \begin{bmatrix} P_2(\mathbf{x}^0) \\ P_3(\mathbf{x}^0) \\ Q_3(\mathbf{x}^0) \end{bmatrix} = \begin{bmatrix} 0.6661 \\ -2.8653 \\ -1.7244 \end{bmatrix} - \begin{bmatrix} 0 \\ 0 \\ -0.52 \end{bmatrix} = \begin{bmatrix} 0.6661 \\ -2.8653 \\ -1.2044 \end{bmatrix}$$

As expected for such a crude guess, the mismatch is big. Next we calculate $\mathbf{J}^0$:

$$\mathbf{J}^0 = \begin{bmatrix} 21 & -10.5 & 0 \\ -10.5 & 20.5 & 0 \\ 0 & 0 & 19.46 \end{bmatrix} \tag{6.42}$$

Note that $\mathbf{J}_{12}^0$ and $\mathbf{J}_{21}^0$ are both zero.

We next solve (6.38) by finding the inverse of $\mathbf{J}^0$, taking advantage of the block diagonal structure.

$$\mathbf{J}_0^{-1} = \begin{bmatrix} \mathbf{J}_{11} & \mathbf{0} \\ \mathbf{0} & \mathbf{J}_{22} \end{bmatrix}^{-1} = \begin{bmatrix} \mathbf{J}_{11}^{-1} & \mathbf{0} \\ \mathbf{0} & \mathbf{J}_{22}^{-1} \end{bmatrix} = \begin{bmatrix} 0.0640 & 0.0328 & 0 \\ 0.0328 & 0.0656 & 0 \\ 0 & 0 & 0.0514 \end{bmatrix} \tag{6.43}$$

Substituting in (6.38), we get

$$\Delta\mathbf{x}^0 = \begin{bmatrix} \Delta\theta_2 \\ \Delta\theta_3 \\ \Delta|V_3| \end{bmatrix}^0 = \begin{bmatrix} -0.0513 \text{ rad} \\ -0.1660 \text{ rad} \\ -0.0619 \end{bmatrix} = \begin{bmatrix} -2.9396° \\ -9.5139° \\ -0.0619 \end{bmatrix}$$

We now find $\mathbf{x}^1$ by using (6.26):

$$\mathbf{x}^1 = \mathbf{x}^0 + \Delta\mathbf{x}^0 = \begin{bmatrix} 0 \\ 0 \\ 1 \end{bmatrix} + \begin{bmatrix} -2.9396° \\ -9.5139° \\ -0.0619 \end{bmatrix} = \begin{bmatrix} -2.9396° \\ -9.5139° \\ 0.9381 \end{bmatrix}$$

We note that the exact solution is $\begin{bmatrix} -3 \\ -10 \\ 0.95 \end{bmatrix}$, so this is pretty good progress for one iteration!

We proceed to the next iteration using the new values $\theta_2^1 = -2.9396°$, $\theta_3^1 = -9.5139°$, and $|V_3|^1 = 0.9381$. Substituting in (6.41a), we get $P_2(\mathbf{x}^1) = 0.5893$, and thus $\Delta P_2^1 = 0.6661 - 0.5893 = 0.0768$. Similarly, using (6.41b) and (6.41c), we get an updated mismatch vector:

$$\begin{bmatrix} \Delta P_2 \\ \Delta P_3 \\ \Delta Q_3 \end{bmatrix}^1 = \begin{bmatrix} 0.0768 \\ -0.1870 \\ 0.2298 \end{bmatrix} \qquad (6.44)$$

Note: In one iteration the mismatch vector has been reduced by a factor of about 10. Calculating $\mathbf{J}^1$, we find that

$$\mathbf{J}^1 = \begin{bmatrix} 20.2715 & -9.7853 & 1.2022 \\ -9.7853 & 19.0372 & -2.8550 \\ 1.1277 & -2.6783 & 17.1931 \end{bmatrix} \qquad (6.45)$$

The matrix should be compared with $\mathbf{J}^0$ in (6.42). It has not changed much. The off-diagonal matrices are no longer zero, but their elements are small compared to the terms in the diagonal matrices. The diagonal matrices themselves have not changed much. The same observation is true about the inverses. The updated inverse is

$$\mathbf{J}_1^{-1} = \begin{bmatrix} 0.0656 & 0.0339 & 0.0010 \\ 0.0339 & 0.0713 & 0.0095 \\ 0.0009 & 0.0089 & 0.0596 \end{bmatrix} \qquad (6.46)$$

Comparing (6.46) with (6.43), we do not see much change. Using (6.38) and (6.26), we find that

$$\mathbf{x}^2 = \begin{bmatrix} \theta_2 \\ \theta_3 \\ |V_3| \end{bmatrix}^2 = \begin{bmatrix} -3.0010 \\ -10.0036 \\ 0.9502 \end{bmatrix}$$

This is very close to the correct answer. The largest error is only about 0.036%. Of course in the usual problem we do not know the answer and would continue into the next iteration. We would find

$$\begin{bmatrix} \Delta P_2 \\ \Delta P_3 \\ \Delta Q_3 \end{bmatrix}^2 = \begin{bmatrix} -0.0005 \\ 0.0016 \\ -0.0037 \end{bmatrix}$$

The mismatch has been reduced from (6.44) by about a factor of 100 and is small enough. On that basis we could stop here. Or we could calculate $\mathbf{x}^3$ and show convergence by noting how very close $\mathbf{x}^2$ and $\mathbf{x}^3$ are. So we stop with the values $\theta_2 = -3.001°$, $\theta_3 = -10.0036°$, and $|V_3| = 0.9502$. It remains to calculate S_{G1} and Q_{G2} using the calculated values of θ_2, θ_3, and $|V_3|$. From (6.29), for the case at hand, we get

$$P_{G1} = P_1 = |V_1||V_2|b_{12}\sin(\theta_1 - \theta_2) + |V_1||V_3|b_{13}\sin(\theta_1 - \theta_3)$$

$$= 10.5\sin 3.001° + 9.502\sin 10.0036° = 2.2003$$

$$Q_{G1} = Q_1 = -[|V_1||V_2|b_{12}\cos(\theta_1 - \theta_2) + |V_1||V_3|b_{13}\cos(\theta_1 - \theta_3) + |V_1|^2 b_{11}]$$

$$= -[10.5\cos 3.001° + 9.502\cos 10.0036° - 19.98]$$

$$= 0.1369$$

$$Q_{G2} = Q_2 = -[|V_2||V_1|b_{21}\cos(\theta_2 - \theta_1) + |V_2||V_3|b_{23}\cos(\theta_2 - \theta_3) + |V_2|^2 b_{22}]$$

$$= -[10.5\cos(-3.001°) + 9.997\cos(7.0025°) - 22.028]$$

$$= 1.630Y$$

This completes the example.

6.7 DECOUPLED POWER FLOW

In Example 6.8 we notice a potentially very useful property of the Jacobian matrix: The off-diagonal submatrices $\mathbf{J}_{12}$ and $\mathbf{J}_{21}$ were quite small. In the general case, in calculations for power systems under the usual operating conditions, we find a similar situation. The reason may be seen by considering some typical terms. For example, referring to (6.40), consider a typical term in $\mathbf{J}_{12}$:

$$\frac{\partial P_2}{\partial|V_3|} = |V_2|[g_{23}\cos(\theta_2 - \theta_3) + b_{23}\sin(\theta_2 - \theta_3)]$$

Since transmission links are mostly reactive, the conductance g_{23} is quite small. Since under normal operating conditions the angle $\theta_2 - \theta_3$ is reasonably small (typically less than about 10°), the term involving b_{23} is also quite small. The same reasoning applies to the typical term in $\mathbf{J}_{21}$, for example, the term

$$\frac{\partial Q_2}{\partial\theta_3} = -|V_2||V_3|[g_{23}\cos(\theta_2 - \theta_3) + b_{23}\sin(\theta_2 - \theta_3)]$$

However, if we look at the typical nonzero terms in the diagonal submatrices $\mathbf{J}_{11}$ and $\mathbf{J}_{22}$, they are not small. The implication is that active power flow depends mostly on the θ_i (and not very much on the $|V_i|$), while reactive power flow depends mostly on the $|V_i|$ (and not very much on the θ_i). Stated differently there is fairly good decoupling between the equations for active power and reactive power. This confirms an observation made in Chapter 2.

This decoupling feature is used in simplifying the N-R algorithm for solving the power flow equations. In the general iteration formula (6.17), instead of picking

$$\mathbf{A}(\mathbf{x}^\nu) = \begin{bmatrix} \mathbf{J}_{11}^\nu & \mathbf{J}_{12}^\nu \\ \mathbf{J}_{21}^\nu & \mathbf{J}_{22}^\nu \end{bmatrix} \qquad \text{we will pick} \qquad \mathbf{A}(\mathbf{x}^\nu) = \begin{bmatrix} \mathbf{J}_{11}^\nu & 0 \\ 0 & \mathbf{J}_{22}^\nu \end{bmatrix}$$

which, in the light of the previous discussion, should work nearly as well. We remind the reader that changing $\mathbf{A}(\mathbf{x})$ changes the rate of convergence, but provided that numerical stability is maintained, the final answer, which is associated with zero mismatch, is unaffected. With our new choice we can replace (6.38) with two sets of equations:

$$\mathbf{J}_{11}^\nu \, \Delta\boldsymbol{\theta}^\nu = \Delta\mathbf{P}(\mathbf{x}^\nu) \tag{6.47a}$$

$$\mathbf{J}_{22}^\nu \, \Delta|V|^\nu = \Delta\mathbf{Q}(\mathbf{x}^\nu) \tag{6.47b}$$

We can simplify matters even further by making approximations for the terms in $\mathbf{J}_{11}$ and $\mathbf{J}_{22}$. For example, assuming small g_{ik} and $\theta_i - \theta_k$ and assuming all the (per unit) bus voltage magnitudes approximately equal, (6.39) may be approximated by

$$\frac{\partial P_2}{\partial \theta_2} \approx \sum_{\substack{k=1 \\ k\neq 2}}^{n} |V_2|\,|V_k|\,b_{2n} = \sum_{k=1}^{n} |V_2|\,|V_k|\,b_{2k} - |V_2|^2 b_{22} \approx -|V_2|^2 b_{22} \tag{6.48}$$

In obtaining this result we make use of the following observations:

1. With all the $|V_k|$ approximately equal, $\sum_{k=1}^{n} |V_2|\,|V_k|\,b_{2k} \approx |V_2|^2 \sum_{k=1}^{n} b_{2k}$.
2. b_{22} = sum of susceptances of *all* the elements of the Π-equivalent circuits incident to bus 2.
3. For $k \neq 2$, b_{2k} = −susceptance of the *bridging* element from bus 2 to bus k.
4. Because of observations 2 and 3, in $\sum_{k=1}^{n} b_{2k}$, all the bridging elements cancel, leaving only the sum of (small) shunt element susceptances (capacitive). Thus we have $|\sum_{k=1}^{n} b_{2k}| \ll |b_{22}|$.

Observations 1 and 4 justify the final approximation in (6.48). To verify this in a numerical example the reader should check the terms in $\mathbf{Y}_{\text{bus}}$ in Example 6.2

Note: If all the (per unit) $|V_i|$ are equal, we have a so-called "flat profile"; under normal operating conditions this is a reasonable approximation.

In similar fashion, with the same assumptions, we find from (6.40),

$$\frac{\partial Q_2}{\partial |V_2|} \approx -\sum_{k=1}^{n} |V_k|\,b_{2k} - |V_2|\,b_{22} \approx -|V_2|\,b_{22} \tag{6.49}$$

Equations (6.48) and (6.49) give the pattern for the diagonal terms. For the off-diagonal terms, from (6.40), we obtain the following approximations:

$$\frac{\partial P_2}{\partial \theta_3} = -|V_2|\,|V_3|\,b_{23}$$

$$\frac{\partial Q_2}{\partial |V_3|} = -|V_2|\,b_{23} \tag{6.50}$$

from which we may deduce the general term.

To complete this analysis it is convenient to reintroduce vector notations. Let

$$\mathbf{B} \triangleq \begin{bmatrix} b_{22} & b_{23} & \cdots & b_{2n} \\ \vdots & & & \vdots \\ b_{n2} & \cdots & \cdots & b_{nn} \end{bmatrix} \qquad [\mathbf{V}] \triangleq \begin{bmatrix} |V_2| & 0 & \cdots & 0 \\ \vdots & |V_3| & & \vdots \\ 0 & \cdots & \cdots & |V_n| \end{bmatrix} \tag{6.51}$$

We can check that $\mathbf{B}$ may be obtained from $\mathbf{Y}_{bus}$ by stripping away the first row and column and then taking the imaginary part. $[\mathbf{V}]$ is a diagonal matrix; we note that $|\mathbf{V}|$, previously defined, is a column vector. Noting the general terms of $\mathbf{J}_{11}$ and $\mathbf{J}_{22}$ as suggested by (6.48) to (6.50), the reader can check that

$$\mathbf{J}_{11} = -[\mathbf{V}]\mathbf{B}[\mathbf{V}] \tag{6.52a}$$

$$\mathbf{J}_{22} = -[\mathbf{V}]\mathbf{B} \tag{6.52b}$$

Then we can replace (6.47a) and (6.47b) by

$$-[\mathbf{V}^\nu]\mathbf{B}[\mathbf{V}^\nu]\,\Delta\boldsymbol{\theta}^\nu = \Delta\mathbf{P}(\mathbf{x}^\nu) \tag{6.53a}$$

$$-[\mathbf{V}^\nu]\mathbf{B}\,\Delta|\mathbf{V}|^\nu = \Delta\mathbf{Q}(\mathbf{x}^\nu) \tag{6.53b}$$

One final approximation: Consistent with the assumption of an approximately flat voltage profile, we replace the second $[\mathbf{V}^\nu]$ in (6.53a) by the identity matrix. Then premultiplying each equation by $[\mathbf{V}^\nu]^{-1}$, we get

$$-\mathbf{B}\,\Delta\boldsymbol{\theta}^\nu = \Delta\tilde{\mathbf{P}}(\mathbf{x}^\nu) \tag{6.54a}$$

$$-\mathbf{B}\,\Delta|\mathbf{V}|^\nu = \Delta\tilde{\mathbf{Q}}(\mathbf{x}^\nu) \tag{6.54b}$$

where $\Delta\tilde{\mathbf{P}} = [\mathbf{V}]^{-1}\,\Delta\mathbf{P}$ and $\Delta\tilde{\mathbf{Q}} = [\mathbf{V}]^{-1}\,\Delta\mathbf{Q}$, respectively. Since $[\mathbf{V}]$ is diagonal, the right-hand sides of (6.54a) and (6.54b) are simply scaled versions of $\Delta\mathbf{P}$ and $\Delta\mathbf{Q}$, respectively.

An important observation is that $\mathbf{B}$ is a constant matrix, independent of iteration count. The advantage is that if matrix inversion of $\mathbf{B}$ is used, only one inversion is necessary. If LU decomposition of $\mathbf{B}$ is used, it also need be done only once.

These equations or slight modifications of them are called *fast-decoupled* power flow equations. In one study, convergence was reached in only four to seven iterations, each iteration taking about as much time as 1.5 Gauss–Seidel iterations but requiring only about 0.2 of the time required for a Newton–Raphson iteration.

Finally, when voltage control buses are present we modify (6.54b) by deleting the corresponding terms from $\Delta\tilde{\mathbf{Q}}$, $\Delta|\mathbf{V}|$, and $\mathbf{B}$. In particular, with our numbering system we delete the first $m - 1$ rows of $\Delta\tilde{\mathbf{Q}}$ and $\Delta|\mathbf{V}|$, and the first $m - 1$ rows and columns of $\mathbf{B}$. We do not change anything in (6.54a).

Example 6.9

Consider the circuit shown in Fig. E6.9. Using the data in Example 6.8, use fast-decoupled power flow to find θ_2, θ_3, and $|V_3|$. For convenience the data of the problem are summarized on page 182.

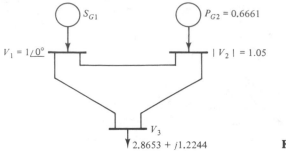

$V_1 = 1\angle 0°$ $|V_2| = 1.05$

V_3
$2.8653 + j1.2244$ **Figure E6.9**

$$P_{G2} = 0.6661$$

$$|V_2| = 1.05$$

$$b_{ij} = 10, \quad i \neq j$$

$$b_{ii} = -19.98$$

$$P_2 = 0.6661$$

$$P_3 = -2.8653$$

$$Q_3 = -1.2244$$

Solution We first find **B** and then use (6.54).

$$\mathbf{B} = \begin{bmatrix} b_{22} & b_{23} \\ b_{32} & b_{33} \end{bmatrix} = \begin{bmatrix} -19.98 & 10 \\ 10 & -19.98 \end{bmatrix}$$

thus (6.54a) becomes

$$-\begin{bmatrix} -19.98 & 10 \\ 10 & -19.98 \end{bmatrix} \begin{bmatrix} \Delta\theta_2 \\ \Delta\theta_3 \end{bmatrix}^\nu = \begin{bmatrix} \dfrac{\Delta P_2}{1.05} \\ \dfrac{\Delta P_3}{|V_3|} \end{bmatrix}^\nu$$

and (6.54b), with the $|V_2|$ equation stripped away, becomes

$$-(-19.98)\,\Delta|V_3|^\nu = \left(\frac{\Delta Q_3}{|V_3|}\right)^\nu$$

In this simple case we can use matrix inversion to get the explicit iteration formulas,

$$\begin{bmatrix} \Delta\theta_2 \\ \Delta\theta_3 \end{bmatrix}^\nu = \begin{bmatrix} 0.0668 & 0.0334 \\ 0.0334 & 0.0668 \end{bmatrix} \begin{bmatrix} \dfrac{\Delta P_2}{1.05} \\ \dfrac{\Delta P_3}{|V_3|} \end{bmatrix}^\nu \tag{6.55a}$$

$$\Delta|V_3|^\nu = 0.0501 \left(\frac{\Delta Q_3}{|V_3|}\right)^\nu \tag{6.55b}$$

We still need to use (6.36) and (6.41) to calculate the mismatch terms. The results of the iterations are tabulated in Table E6.9 starting with initial values of $\theta_2 = \theta_3 = 0$ and $|V_3| = 1.0$. The tabulated values of θ_2 and θ_3 are in degrees. With such small mismatches

TABLE E6.9

Iteration number	θ_2	θ_3	$|V_3|$	$\Delta \tilde{P}_2$	$\Delta \tilde{P}_3$	$\Delta \tilde{Q}_3$
0	0	0	1	0.6344	−2.8653	−1.2044
1	−3.0552	−9.7525	0.9397	0.0714	−0.1307	0.2057
2	−3.0320	−10.1161	0.9500	−0.0083	0.0353	−0.0055
3	−2.9962	−9.9970	0.9497	−0.0004	−0.0013	0.0056

we stop after three iterations. As compared with the results of Example 6.8, it has taken three iterations to reduce the mismatches to the levels reached in only two iterations using Newton–Raphson. Still, the simplification is very great and helps explain the increased popularity of the fast-decoupled power flow methods.

6.8 CONTROL IMPLICATIONS

The Jacobian matrix comes up in another connection related to system control. Consider (6.38). What we have at each stage of the iteration process is a relationship between desired increments in $\mathbf{P}(\mathbf{x})$ and $\mathbf{Q}(\mathbf{x})$ and the increments in $\mathbf{x}$ to bring it about. We have been using this relationship to solve the power flow equations, but it is also directly applicable to the problem of control.

Suppose that the system is in a particular operating state or condition $\mathbf{x}^0$ with corresponding bus powers $\mathbf{P}(\mathbf{x}^0)$ and $\mathbf{Q}(\mathbf{x}^0)$. Suppose we now wish to make a small change in the bus powers by exercising control at the generator buses (i.e., by changing some of the components of $\mathbf{x}$). We then need to consider how changes in $\mathbf{x}$ affect changes in $\mathbf{P}(\mathbf{x})$ and $\mathbf{Q}(\mathbf{x})$; for small increments the relationship is linear and is given by (6.38), where the Jacobian matrix is evaluated at the operating state $\mathbf{x}^0$. In using (6.38) we can simply ignore the notation ν.

Just as discussed in Section 6.8, we can take advantage of certain properties of $\mathbf{J}(\mathbf{x})$. In particular, we note that under the usual operating conditions $\mathbf{J}_{12}$ and $\mathbf{J}_{21}$ are small terms. Thus there is reasonably good decoupling between the control of active and reactive power; to control $\mathbf{P}$ we vary $\boldsymbol{\theta}$ and to control $\mathbf{Q}$ we vary $|\mathbf{V}|$. As an approximation we can substitute (6.47) or (6.54) for (6.38), which makes the decoupling explicit. Again, in using (6.47) or (6.54) we can ignore the notation ν.

In part (d) of the following example we have a control application of (6.54).

Example 6.10

Consider the power system in Fig. E6.10. Assume that $Z_L = R_L + jX_L = 0.0099 + j0.099 = 0.0995\ \underline{/84.2894°}$.

(a) Verify that a solution of the power flow equations is given by

$$\boldsymbol{\theta} = \begin{bmatrix} \theta_2 \\ \theta_3 \\ \theta_4 \\ \theta_5 \end{bmatrix} = \begin{bmatrix} -5° \\ -10° \\ -10° \\ -15° \end{bmatrix} \qquad |\mathbf{V}| = \begin{bmatrix} |V_4| \\ |V_5| \end{bmatrix} = \begin{bmatrix} 1.0 \\ 1.0 \end{bmatrix}$$

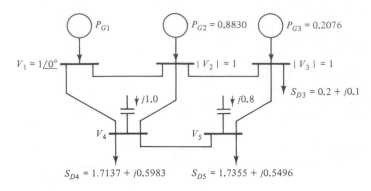

Figure E6.10

(b) Calculate the slack bus power $S_1 = S_{G1}$.

(c) Calculate the total line losses.

(d) Show that the (complex) load demand may be met with lower line losses by "shifting" generation to generator 3.

Solution   (a) We simply plug the solution into the appropriate power flow equations and verify that they are satisfied. The appropriate equations are the ones that would be solved for the unknown (implicit) variables (i.e., the equations with P_2, P_3, P_4, P_5, Q_4, and Q_5 on the left side). In fact, these are the only equations that can be checked since P_1, Q_1, Q_2, and Q_3 are unknown.

We can use either (6.29) or (6.5) as convenient. In either case we need the elements of $\mathbf{Y}_{\text{bus}}$ (i.e., $y_{ik} = g_{ik} + jb_{ik}$). By the method described in Section 6.1 and noting that $Y_L = Z_L^{-1} = 10.503 \, \underline{/-84.2894°} = 1 - j10$, we have

$$\mathbf{Y}_{bus} = \begin{bmatrix} 2 - j20 & -1 + j10 & 0 & -1 + j10 & 0 \\ -1 + j10 & 3 - j30 & -1 + j10 & -1 + j10 & 0 \\ 0 & -1 + j10 & 2 - j20 & 0 & -1 + j10 \\ -1 + j10 & -1 + j10 & 0 & 3 - j30 & -1 + j10 \\ 0 & 0 & -1 + j10 & -1 + j10 & 2 - j20 \end{bmatrix} \quad (6.56)$$

Using (6.5) for bus 4, for example, we get for the bus power into the network

$$S_4^* = V_4^* \sum_{k=1}^{5} y_{4k} V_k$$

$$= 3 - j30 + 1\angle 10°(-1 + j10)(1\angle 0° + 1\angle -5° + 1\angle -15°$$

$$= -1.7137 - j0.4017$$

Thus

$$S_4 = -1.7137 + j0.4017$$

This checks the value derived from power terminal constraints:

$$S_4 = S_{G4} - S_{D4} = j1.0 - (1.7137 + j0.5983)$$

$$= -1.7137 + j0.4017$$

Similarly, we verify the complex power balances at bus 5 and the active power balances at buses 2 and 3.

(b) With all the complex voltages known, we can calculate the slack bus power using (6.3a). We get

$$S_1 = V_1 \sum_{k=1}^{4} y_{1k}^* V_k^*$$

$$= 2 + j20 + 1\angle 0°(-1 - j10)(1\angle 5° + 1\angle 10°)$$

$$= 2.6270 - j0.0709$$

(c) In a lossless transmission system, $\sum_{i=1}^{n} P_i = 0$. In our lossy transmission system, consideration of conservation of power gives the following formula for the line losses:

$$P_L = \sum_{i=1}^{5} P_i = 2.6270 + 0.8830 + 0.0076 - 1.7137 - 1.7355$$

$$= 0.0684$$

This amounts to about 1.8% of the generated power and represents power lost as heat in the transmission system. We note that we would get exactly the same result if we summed the individual power losses in the six transmission links.

(d) Since the active power demand at bus 5 is so large, it seems reasonable that line losses can be reduced by increasing generation at bus 3, which is the closest generator bus. Let us consider the effect of raising P_{G3} a small amount, say 0.1, to see if we can reduce the line losses. If the small increment works, we can try a larger increment later. Since the increment is relatively small we can use linearization techniques to estimate the new operating state. With small θ_{ik}, small g_{ik} (compared to the b_{ik}), and with a flat voltage (magnitude) profile, the use of (6.54) is justified. Using (6.56), we strip away the first row and column, and then, taking imaginary parts, we find **B** directly and substitute in (6.54a). We get

$$-\begin{bmatrix} -30 & 10 & 10 & 0 \\ 10 & -20 & 0 & 10 \\ 10 & 0 & -30 & 10 \\ 0 & 10 & 10 & -20 \end{bmatrix} \begin{bmatrix} \Delta\theta_2 \\ \Delta\theta_3 \\ \Delta\theta_4 \\ \Delta\theta_5 \end{bmatrix} = \begin{bmatrix} \Delta P_2 \\ \Delta P_3 \\ \Delta P_4 \\ \Delta P_5 \end{bmatrix} = \begin{bmatrix} 0 \\ 0.1 \\ 0 \\ 0 \end{bmatrix} \quad (6.57)$$

For the voltage variables we strip away the first two rows and columns of **B** because the voltages at buses 2 and 3 are fixed. Substituting in (6.54b), we get

$$-\begin{bmatrix} -30 & 10 \\ 10 & -30 \end{bmatrix} \begin{bmatrix} \Delta|V_4| \\ \Delta|V_5| \end{bmatrix} = \begin{bmatrix} \Delta Q_4 \\ \Delta Q_5 \end{bmatrix} = \begin{bmatrix} 0 \\ 0 \end{bmatrix} \quad (6.58)$$

The right side is zero because we are not making any change in the reactive power injections at buses 4 and 5. Equation (6.58) predicts that $|V_4|$ and $|V_5|$ remain unchanged at their original value 1.

Solving (6.57), we get

$$\Delta\boldsymbol{\theta} = \begin{bmatrix} \Delta\theta_2 \\ \Delta\theta_3 \\ \Delta\theta_4 \\ \Delta\theta_5 \end{bmatrix} = \begin{bmatrix} 0.00545 \\ 0.01182 \\ 0.00455 \\ 0.00818 \end{bmatrix} \text{rad} = \begin{bmatrix} 0.312 \\ 0.677 \\ 0.261 \\ 0.469 \end{bmatrix} \text{deg}$$

and thus the new bus angles are

$$\boldsymbol{\theta} = \boldsymbol{\theta}^0 + \Delta\boldsymbol{\theta} = \begin{bmatrix} -4.688 \\ -9.323 \\ -9.739 \\ -14.531 \end{bmatrix} \text{deg}$$

We note that the power angle $\theta_3 - \theta_5$ has increased, which shows that more power is being supplied to bus 5 from generator 3.

To find the new line losses corresponding to these angles, we can recalculate all the bus powers and take their sum. It is simpler to add the individual line losses directly. From (2.26) and (2.27) we have for the losses in the link between buses 1 and 2,

$$P_{\text{losses}} = \text{Re}(S_{12} + S_{21}) = \frac{R_L}{|Z_L|^2}[|V_1|^2 + |V_2|^2 - 2|V_1|\,|V_2| \cos\theta_{12}]$$

$$= 2(1 - \cos\theta_{12}) = 2(1 - \cos 4.6888°) = 0.0067$$

In this way, adding up the losses for all six lines, we get the total losses

$$P_L = 2(6 - \cos 4.688° - \cos 4.635° - \cos 9.739° - \cos 5.051° - \cos 4.792°$$
$$\quad - \cos 5.208°)$$

$$= 0.0651$$

The line losses have decreased slightly from the original value of 0.0684. A further increment in P_{G3} is indicated if reduction of transmission losses is desired.

Ordinarily, minimizing transmission line losses is not an objective in itself, but is a consideration in the larger problem of reducing the cost of serving the load demand. Clearly, the cost of producing the power also needs to be considered. If the cost of producing additional power at generator 3 is excessive, it may be cheaper to produce it elsewhere even though the transmission losses are thereby increased. These questions will be addressed in Chapter 7.

Example 6.10 illustrates a further simplification of (5.54) used in a control application. In the example, since the Q_{Di} at the load buses (buses 4 and 5) are constant, we find from (6.58) that $|V_4|$ and $|V_5|$ also remain constant. Since $|V_1|$, $|V_2|$, and $|V_3|$ (at generator buses) are also constant, we have all bus voltage magnitudes constant. Thus we do not have to consider (6.58) at all and can limit our consideration to the active power equations in (6.57).

More generally, the same thing happens in the control applications of (6.54) if the Q_{Di} at the load buses are not changed. In this case we can use (6.54a) (with the $|V_i|$ constant at their original values) to investigate the effect of (small) changes in active power injections. On the other hand, if the active power injections are not changed, we can use (6.54b) to investigate the effect of (small) changes in the Q_{Di} at the load buses. Thus in these cases we obtain *complete* decoupling between the equations involving active and reactive power.

With complete decoupling we can also greatly simplify the system model. Consider the five-bus system in Fig. E6.10. We find, by an exact (complex) power flow analysis, that $|V_4| = |V_5| = 1.0$. Thereafter in determining the effect of (small) changes in active power injections, all the bus voltage magnitudes remain (approximately) constant. In determining the effect of the changes in active power injections, we can therefore use Fig. 6.6, in which all the reactive power injections have been left out.

Taking a model such as in Fig. 6.6 as a starting point rather than Fig. E6.10 is a significant simplification which we will use frequently henceforth. Finally, we note the use of this (active power) model, as an approximation, even in cases in which (almost) complete decoupling would not be expected.

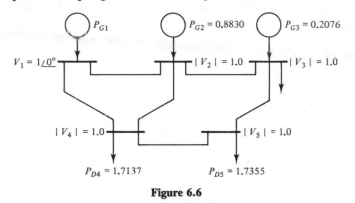

P_{G1} $P_{G2} = 0.8830$ $P_{G3} = 0.2076$

$V_1 = 1\angle 0°$ $|V_2| = 1.0$ $|V_3| = 1.0$

$|V_4| = 1.0$ $|V_5| = 1.0$

$P_{D4} = 1.7137$ $P_{D5} = 1.7355$

Figure 6.6

6.9 SUMMARY

The power flow equations give the relationships between bus powers and bus voltages in terms of the admittance parameters of the transmission system. Operational and mathematical considerations lead us to define the following types of buses. There are load (P, Q) buses, generator $(P, |V|)$ buses, and a slack or swing bus. A bus with reactive power injections from a capacitor bank is a P, Q bus if the Q injection is fixed but is a $P, |V|$ bus if the Q injection is varied to keep $|V|$ fixed.

At the P, Q buses we do not know the (complex) voltages (i.e., we do not know $|V|, \theta$). At the $P, |V|$ buses we do not know Q, θ. The $|V|$ and θ variables are implicit variables in the (nonlinear) power flow equations and iterative solution methods are required.

One method of solution is Gauss iteration or its minor variation, Gauss–Seidel iteration. The complex form of the power flow equations is used. The iteration formula remains unchanged through the entire calculation. No advantage is taken of the presence of $P, |V|$ buses (which reduces the number of unknown implicit variables).

Another method of solution uses Newton–Raphson iteration. The real form of the power flow equations is used. The iteration formula involves a Jacobian matrix that changes as the iterations proceed. Advantage is taken of the presence of $P, |V|$ buses; the number of equations that need to be considered is reduced by the number of $P, |V|$ buses.

Comparing the computational burden of the two methods, each Newton–Raphson iteration takes longer than the corresponding Gauss (or Gauss–Seidel) iteration, but convergence is obtained with fewer iterations so, overall, there is usually a saving in computation.

For computations involving power systems under the usual operating conditions, some simplifications of the Newton–Raphson scheme are usually possible. One of these modifications is called decoupled power flow. It still requires the updating of Jacobian matrices for each iteration, but the dimensionality of the computation is reduced. Another modification is called fast-decoupled power flow. In this case, the updating of matrices is no longer required and the computational burden is greatly reduced.

PROBLEMS

6.1. Find $\mathbf{Y}_{bus}$ for Fig. 6.1 assuming that all the transmission links have identical Π-equivalent circuits; the series element impedance is $0.01 + j0.1$ and each (of the two) shunt element admittances is $j0.8$.

6.2. Write the power flow equations corresponding to Problem 6.1. Write the equations in the form (6.3).

6.3. In Fig. P6.3, all the transmission links are the same and each is modeled by the Π-equivalent circuit shown; the element values are impedances. Find $\mathbf{Y}_{bus}$.

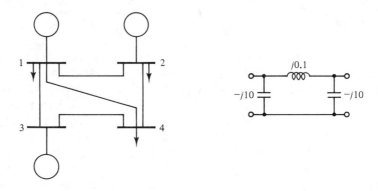

Figure P6.3

6.4. Write the power flow equations corresponding to Problem 6.3 in the form (6.3).

6.5. Refer to Fig. P6.5.
(a) Find V_2 exactly (take the larger of two possible values).

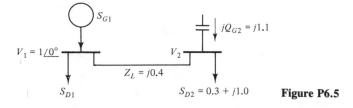

Figure P6.5

(b) Find V_2 by Gauss iteration starting with $V_2^0 = 1 \underline{/0°}$. If you use hand calculation, stop after one iteration.

(c) Find $S_1 = S_{G1} - S_{D1}$.

6.6. We are given the system shown in Fig. P6.6 and the following equations for bus powers:

$$S_1 = j19.98 \, |V_1|^2 - j10V_1V_2^* - j10V_1V_3^*$$

$$S_2 = -j10V_2V_1^* + j19.98 \, |V_2|^2 - j10V_2V_3^*$$

$$S_3 = -j10V_3V_1^* - j10V_3V_2^* + j19.98 \, |V_3|^2$$

Do one step of Gauss iteration to find V_2^1 and V_3^1. Start with $V_2^0 = V_3^0 = 1 \underline{/0°}$.

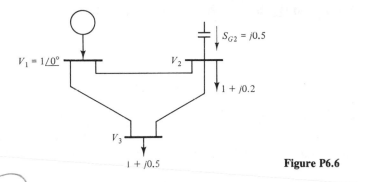

Figure P6.6

6.7. Repeat Problem 6.6 using Gauss–Seidel iteration. Do only one iteration.

6.8. In Fig. P6.8, assume that

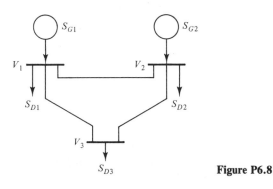

Figure P6.8

$$S_{D1} = 1.0 \qquad\qquad V_1 = 1\angle 0°$$

$$S_{D2} = 1.0 - j0.8 \qquad Q_{G2} = -0.3$$

$$S_{D3} = 1.0 + j0.6 \qquad P_{G2} = 0.8$$

$$Z_L = j0.4, \text{ all lines}$$

Use Gauss iteration to find V_2 and V_3. Start with $V_2^0 = V_3^0 = 1\angle 0°$. Do one iteration only (i.e., calculate V_2^1 and V_3^1). *Note:* Bus 2 is a P, Q bus because Q_{G2} is specified rather than $|V_2|$.

6.9. Repeat Problem 6.8 with one change. We specify $|V_2| = 1.0$ and do not specify Q_{G2}. In this case bus 2 is a P, $|V|$ bus.

6.10. In Fig. P6.10, assume that

$$Z_L = j0.5 \text{ for all lines}$$

$$P_{G2} = 0.2, \quad |V_2| = 1$$

$$V_1 = 1\angle 0°$$

$$S_{D3} = 0.3 + j0.1$$

Find $\angle V_2$, $|V_3|$ and $\angle V_3$, by using small-angle approximation. More specifically, in the equations

$$S_2^* = V_2^* \sum y_{2k} V_k \tag{1}$$

$$S_3^* = V_3^* \sum y_{3k} V_k \tag{2}$$

1. Approximate $e^{j\theta_{ik}}$ by $1 + j\theta_{ik}$,
2. Equate imaginary parts of (2) to find $|V_3|$ (use the larger of the solutions);
3. Then use the real parts of (1) and (2) to find $\angle V_2$ and $\angle V_3$.

Do you think the approximation is reasonable in this problem?

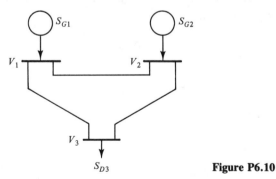

Figure P6.10

6.11. Make up a simple (scalar) example showing the graphical solution of $f(x) = 0$ by use of the iteration formula

$$x^{\nu+1} = x^\nu - \alpha f(x^\nu)$$

Show typical solutions for (a) $\alpha = 1$, (b) $\alpha > 1$, (c) < 1, and (d) $\alpha = [f'(x^{\nu})]^{-1}$.

6.12. Use the Newton–Raphson method to solve

$$f_1(\mathbf{x}) = x_1^2 - x_2 - 1 = 0$$

$$f_2(\mathbf{x}) = x_2^2 - x_1 - 1 = 0$$

$x_1^0 = x_2^0 = 1$. Do *two* iterations only. *Note:* Exact solution is $x_1 = x_2 = 1.618$. (Another solution is $x_1 = x_2 = -0.618$).

6.13. Repeat Problem 6.12 with initial condition $x_1^0 = x_2^0 = -1$.

6.14. Use the Newton–Raphson method to solve

$$f_1(\mathbf{x}) = x_1^2 + x_2^2 - 1 = 0$$

$$f_2(\mathbf{x}) = x_1 + x_2 = 0$$

with an initial guess $x_1^0 = 1$ and $x_2^0 = 0$. Do *two* iterations. *Note:* Exact solution is $x_1 = -x_2 = 1/\sqrt{2}$.

6.15. For the system in Example 6.8, assume that $P_{G2} = 0.3$, $|V_2| = 0.95$, and $S_{D3} = 0.5 + j0.2$. Starting with $\theta_2^0 = \theta_3^0 = 0$ and $|V_3|^0 = 1.0$, do one N-R iteration and find θ_2^1, θ_3^1, and $|V_3|^1$. Evaluate the mismatch vector after this single iteration.

6.16. Use the fast-decoupled power flow method in Problem 6.15. Do two iterations. Tabulate the values of θ_2, θ_3, $|V_3|$, $\Delta\tilde{P}_2$, $\Delta\tilde{P}_3$, and $\Delta\tilde{Q}_3$ obtained as the iterations proceed.

Economic Operation of Power Systems

chapter 7

7.0 INTRODUCTION

In this chapter we consider the important problem of economic operation of power systems: how to operate a power system to supply all the (complex) loads at minimum cost. Here we assume that we have some flexibility in adjusting the power delivered by each generator. Of course, if we have a "peak" demand for power which is so large that all the available generator capacity must be used, there are no options. But usually the total load is less than the available generator capacity and there are many possible generation assignments.

In this case it is important to consider the cost of generating the power, and to pick the P_{Gi} to minimize these "production" costs while "satisfying" the loads and the losses in the transmission links. When hydro generation is not considered, it is reasonable to choose the P_{Gi} on an instantaneous basis (i.e., always to minimize the present production cost rate). With hydro, however, in dry periods, the replenishment of the water supply may be a problem. The water used today may not be available in the future when its use might be more advantageous. Even without the element of prediction involved, the problem of minimizing production cost over time becomes much more complicated and we will not consider this problem. We will consider the simpler problem of minimization of instantaneous production costs; this is called the *economic dispatch problem*.

It should be mentioned that economy of operation is not the only possible consideration. If the "optimal" economic dispatch requires all the power to be imported from a neighboring utility through a single transmission link, considerations of system security might preclude that solution. When water used for hydro generation is also used for irrigation, nonoptimal releases of water may be required. Under adverse atmospheric conditions it may be necessary to limit generation at certain fossil-fuel plants to reduce emissions.

In general, costs, security, and emissions are all areas of concern in power plant operation, and in practice, the system is operated to effect a compromise between the frequently conflicting requirements. We will, however, limit ourselves to economic considerations.

It is important to emphasize that the problem we will be considering is the problem of minimizing production costs in real time under the assumption that the generators available to us have been specified (i.e., we know which generators are "on-line" or "committed" at a given moment). Our optimization then concerns this particular set of generators.

We will not consider, in detail, the interesting related problem of scheduling the startup and shutdown of generators to meet the anticipated variations in demand for power over a 24-hour period. We note that the small variations in power demand are taken care of by adjusting the generators already on-line. The larger variations are accommodated, basically, by starting up generating units when the loads are on the upswing and shutting them down when the loads decrease. The problem is complicated by considerations of the long lead time required (6 to 8 hours) for preparing a "cold" thermal unit for service, startup and shutdown costs, and the requirement that enough spare generating capacity (also called "spinning reverse") be available on-line in the event of a random generator failure.

We turn next to a formulation of the economic dispatch problem. We remind the reader that in Chapter 6 we assumed that the P_i were specified at the P, $|V|$ buses. Since we assume that the S_{Di} are given, this is equivalent to specification of the P_{Gi}. In this chapter we consider how the P_{Gi} are specified by economic considerations.

7.1 FORMULATION OF THE ECONOMIC DISPATCH PROBLEM

The total cost of operation includes fuel, labor, and maintenance costs, but for simplicity we will assume that the only variable costs we need to consider are fuel costs. For the fuel costs we assume that we are given the fuel-cost curves for each generating unit, specifying the cost of the fuel used per hour as a function of the generator power output. For simplicity we assume that each generating unit consists of a generator, turbine, steam generating unit (boiler-furnace), and associated auxiliary equipment. An approximation to a typical fuel-cost curve is shown in Fig. 7.1. Note the use of the symbol P_{Gi} for 3ϕ power output instead of $P_{Gi}^{3\phi}$. For most of the chapter (until Section 7.5) only 3ϕ power is considered and the less burdensome notation is preferable.

The shape of the fuel-cost curve (concave upward) may be understood in terms of the heat-rate curve, which is determined by field testing the generating units. The

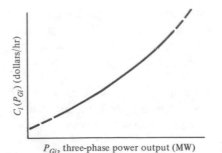

     P_{Gi}, three-phase power output (MW)   **Figure 7.1**   Fuel-cost curve.

approximate shape of this curve is given in Fig. 7.2. The curve gives $H_i(P_{Gi})$, the MBtu (Btu is British thermal units) of heat energy supplied by burning the fuel, per MWh of electric energy. At the minimum point the generating unit is most efficient. The curve reflects the typical drop in efficiency of most energy conversion machines at the low and high ends. At peak efficiency the heat rates of modern fossil-fuel plants are in the range 8.6 to 10 MBtu/MWh. If the conversion from chemical energy in the fuel to electrical energy were 100% efficient, the heat rate would be approximately 3.41 MBtu/MWh. Thus we are obtaining typical maximum overall efficiencies in the range 34 to 39%.

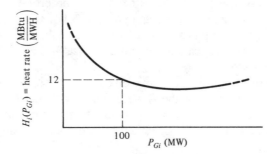

      100   P_{Gi} (MW)   **Figure 7.2**   Heat-rate curve.

From the heat-rate curve we can determine the input energy rate as a function of output power. For example, suppose that in Fig. 7.2, $P_{Gi} = 100$ MW. From the figure we see that $H_i(P_{Gi})$, the corresponding heat rate, is 12. This means that there is 12 MBtu of heat energy required for each 1 MWh of electrical energy output. In an hour the electrical energy output would be 100 MWh while the heat energy required would be 100×12 MBtu = 1200 MBtu. Thus a heat input energy rate of 1200 MBtu/hr is required to sustain a power output of 100 MW; we get the figure by multiplying P_{Gi} (3ϕ power) by the corresponding heat rate $H_i(P_{Gi})$. More generally, the heat input energy rate, $F_i(P_{Gi})$, is found by the formula

$$F_i(P_{Gi}) = P_{Gi}H_i(P_{Gi}) \tag{7.1}$$

where P_{Gi} is 3ϕ power in MW, H_i is the heat rate in MBtu/MWh, and F_i is in MBtu/hr. The graph of $F_i(P_{Gi})$ is called an *input-output curve*.

Suppose now that it is specified that the cost of the fuel is K dollars/MBtu. Then

$$C_i(P_{Gi}) = KF_i(P_{Gi}) \tag{7.2}$$

is the fuel cost in dollars per hour to supply P_{Gi} MW of electrical power.

The general shape of the fuel-cost curve may thus be inferred from the shape of the heat-rate curve. It is found that the heat-rate curves may be approximated in the form $H(P_G) = (\alpha'/P_G) + \beta' + \gamma'P_G$ with all coefficients positive. Then using (7.1) and (7.2), we find that the corresponding $C(P_G)$ is a quadratic expression with positive coefficients in the form

$$C(P_G) = \alpha + \beta P_G + \gamma P_G^2 \qquad \text{dollars/hr} \tag{7.3}$$

where P_G is expressed in MW.

Example 7.1

Assume that the heat rate of a 50-MW gas-fired generator unit is measured as follows:

25% of rating:	14.26 MBtu/MWh
40% of rating:	12.94 MBtu/MWh
100% of rating:	11.70 MBtu/MW

Assume that the cost of gas is $5 per MBtu.

(a) Find $C(P_G)$ in the form (7.3).
(b) Find the fuel cost in cents/kWh when 100% loaded.
(c) Find the fuel cost when 50% loaded.
(d) Find the fuel cost when 25% loaded.

Solution   (a) The assumed polynomial form of $C(P_G)$ is equivalent to a heat-rate curve of the form

$$H(P_G) = \frac{\alpha'}{P_G} + \beta' + \gamma'P_G$$

The three test measurements give us three points on the curve, and hence we can solve for the three unknown coefficients, α', β', and γ'.

$$\frac{\alpha'}{12.5} + \beta' + \gamma'12.5 = 14.26$$

$$\frac{\alpha'}{20} + \beta' + \gamma'20 = 12.94$$

$$\frac{\alpha'}{50} + \beta' + \gamma'50 = 11.70$$

The solution is $\alpha' = 44.89$, $\beta' = 10.62$, and $\gamma' = 0.0036$. Thus

$$H(P_G) = \frac{44.89}{P_G} + 10.62 + 0.0036P_G \qquad \text{MBtu/MWh}$$

If we multiply by P_G, we get the fuel input rate in MBtu/hr. Multiplying by $5, the cost of fuel per MBtu, we get

$$C(P_G) = 224.5 + 53.1P_G + 0.0180P_G^2 \qquad \text{dollars/hr}$$

(b) When fully loaded

$$C(P_G) = \$2924.5/\text{hr (for 50 MW)}$$

$$= 5.85 \text{ cents/hr (for 1 kW)}$$

This is usually expressed as 5.85¢/kWh.

(c) At 40% load

$$C(P_G) = \$1293.7/\text{hr (for 20 MW)}$$

$$= 6.47 \text{ cents/kWH}$$

(d) At 25% load

$$C(P_G) = \$891.1/\text{hr (for 12.5 MW)}$$

$$= 7.13 \text{ cents/kWh}$$

Exercise 1. What shape of heat-rate curve would yield a linear fuel-cost curve?

Assume henceforth that the $C_i(P_{Gi})$ curves are given. We may then formulate a general problem of optimum economic dispatch, or optimum power flow.

General Problem Formulation

Given a system with m generators committed and all the S_{Di} given, pick the P_{Gi} and $|V_i|$, $i = 1, 2, 3, \ldots, m$, to minimize the total cost

$$C_T \triangleq \sum_{i=1}^{m} C_i(P_{Gi})$$

subject to the satisfaction of the power flow equations and to the following inequality constraints on generator power, line power flow, and voltage magnitude.

1. $P_{Gi}^{\min} \leq P_{Gi} \leq P_{Gi}^{\max}, \qquad i = 1, 2, \ldots, m$
2. $|P_{ij}| \leq P_{ij}^{\max}, \qquad \text{all lines}$
3. $|V_i|^{\min} \leq |V_i| \leq |V_i|^{\max}, \qquad i = 1, 2, \ldots, m, \ldots, n$

We next comment briefly on the formulation.

1. The power flow equations must be satisfied (i.e., they are an equality constraint on the optimization).
2. The upper limit on P_{Gi} is set by thermal limits on the turbine generator unit, while the lower limit is set by boiler and/or other thermodynamic considerations. A

certain minimum flow of water and/or steam is required in the boiler to prevent "hot spots" from developing. The fuel burning rate must also be sufficient to keep the flame from going out ("flame out").

3. The constraints on voltage keep the system voltages from varying too far from their rated or nominal values. The objective is to help maintain the consumer's voltage; the voltage should neither be too high or too low.

4. The reasons for constraints on the transmission-line powers relate to thermal and stability limits and have been discussed in Section 4.6.

5. The formulation of the problem is consistent with the availability of injected active power and bus voltage magnitude as control variables at each generator. It can be extended to deal with other control variables, such as the phase angles across phase-shifting transformers, the turns ratios of tap-changing transformers, and the admittances of variable (and controllable) shunt and series inductors and capacitors.

6. The minimization of a cost function subject to equality and inequality constraints is a problem in optimization which is treated by a branch of applied mathematics called *nonlinear programming*.

To help fix ideas, consider next the formulation as applied to the simple system in Fig. 7.3. Assume that the system is specified; in particular, all the bus admittance parameters are given, the reference voltage angle is given, and all the S_{Di} are given. We can pick a particular set of $|V_1|$, P_{G2}, $|V_2|$, P_{G3}, and $|V_3|$ (within the constraint set) and by power flow analysis find P_{G1} and the transmission-line power angles. If P_{G1} and the transmission-line powers satisfy the inequality constraints, our choice is "feasible" and we calculate the total cost C_T. The problem is to minimize C_T over the set of feasible independent control variables $|V_1|$, P_{G2}, $|V_2|$, P_{G3}, and $|V_3|$.

The general nonlinear programming methods for solving this problem are computationally expensive and they do not easily give insights into the nature of the optimal solutions. We therefore make some approximations which simplify the calculations and give physical insights into the problem of economic dispatch.

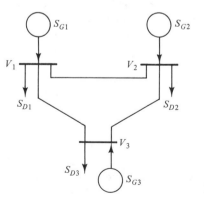

Figure 7.3 System to be optimized.

7.2 CLASSICAL ECONOMIC DISPATCH (Line Losses Neglected)

As shown in Section 6.7, under normal operating conditions there is relatively little coupling between active power flows and power angles on the one hand, and reactive power flows and voltage magnitudes on the other. Thus in the general problem formulation in Section 7.1 we expect that the cost C_T will be critically dependent on the P_{Gi} and P_{Di} but only marginally dependent on the $|V_i|$ and Q_{Di}. For this reason we reformulate the problem entirely in terms of active power flows, setting the $|V_i|$ at their nominal values.

Suppose that, in addition, we neglect the line power flow constraints and for the moment, neglect line losses as well. We can then replace the general problem formulation with a much simplified version.

Simplified Problem Formulation

Pick the P_{Gi} to minimize

$$C_T = \sum_{i=1}^{m} C_i(P_{Gi}) \tag{7.4}$$

such that

$$\sum_{i=1}^{m} P_{Gi} = P_D \triangleq \sum_{i=1}^{n} P_{Di} \tag{7.5}$$

and

$$P_{Gi}^{\min} \leq P_{Gi} \leq P_{Gi}^{\max} \qquad i = 1, 2, \ldots, m \tag{7.6}$$

Note that (7.5), the equality constraint, is simply a statement of conservation of active power in the case of a lossless transmission system. It takes the place of the power flow equations, as an equality constraint, in the previous formulation. Note that the (active) power flow equations might still be needed if the line power flow constraints were to be checked, but in the simplified problem formulation there is no need whatever to consider the power flow equations.

The formulation may be considered as a first approximation; we will soon be considering modifications that permit line losses to be considered, and in this case we will need to consider the power flow equations.

Returning to the simplified problem formulation, we note that there is a simple and elegant solution in terms of so-called *incremental costs* (ICs). First we need a definition:

$$IC_i = \frac{dC_i(P_{Gi})}{dP_{Gi}} = \text{slope of fuel-cost curve} \tag{7.7}$$

If the units of $C_i(P_{Gi})$ are dollars/hr, the units of the IC_i are dollars/hr/MW or dollars/MWh. The IC_i figure represents the increase in cost rate per increase in MW output or equivalently the increase in cost per increase in MWh.

If as in (7.4) we assume that the fuel-cost curves are quadratic (with positive coefficients), the incremental cost curves are linear (with positive coefficients):

$$IC_i = \beta_i + 2\gamma_i P_{Gi} \qquad (7.8)$$

We note that the curves are strictly monotonically increasing, a property we will be assuming even if we do not use the linear form of (7.8).

Example 7.2

With the data of Example 7.1:

(a) Find an expression for IC $= dC/dP_G$.
(b) Find the incremental cost when 100% loaded.
(c) Find the incremental cost when 25% loaded.

Solution   We use (7.8) in Example 7.1.
(a) IC $= 53.1 + 0.0360P_G$.
(b) At 100% loading, IC $=$ \$54.90/MWh.
(c) At 25% loading, IC $=$ \$53.55/MWh.

Note. In this example there is very little change in incremental costs as a function of loading.

We now return to the solution of the simplified optimization problem posed in (7.4) to (7.6). It is convenient to consider first the case without the inequality constraints [i.e., without the generator limits given in (7.6)]. In this special case we get the following simple rule.

Optimal Dispatch Rule (No Losses—No Generator Limits).   Operate every generator at the same incremental cost.

Note 1: The rule gives only a property of the solution. But with this property we reduce an *m*-dimensional search (for the best $P_{G1}, P_{G2}, \ldots, P_{Gm}$), to a one-dimensional search for the common incremental cost.

Note 2: In the statement of the rule there is an implicit assumption that it is possible to operate every generator at the same incremental cost.

Intuitively, it is clear why it cannot possibly be optimal to operate generators at different ICs. Suppose that we have two generators with, say, $P_{G1} = 100$ MW and $P_{G2} = 50$ MW and the corresponding IC_i (i.e., the slopes of the cost curves) are not equal. The situation is shown in Fig. 7.4. As shown in the figure, IC_1 is greater than IC_2. Since IC_1 is greater than IC_2, we notice the desirability of the following adjustment of generator powers. If we reduce P_{G1} by (say) 10 MW, we save quite a lot of cost per hour (because the slope of the cost curve is large). If we add the 10 MW to P_{G2}, the cost goes up less (because the slope is smaller). Thus we can deliver the same total power (150 MW) at less cost. In general, it pays to reduce the power output of the generator with the higher incremental cost. The slope of the curves (strictly monotonically increasing slope) is such that a continuation of this process ultimately leads to equal incremental costs.

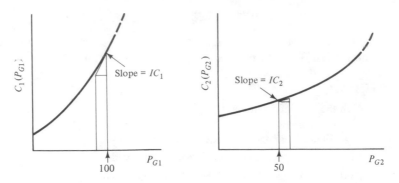

Figure 7.4 Two generators operating at different incremental costs.

This line of reasoning obviously extends to any number of generators, considering them in pairs, and leads to the conclusion that all the ICs must be equal at the minimum cost point. A more formal proof will be offered later in this section.

Exercise 2. To understand better why the procedure above leads to a minimum, consider some fictitious cost curves: for example, concave upward but with negative slope or concave downward with positive (or negative) slope. Do we get convergence? If so, do we get a minimum or a maximum cost when the two ICs are equal?

Example 7.3

The fuel-cost curves of two generators are given as follows:

$$C_1(P_{G1}) = 900 + 45P_{G1} + 0.01P_{G1}^2$$

$$C_2(P_{G2}) = 2500 + 43P_{G2} + 0.003P_{G2}^2$$

The total load to be supplied is $P_D = P_{D1} + P_{D2} = 700$ MW. Use the optimal dispatch rule (special case) to find P_{G1} and P_{G2}.

Solution The incremental costs are

$$IC_1 = \frac{dC_1}{dP_{G1}} = 45 + 0.02P_{G1}$$

$$IC_2 = \frac{dC_2}{dP_{G2}} = 43 + 0.006P_{G2}$$

It is instructive to plot the IC curves (Fig. E7.3). We see that IC_2 is below IC_1, which implies that generator 2 will carry more load than generator 1. Invoking the condition $IC_1 = IC_2$, we can see graphically that G_2 will carry almost all the load. We can find the exact division iteratively by guessing a value for $IC_1 = IC_2$, solving for P_{G1} and P_{G2}, checking if $P_{G1} + P_{G2} = P_D = 700$ MW, and increasing (decreasing) the incremental cost if $P_{G1} + P_{G2}$ is too low (high). This method works even if the IC_i curves are not linear. Alternatively, in the present case we can easily solve for the exact values analytically.

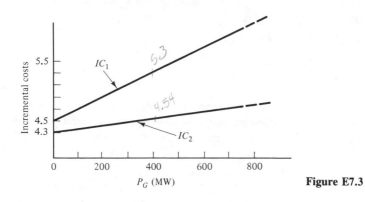

Figure E7.3

$$IC_1 = 45 + 0.02P_{G1} = IC_2 = 43 + 0.006(700 - P_{G1})$$

Solving, we get

$$P_{G1} = 84.6 \text{ MW} \qquad P_{G2} = 615.4 \text{ MW} \qquad IC_1 = IC_2 = \$46.69/\text{hr/MW}$$

Note that if P_D is raised from 700 MW to 701 MW (i.e., increased by 1 MW), the fuel cost rate will go up (approximately) \$46.69/hr (i.e., by \$46.69/MWh or 4.669 cents/kWh). This incremental cost is to be contrasted with the *average* cost of producing the 700 MW. Plugging the optimal values of P_{G1} and P_{G2} into $C_1(P_{G1})$ and $C_2(P_{G2})$, we get a total cost $C_T = \$34,877/\text{hr}$. This translates into an average cost of $(34877 \times 100)/(700 \times 1000) = 4.98$ cents/kWh.

Exercise 3.   Consider some alternative generator assignments or "dispatches" in Example 7.3 and convince yourself that the cost of producing 700 MW goes up.

We turn next to a formal proof of the result on equal incremental costs.

Proof of Optimal Dispatch Rule (Special Case).   Referring to the simplified problem formulation (without inequality constraints), the problem is to pick the m values of P_{Gi} to minimize $C_T = C_1(P_{G1}) + C_2(P_{G2}) + \cdots + C_m(P_{Gm})$ subject to the constraint $P_{G1} + P_{G2} + \cdots + P_{Gm} = P_D$. The constraint permits us to express one of the variables, say P_{Gm}, in terms of the rest; in this case $P_{Gm} = P_D - P_{G1} - \cdots - P_{G,m-1}$. Thus the problem reduces to the following: Pick the $m - 1$ independent variables P_{G1}, $P_{G2}, \ldots, P_{G,m-1}$ to minimize

$$C_T = C_1(P_{G1}) + C_2(P_{G2}) + \cdots + C_{m-1}(P_{G,m-1})$$
$$+ C_m(P_D - P_{G1} - \cdots - P_{G,m-1}) \qquad (7.9)$$

In this way we have converted the constrained problem to an unconstrained problem of lower dimension.

The solution to the unconstrained problem may be found in the usual way by the differential calculus. The necessary condition for a stationary point is that the $m - 1$ partial derivatives, $\partial C_T/\partial P_{Gi}$, are zero. Thus from (7.9), by using the "chain rule" of differentiation, we get

$$\frac{\partial C_T}{\partial P_{Gi}} = \frac{dC_i}{dP_{Gi}} + \frac{dC_m}{dP_{Gm}} \frac{\partial P_{Gm}}{\partial P_{Gi}}$$

$$= \frac{dC_i}{dP_{Gi}} - \frac{dC_m}{dP_{Gm}} = 0 \qquad i = 1, 2, \ldots, m - 1 \qquad (7.10)$$

The result is that all the incremental costs must be equal to the incremental cost of the mth generating unit (i.e., that they are all equal). We note (without proof) that the shape of the C_i curves is such (concave upward) that the stationary point defined by (7.10) is a minimum. Also, the assumption that the IC curves are strictly monotonic implies a unique solution for the corresponding P_{Gi}. Thus, to summarize, there is only one possible set of P_{Gi} which satisfies the equal IC condition, and thus the corresponding cost function must be at a global minimum. Stated differently, the equal IC condition is both a necessary and a sufficient condition for optimization.

There is an alternative approach to the one above which is frequently more convenient to use. It is the classical method using Lagrange multipliers, for minimizing (or maximizing) a function with equality constraints as side conditions. For a brief introduction to the method, the reader is referred to Appendix 3.

Using the method, we replace the cost function $\tilde{C}_T$ by an "augmented" cost function $\tilde{C}_T$.

$$\tilde{C}_T \triangleq C_T - \lambda\left(\sum_{i=1}^{m} P_{Gi} - P_D\right) \qquad (7.11)$$

where λ is the Lagrange multiplier. We then find the stationary points of C_T with respect to λ and the P_{Gi} as variables. If we satisfy

$$\frac{\partial \tilde{C}_T}{\partial P_{Gi}} = 0 \qquad i = 1, 2, 3, \ldots, m$$

$$\frac{\partial \tilde{C}_T}{\partial \lambda} = 0$$

then we satisfy the equality constraint, $\sum_{i=1}^{m} P_{G1} - P_D = 0$ and simultaneously find a stationary point of C_T taking the constraint into consideration.

Applied to our problem, with concave-upward cost curves, we get a single stationary point (a minimum), and

$$\frac{dC_i(P_{Gi})}{dP_{Gi}} = \lambda \qquad i = 1, 2, \ldots, m \qquad (7.12a)$$

$$\sum_{i=1}^{m} P_{Gi} - P_D = 0 \qquad (7.12b)$$

Note that the result in (7.12) is the same as implied by (7.10), with λ the common value of incremental cost. For this reason we refer to λ as the system incremental cost, or system λ.

Just as in Example 7.3, λ, which is the common value of incremental cost, relates increased fuel cost rate (in dollars/hr) to increased demand (in MW). It is instructive to derive this result formally. Suppose that with a given demand P_D^0, we find an optimal set of P_{Gi}^0 and the corresponding cost is C_T^0. Suppose that now the load increases (incrementally) to $P_D = P_D^0 + \Delta P_D$ and we wish to find the new cost C_T. We may use a two-term Taylor series:

$$C_T = C_T^0 + \Delta C_T = \sum_{i=1}^{m}\left[C_i(P_{Gi}^0) + \frac{dC_i(P_{Gi}^0)}{dP_{Gi}}\Delta P_{Gi}\right] \qquad (7.13)$$

Relating increments, we get

$$\Delta C_T = \sum_{i=1}^{m} \frac{dC_i(P_{Gi}^0)}{dP_{Gi}}\Delta P_{Gi} \qquad (7.14)$$

Using (7.12a), we get

$$\Delta C_T = \lambda \sum_{i=1}^{m} \Delta P_{Gi} = \lambda \Delta P_D \qquad (7.15)$$

Thus λ is the constant of proportionality relating cost-rate increase (in dollars/hr) to increase in system power demand (in MW). Equivalently, λ gives us the additional cost of additional (incremental) energy in dollars/MWh.

Finally, we consider how to find λ in the present case. If the cost curves are quadratic, the IC_i curves are straight lines and the problem reduces to a linear problem. The solution is considered in a homework problem. In more general cases, the problem of finding λ can be determined by an iterative process. The procedure may be explained by considering Fig. 7.5, where the incremental costs are given graphically. The basic iterative procedure is:

1. Pick an initial λ.
2. Find the corresponding $P_{G1}(\lambda)$, $P_{G2}(\lambda)$, and $P_{G3}(\lambda)$.
3. If $\sum_{i=1}^{3} P_{Gi}(\lambda) - P_D < 0$, increase λ and go to 2. If $\sum_{i=1}^{3} P_{Gi}(\lambda) - P_D > 0$, decrease λ and go to 2. If $\sum_{i=1}^{3} P_{Gi}(\lambda) - P_D = 0$, stop.

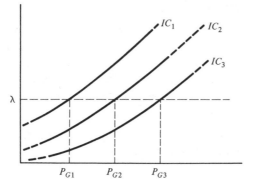

Figure 7.5 Use of incremental cost curves.

7.3 GENERATOR LIMITS INCLUDED

The case we have been considering is only an introduction to the more practical case in which there are generator limits. Returning to the simplified problem formulation, we now add the inequality constraints (7.6). Consider, then, the case with generator limits. We can show these constraints on the IC curves, as is done in Fig. 7.6.

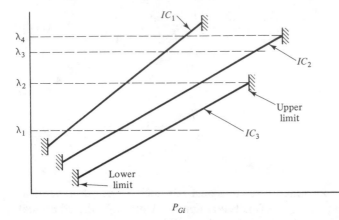

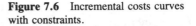

Figure 7.6 Incremental costs curves with constraints.

Consider the optimal dispatch in this case. Suppose that for a given P_D the system λ is λ_1. All three generators are operating in accordance with the optimal dispatch rule and the question of generator limits is moot since each generator is operating away from any limit. Now suppose that P_D increases and we increase λ to provide more generation. Continuing the process in this way we reach λ_2. What if P_D increases further? P_{G3} has reached a limit and cannot be increased further. The increased load must be taken by P_{G1} and P_{G2}. Clearly, they should operate at equal IC, say λ_3. Further increases in load can be taken by P_{G1} and P_{G2} operating at equal IC until P_{G2} reaches its upper limit, and $\lambda = \lambda_4$. Beyond this point, only P_{G1} is available to take an increased load. Considerations of this type lead to the more general rule that follows.

Optimal Dispatch Rule (No Losses). Operate all generators, not at their limits, at equal incremental cost λ.

Procedure for Finding λ. Pick an initial λ such that all the generators operate at equal incremental cost and within their constraints. If this choice of λ is not consistent with satisfaction of the load demand, adjust λ just as in the nonconstrained cases. If in this process a generating unit reaches a maximum or minimum value, fix the unit at the limit and continue the process of adjustment of λ with the remaining units.

Note 1: In applying the rule it is helpful to proceed graphically, at least to determine which generators will not be at their limits.

Note 2: We remind the reader that we are considering a given set of committed generators. This set of generators once specified is always kept on-line. We don't "shut down" a particular generator which may be operating at its lower limit.

Note 3: We can interpret the rule as a requirement that we operate as *closely as possible* to equality in the incremental costs.

Outline of Proof. It should be clear from the unconstrained problem that generators not at their limits should operate at the same λ. Suppose, then, that we have a case such as that in Fig. 7.6, with unit 3 at its upper limit (with incremental cost λ_2) and units 1 and 2 operating at equal incremental costs λ_3. We will show that we cannot lower the cost by backing away from the limit on unit 3 and making up for it by increasing the output on units 1 and 2. Assume that $\Delta P_{G3} = -\epsilon$, where $\epsilon > 0$ (i.e., we decrease P_{G3}). Then using Taylor's series as in (7.13), we get

$$\Delta C_T = \sum_{i-1}^{3} \lambda_i \, \Delta P_{Gi} = -\lambda_2 \epsilon + \lambda_3 \frac{\epsilon}{2} + \lambda_3 \frac{\epsilon}{2} = \epsilon(\lambda_3 - \lambda_2) > 0$$

Here we assume that P_{G1} and P_{G2} have each been increased by $\epsilon/2$, but it does not matter how the small increment ϵ is distributed. Since $\lambda_3 > \lambda_2$, the result is that the cost goes up. The same thing happens if we consider the case of backing away from a lower limit; the cost goes up. Although we have considered the special case of three machines, the result in the general case is the same. Since no improvement results by any perturbation, the dispatch rule is optimal. It is easy to check that (7.15) still gives the sensitivity of cost rate increase to increase in power demand, where the summation is only over the units not at their limits and the λ is the common value for these units. We still call λ the system λ, or system incremental cost.

Example 7.4

Suppose that in Example 7.3 the generator limits are

$$50 \text{ MW} \leq P_{G1} \leq 200 \text{ MW}$$

$$50 \text{ MW} \leq P_{G2} \leq 600 \text{ MW}$$

Find the optimal dispatch for $P_D = 700$ MW. Find the system incremental cost and compare with the result in Example 7.3.

Solution   Following the suggested procedure for finding λ, we can use the graph of ICs in Example 7.3. With $\lambda = 45.1$, say, both generators operate within their constraints but the total generation is less than the required 700 MW. Raising λ, we find that generator 2 reaches its upper limit (600 MW) before the total generation has reached 700 MW. Fixing $P_{G2} = 600$ MW, we immediately get $P_{G1} = 700 - P_{G2} = 100$ MW.

In this example there is a simpler alternative solution. First solve the problem without the limits; this was done in Example 7.3 with the result $P_{G1} = 84.6$ MW and $P_{G2} = 615.4$ MW. Second, noting that P_{G2} has exceeded its upper limit (600 MW), we set $P_{G2} = 600$ MW. Then it immediately follows that $P_{G1} = 100$ MW.

We note that we can extend the alternative solution method to more general problems, but occasionally errors may result in those cases where both upper and lower limits are violated simultaneously. If only upper limits, or only lower limits, are violated, the alternative method works fine.

Finally, we can calculate the system incremental cost. This is the (common) λ for the units not at their limits. In our case generator 1 is the only such unit. From Example 7.3 we have $IC_1 = 45 + 0.02P_{G1}$. Evaluated for $P_{G1} = 100$ MW, we get $\lambda = 47$. This is higher than the value (46.69) found in Example 7.3.

Example 7.5

For the generating units in Example 7.3 subject to the limits in Example 7.4, consider a range of possible values for P_D rather than the single fixed value 700 MW.

(a) Plot P_{G1} and P_{G2} versus P_D for optimal dispatch.
(b) Find the system incremental cost λ as a function of total generator output, $P_D = P_{G1} + P_{G2}$.

Solution (a) We plot the IC curves as in Example 7.3 but this time showing the limits [Fig. E7.5(a)]. For values of λ up to 46, $P_{G1} = 50$ MW (at its lower limit), while $P_{G2} = P_D - 50$ MW supplies the remainder of the load. For values of $46 < \lambda < 46.6$, neither unit is limited and we can find P_{Gi} by using the IC formulas in Example 7.3. For values of λ greater than 46.6, $P_{G2} = 600$ (at its upper limit) and $P_{G1} = P_D - 600$. The results are shown in the curves in Fig. E7.5(b). As expected, since the incremental costs of unit 1 are so high, it does not come away from its lower limit until unit 2 is heavily loaded (500 MW).

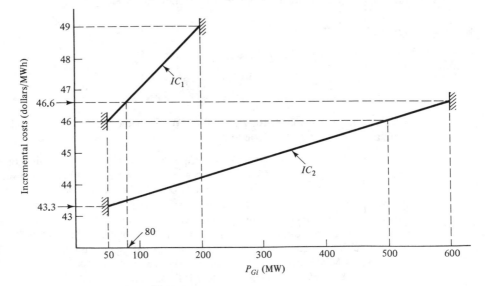

Figure E7.5(a)   Generator incremental costs.

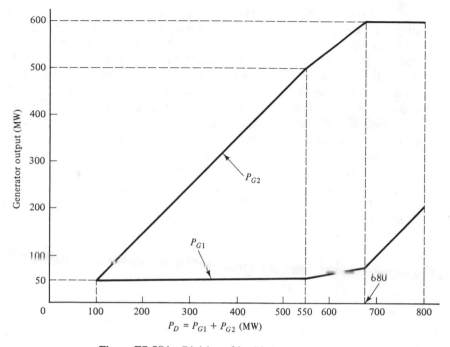

Figure E7.5(b)   Division of load between generators.

(b) While unit 1 is at its lower limit (i.e., for $\lambda < 46$), $\lambda = 43 + 0.006P_{G2} = 43 + 0.006(P_D - 50) = 42.7 + 0.006P_D$. While unit 2 is at its upper limit (i.e., for $\lambda > 46.6$), $\lambda = 45 + 0.02P_{G1} = 45 + 0.02(P_D - 600) = 33 + 0.02P_D$. When both generators are away from their limits, $\lambda = 43.46 + 0.0046P_D$. The resulting system λ is shown in Fig. E7.5(c).

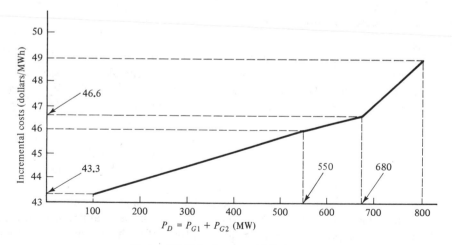

Figure E7.5(c)   Incremental fuel costs.

We can interpret the curve of λ versus P_D as that of an "equivalent generator" supplying the power $P_{G1} + P_{G2}$. This viewpoint finds some advantages when considering all the (on-line) generators at a given location (i.e., at a power plant). A curve such as Fig. E7.5(c) adequately characterizes the plant for finding the economic dispatch with respect to the rest of the system. Of course, the plant operator needs to be aware of the optimal division of generation among the various local generating units. This is obtained from curves such as Fig. E7.5(b).

Finally, we consider a case with a particularly simple solution. Suppose that the IC curves look as shown in Fig. 7.7. We can easily show that the following procedure for loading up the units is optimal. We first load up unit 1 until it reaches its upper limit, then load up unit 2 similarly, and finally unit 3. We do not operate at equal IC because only one generator at a time is not at a limit. We need only decide in which order to load the generators. Unit 1 is "best," so it is first on our list, followed by unit 2, and then unit 3. This is the "order of merit" method of dispatch.

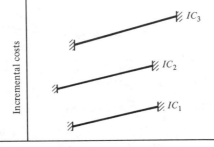

Figure 7.7 Order-of-merit dispatch.

Exercise 4.   For a case similar to Fig. 7.7, sketch the system λ as a function of $P_D = P_{G1} + P_{G2} + P_{G3}$.

7.4 LINE LOSSES CONSIDERED

If all the generators are located in one plant or are otherwise very close geographically, it is physically reasonable to neglect line losses in calculating the optimal dispatch. On the other hand, if the power stations are spread out geographically, the transmission-line (link) losses must usually be considered, and this will modify the optimal generation assignments as determined in the preceding section.

To take a simple case, suppose that all the generating units on the system are identical. Then, considering line losses, we expect it will be cheaper to draw most heavily from the generators closest to the loads, and this expectation is confirmed in analyses which include the line losses.

The standard approach to the problem of considering line losses can be understood by assuming that we have an expression for the total line losses, P_L, in terms of generator outputs. We next consider the form of such an expression. From the law of conservation of power we know that

$$P_L = \sum_{i=1}^{n} P_i = \sum_{i=1}^{m} P_{Gi} - \sum_{i=1}^{n} P_{Di} \qquad (7.16)$$

where P_L is the total line loss and the P_i are the bus powers. We assume that the n P_{Di} are specified and fixed but the P_{Gi} are variables. With the P_{Di} fixed, it can be seen from (7.16) that P_L depends only on the P_{Gi}. In fact, however, only $m - 1$ of the P_{Gi} are independent variables. As discussed in Chapter 6, bus 1 is a slack or swing bus and the bus power P_1 (and hence $P_{G1} = P_1 + P_{D1}$) is a dependent variable found by solving the power flow equations. Thus for a given system and a given "base case" (i.e., specific choice of $S_{Di} = P_{Di} + jQ_{Di}$ at all buses and $|V_i|$ specified at buses 1, 2, 3, . . . , m), the functional dependence of P_L on the generator outputs may be written

$$P_L = P_L(\mathbf{P}_G) \triangleq P_L(P_{G2}, P_{G3}, \ldots, P_{Gm}) \qquad (7.17)$$

Two points should be noted. The first is that the function in (7.17) depends on the choice of base case. The second is that (7.17) is not an explicit formula but depends on the solution of implicit equations (i.e., the power flow equations).

We now can state a version of the optimization problem involving line losses.

Problem Statement.   Pick the P_{Gi} to minimize

$$C_T = \sum_{i=1}^{m} C_i(P_{Gi})$$

subject to

$$\sum_{i=1}^{m} P_{Gi} - P_L(P_{G2}, \ldots, P_{Gm}) - P_D = 0$$

and

$$P_{Gi}^{\min} \le P_{Gi} \le P_{Gi}^{\max} \qquad i = 1, 2, \ldots, m$$

Note: The equality constraint is simply a rearrangement of (7.16) and is once again a statement of conservation of (active) power.

Just as in the lossless case, we will first consider the case without the generator limits. To obtain at least a formal solution, we can use the method of Lagrange multipliers. Define the augmented cost function

$$\tilde{C}_T = \sum_{i=1}^{m} C_i(P_{Gi}) - \lambda \left(\sum_{i=1}^{m} P_{Gi} - P_L(P_{G2}, \ldots, P_{Gm}) - P_D \right)$$

We next find a stationary point of $\tilde{C}_T$ with respect to λ and the P_{Gi}:

$$\frac{d\tilde{C}_T}{d\lambda} = \sum_{i=1}^{m} P_{Gi} - P_L - P_D = 0 \qquad (7.18)$$

$$\frac{d\tilde{C}_T}{dP_{G1}} = \frac{dC_1}{dP_{G1}} - \lambda = 0 \qquad (7.19)$$

$$\frac{d\tilde{C}_T}{dP_{Gi}} = \frac{dC_i(P_{Gi})}{dP_{Gi}} - \lambda\left(1 - \frac{\partial P_L}{\partial P_{Gi}}\right) = 0 \qquad i = 2, \ldots, m \tag{7.20}$$

An alternative expression for (7.20) is

$$\frac{1}{1 - \dfrac{\partial P_L}{\partial P_{Gi}}} \cdot \frac{dC_i(P_{Gi})}{dP_{Gi}} = \lambda \qquad i = 2, \ldots, m \tag{7.21}$$

Next define the *penalty factor* L_i for the ith generator.

$$L_1 = 1 \tag{7.22a}$$

$$L_i = \frac{1}{1 - \dfrac{\partial P_L}{\partial P_{Gi}}} \qquad i = 2, 3, \ldots, m \tag{7.22b}$$

Substituting (7.22b) in (7.21) and (7.22a) in (7.19) the necessary conditions for optimization given in (7.19) and (7.20) may be replaced by

$$L_1\frac{dC_1}{dP_{G1}} = L_2\frac{dC_2}{dP_{G2}} = \cdots = L_m\frac{dC_m}{dP_{Gm}} = \lambda \tag{7.23}$$

Recalling that dC_i/dP_{Gi} are the incremental costs, we now get the following rule.

Optimal Dispatch Rule (Line Losses Considered—No Generator Limits). Operate all generators so that the product $L_i \times IC_i = \lambda$ for every generator.

We see that it is no longer optimal to operate each generator at the same incremental cost. The ICs now must be weighted by the penalty factors L_i; a large penalty factor makes the corresponding plant less attractive and a smaller IC from that plant is required. We can expect that plants remote from load centers will have larger penalty factors than plants which are close.

If we now reintroduce the generator limits we find a rule analogous to the one in the lossless case.

Optimal Dispatch Rule (Line Losses Considered).   Operate all generators, not at their limits, such as to satisfy

$$L_i \times IC_i = \lambda \tag{7.24}$$

We next illustrate the use of the optimal dispatch rule with a simple two-bus example in which an explicit expression for $P_L(P_{G2})$ is given. We defer the important question of how to obtain such an expression for the losses.

Example 7.6

Consider a system without generator limits (Fig. E7.6). Assume that

$$IC_1 = 0.007P_{G1} + 4.1 \ \text{dollars/MWh}$$

$$IC_2 = 0.007P_{G2} + 4.1 \ \text{dollars/MWh}$$

$$P_L = 0.001(P_{G2} - 50)^2 \text{ MW}$$

Find the optimal generation for each plant and the power loss in the transmission link.

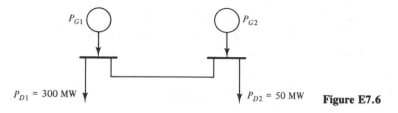

P_{G1} P_{G2}

$P_{D1} = 300 \text{ MW}$ $P_{D2} = 50 \text{ MW}$ **Figure E7.6**

Solution   We first calculate

$$\frac{\partial P_L}{\partial P_{G2}} = 0.002(P_{G2} - 50) = 0.002P_{G2} - 0.1$$

Using (7.22), we find the penalty factors

$$L_1 = 1.0 \qquad L_2 = \frac{1}{1.1 - 0.002P_{G2}}$$

Applying the appropriate optimal dispatch rule, we get two equations:

$$L_1 \frac{dC_1}{dP_{G1}} = 0.007P_{G1} + 4.1 = \lambda$$

$$L_2 \frac{dC_2}{dP_{G2}} = \frac{1}{1.1 - 0.002P_{G2}}(0.007P_{G2} + 4.1) = \lambda$$

We next need to pick an initial value of λ and solve for P_{G1} and P_{G2}. For this purpose a more convenient form of the equations is

$$P_{G1} = \frac{\lambda - 4.1}{0.007}$$

$$P_{G2} = \frac{1.1\lambda - 4.1}{0.007 + 0.002\lambda}$$

Starting with $\lambda = 5.0$, we calculate

$$P_{G1} = 128.6 \text{ MW} \qquad P_{G2} = 82.4 \text{ MW} \qquad P_L = 1.0 \text{ MW}$$

Since $P_{G1} + P_{G2} - P_L = 210$ MW is less than $P_D = 350$ MW, we need to increase λ. $\lambda = 6$ turns out to be too large. After a few iterations we find $\lambda = 5.694$, which gives $P_{G1} = 227.72$, $P_{G2} = 117.65$, $P_L = 4.58$, and $P_{G1} + P_{G2} - P_L = 349.9$. This last value is close enough to $P_D = 350$.

Example 7.7

Consider the same system as in Example 7.6. Suppose that we wish to supply the same load (i.e., $P_D = 350$ MW) but neglect line losses in carrying out the optimization. Find P_{G1}, P_{G2}, P_L, and the increased cost per hour in delivering the load under the nonoptimal condition.

Solution   Since the generator IC curves are identical, if we neglect line losses, we should pick $P_{G1} = P_{G2}$. Then $P_{G1} + P_{G2} - P_L = 2P_{G2} - 0.001(P_{G2} - 50)^2 = 350$. Solving the quadratic, we get two solutions. The physically reasonable value is $P_{G1} = P_{G2} = 183.97$ MW. Then $P_L = 17.95$ MW. Note how much higher the losses are compared with Example 7.6.

To calculate the increased cost, we note that the given IC characteristic implies the following fuel-cost curve:

$$C_i(P_{Gi}) = \alpha + 4.1P_{Gi} + 0.0035P_{Gi}^2 \qquad \text{dollars/hr}$$

The constant term α is not specified nor is it needed in calculating the *increased* cost. When $P_{G1} = P_{G2} = 183.97$ MW.

$$C_T = C_1(P_{G1}) = C_2(P_{G2}) = 2\alpha + 1745.47 \text{ dollars/hr}$$

With the optimal dispatch (from Example 7.6), $P_{G1} = 227.71$ MW and $P_{G2} = 117.65$ MW. We get

$$C_T = C_1(P_{G1}) + C_2(P_{G2}) = 2\alpha + 1645.90 \text{ dollars/hr}$$

The increased cost is therefore \$99.57/hr.

Exercise 5.   In Example 7.6, $\partial P_L/\partial P_{G2} = 0.002(P_{G2} - 50)$ can be a positive or negative number depending on whether P_{G2} is greater or less than 50 MW. In this case the penalty factor L_2 can be less than or greater than 1.0. Consider the effect on the optimal dispatch and decide whether the result seems physically reasonable. *Hint:* Note that $P_{D2} = 50$ MW, so $P_{G2} - 50$ represents active power transmitted down the line.

Finally, we consider the physical significance of λ in the present case. In the lossless case we found that λ related increase in cost rate (dollars/hr) to increase in power demand (MW). Is this still true with losses?

Suppose that for a given set of P_{Di}^0, we find the optimal set of P_{Gi}^0 and the corresponding C_T^0. Now increase the total load incrementally from P_D^0 to $P_D = P_D^0 + \Delta P_D$; it does not matter how the increment of increased demand is distributed. We wish to find the new cost C_T. Using a two-term Taylor series as in (7.13), we get the same result:

$$\Delta C_T = \sum_{i=1}^{m} \frac{dC_i(P_{Gi}^0)}{dP_{Gi}} \Delta P_{Gi} \tag{7.25}$$

The increments in generator powers must satisfy a power balance equation,

$$\sum_{i=1}^{m} \Delta P_{Gi} - \Delta P_L = \Delta P_D \tag{7.26}$$

Now we also have, for the new line losses, using a two-term Taylor series,

$$P_L = P_L(P_{G2}^0 + \Delta P_{G2}, P_{G3}^0 + \Delta P_{G3}, \ldots, P_{Gm}^0 + \Delta P_{Gm})$$

$$= P_L^0 + \sum_{i=2}^{m} \frac{\partial P_L(\mathbf{P}_G^0)}{\partial P_{Gi}} \Delta P_{Gi} \tag{7.27}$$

Then

$$\Delta P_L = \sum_{i=2}^{m} \frac{\partial P_L(\mathbf{P}_G^0)}{\partial P_{Gi}} \Delta P_{Gi} \tag{7.28}$$

Using (7.28) in (7.26), we get

$$\Delta P_{G1} + \sum_{i=2}^{m} \left(1 - \frac{\partial P_L(\mathbf{P}_G^0)}{\partial P_{Gi}} \right) \Delta P_{Gi} = \Delta P_D \tag{7.29}$$

We note that the quantities in brackets are the reciprocals of the penalty factors L_i. Using also the definition $L_1 = 1$, we may replace (7.29) by

$$\sum_{i=1}^{m} L_i^{-1} \Delta P_{Gi} = \Delta P_D \tag{7.30}$$

Since we have been assuming optimal dispatch, $L_i \times \text{IC}_i = \lambda$, $i = 1, 2, \ldots, m$. Then from (7.25) and (7.30), we find

$$\Delta C_T = \lambda \sum_{i=1}^{m} L_i^{-1} \Delta P_{Gi} = \lambda \Delta P_D \tag{7.31}$$

The end result is exactly the same as in (7.15). Just as previously, if some generators are at their limits we exclude them and sum only over the generators associated with the common value λ.

7.5 CALCULATION OF PENALTY FACTORS (METHOD 1)

In Example 7.6 we were given an explicit expression for P_L in terms of P_{G2} from which L_2 could be found. Suppose that the expression is not given. How can we find the penalty factors? We turn next to this question.

We will need to consider the transmission network in more detail. For this purpose it is convenient to go back to the per unit system. We can rewrite (7.3) using $P_{Gi} = P_{Gip.u.} \times S_B^{3\phi}$ with $S_B^{3\phi}$ in MW.

$$C_i = \alpha_i + \beta_i S_B^{3\phi} P_{Gip.u.} + \gamma_i (S_B^{3\phi})^2 P_{Gip.u.}^2 \qquad \text{dollars/hr} \tag{7.32}$$

We can replace (7.8) by

$$\text{IC}_i = \frac{dC_i}{dP_{Gi}} = \beta_i + 2\gamma_i (S_B^{3\phi}) P_{Gip.u.} \qquad \text{dollars/MWh} \tag{7.33}$$

Thus the conversion to the p.u. system is very simple. In (7.33) instead of γ_i we have $\gamma_i S_B^{3\phi}$; in (7.32) we have $\beta_i S_B^{3\phi}$ instead of β_i and $\gamma_i (S_B^{3\phi})^2$ instead of γ_i. We end up with the same expressions but with different numerical values for the constants. For the rest of the chapter except in specific instances which will be identified, we will be using the per unit system and it is usually unambiguous and simpler to drop the p.u. notation. Also, it is to be noted that, in a sense, the units do the talking; in dealing with a practical system, if we indicate $P_{Gi} = 1$, it is almost a certainty that it is 1 p.u.

We return now to the problem of calculating penalty factors and consider an example which suggests how they may be found in general.

Example 7.8

Consider a system similar to the one in Example 7.6, but without an explicit formula for P_L in terms of P_{G2} (Fig. E7.8). Assume that the generator ICs are the same as in Example 7.6. Assume that $S_B^{3\phi} = 100$ MW. Then, using (7.33),

$$IC_1 = 4.1 + 0.7P_{G1\text{p.u.}} \quad \text{dollars/MWh} \tag{7.34}$$

$$IC_2 = 4.1 + 0.7P_{G2\text{p.u.}} \quad \text{dollars/MWh} \tag{7.35}$$

To check, note that if $P_{G1} = 100$ MW, then $P_{G1\text{p.u.}} = 1$ and $IC_1 = 4.8$ dollars/MWh, just as in the previous formula. In (7.34) and (7.35) the per unit notation was retained but in the rest of the example we will simplify and drop the p.u. subscript.

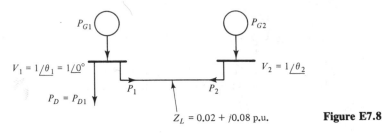

$V_1 = 1\underline{/\theta_1} = 1\underline{/0°}$

$V_2 = 1\underline{/\theta_2}$

$P_D = P_{D1}$

P_1 P_2

$Z_L = 0.02 + j0.08$ p.u.   **Figure E7.8**

Next we calculate $Y_L = Z_L^{-1} = 2.94 - j11.76$. From this the admittance parameters are seen to be

$$g_{11} = g_{22} = 2.94 \qquad g_{12} = g_{21} = -2.94$$
$$b_{11} = b_{22} = -11.76 \qquad b_{12} = b_{21} = 11.76$$

From (6.29) we get

$$P_{G1} - P_{D1} = P_1 = 2.94(1 - \cos\theta_{12}) + 11.76\sin\theta_{12} \tag{7.36}$$

$$P_{G2} = P_2 = 2.94(1 - \cos\theta_{12}) - 11.76\sin\theta_{12} \tag{7.37}$$

By the law of conservation of power, we then get

$$P_L = P_1 + P_2 = 5.88(1 - \cos\theta_{12}) \tag{7.38}$$

Note that this very simple expression for power loss is *explicit* in terms of θ_{12}. Note that P_{G2} is also an explicit function of θ_{12}. This suggests using θ_{12} as an intermediary to avoid the problem of the nonexplicit dependence of P_L on P_{G2}. Following this lead we get, by the chain rule of differentiation,

$$\frac{dP_L}{d\theta_{12}} = \frac{dP_L}{dP_{G2}}\frac{dP_{G2}}{d\theta_{12}} \tag{7.39}$$

We remind the reader that consistent with our formulation, P_L depends only on P_{G2}; P_{G1}, specified by the slack variable P_1, is not an independent variable. Using (7.38), we get

$$\frac{dP_L}{d\theta_{12}} = 5.88 \sin \theta_{12} \tag{7.40}$$

Using (7.37) yields

$$\frac{dP_{G2}}{d\theta_{12}} = \frac{dP_2}{d\theta_{12}} = 2.94 \sin \theta_{12} - 11.76 \cos \theta_{12} \tag{7.41}$$

Inserting (7.40) and (7.41) in (7.39), we get

$$\frac{dP_L}{dP_{G2}} = \frac{5.88 \sin \theta_{12}}{2.94 \sin \theta_{12} - 11.76 \cos \theta_{12}} \tag{7.42}$$

Thus we can calculate the required penalty factor:

$$L_2 = \frac{1}{1 - dP_L/dP_{G2}} = \frac{11.76 \cos \theta_{12} - 2.94 \sin \theta_{12}}{11.76 \cos \theta_{12} + 2.94 \sin \theta_{12}} = \frac{1 - 0.25 \tan \theta_{12}}{1 + 0.25 \tan \theta_{12}} \tag{7.43}$$

Note that L_2 is an explicit function of θ_{12} even though, through (7.37), it is an implicit function of P_{G2}. We next need to choose P_{D1} so that we can proceed numerically. Suppose that we choose $P_{D1} = 3.0$ p.u. (which corresponds to 300 MW). We next wish to illustrate the technique of finding the optimal dispatch.

The basic underlying technique is to pick P_{G2}, and in succession, find θ_{12}, P_{G1}, L_2, and check (7.24). If (7.24) is not satisfied, we pick a better P_{G2} and continue to iterate until convergence is obtained. In fact, it is much simpler to iterate on the choice of θ_{12} illustrated as follows. Initial guess: $\theta_{12} = -5°$. Then we calculate,

$$P_1 = -1.0138 \Rightarrow P_{G1} = 1.9862 \quad \text{from (7.36)}$$

$$P_2 = 1.0361 \Rightarrow P_{G2} = 1.0361 \quad \text{from (7.37)}$$

$$IC_1 = 5.4903 \quad IC_2 = 4.8253 \quad \text{from (7.34) and (7.35)}$$

$$L_1 = 1.0 \quad L_2 = 1.0447 \quad \text{from (7.43)}$$

$$L_1 \times IC_1 = 5.4903 \quad L_2 \times IC_2 = 5.0410$$

We see that (7.24) is not satisfied and P_{G2} should be increased. If P_{G2} is increased, IC_2 and L_2 will increase while P_{G1} will decrease, causing IC_1 to drop. A few steps of the iteration bring us to the value $\theta_{12} = -6.325°$ and

$$P_1 = -1.2777 \Rightarrow P_{G1} = 1.7223$$

$$P_2 = 1.3135 \Rightarrow P_{G2} = 1.3135$$

$$IC_1 = 5.3056 \quad IC_2 = 5.0194$$

$$L_1 = 1.0 \quad L_2 = 1.0570$$

$$L_1 \times IC_1 = 5.3056 \quad L_2 \times IC_2 = 5.3055$$

This satisfies (7.24) close enough and we stop here. Converting to MW values, we get

$$P_{G1} = 172.23 \text{ MW} \quad P_{G2} = 131.35 \text{ MW}$$

$$P_{D1} = 300.00 \text{ MW} \quad P_L = 3.58 \text{ MW}$$

Note: Neglecting line losses, we would have picked $P_{G1} = P_{G2}$.

Exercise 6. From (7.32) we can deduce that C_i in Example 7.8 has the form

$$C_i = \alpha_i + 410 P_{Gip.u.} + 35 P_{Gip.u.}^2$$

Thus we can calculate C_T to within a constant. Convince yourself that the values of P_{G1} and P_{G2} corresponding to $\theta_{12} = -6.325°$ are optimal. (Try other values of θ_{12}, find the corresponding P_{G1} and P_{G2}, and calculate C_T.)

The example offers a clue to a general method for finding the penalty factors. The clue is to use the system θ_i's as the intermediaries. This is bound to be helpful since we can calculate the $\partial P_{Gi}/\partial \theta_k$ explicitly from the power flow equations. We have, first of all,

$$P_L = \sum_{i=1}^{n} P_i = \sum_{i=1}^{m} P_{Gi} - \sum_{i=1}^{n} P_{Di} \qquad (7.16)$$

where the P_i are the bus powers. Using (7.16), we can calculate (using $P_L = \sum_{i=1}^{n} P_i$)

$$\frac{\partial P_L}{\partial \theta_k} = \frac{\partial P_1}{\partial \theta_k} + \cdots + \frac{\partial P_m}{\partial \theta_k} + \frac{\partial P_{m+1}}{\partial \theta_k} + \cdots + \frac{\partial P_n}{\partial \theta_k} \qquad k = 2, 3, \ldots, n \quad (7.44)$$

Since $\theta_1 = $ constant $(= 0°)$, we don't include $k = 1$.

Note that we sum contributions from all n buses, not just the m generator buses. Since we are including load buses (without voltage support), the load voltage magnitudes may vary with θ, and strictly speaking, this dependence should be included in evaluating the $\partial P_i/\partial \theta_k$. For simplicity, however, we will assume the $|V_i|$ are constant at *all* buses. The assumption is not an unreasonable approximation. The reader may recognize this as an application of the active power model discussed in Section 6.8.

Now, with any given $\boldsymbol{\theta}$, with components $\theta_2, \ldots, \theta_n$, we can calculate the $\partial P_i/\partial \theta_k$ *explicitly* from the power flow equations (6.29). Taking account of the relation $P_{Gi} = P_i - P_{Di}$, we note also that with load powers fixed, $\partial P_i/\partial \theta_k = \partial P_{Gi}/\partial \theta_k$ for $i = 1, 2, \ldots, m$.

Equation (7.44) gives one expression for $\partial P_L/\partial \theta_k$; we next derive an alternative expression using (7.17) which involves quantities we need in calculating the penalty factors. Thus, using the chain rule differentiation,

$$\frac{\partial P_L}{\partial \theta_k} = \frac{\partial P_L(\mathbf{P}_G)}{\partial P_{G2}} \frac{\partial P_2}{\partial \theta_k} + \cdots + \frac{\partial P_L(\mathbf{P}_G)}{\partial P_{Gm}} \frac{\partial P_m}{\partial \theta_k} \qquad k = 2, 3, \ldots, n \quad (7.45)$$

In (7.45) we made use of the fact that at generator buses $\partial P_{Gi}/\partial \theta_k = \partial P_i/\partial \theta_k$. We also used the notation $P_L(\mathbf{P}_G)$ to emphasize that the partial derivatives were of the function P_L in (7.17). Since the function P_L does not include P_{G1} among its arguments, $\partial P_L/\partial P_{G1} = 0$ and is not included in (7.45).

We can proceed as follows. Subtract (7.45) from (7.44) to get

$$\frac{\partial P_1}{\partial \theta_k} + \frac{\partial P_2}{\partial \theta_k}\left(1 - \frac{\partial P_L}{\partial P_{G2}}\right) + \cdots + \frac{\partial P_m}{\partial \theta_k}\left(1 - \frac{\partial P_L}{\partial P_{Gm}}\right) + \frac{\partial P_{m+1}}{\partial \theta_k} + \cdots + \frac{\partial P_n}{\partial \theta_k} = 0$$

$$k = 2, 3, \ldots, n \qquad (7.46)$$

In matrix form we can replace (7.46) by

$$
\begin{bmatrix}
\dfrac{\partial P_2}{\partial \theta_2} \cdots \dfrac{\partial P_m}{\partial \theta_2} \cdots \dfrac{\partial P_n}{\partial \theta_2} \\
 \\
\cdot \qquad \cdot \qquad \cdot \\
\cdot \qquad \cdot \qquad \cdot \\
\cdot \qquad \cdot \qquad \cdot \\
\dfrac{\partial P_2}{\partial \theta_n} \cdots \dfrac{\partial P_m}{\partial \theta_n} \cdots \dfrac{\partial P_n}{\partial \theta_n}
\end{bmatrix}
\begin{bmatrix}
1 - \dfrac{\partial P_L}{\partial P_{G2}} \\
\cdot \\
\cdot \\
\cdot \\
1 - \dfrac{\partial P_L}{\partial P_{Gm}} \\
1 \\
\cdot \\
\cdot \\
\cdot \\
1
\end{bmatrix}
= -
\begin{bmatrix}
\dfrac{\partial P_1}{\partial \theta_2} \\
\cdot \\
\cdot \\
\cdot \\
\dfrac{\partial P_1}{\partial \theta_n}
\end{bmatrix}
\qquad (7.47)
$$

Since the terms in the matrix in (7.47) relate the bus powers P_i to the angles θ_k, it is not surprising that the matrix is just the transpose of $\mathbf{J}_{11}$ found in Section 6.6. Some typical elements of $\mathbf{J}_{11}$ are given in (6.39) and the top row of (6.40).

Since $\mathbf{J}_{11}$ is used in a similar manner in the iterative solution of power flow equations by the Newton–Raphson method, computer programs for the solution of (7.47) are readily available. We can then obtain the $m - 1$ unknown values of $1 - \partial P_L / \partial P_{Gi}$.

An expression for $\partial P_1 / \partial \theta_i$ on the right side of (7.47) is needed and may easily be found from (6.29). It is

$$\frac{\partial P_1}{\partial \theta_i} = |V_1||V_i|\left[g_{1i} \sin(\theta_1 - \theta_i) - b_{1i} \cos(\theta_1 - \theta_i)\right] \qquad i = 2, 3, \ldots, n$$

$$(7.48)$$

We next consider in outline the solution for the optimal dispatch. For simplicity assume that none of the generators are at their limits (i.e., all the committed generators are available for adjustment). Figure 7.8 is a flowchart that describes the solution, where:

1. The single lines represent scalars. The double lines represent vectors with components over the index set $i = 2, 3, \ldots, n$, except in the case of the vectors involving the P_{Gi} and L_i, which have components over the index set $i = 2, 3, \ldots, m$.

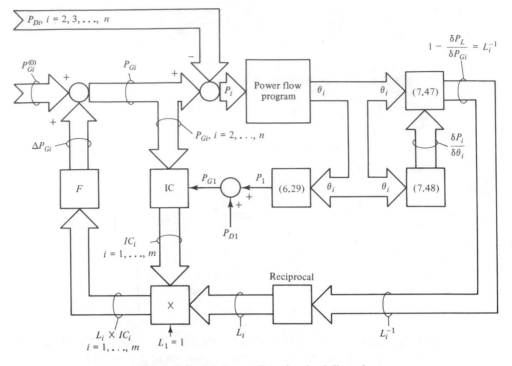

Figure 7.8   Solution outline of optimal dispatch.

2. F generates corrections to P_{Gi} in the event that (7.24) is not satisfied. If $L_i \times IC_i$ is too low (high) compared with $L_1 \times IC_1$, the corresponding P_{Gi} should be raised (lowered).

3. The equation numbers in the blocks describe the operation.

The steps in the solution are the following.

1. Starting with a set of P_{Di}, $i = 1, 2, \ldots, n$ and an initial set of $P_{Gi}^{(0)}$, $i = 2$, $3, \ldots, m$, we calculate $P_i = P_{Gi} - P_{Di}$, $i = 2, 3, \ldots, n$, and enter the power flow program and compute θ_i, $i = 2, 3, \ldots, n$. Since we are using the active power model (all the $|V_i|$ assumed known), it is appropriate to use (6.47a).

2. We can now calculate the slack bus power P_1 using (6.29) and then $P_{G1} = P_1 + P_{D1}$.

3. Using (7.48), we also calculate $\partial P_1 / \partial \theta_i$, $i = 2, 3, \ldots, m$, and with $\mathbf{J}_{11}(\theta)$ in (7.47) solve for $1 - \partial P_L / \partial P_{Gi}$, $i = 2, 3, \ldots, m$. Taking reciprocals, we have the L_i, $i = 2, 3, \ldots, m$.

4. Since $L_1 = 1$, we now have all the L_i. We also have all the P_{Gi}, so we can calculate the IC_i. Multiplying, we get the products $L_i \times IC_i$. If they are all equal, we have the optimal dispatch. If not, we need to change the P_{Gi}, $i = 2, 3, \ldots$,

m, and iterate. If a particular product is too low (high) compared with $L_1 \times IC_1$, the corresponding P_{Gi} should be raised (lowered).

This outline concludes the various solution steps. Although the outline does not constitute an algorithm, it is perhaps easier to follow because it is divorced from detailed computational considerations.

It is important to note that in Fig. 7.8 the basic input is the set of P_{Di}, $i = 1$, 2, . . . , n, while the output is the set of optimal P_{Gi}, $i = 1, 2, \ldots, m$. In practice, since the load varies, it is desirable to recalculate the P_{Gi}. This is done "on-line" at 3- to 5-minute intervals at least during times of variable load.

We consider next a very simple example which illustrates the calculation outlined in Fig. 7.8.

Example 7.9

Consider the system shown in Fig. E7.9. Assume that

$$IC_i = 4.1 + 0.1P_{Gi} \qquad i = 1, 2$$

$$Z_{L1} = 0.0099 + j0.099$$

$$Z_{L2} = 0.0198 + j0.198$$

For simplicity we have picked values such that $Z_{L2} = 2Z_{L1}$. Although the generators are identical, line 2 has more resistance than line 1 and we expect more of the load will be carried by unit 1. We next find

$$Y_{L1} = Z_{L1}^{-1} = 1 - j10$$

$$Y_{L2} = Z_{L2}^{-1} = 0.5 - j5.0$$

and assemble $\mathbf{Y}_{bus}$.

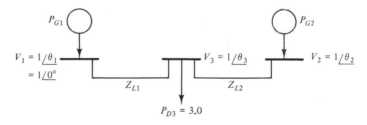

$$P_{D3} = 3.0$$

Figure E7.9

$$\mathbf{Y}_{bus} = \begin{bmatrix} 1 - j10 & 0 & -1 + j10 \\ 0 & 0.5 - j5.0 & -0.5 + j5.0 \\ -1 + j10 & -0.5 + j5.0 & 1.5 - j15 \end{bmatrix}$$

From $\mathbf{Y}_{bus}$ we get the values of $y_{ik} = g_{ik} + jb_{ik}$. From (6.31) we get

$$P_1(\boldsymbol{\theta}) = 1 - \cos \theta_{13} + 10 \sin \theta_{13} \tag{7.49}$$

$$P_2(\boldsymbol{\theta}) = 0.5 - 0.5 \cos \theta_{23} + 5 \sin \theta_{23} \tag{7.50}$$

$$P_3(\boldsymbol{\theta}) = 1.5 - \cos \theta_{31} + 10 \sin \theta_{31} - 0.5 \cos \theta_{32} + 5 \sin \theta_{32} \qquad (7.51)$$

We can also find

$$P_L = \sum_{i=1}^{3} P_i(\boldsymbol{\theta}) = 3 - 2 \cos \theta_{13} - \cos \theta_{23} \qquad (7.52)$$

and note as a check that with $\boldsymbol{\theta} = \mathbf{0}$, $P_L = 0$, as expected.

To find $\mathbf{J}_{11}$ we take the required partial derivatives $\partial P_i / \partial \theta_j$, $i, j = 2, 3$.

$$\frac{\partial P_2}{\partial \theta_2} = 0.5 \sin \theta_{23} + 5 \cos \theta_{23}$$

$$\frac{\partial P_2}{\partial \theta_3} = -0.5 \sin \theta_{23} - 5 \cos \theta_{23}$$

$$\frac{\partial P_3}{\partial \theta_2} = -0.5 \sin \theta_{32} - 5 \cos \theta_{32}$$

$$\frac{\partial P_3}{\partial \theta_3} = \sin \theta_{31} + 10 \cos \theta_{31} + 0.5 \sin \theta_{32} + 5 \cos \theta_{32} \qquad (7.53)$$

Using (7.48), we calculate

$$\frac{\partial P_1}{\partial \theta_2} = 0$$

$$\frac{\partial P_1}{\partial \theta_3} = -\sin \theta_{13} - 10 \cos \theta_{13} \qquad (7.54)$$

In this case (7.47) reduces to the form

$$
\begin{bmatrix} \dfrac{\partial P_2}{\partial \theta_2} & \dfrac{\partial P_3}{\partial \theta_2} \\[2mm] \dfrac{\partial P_2}{\partial \theta_3} & \dfrac{\partial P_3}{\partial \theta_3} \end{bmatrix}
\begin{bmatrix} 1 - \dfrac{\partial P_L}{\partial P_{G2}} \\[2mm] 1 \end{bmatrix}
= - \begin{bmatrix} 0 \\[2mm] \dfrac{\partial P_1}{\partial \theta_3} \end{bmatrix}
$$

and it is easy to solve for $1 - \partial P_L / \partial P_{G2}$ by using Cramer's rule. We get

$$
L_2^{-1} = 1 - \frac{\partial P_L}{\partial P_{G2}} = \frac{\begin{bmatrix} 0 & \dfrac{\partial P_3}{\partial \theta_2} \\[2mm] -\dfrac{\partial P_1}{\partial \theta_3} & \dfrac{\partial P_3}{\partial \theta_3} \end{bmatrix}}{\begin{bmatrix} \dfrac{\partial P_2}{\partial \theta_2} & \dfrac{\partial P_3}{\partial \theta_2} \\[2mm] \dfrac{\partial P_2}{\partial \theta_3} & \dfrac{\partial P_3}{\partial \theta_3} \end{bmatrix}}
$$

Substituting from (7.53) and (7.54), we get

$$L_2^{-1} = \frac{10 \cos \theta_{13} + \sin \theta_{13}}{10 \cos \theta_{13} - \sin \theta_{13}} \frac{10 \cos \theta_{23} - \sin \theta_{23}}{10 \cos \theta_{23} + \sin \theta_{23}}$$

Thus

$$L_2 = \frac{1 - 0.1 \tan \theta_{13}}{1 + 0.1 \tan \theta_{13}} \frac{1 + 0.1 \tan \theta_{23}}{1 - 0.1 \tan \theta_{23}} \tag{7.55}$$

which is a very nice closed-form solution in this particular example.

We next need to calculate the θ_{ij} corresponding to a particular P_{G2}. For the initial choice we pick $P_{G2} = 1.5$. Then $P_2 = 1.5$. We now do a simple power flow calculation. From (7.50) we solve (iteratively) for $\theta_{23} = 17.19°$. From (7.51), with $P_3 = -P_{D3} = -3.0$, we solve for $\theta_{13} = -\theta_{31} = 8.956°$. From (7.49) we find $P_1 = P_{G1} = 1.569$. Now we find

$$IC_1 = 5.198$$

$$IC_2 = 5.150$$

Using the angles $\theta_{13} = 8.956°$ and $\theta_{23} = 17.19°$ in (7.55),

$$L_2 = 1.0309$$

L_1 is always 1, so

$$L_1 \times IC_1 = 5.198$$

$$L_2 \times IC_2 = 5.309$$

The second product is too large relative to the first, so we need to decrease P_{G2}.

Pick $P_{G2} = 1.4$ and repeat the iteration. We find successively $\theta_{23} = 16.028°$, $\theta_{13} = 9.513°$, $P_{G1} = 1.667$, $IC_1 = 5.267$, $IC_2 = 5.080$, and $L_2 = 1.0242$. Then

$$L_1 \times IC_1 = 5.269$$

$$L_2 \times IC_2 = 5.203$$

We need to bring up the second product. In this manner we can continue the iterations and soon reach the values $P_{G2} = 1.4365$, $\theta_{23} = 16.451°$, $\theta_{13} = 9.309°$, $P_{G1} = 1.6308$, $IC_1 = 5.2415$, $IC_2 = 5.1056$, and $L_2 = 1.0266$. Then

$$L_1 \times IC_1 = 5.2415$$

$$L_2 \times IC_2 = 5.2416$$

This is certainly close enough, so we stop here. In fact, in practice we would have stopped before this point because the accuracy of the source data and the intended application do not warrant the precision in the calculation. As a check of the calculation we can calculate the variable part of the total cost C_T and verify that the last iteration gives the smallest C_T among the various trials.

We turn next to an alternative method for calculating the penalty factors.

7.6 CALCULATION OF PENALTY FACTORS (METHOD 2)

A well-established method in common use by the power utility industry is based on the use of an approximate expression for the line losses in terms of generator bus powers. The usual approximation is that for a given operating condition (or base case), the transmission system losses are quadratic in the bus powers at the generator buses as follows:

$$P_L = \sum_{i=1}^{m} \sum_{j=1}^{m} B_{ij} P_i P_j \tag{7.56}$$

Here the B_{ij} are constants called the B coefficients or loss coefficients. Without loss of generality it may be assumed that $B_{ij} = B_{ji}$. With a significantly different operating condition a different set of B coefficients would be used. Note that the summation is only over the generator buses. In some loss formulas additional constant and linear terms are added.

The advantage of an explicit loss formula like (7.56) is that the calculation of $\partial P_L / \partial P_i$ is so very simple. It is easy to show that

$$\frac{\partial P_L}{\partial P_i} = 2 \sum_{j=1}^{m} B_{ij} P_j \qquad i = 1, 2, \ldots, m \tag{7.57}$$

Since by the chain rule of differentiation,

$$\frac{\partial P_L}{\partial P_{Gi}} = \frac{\partial P_L}{\partial P_i} \frac{\partial P_i}{\partial P_{Gi}} = \frac{\partial P_L}{\partial P_i}$$

the penalty factors may be found immediately from (7.57).

We turn next to an interpretation of the B_{ij} and a possible way to determine them experimentally. If in (7.57) we take the derivative of $\partial P_L / \partial P_i$ with respect to P_j, we get a second (mixed) partial derivative and find that

$$B_{ij} = \frac{1}{2} \frac{\partial^2 P_L}{\partial P_i \partial P_j} \qquad i, j = 1, 2, \ldots, m \tag{7.58}$$

Thus B_{ij} quantifies the dependence of $\partial P_L / \partial P_i$ on P_j, with all other P_k held constant.

Equation (7.58) suggests a way to find the B_{ij} needed for use in (7.56). We can think of the right sides of (7.58) being found by experiments on the actual power system, or more reasonably, by using the power flow equations, which appropriately model the power system.

In such a study it is found that the second (mixed) partial derivatives depend on the particular operating point at which the derivatives are evaluated. Thus the B_{ij} evaluated by (7.58) will depend on operating point. Fortunately, the dependence is not critical.

It should be noted that P_1 appears in the line-loss formula (7.56) (i.e., P_1 is treated as an independent variable). This formulation has the advantage of complete symmetry in the treatment of the different generator outputs. Implied in the formulation is the notion that bus 1 is no longer the slack bus. One consequence of this change is that L_1 is no longer necessarily equal to 1. Despite the change, it is easy to modify the proof and check that (7.24) remains the condition for optimal dispatch. In general, we can expect somewhat different generator dispatches using the B coefficients rather than the more accurate (but more involved) calculations described in Section 7.5.

Next we consider an example similar to Example 7.9 which we will solve using B coefficients.

Example 7.10

For the system in Example 7.9:

(a) Find B_{11}, B_{12}, and B_{22}.
(b) Determine the optimal dispatch.

Solution   (a) The bus powers and line losses are described by (7.49) to (7.52). Using these equations and (7.58), we can find the B_{ij}. From the form of the equations and/or the physical arrangement of lines, it is apparent that $B_{12} = 0$. However, we can check this result by direct calculation.

$$B_{12} = \frac{1}{2} \frac{\partial^2 P_L}{\partial P_1 \, \partial P_2} = \frac{1}{2} \frac{\partial}{\partial P_1} \left(\frac{\partial P_L}{\partial P_2} \right) = \frac{1}{2} \frac{\partial f_{L2}}{\partial P_1}$$

where $f_{L2} \triangleq \partial P_L / \partial P_2$. From (7.50) and (7.52) we can see that P_L depends on P_2 through θ_{23}. Thus using the chain rule again,

$$\frac{\partial P_L}{\partial \theta_{23}} = \frac{\partial P_L}{\partial P_2} \frac{\partial P_2}{\partial \theta_{23}}$$

Using (7.52) and (7.50), we get

$$\frac{\partial P_L}{\partial \theta_{23}} = \sin \theta_{23}$$

$$\frac{\partial P_2}{\partial \theta_{23}} = 0.5 \sin \theta_{23} + 5 \cos \theta_{23}$$

Thus

$$f_{L2} \triangleq \frac{\partial P_L}{\partial P_2} = \frac{2 \sin \theta_{23}}{\sin \theta_{23} + 10 \cos \theta_{23}}$$

We can now evaluate $B_{12} = \frac{1}{2}(\partial f_{L2} / \partial P_1)$ by using the chain rule again.

$$\frac{\partial f_{L2}}{\partial \theta_{13}} = \frac{\partial f_{L2}}{\partial P_1} \frac{\partial P_1}{\partial \theta_{13}} = 2B_{12} \frac{\partial P_1}{\partial \theta_{13}}$$

But f_{L2} is not a function of θ_{13}, so $\partial f_{L2} / \partial \theta_{13} = 0$, which implies that $B_{12} = \frac{1}{2}(\partial f_{L2} / \partial P_1) = 0$.

Turning next to the calculation of B_{11}, we find that

$$B_{11} = \frac{1}{2} \frac{\partial^2 P_L}{\partial P_1^2} = \frac{1}{2} \frac{\partial}{\partial P_1} \left(\frac{\partial P_L}{\partial P_1} \right) = \frac{1}{2} \frac{\partial f_{L1}}{\partial P_1}$$

where $f_{L1} \triangleq \partial P_L / \partial P_1$. In evaluating $f_{L1} = \partial P_L / \partial P_1$ we use

$$\frac{\partial P_L}{\partial \theta_{13}} = \frac{\partial P_L}{\partial P_1} \frac{\partial P_1}{\partial \theta_{13}}$$

Using (7.52) and (7.49), we find that

$$\frac{\partial P_L}{\partial \theta_{13}} = 2 \sin \theta_{13}$$

$$\frac{\partial P_1}{\partial \theta_{13}} = \sin \theta_{13} + 10 \cos \theta_{13}$$

Thus

$$f_{L1} \triangleq \frac{\partial P_L}{\partial P_1} = \frac{2 \sin \theta_{13}}{\sin \theta_{13} + 10 \cos \theta_{13}}$$

Continuing the calculation, we find that

$$\frac{\partial f_{L1}}{\partial \theta_{13}} = \frac{\partial f_{L1}}{\partial P_1} \frac{\partial P_1}{\partial \theta_{13}} = 2B_{11} \frac{\partial P_1}{\partial \theta_3}$$

Since we have an expression for f_{L1}, we can find $\partial f_{L1}/\partial \theta_{13}$; we have already found $\partial P_1/\partial \theta_{13}$. Thus

$$B_{11} = \frac{1}{2} \frac{\partial^2 P_L}{\partial P_1^2} = \frac{10}{(\sin \theta_{13} + 10 \cos \theta_{13})^3} \tag{7.59}$$

Similarly, we find that

$$B_{22} = \frac{1}{2} \frac{\partial^2 P_L}{\partial P_2^2} = \frac{20}{(\sin \theta_{23} + 10 \cos \theta_{23})^3} \tag{7.60}$$

The dependence of B_{11} and B_{22} on θ_{13} and θ_{23}, respectively, may be replaced by the dependence on P_1 and P_2. We simply pick a value of θ_{13} and calculate corresponding values of P_i [from (7.49) or (7.50)] and B_{ii}. These values are plotted in Fig. E7.10. It may be seen that the curves are reasonably constant for values of P_i between about -1 and 2, corresponding to θ_{ij} between about -10 to $+20°$; this is the range of interest in this

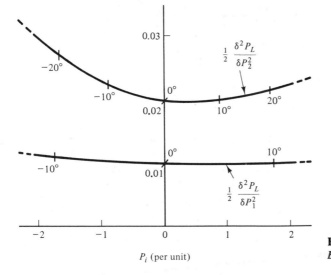

P_i (per unit)

Figure E7.10 Calculation of B_{11} and B_{22}.

problem. As an approximation we can set B_{ii} to constant values corresponding to $P_i = 0$ and get $B_{11} = 0.01$ and $B_{22} = 0.02$. The loss formula is then

$$P_L = 0.01P_1^2 + 0.02P_2^2 \qquad (7.61)$$

Note: In (7.61) we depart from our previous practice and write P_L in terms of P_i rather than P_{Gi}. In the present example $P_i = P_{Gi}$, but even if not, it is logical since line losses depend on the P_i. Of course, if desired, we can make the substitution $P_i = P_{Gi} - P_{Di}$ and obtain the usual expression for P_L in terms of P_{Gi}.

(b) We next find the optimal dispatch. In the formula for the penalty factors we have the expression $\partial P_L/\partial P_{Gi}$, but since $\partial P_i/\partial P_{Gi} = 1$ we can substitute $\partial P_L/\partial P_i$ instead, even if $P_i \neq P_{Gi}$. From (7.61) we get

$$L_1 = \frac{1}{1 - 0.02P_1} \qquad L_2 = \frac{1}{1 - 0.04P_2}$$

Thus, to satisfy (7.24),

$$\frac{4.1 + 0.7P_1}{1 - 0.02P_1} = \frac{4.1 + 0.7P_2}{1 - 0.04P_2} = \lambda \qquad (7.62)$$

subject to the power balance constraint

$$P_1 + P_2 - (0.01P_1^2 + 0.02P_2^2) = 3.0 \qquad (7.63)$$

A good solution technique that extends beyond the two-dimensional case is to pick λ, solve for P_1 and P_2 in (7.62), insert the values in (7.63), and iterate on λ until convergence is obtained. Of course, it is convenient to rearrange (7.62) to express the P_i explicitly in terms of λ. The result of the iteration is

$$P_{G1} = P_1 = 1.6303$$

$$P_{G2} = P_2 = 1.4376$$

These values fall within the range of applicability of the loss formula (7.61). The results should be compared with the results of Example 7.9, where the optimal solution was found to be

$$P_{G1} = 1.6308 \qquad P_{G2} = 1.4365$$

From a practical standpoint the two sets of results are indistinguishable. We can also compare the values of P_L. By use of (7.61) we get $P_L = 0.0679$; by use of (7.52) we get $P_L = 0.0673$.

The reader may wish to consider why we were able to dispense with the slack bus in this example. The reason was the availability of an explicit expression for line losses for use in (7.63), the power balance constraint.

Finally, we note that the example illustrates what has been observed in practice—that very good results may be obtained by properly using the B coefficients. We note that the method used in the example for finding the B coefficients can be extended to the general case and is one of the methods used by the industry.

We conclude this section with another example.

Example 7.11

Consider the power system shown in Fig. E7.11. Assume that

$$B_{11} = 0.04$$

$$B_{12} = 0.02 = B_{21}$$

$$B_{22} = 0.03$$

$$IC_1 = 2.0 + P_{G1}$$

$$IC_2 = 2.0 + 2P_{G2}$$

(a) Find the optimal dispatch neglecting losses.
(b) Find the optimal dispatch considering losses.

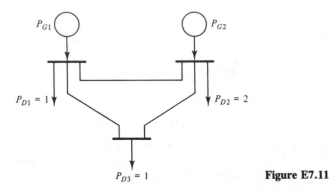

P_{G1} P_{G2}

$P_{D1} = 1$ $P_{D2} = 2$

$P_{D3} = 1$ **Figure E7.11**

Solution   (a) Without losses, $IC_1 = IC_2$. The power balance equation gives
$P_{G1} + P_{G2} = P_{D1} + P_{D2} + P_{D3} = 4$. Thus

$$2.0 + P_{G1} = 2.0 + 2(4 - P_{G1})$$

and optimal dispatch is

$$P_{G1} = 2.667 \qquad P_{G2} = 1.333$$

Generator 1 carries more load than generator 2 because of its more favorable IC characteristic.

(b) We now consider losses using the specified B coefficients in (7.56). We get

$$P_L = 0.04P_1^2 + 2(0.02)P_1P_2 + 0.03P_2^2$$

Thus the penalty factors (in terms of the P_i) are

$$L_1 = \frac{1}{1 - 2(0.04P_1 + 0.02P_2)}$$

$$L_2 = \frac{1}{1 - 2(0.02P_1 + 0.03P_2)}$$

As in Example 7.10, we must satisfy (7.24). Expressing the IC_i in terms of the P_i, we get

$$\frac{2.0 + (P_1 + 1)}{1 - 0.08P_1 - 0.04P_2} = \frac{2.0 + 2(P_2 + 2)}{1 - 0.04P_1 - 0.06P_2} = \lambda \qquad (7.64)$$

subject to the power balance constraint

$$P_1 + P_2 - (0.04P_1^2 + 0.04P_1P_2 + 0.03P_2^2) = P_{D3} = 1 \qquad (7.65)$$

In (7.64) we used $P_{G1} = P_1 + 1$ and $P_{G2} = P_2 + 2$. In this way (7.64) and (7.65) are both expressed in terms of bus powers.

The calculation is more complicated than in Example 7.10 because of the presence of a nonzero B_{12} term, but the basic solution technique is the same. We pick λ, solve (7.64) for P_1 and P_2, insert the values in (7.65), and iterate on λ until convergence is obtained. In this connection note that for each fixed λ we may easily obtain two linear algebraic equations in P_1 and P_2 by clearing the terms in the denominator of (7.64).

As an initial value for λ we can choose the λ corresponding to the solution of part (a). This is

$$\lambda = 2.0 + P_{G1} = 2.0 + 2.667 = 4.667$$

Inserting this value in (7.64) and clearing the terms in the denominator, we get

$$1.3734P_1 + 0.1867P_2 = 1.667$$

$$0.1867P_1 + 2.2800P_2 = -1.333$$

Solving we get $P_1 = 1.30078$ and $P_2 = -0.6917$. Using these values in (7.65), we get

$$P_1 + P_2 - (0.04P_1^2 + 0.04P_1P_2 + 0.03P_2^2) = 0.57$$

This value is too low; to satisfy the power balance constraint we need to increase the left side of (7.65). Since P_L is small compared with $P_1 + P_2$, our objective should be to increase P_1 and P_2, and this can be accomplished by increasing λ. Thus in the next iteration we try $\lambda = 5.0$.

Proceeding in this way, after a few iterations we find $\lambda = 5.14$. With this value we get $P_1 = 1.5914$ and $P_2 = -0.5143$, and the left side of (7.65) equals 1.0006, which is close enough to 1. Thus the end result is

$$P_{G1} = P_1 + 1 = 2.591$$

$$P_{G2} = P_2 + 2 = 1.486$$

Compared with the result in part (a), the lossless case, we see a shift of generation to generator 2. The physical reason is that the transmission losses are reduced by a more local source of power for the relatively large P_{D2}.

7.7 SUMMARY

The important problem of minimizing fuel costs in supplying a given load is called the economic dispatch problem. We assume that fuel-cost curves are available for all the (committed) generating units. In the classic economic dispatch problem, neglecting losses in the transmission links, we find a simple optimal dispatch rule: Operate all generators that are not at their limits at equal incremental cost λ.

With transmission-line (link) losses considered, the rule is modified as follows: Operate all generators that are not at their limits such that for each generator the product of penalty factor and incremental cost is λ. In calculating the penalty factors we may use an explicit loss formula which is quadratic in the bus powers (at the generator buses) and involves the B coefficients. Alternatively, we may calculate the penalty factors (and the optimal dispatch) by using iterative techniques as described in Section 7.5.

In this chapter we have discussed the importance of controlling generator output to reduce fuel costs. In the next chapter we study the means by which this control is exercised.

PROBLEMS

7.1. The fuel input rate of a gas-fired generating unit is approximated as follows:

$$\text{fuel input rate} = 175 + 8.7P_G + 0.0022P_G^2 \qquad \text{MBtu/hr}$$

where P_G is the 3ϕ power output in MW. If the cost of gas is \$5 per MBtu, find:
   (a) The fuel input rate when the unit is delivering 100 MW.
   (b) The corresponding fuel cost in dollars/hr.
   (c) The incremental cost in dollars/MWh.
   (d) The approximate cost in dollars/hr to deliver 101 MW.

7.2. Assume that we have the following fuel-cost curves for three generating units:

$$C_1(P_{G1}) = 300 + 8.0P_{G1} + 0.0015P_{G1}^2$$

$$C_2(P_{G2}) = 450 + 8.0P_{G2} + 0.0005P_{G2}^2$$

$$C_3(P_{G3}) = 700 + 7.5P_{G3} + 0.0010P_{G3}^2$$

Neglecting line losses and generator limits, find the optimal dispatch and the total cost in dollars/hr when the total load, P_D, is **(a)** 500 MW, **(b)** 1000 MW, and **(c)** 2000 MW.

7.3. Suppose that instead of operating optimally, the three generators in Problem 7.2 share the load equally. Find the additional cost per hour in cases (a), (b), and (c).

7.4. Repeat Problem 7.2, but this time introduce the following generator limits (in MW):

$$50 \le P_{G1} \le 400$$

$$50 \le P_{G2} \le 800$$

$$50 \le P_{G3} \le 1000$$

7.5. The three generating units in Problem 7.2 (subject to the generator constraints in Problem 7.4) are connected to the same plant bus. Plot the graph of plant incremental cost versus total load P_D when the generators are sharing the load optimally. Plot the graph for $150 \le P_D \le 2200$.

7.6. For the system in Problem 7.4 suppose that $P_D = 1800$ MW and calculate the approximate additional cost per hour of supplying one additional megawatt (i.e., 1801 MW).

7.7. Two generating units supply a system.

$$G_1: \quad IC_1 = 0.012P_{G1} + 8.0 \quad \text{dollars/MWh}$$
$$G_2: \quad IC_2 = 0.018P_{G2} + 7.0 \quad \text{dollars/MWh}$$

$$G_1: \quad 100 \text{ MW} \leq P_{G1} \leq 650 \text{ MW}$$
$$G_2: \quad 50 \text{ MW} \leq P_{G2} \leq 500 \text{ MW}$$

(a) Find the system λ for optimal operation when $P_{G1} + P_{G2} = P_D = 600$ MW. Find P_{G1} and P_{G2}.

(b) Suppose that P_D increases by 1 MW (to 601 MW). Find the extra cost in dollars/hr.

7.8. Suppose that there are n generating units on economic dispatch supplying a total load P_D. The incremental cost of each generating unit is $IC_i = \beta_i + 2\gamma_i P_{Gi}$. Neglecting line losses and generator limits, show that the relationship between system λ and P_D is linear, in fact,

$$\lambda = \left(P_D + \sum_{i=1}^{n} \frac{\beta_i}{2\gamma_i} \right) \bigg/ \sum_{i=1}^{n} \frac{1}{2\gamma_i}$$

Hint: Operating on economic dispatch, $\beta_i + 2\gamma_i P_{Gi} = \lambda$, for $i = 1, 2, \ldots, n$. Solve for P_{Gi}, $i = 1, 2, \ldots, n$ and add to get P_D in terms of λ and the β_i and γ_i parameters.

7.9. Jack and Jill are dispatchers at the Ideal Power Company, which has lossless lines and three generating units. Jack claims that the system is on economic dispatch. In the next few minutes Jill observes the following sets of data representing incremental changes in generator MW outputs and total production cost rates:

ΔC_T(dollars/hr)	ΔP_{G1}	ΔP_{G2}	ΔP_{G3}
0	1	1	-2
30	1	1	1
-20	-3	1	1

After the third reading she says, "Jack you're wrong." How does she know?

7.10. Consider a short transmission line with (per phase) series impedance $Z = R + jX$. Show that if $|V_1| = |V_2|$, $R \ll X$, and θ_{12} is small (so that $1 - \cos \theta_{12} \approx \theta_{12}^2/2$, and $\sin \theta_{12} \approx \theta_{12}$), then, as an approximation, the line losses vary quadratically with the transmitted (active) power. In fact,

$$P_L = \frac{R}{|V_1|^2} P_{12}^2$$

7.11. Repeat Example 7.6 but assume that $IC_i = 0.002P_{Gi} + 7.0$, $P_L = 0.0008(P_{G2} - 100)^2$, $P_{D1} = 300$ MW, and $P_{D2} = 100$ MW. Suppose that P_D increases from 400 MW to 401 MW. Find the increased cost per hour to supply the additional 1 MW.

7.12. Suppose that in Problem 7.11 line losses were ignored in calculating the optimal dispatch. In this case we would pick $P_{G1} = P_{G2}$. With this assignment, calculate P_L and the increased cost of supplying the total load power $P_D = 400$ MW.

7.13. In Example 7.8 suppose that $IC_i = 7.0 + 0.02P_{Gi}$. Do a few iterations to estimate the optimal P_{G1}, P_{G2}, and the resulting P_L.

7.14. Suppose that in Example 7.11, $IC_1 = 7.0 + 0.5P_{G1}$, $IC_2 = 8.0 + 0.4P_{G2}$.

(a) Find P_{G1} and P_{G2} (in p.u.) if line losses are ignored.

(b) Find P_{G1}, P_{G2}, and P_L (in p.u.) taking line losses into account.

7.15. The numerical data in Example 7.11 are appropriate for powers expressed in per unit. Suppose that $S_B^{3\phi}$ is 200 MW. Convert the given data so that the problem can be phrased in terms of actual (active) powers in MW.

7.16. Consider the system shown in Fig. P7.16, given that

$$IC_1 = 0.007 P_{G1} + 4.1 \text{ dollars/MWh}$$

$$IC_2 = 0.007 P_{G2} + 4.1 \text{ dollars/MWh}$$

$$B_{11} = 0.1 \times 10^{-2} \text{ MW}^{-1}$$

$$B_{12} = B_{21} = -0.005 \times 10^{-2} \text{ MW}^{-1}$$

$$B_{22} = 0.13 \times 10^{-2} \text{ MW}^{-1}$$

(a) Find the optimal dispatch P_{G1} and P_{G2} in MW.

(b) Suppose that with optimal dispatch, $P_D = P_{D1} + P_{D2}$ is increased by 1 MW. Find the additional cost per hour.

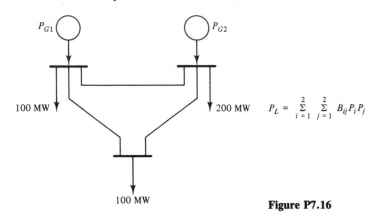

$$P_L = \sum_{i=1}^{2} \sum_{j=1}^{2} B_{ij} P_i P_j$$

Figure P7.16

Generator Modeling I (Machine Viewpoint)

chapter 8

8.0 INTRODUCTION

We now wish to turn our attention to the generators which supply complex power to the loads through the transmission system. We will consider the means by which the active and reactive power supplied to each generator bus may be adjusted and gain an appreciation for the limitations on the available complex power. For this purpose it is appropriate to consider a steady-state model of the generator.

In this chapter we derive a steady-state generator model based on consideration of the rotating fields in the generator due to field and stator currents. This emphasis on the fields is the heart of the classical "machine" viewpoint and is very successful in giving clear physical descriptions of the basic voltage generation mechanism, of "armature reaction," and of the physical significance of "direct-axis" and "quadrature-axis" reactances.

The machine viewpoint is not quite as clear when transients or nonpositive-sequence operating conditions are considered. In Chapter 10, therefore, when we consider these more general cases, we will adopt a "circuit" point of view as an alternative. This provides an easier development of a more general generator model.

We limit our discussion to ac generators (synchronous generators, alternators) driven by steam, hydro, or gas turbines or possibly diesel engines. At this time these are the only significant sources of electrical energy supplied to power systems.

It should be noted that the theory of synchronous generators and synchronous motors is the same. For this reason we sometimes use the terminology "synchronous machine" to cover both motors and generators.

8.1 CLASSICAL MACHINE DESCRIPTION

The single phase ac generator described in elementary physics textbooks consists of a coil (loop of wire) rotating in a uniform field. The ac voltage is brought out through "slip rings." In practical large ac generators the field rotates and the ac coils are stationary. A consideration of the higher currents, voltages, and winding complexity on the ac side makes this the more feasible arrangement. Thus we have a rotating part called a *rotor* or *field* and a stationary part called a *stator*. A schematic of the cross section of such a machine is shown in Fig. 8.1. In the figure the two sides of each coil are designated by a letter and its prime. For example, the *a* phase coil in the horizontal plane is designated by the letters *a* and *a'*. The reference directions for the currents are indicated in the usual way by arrows. In this case, in cross section we see the point of the arrow coming out of the page, and the feathers of the arrow going into the page. We may then find the associated reference directions for voltage and flux.

The rotor and stator are made of high-permeability iron to achieve a high ratio of flux density to mmf. The figure is highly schematic. We show each phase winding as if it were a single turn (the wires *aa'*, *bb'*, *cc'* are connected at the back) placed in a single pair of stator slots. Actually, it is a multiturn coil physically distributed in a number of stator slots. Similarly, the field winding mounted on the rotor is a multiturn winding.

The two-pole rotor shown rotates at 3600 revolutions per minute to generate 60 Hz. At this speed the surface speed of a 3.5-ft-diameter rotor is about 450 mi/hr

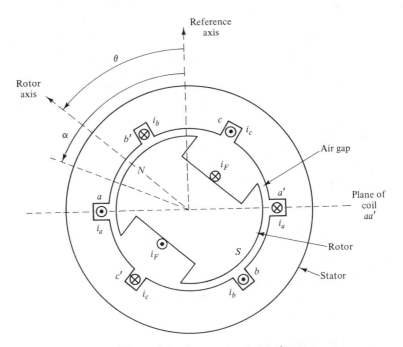

Figure 8.1 Generator cross section.

and the wind resistance is significant even in the low-density hydrogen-gas atmosphere which is used. For this reason (among others), in practice, a two-pole rotor would have a circular cross section. Such a rotor is called a *smooth* or *round* or *cylindrical rotor*.

The rotor shown in Fig. 8.1 is far from smooth. It is of the "salient pole" type, a type which is used in lower-speed multipole generators, for example, those driven by water turbines. Since multipole machines may be analyzed using a two-pole model, while even smooth rotor two-pole machines have slightly different magnetic properties in the rotor axis v-s cross rotor axis directions (because of the presence of the field windings), we may take the rotor in Fig. 8.1 as the generic rotor.

8.2 VOLTAGE GENERATION

We next wish to consider the generator terminal voltages. Neglecting resistances for the moment, the terminal voltages depend on the time rate of change of the corresponding flux linkages. Almost all of the flux linking each coil crosses the air gap, from rotor to stator and back again from stator to rotor; we call this the air-gap flux and the corresponding voltage is called the "air-gap" voltage. There is also a small "leakage" flux, which does not cross the air gap; the flux surrounding the "end turns" (i.e., the connections at the back of the conductors) is a part of this leakage flux. We can use a leakage inductance to model the effect of the leakage flux.

Considering only the air-gap flux for now, we note that this flux depends collectively on the field current i_F and the stator currents i_a, i_b, and i_c. We now make some assumptions that greatly simplify the analysis.

Assumptions

1. Magnetic circuit is linear.
2. Flux density, **B**, in air gap, due to i_F alone, is radial, and has a spatial distribution as a function of angle given by

$$B = B_{\max} \cos (\alpha - \theta) \tag{8.1}$$

   where B is the scalar flux density directed outward. The angles α and θ are defined in Fig. 8.1.
3. Each coil consists of N turns concentrated in a single slot.

Assumption 1 is crucial. It permits us to use superposition to find the total air-gap flux linkages by considering separately the flux linkages due to i_F alone (i.e., with $i_a = i_b = i_c = 0$), and of i_a, i_b, and i_c alone (i.e., with $i_F = 0$). Corresponding to these two flux linkage components there are two voltage components. The first is the open-circuit voltage. The second voltage is less familiar; it is called the *armature reaction voltage*. The sum of these two voltages gives the air-gap voltage due to the total air-gap flux.

Assumption 2 provides for a spatial distribution of flux density which is a maximum (outward) along the rotor axis (i.e., with $\alpha = \theta$) and drops off in a sinusoidal fashion with angle $\alpha - \theta$. This condition is an objective in generator design and can be approximated closely by distributing the field windings appropriately and/or by shaping the pole faces. Note the distinction between the angles θ and α. θ is the angle the rotor axis (north pole) is "looking" at; α is the more general angle *we* are looking at.

Assumption 3 is a step in the direction of considering the realistic case of N turns per phase winding. For simplicity we assume that all N turns are in the same slot. This assumption can be relaxed in a more detailed treatment, to account for a more realistic spatial distribution of conductors.

In Section 8.3 we find the open-circuit flux linkages and voltages. In Section 8.4 we find the effect of the stator currents. In Section 8.5 we find the combined effect and then finally the generator terminal voltages.

8.3 OPEN-CIRCUIT VOLTAGE

We assume that $i_a = i_b = i_c = 0$, and consider the flux linkages $\lambda_{aa'}$ in coil aa' due to i_F alone. From Fig. 8.1, using the right-hand rule, we see that $\lambda_{aa'}$ is positive if the flux is upward through the coil aa'. Next, using the flux density pattern given in (8.1) we draw a half-cylindrical Gaussian surface in the air gap and integrate B to find the total flux ϕ crossing the surface. This is shown in Fig. 8.2.

The flux in the crosshatched rectangular slot at an angle α is

$$d\phi = lrB_{\max} \cos (\alpha - \theta) \, d\alpha$$

Thus over the whole Gaussian surface,

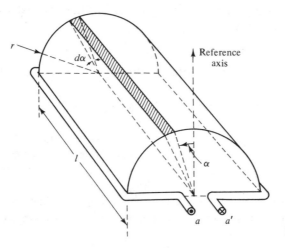

Figure 8.2 Gaussian surface.

$$\phi = \int \mathbf{B} \cdot \mathbf{ds} = lr \int_{-\pi/2}^{\pi/2} B_{max} \cos(\alpha - \theta)\, d\alpha$$

$$= 2lrB_{max} \cos \theta \tag{8.2}$$

$$= \phi_{max} \cos \theta$$

where $\phi_{max} \triangleq 2lrB_{max}$. This confirms the physically obvious fact that the maximum flux crossing the surface will occur for $\theta = 0$ (i.e., with the rotor in Fig. 8.1 pointed straight up).

For this angle $\lambda_{aa'}$ will also be a maximum. For an N-turn concentrated coil (assumption 3), the expression for flux linkages of coil aa' is

$$\lambda_{aa'} = N\phi_{max} \cos \theta = \lambda_{max} \cos \theta \tag{8.3a}$$

where $\lambda_{max} \triangleq N\phi_{max}$.

We next find $\lambda_{bb'}$ and $\lambda_{cc'}$ by noting that the flux linkages of coil bb' are a maximum when $\theta = 120°$, while coil cc' reaches its maximum at an angle of $240°$ or equivalently $-120°$. Thus

$$\lambda_{bb'} = \lambda_{max} \cos \left(\theta - \frac{2\pi}{3} \right) \tag{8.3b}$$

$$\lambda_{cc'} = \lambda_{max} \cos \left(\theta + \frac{2\pi}{3} \right) \tag{8.3c}$$

Returning to consider coil aa' and assuming a uniform rate of rotation of the rotor (i.e., $\theta = \omega_0 t + \theta_0$) we get

$$\lambda_{aa'} = \lambda_{max} \cos (\omega_0 t + \theta_0) \tag{8.4}$$

Using the circuits convention on associated reference directions, we find the voltage

$$e_{a'a} = \frac{d\lambda_{aa'}}{dt} = -\omega_0 \lambda_{max} \sin (\omega_0 t + \theta_0) \tag{8.5}$$

by use of (8.4). The reader may recall from circuit theory that in dealing with generators or sources it is conventional to define the reference direction for the sources differently. Accordingly we use $e_{aa'} = -e_{a'a}$, and instead of (8.5) we get

$$e_{aa'} = -\frac{d\lambda_{aa'}}{dt}$$

$$= \omega_0 \lambda_{max} \sin (\omega_0 t + \theta_0) \tag{8.6}$$

$$= E_{max} \cos \left(\omega_0 t + \theta_0 - \frac{\pi}{2} \right)$$

where $E_{max} \triangleq \omega_0 \lambda_{max}$. The assumption of linearity (assumption 1) implies that E_{max} is proportional to i_F. We also note that the voltage $e_{aa'}(t)$ reaches its maximum value when $\theta = \omega_0 t + \theta_0 = 90°$ (i.e., at the instant the rotor axis in Fig. 8.1 is pointed to the left). We can infer the same result from the phasor diagram relating the (effective) phasor representations of $\lambda_{aa'}(t)$ and $e_{aa'}(t)$. Using capital letters for effective phasors, from (8.4) and (8.6) we have

$$\Lambda_{aa'} = \frac{\lambda_{max}}{\sqrt{2}} e^{j\theta_0} \tag{8.7}$$

$$E_{aa'} = \frac{E_{max}}{\sqrt{2}} e^{j(\theta_0 - \pi/2)} = -j\frac{\omega_0 \lambda_{max}}{\sqrt{2}} e^{j\theta_0} = -j\omega_0 \Lambda_{aa'} \tag{8.8}$$

Thus the phasor $E_{aa'}$ lags $\Lambda_{aa'}$ by $90°$. From this it is clear that $e_{aa'}$ reaches its maximum value $\frac{1}{4}$ cycle after $\lambda_{aa'}$ has reached its maximum (i.e., with the rotor pointed straight up). But in $\frac{1}{4}$ cycle the rotor turns $\frac{1}{4}$ revolution; hence we get the same result as before.

Example 8.1

Suppose that $\theta_0 = \pi/4$. Show the directions of $\Lambda_{aa'}$, $E_{aa'}$, and the rotor position at $t = 0$.

Solution   Using (8.7), (8.8), and $\theta = \omega_0 t + \theta_0$, we get the results shown in Fig. E8.1.

Note: To obtain the instantaneous $e_{aa'}(t)$ and $\lambda_{aa'}t$, we multiply the corresponding phasors by $\sqrt{2}e^{j\omega_0 t}$ [which sets them into a counterclockwise (CCW) rotation with angular velocity ω_0] and take the real parts. Thus, corresponding to the CCW rotation of the rotor (with angular velocity ω_0), we have the CCW rotation of the "activated" phasors. Figure E8.1 can then be viewed as the motion frozen at $t = 0$.

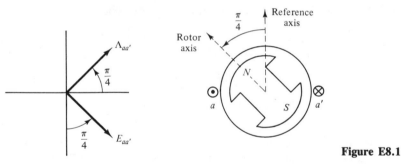

Figure E8.1

Exercise 1.   To establish firmly the connection between field (rotor) rotation, field flux linkages, and open-circuit voltage, consider several values of θ_0 (in $\theta = \omega_0 t + \theta_0$) and sketch the sinusoidal waveforms of $\lambda_{aa'}(t)$ and $e_{aa'}(t)$.

Returning to the main development, if we now consider the voltages $E_{bb'}$ and $E_{cc'}$, then as a consequence of (8.3) we get a balanced (positive sequence) set of voltages. We assume that a', b', c' are always connected to a neutral point n and then can dispense with the double-subscript notation. Thus we will use the single subscript notation E_a, E_b, E_c.

Exercise 2.   Suppose that the rotor is turning clockwise. Show that the voltages E_a, E_b, E_c are now a negative sequence set.

8.4 ARMATURE REACTION

In the preceding section we assumed that $i_a = i_b = i_c = 0$. In this case the (open-circuit) terminal voltages are E_a, E_b, and E_c, where $E_a = E_{aa'}$ is given by (8.8). If stator currents flow, they will modify the flux linkages and consequently the terminal voltages will no longer be E_a, E_b, and E_c. As a step in finding the air-gap voltage in this case, we consider the effect of armature reaction (i.e., the effect of these stator currents, acting alone). Thus suppose that i_a, i_b, and i_c are a balanced (positive-sequence) set of sinusoidal currents and $i_F = 0$. Our initial objective is to find the air-gap flux and the flux linkages of coil aa' associated with these currents. To simplify the preliminary discussion, assume a round rotor. In this case, since $i_F = 0$, θ does not matter (i.e., it does not matter in which direction the rotor is pointed).

Consider the flux due to i_a alone. The coil aa' and typical flux lines are shown in Fig. 8.3. The flux lines are symmetrical with respect to the plane of aa'. First let us find the air-gap flux density B as a function of α. Using Ampère's circuital law, with the path Γ coinciding with a flux line as shown, we get

$$F = \oint_\Gamma \mathbf{H} \cdot d\mathbf{l} = \frac{1}{\mu_r \mu_0} \int_{\Gamma_{\text{iron}}} \mathbf{B} \cdot d\mathbf{l} + \frac{1}{\mu_0} \int_{\Gamma_{\text{air}}} \mathbf{B} \cdot d\mathbf{l} = Ni_a \qquad (8.9)$$

In (8.9) the closed path Γ has been separated into the iron and air-gap portions. The

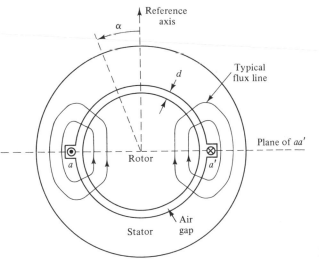

Figure 8.3 Air-gap flux due to i_a.

relative permeability in iron is very high, on the order of 1000, and as an approximation we will therefore neglect the integral representing the mmf drop in the iron. In evaluating the air-gap contribution we make use of the fact that the vector flux density in the air gap is approximately radial. Then over the small air-gap distances we get

$$\frac{1}{\mu_0} \int_{\Gamma_{air}} \mathbf{B} \cdot d\mathbf{l} \approx 2\frac{Bd}{\mu_0} = Ni_a \tag{8.10}$$

where d is the air-gap width and there is a factor of 2 because of the two symmetrically disposed air-gap crossings. Solving for B in (8.10), we get

$$B = \frac{\mu_0 Ni_a}{2d} = Ki_a \tag{8.11}$$

as the (radially outward) scalar flux density in the air gap above the coil aa' and the (radially inward) scalar flux density in the air gap below the coil. We note that B is proportional to i_a with $K \triangleq \mu_0 N/2d$ as the (positive) constant of proportionality. Note, also, that, at least for $0 < \alpha < \pi/2$, B does not depend on α (i.e., on which particular flux line coincides with the path Γ). Checking further, as a function of α, we get the spatial flux distribution shown in Fig. 8.4. Positive (negative) values of B correspond to radially outward (inward) flux density. The waveform is (approximately) a square wave. Also shown in Fig. 8.4 is the fundamental component obtained by Fourier analysis. This component is given by

$$B_a = \frac{4}{\pi}Ki_a \cos \alpha \tag{8.12}$$

We have assumed a concentrated coil. If, as in the realistic case, the coil is distributed, the flux density wave more nearly approximates a truncated triangular wave with maximum at $\alpha = 0$. In either case, in a simplified analysis, we will consider only the fundamental component of the spatial flux density wave, as indicated in Fig. 8.4.

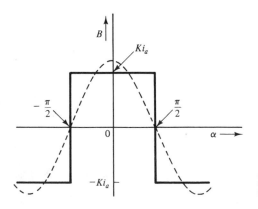

Figure 8.4 Spacial flux density due to to i_a.

Now consider (8.12) with i_a a sinusoidal function of t. We may assume that

$$i_a(t) = \sqrt{2}\,|I_a|\cos(\omega_0 t + \underline{/I_a})\qquad(8.13)$$

then using (8.12) we get

$$B_a = \frac{4}{\pi}Ki_a(t)\cos\alpha$$

$$= B'_{max}\cos\alpha\cos(\omega_0 t + \underline{/I_a})\qquad(8.14)$$

B'_{max} absorbs many constants; we note that it is proportional to $|I_a|$. Equation (8.14) gives both the spatial and temporal distribution of B_a. We note that the B_a distribution is stationary in space (with the centerline straight up) and varies sinusoidally with t.

If we now repeat the calculation to get the contributions of the remaining two phases (i.e., of i_b and i_c), we get a result similar to (8.14) but with appropriate shifts in the spatial and time arguments. Because of the placement of the coils bb' and cc', the spatial distributions of B_b and B_c are shifted by 120° and 240°, respectively; because i_a, i_b, and i_c are assumed to be a positive sequence, there are corresponding displacements of 120° and 240° in the time behavior as well.

Consider next the flux density with all three (positive sequence) currents i_a, i_b, and i_c present simultaneously. Using superposition and (8.14) but taking the spatial and temporal shifts into account, we get

$$B_{abc} = B'_{max}\left[\cos\alpha\cos(\omega_0 + \underline{/I_a}) + \cos\left(\alpha - \frac{2\pi}{3}\right)\cos\left(\omega_0 t + \underline{/I_a} - \frac{2\pi}{3}\right)\right.$$

$$\left. + \cos\left(\alpha + \frac{2\pi}{3}\right)\cos\left(\omega_0 t + \underline{/I_a} + \frac{2\pi}{3}\right)\right]\qquad(8.15)$$

$$= \tfrac{3}{2}B'_{max}\cos(\alpha - \omega_0 t - \underline{/I_a})$$

Physically, this represents a sinusoidal traveling wave in the air gap rotating in a CCW direction at an angular velocity ω_0. We note that the ability to generate such rotating fields is a valuable additional advantage to the use of three-phase (over single-phase) supply.

Thus, to reiterate, we find that positive-sequence stator currents of frequency ω_0 give rise to a spatial armature reaction flux wave rotating in the air gap with the same angular velocity as the rotor. Note that at $t = 0$, the maximum of the flux density wave occurs for $\alpha = \underline{/I_a}$. Thus, at $t = 0$, the centerline or north pole of the wave is at an angle $\underline{/I_a}$ with respect to the reference axis in Fig. 8.1. We will make frequent use of this result in Section 8.5.

In describing the rotating fluxes in the air gap it is frequently useful to imagine that we are viewing the fluxes from a "synchronously rotating reference frame." Suppose that the reference frame angle is $\omega_0 t$ and we are observing the air-gap fluxes relative to the frame. In the steady-state case we have been considering, the rotor angle (relative to the synchronously rotating reference frame) is θ_0, a constant; the centerline of the armature reaction flux is also at a fixed angle $\underline{/I_a}$. The merits of using the synchronously rotating reference frame will be clearer in the context of the application to rotor angle transients to be considered in Chapter 9.

Exercise 3.   We can get a good feeling for the rotating armature reaction flux by sketching the centerlines of flux directions at different times. Suppose that i_a, i_b, and i_c form a positive-sequence set, with $i_a(t) = 10 \cos \omega_0 t$. Use the right-hand rule, and symmetry, to show the flux directions at $\omega_0 t = 0$, $\omega_0 t = 2\pi/3$, and $\omega_0 t = 4\pi/3$. What happens if the currents form a negative-sequence set?

We have been considering the rotating armature reaction air-gap flux wave due to the currents i_a, i_b, and i_c. We next consider the corresponding flux linkages with coil aa'. Let these armature reaction flux linkages be designated by λ_{ar}. We use the same Gaussian surface used earlier to calculate $\lambda_{aa'}$ (see Fig. 8.2), and find, using (8.15),

$$
\begin{aligned}
\lambda_{ar}(t) &= N\phi(t) = Nlr \int_{-\pi/2}^{\pi/2} \tfrac{3}{2} B'_{max} \cos(\alpha - \omega_0 t - \underline{/I_a})\, d\alpha \\
&= 3Nlr\, B'_{max} \cos(\omega_0 t + \underline{/I_a}) \qquad\qquad (8.16) \\
&= \sqrt{2} L_{s1} |I_a| \cos(\omega_0 t + \underline{/I_a}) \\
&= L_{s1} i_a(t)
\end{aligned}
$$

In going from line 2 to line 3 above, we make use of the fact that B'_{max} is proportional to $|I_a|$ and absorb many constants into L_{s1}. The interesting result in (8.16) is that *instantaneously* λ_{ar} is proportional to i_a just as if i_b and i_c were not present. Their presence changes only the constant of proportionality. Since i_a is a sinusoid, we can also represent λ_{ar} and i_a by phasors. We get

$$
\Lambda_{ar} = L_{s1} I_a \qquad\qquad (8.17)
$$

Equation (8.16) or (8.17) gives us the armature reaction effect for phase a. Similarly, the flux linkages of coils bb' and cc' are related to i_b and i_c, respectively, by the same constant L_{s1}.

8.5 TERMINAL VOLTAGE

We now use superposition to combine the results of Sections 8.3 and 8.4. We assume that *all* the currents are present. The total air-gap flux wave then consists of the sum of field flux wave plus armature reaction flux wave. At $t = 0$, the centerline of the field flux wave is at an angle θ_0, while that of the armature reaction flux wave is at an angle $\underline{/I_a}$. Both angles are measured with respect to the reference direction in Fig. 8.1. Since we assume sinusoidal distribution in space for both flux densities, we can add them (as we add phasors) to find a single resultant air-gap flux wave. This resultant flux wave rotates in the air gap at an angular rate ω_0 radians per second.

The flux linkages also add. Defining λ_{ag} as the total air-gap flux linkages of coil aa', we have

$$\lambda_{ag} = \lambda_{aa'} + \lambda_{ar} \qquad (8.18)$$

Since these are sinusoidal functions of time, we also get the corresponding relation between phasors:

$$\Lambda_{ag} = \Lambda_{aa'} + \Lambda_{ar} \qquad (8.19)$$

The actual generated voltage (the air-gap voltage) is

$$v_{ag} = -\frac{d\lambda_{ag}}{dt} \qquad (8.20)$$

or in terms of phasors,

$$V_{ag} = -j\omega_0 \Lambda_{ag} \qquad (8.21)$$

The reason for the minus sign was discussed just after (8.5). Differentiating (8.18) and using (8.6) and (8.16), we get successively

$$-\frac{d\lambda_{ag}}{dt} = -\frac{d\lambda_{aa'}}{dt} - \frac{d\lambda_{ar}}{dt}$$

$$v_{ag} = e_a - L_{s1}\frac{di_a}{dt} \qquad (8.22)$$

Using phasors, we can replace (8.22) by

$$V_{ag} = E_a - j\omega_0 L_{s1} I_a \qquad (8.23)$$

It is remarkable that we are able to account for the difference between the open-circuit voltage E_a and the actual generated voltage V_{ag} by the voltage drop across a fictitious inductance L_{s1}.

V_{ag} is not quite the terminal voltage. It would be if there were no resistance or leakage reactance in the windings. We account for these effects by adding circuit elements r and X_l and obtain the per phase equivalent circuit in Fig. 8.5. The added elements are relatively small. X_l is in the order of 10% of X_{s1}; r is less than 1% of X_{s1}. As indicated in the figure, we define the *synchronous reactance* $X_s \triangleq X_{s1} + X_l$ and get

$$V_a = E_a - rI_a - jX_s I_a \qquad (8.24)$$

The relationships between various voltage and the current may be seen in the phasor

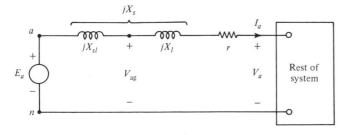

Figure 8.5 Per phase circuit diagram (round-rotor machine).

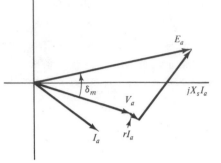

Figure 8.6 Phasor diagram corresponding to (8.24).

diagram in Fig. 8.6. We identify a power angle $\delta_m = \underline{/E_a} - \underline{/V_a}$. Later we will discuss
the physical significance of this angle.

Example 8.2

Given a round-rotor generator with $V_a = 1.0$, $X_s = 1.6$, $r = 0.004$, and $I_a = 1\underline{/-60°}$
find E_a. Draw a phasor diagram.

Solution   The problem reduces to a simple circuits problem.

$$E_a = V_a + rI_a + jX_sI_a$$
$$= 1 + 0.004 \underline{/-60°} + 1.6 \underline{/30°}$$
$$= 2.517 \underline{/18.45°}$$

If we had neglected r, the result would have been $E_a = 2.516 \underline{/18.54°}$, which is close
enough.

The phasor diagram is shown in Fig. E8.2. The voltage drop in r is exaggerated
to make it visible.

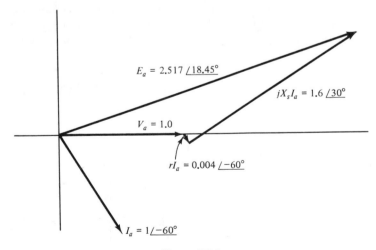

Figure E8.2

Example 8.3

Suppose that in Example 8.2 there is a symmetric 3ϕ short circuit at the generator terminals. Find the steady-state short-circuit current magnitude assuming that i_F does not change from its prefault value.

Solution   If i_F does not change, neither does $|E_a|$, which is proportional to it. Then, in the steady state,

$$I_a = \frac{E_a}{r + jX_s} \Rightarrow |I_a| = \frac{2.517}{1.6000} = 1.573$$

I_a lags E_a by 89.86°, which we may approximate by 90°.

Example 8.4

Suppose that in Example 8.3, in the steady state, the rotor is turning at a uniform rate with $\theta = \omega_0 t + \theta_0$. Assume that I_a lags E_a by 90°. Consider the rotating fluxes in the air gap and show that the armature reaction flux and the flux due to i_F are oppositely directed.

Solution   From (8.6) or (8.8) we see that $\underline{/E_a} = \theta_0 - 90°$. Then, since I_a lags E_a by 90°, we have $\underline{/I_a} = \theta_0 - 90° - 90° = \theta_0 - 180°$. Then, as discussed in Section 8.4, at $t = 0$, the centerline of the armature reaction flux wave is at an angle $\theta_0 - 180°$. Thus the two components of the air-gap flux are oppositely directed.

In describing the effect of the short-circuit currents, we say they set up a "demagnetizing" flux which opposes the flux due to i_F.

So far we have been considering the round-rotor case. It remains to consider the salient-pole case. In this case $\underline{/I_a}$ relative to θ_0 must be considered. We now consider some possibilities. Suppose that $\underline{/I_a} = \theta_0$. Then the armature reaction flux is aligned with the rotor. If $\underline{/I_a} = \theta_0 - 180°$, as in Example 8.4, the flux is in the opposite direction (in the "demagnetizing" direction). In either case the air gap "seen" by most of the armature reaction flux is a small one and by an analysis similar to that in Section 8.4, we can expect to get a relatively large inductance parameter relating flux linkages of coil aa' and i_a. On the other hand, suppose that $\underline{/I_a} = \theta_0 \pm 90°$. Then the armature reaction flux is pointed in a direction perpendicular to the rotor axis. The important point is this. Since the air gap seen by most of the armature reaction flux is bigger in this case, we expect i_a, i_b, and i_c to be less effective in producing the flux; equivalently, after a development like that in Section 8.4, we will get a smaller inductance parameter in (8.16).

In the general case we resolve the phasors I_a, I_b, and I_c, into two groups of components. One set of components I_{ad}, I_{bd}, and I_{cd} sets up a rotating flux with centerline aligned with the "direct axis" of the rotor. The other set of components I_{aq}, I_{bq}, and I_{cq} sets up a rotating flux with centerline in the "quadrature axis" direction. We can find the armature reaction flux wave by superposition considering the contributions of direct axis and quadrature axis currents separately. A similar statement can be made about the corresponding flux linkages of coil aa'. Thus, in terms of phasors, we have

$$\Lambda_{ar} = \Lambda_{ad} + \Lambda_{aq} \tag{8.25}$$

where the d refers to the direct axis component and the q to the quadrature axis component. Just as in Section 8.4, Λ_{ad} (which depends on I_{ad}, I_{bd}, and I_{cd}) is proportional to I_{ad} and, similarly, Λ_{aq} is proportional to I_{aq}. The constants of proportionality are different for I_{ad} and I_{aq}. The situation is as shown in Fig. 8.7 for the case $\theta_0 = 90°$. As shown in the figure I_a is resolved into two components I_{aq} and I_{ad} which (in this case) are equal in magnitude. However since I_{aq} is less effective in producing flux linkages than is I_{ad}, the magnitude of Λ_{aq} is much less than that of Λ_{ad}.

If we define fictitious inductance parameters such that

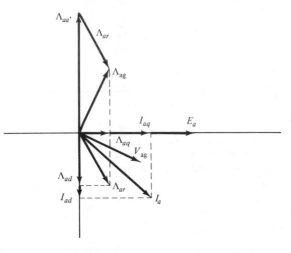

Figure 8.7   Flux linkages.

$$\Lambda_{ad} = L_{d1}I_{ad} \tag{8.26a}$$

$$\Lambda_{aq} = L_{q1}I_{aq} \tag{8.26b}$$

then we can eliminate the flux linkages and obtain expressions involving only voltages, currents, and impedances. We can start by substituting (8.25) in (8.19). We get

$$\Lambda_{ag} = \Lambda_{aa'} + \Lambda_{ad} + \Lambda_{aq} \tag{8.27}$$

Multiplying by $-j\omega_0$, we get

$$-j\omega_0\Lambda_{ag} = -j\omega_0\Lambda_{aa'} - j\omega_0\Lambda_{ad} - j\omega_0\Lambda_{aq}$$

Using (8.8), (8.21), and (8.26), we can evaluate V_{ag}

$$\begin{aligned} V_{ag} &= E_a - j\omega_0 L_{d1}I_{ad} - j\omega_0 L_{q1}I_{aq} \\ &= E_a - jX_{d1}I_{ad} - jX_{q1}I_{aq} \end{aligned} \tag{8.28}$$

where $X_{d1} = \omega_0 L_{d1}$ and $X_{q1} = \omega_0 L_{q1}$. To get the terminal voltage we subtract the voltage drops in r and X_l:

$$\begin{aligned} V_a &= V_{ag} - rI_a - jX_l I_a \\ &= V_{ag} - rI_a - jX_l(I_{ad} + I_{aq}) \end{aligned} \tag{8.29}$$

Substituting for V_{ag} from (8.28), we get

$$V_a = E_a - rI_a - j(X_l + X_{d1})I_{ad} - j(X_l + X_{q1})I_{aq}$$
$$= E_a - rI_a - jX_dI_{ad} - jX_qI_{aq} \tag{8.30}$$

where $X_d \triangleq X_{d1} + X_l$ and $X_q \triangleq X_{q1} + X_l$. X_d is called the *direct axis reactance*. X_q is called the *quadrature axis reactance*. The generator model using X_d and X_q is called the *two-reaction model*.

In per unit, typical values of X_d and X_q are given in Table 8.1. We note that for the two-pole round rotor machines, there is a small difference between X_d and X_q, although as an approximation we can take $X_s = X_d \approx X_q$. Making this approximation, (8.30) reduces to (8.24).

TABLE 8.1   TYPICAL VALUES OF X_d AND X_q

	Turbine-generators* (two-pole)	Salient-pole machines (with dampers)
X_d	1.20	1.25
X_q	1.16	0.70

*Values for air cooled and hydrogen conventionally cooled generators

Unfortunately, we do not get a simple circuit model from (8.30), but we do get the simple phasor diagram in Fig. 8.8. Figures 8.8 and 8.6 should be compared. In particular, note that in the round-rotor case, with $X_d = X_q = X_s$, the voltage $jX_dI_{ad} + jX_qI_{aq}$ in Fig. 8.8 reduces to the voltage jX_sI_a in Fig. 8.6.

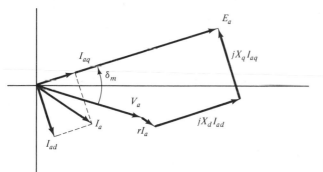

Figure 8.8   Phasor diagram corresponding to (8.30).

Consider next the addition to Fig. 8.8 shown in Fig. 8.9. In Fig. 8.9 we show the result of adding jX_qI_a to $V_a + rI_a$; we claim we get to a point a' in the same direction as E_a.

Proof. The point labeled a' is a complex number described as follows:

$$a' = V_a + rI_a + jX_qI_a = V_a + rI_a + jX_q(I_{ad} + I_{aq})$$

Consider also the point E_a.

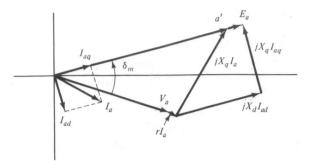

Figure 8.9 Augmented phasor diagram.

$$E_a = V_a + rI_a + jX_dI_{ad} + jX_qI_{aq}$$

If we now take the difference, we get

$$E_a - a' = j(X_d - X_q)I_{ad}$$

From the geometry of Fig. 8.9, jI_{ad}, and thus $E_a - a'$, is parallel to E_a which implies that E_a and a' are collinear.

With the addition of the point a' to Fig. 8.8, we are in a position to solve either of the following two problems.

Problem 1: Given E_a and I_a, find V_a.
Problem 2: Given V_a and I_a, find E_a.

We next consider two examples that illustrate the methods of solution. The first example is an illustration of Problem 1; the second illustrates Problem 2.

Example 8.5

Given $E_a = 1.5 \underline{/30°}$, $I_a = 0.5 \underline{/-30°}$, $X_d = 1.0$, and $X_q = 0.6$, find V_a. Neglect r.

Solution We need to resolve I_a into I_{ad} and I_{aq}. This can be done most easily by sketching I_a and E_a (Fig. E8.5). From the sketch by using simple trigonometry, we find

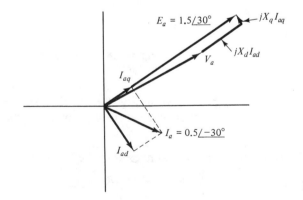

Figure E8.5

$$I_{aq} = 0.25 \,\underline{/30°} \qquad I_{ad} = 0.433 \,\underline{/-60°}$$

Then, using (8.30),

$$V_a = E_a - jX_d I_{ad} - jX_q I_{aq}$$

$$= 1.5 \,\underline{/30°} - j1.0 \times 0.433 \,\underline{/-60°} - j0.6 \times 0.25 \,\underline{/30°}$$

$$= 1.077 \,\underline{/22.00°}$$

As a check, in the phasor diagram, we also show the various voltages that add up to E_a.

Example 8.6

Given $V_a = 1.0 \,\underline{/0°}$, $I_a = 1 \,\underline{/-45°}$, $X_d = 1.0$, and $X_q = 0.6$, find E_a. Neglect r.

Solution   Before we can resolve I_a into components, we need to know the direction of E_a. Thus we find the point a' as shown in Fig. 8.9.

$$a' = V_a + jX_q I_a = 1 + j0.6 \times 1 \,\underline{/-45°} = 1.486 \,\underline{/16.59°}$$

This point is shown in Fig. E8.6 together with the resolution of I_a into I_{aq} and I_{ad}. From the sketch, by using trigonometry, we find

$$I_{aq} = 0.476 \,\underline{/16.59°} \qquad I_{ad} = 0.880 \,\underline{/-73.41°}$$

Then using (8.30) we find that

$$E_a = V_a + jX_d I_{ad} + jX_q I_{aq}$$

$$= 1 + j1.0 \times 0.880 \,\underline{/-73.41°} + j0.6 \times 0.476 \,\underline{/16.59°}$$

$$= 1.838 \,\underline{/16.59°}$$

As a check we also show in the phasor diagram, the various voltages that add up to E_a.

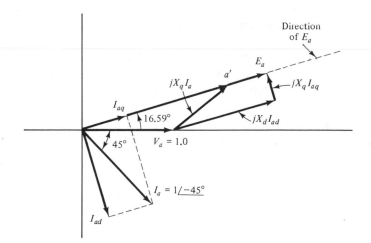

Figure E8.6

Example 8.7

Suppose that in Example 8.5 there is a symmetric three-phase short circuit. Find the steady-state short-circuit current magnitude assuming that i_F (in the steady state) is unchanged from its prefault value.

Solution The assumption regarding i_F implies that $|E_a| = 1.5$ is unchanged. With $V_a = 0$, we have

$$E_a = jX_d I_{ad} + jX_q I_{aq}$$

From Fig. 8.8 we can see that $jX_d I_{ad}$ is parallel to E_a, while $jX_q I_{aq}$ is perpendicular. From this we can conclude that $I_{aq} = 0$ while $I_{ad} = E_a/jX_d$. Thus $I_a = I_{ad} + I_{aq} = I_{ad}$ and $I_a = E_a/jX_d$. Thus we get $|I_a| = |E_a|/X_d = 1.5$.

We emphasize that the short-circuit current does not depend on X_q at all.

Exercise 4. In Example 8.7 in modeling the generator assume that we include a small resistance r. Show that for small r, $|I_a| \approx |E_a|/X_d$. *Hint:* In using (8.30), $rI_a = r(I_{ad} + I_{aq})$. Note that rI_{aq} is parallel to E_a and rI_{ad} is perpendicular to E_a.

Finally, before leaving these generator models, either the round-rotor or two-reaction models, we should note the following about Figs. 8.6 and 8.8.

1. E_a is the internal, or open circuit, or Thévenin equivalent voltage. $|E_a|$ is proportional to the rotor (field) current i_F.
2. In both cases there is a power angle δ_m between the internal voltage and the terminal voltage. As an approximation, if we neglect both r and X_l so that $V_a = V_{ag}$, we get a nice physical interpretation of δ_m. δ_m can be shown to be the angle between the rotor centerline (north pole) and the rotating air-gap flux centerline (north pole). This is suggested in Fig. 8.10 for the salient-pole case. To check the assertion we note that, with our approximation, $v_{ag} = v_a$ lags e_a by the angle δ_m. In terms of the flux linkages of the coil aa' there must be a similar lag of λ_{ag} compared with the open circuit $\lambda_{aa'}$. As a consequence, the centerline of the spatial air-gap flux wave (which fills the coil aa') must lag the rotor north pole by the angle δ_m, as indicated in Fig. 8.10.

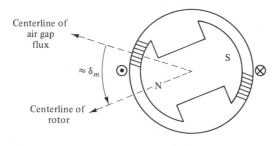

Figure 8.10 Physical significance of δ_m.

3. We can easily extend the use of both models to the case of a generator connected to a bus through a (short) transmission line, by absorbing the (series) line impedance $R_L + jX_L$ into the generator model. Let V_a be the generator terminal voltage and V_t be the receiving-end voltage of the transmission line. Then

$$V_t = V_a - R_L I_a - jX_L I_a \tag{8.31}$$

Using (8.30) and $I_a = I_{ad} + I_{aq}$, we get

$$V_t = E_a - rI_a - jX_d I_{ad} - jX_q I_{aq} - R_L I_a - jX_L(I_{ad} + I_{aq}) \tag{8.32}$$

Grouping terms, we get

$$V_t = E_a - \tilde{r}I_a - j\tilde{X}_d I_{ad} - j\tilde{X}_q I_{aq} \tag{8.33}$$

where $\tilde{r} = r + R_L$, $\tilde{X}_d = X_d + X_L$, and $\tilde{X}_q = X_q + X_L$. Equation (8.33) is in the same form as (8.30) with the line impedances absorbed into the generator model and with V_t as the generator "terminal voltage."

8.6 POWER DELIVERED BY GENERATOR

With the generator models derived in Section 8.5 we may determine the complex power S_G delivered by a generator in terms of its internal voltage E_a and terminal voltage V_a. We consider the round-rotor case first.

Case I: Round rotor.   Using the per phase equivalent circuit in Fig. 8.5, we have

$$S_G = V_a I_a^* = V_a \left(\frac{E_a - V_a}{Z_G}\right)^* \tag{8.34}$$

where $Z_G = r + jX_s$. This is exactly the problem considered in Section 2.6—that of power transmitted between two buses. Thus we can use the power circle diagram to find $S_G = -S_{21}$. This makes it clear that there is an ultimate active power limit just like for a transmission line!

If the generator resistance is neglected, then using (2.31) and (2.33), we get the simple results

$$P_G = \frac{|E_a||V_a|}{X_s} \sin \delta_m \tag{8.35}$$

$$Q_G = \frac{|V_a|(|E_a| \cos \delta_m - |V_a|)}{X_s} \tag{8.36}$$

Case II: Salient-pole generator.   For simplicity we will consider only the case $r = 0$. Also, to simplify the derivation we will assume that

$$E_a = |E_a| \underline{/0^\circ} = |E_a| \tag{8.37}$$

$$V_a = |V_a| \underline{/-\delta_m} = |V_a| e^{-j\delta_m} = |V_a| \cos \delta_m - j|V_a| \sin \delta_m \tag{8.38}$$

This amounts to no more than choosing the time we call $t = 0$ to be the time the rotor angle $\theta = 90^\circ$ (i.e., $\theta_0 = 90^\circ$). The formulas we derive depend on the angle difference θ_m, not the individual angles. Using $I_a = I_{ad} + I_{aq}$, we get

$$S_G = V_a I_a^* = V_a (I_{ad} + I_{aq})^* \tag{8.39}$$

We first determine I_{ad} and I_{aq} as follows using (8.30), (8.37), and (8.38):

$$\begin{aligned} |E_a| &= V_a + jX_d I_{ad} + jX_q I_{aq} \\ &= |V_a| \cos \delta_m - j|V_a| \sin \delta_m + jX_d I_{ad} + jX_q I_{aq} \end{aligned} \tag{8.40}$$

Since, by (8.37), E_a is purely real, so is I_{aq} and $jX_d I_{ad}$. On the other hand, $jX_q I_{aq}$ is purely imaginary. Separating (8.40) into real and imaginary parts, we get two equations:

$$|E_a| = |V_a| \cos \delta_m + jX_d I_{ad} \tag{8.41}$$

$$0 = -|V_a| \sin \delta_m + X_q I_{aq} \tag{8.42}$$

Solving, we get

$$I_{ad} = \frac{|E_a| - |V_a| \cos \delta_m}{jX_d} \qquad I_{aq} = \frac{|V_a \sin \delta_m|}{X_q} \tag{8.43}$$

We next substitute I_{ad} and I_{aq} into (8.39):

$$\begin{aligned} S_G &= |V_a| e^{-j\delta_m} \left(j \frac{|E_a| - |V_a| \cos \delta_m}{X_d} + \frac{|V_a| \sin \delta_m}{X_q} \right) \\ &= |V_a| (\cos \delta_m - j \sin \delta_m) \left(\frac{|V_a| \sin \delta_m}{X_q} + j \frac{|E_a| - |V_a| \cos \delta_m}{X_d} \right) \end{aligned} \tag{8.44}$$

We next find P_G and Q_G. Taking real parts, we get

$$\begin{aligned} P_G &= |V_a| \left(\cos \delta_m \frac{|V_a| \sin \delta_m}{X_q} + \sin \delta_m \frac{|E_a| - |V_a| \cos \delta_m}{X_d} \right) \\ &= \frac{|E_a||V_a|}{X_d} \sin \delta_m + |V_a|^2 \left(\frac{\cos \delta_m \sin \delta_m}{X_q} - \frac{\cos \delta_m \sin \delta_m}{X_d} \right) \\ &= \frac{|E_a||V_a|}{X_d} \sin \delta_m + \frac{|V_a|^2}{2} \left(\frac{1}{X_q} - \frac{1}{X_d} \right) \sin 2\delta_m \end{aligned} \tag{8.45}$$

Note: If $X_d = X_q = X_s$, (8.45) reduces to (8.35).

Next we equate the imaginary parts of (8.44) and after some algebra we get

$$Q_G = \frac{|E_a||V_a|}{X_d} \cos \delta_m - |V_a|^2 \left(\frac{\cos^2 \delta_m}{X_d} + \frac{\sin^2 \delta_m}{X_q} \right) \qquad (8.46)$$

If $X_d = X_q = X_g$, (8.46) reduces to (8.36).

Suppose that with $|E_a|$ and $|V_a|$ fixed we consider the effect of saliency on the curve of P_G versus δ_m. Since we have $X_d > X_q$, then $1/X_q - 1/X_d$ is positive. Thus, sketching (8.45) as in Fig. 8.11, we see that the effect of saliency is to cause the maximum P_G to occur at an angle $\delta_m < 90°$.

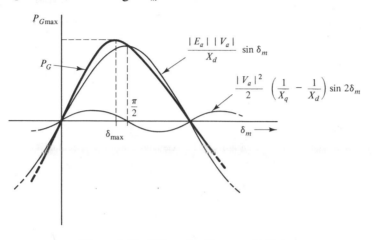

Figure 8.11   Effect of saliency on $P_G(\delta_m)$.

8.7 SYNCHRONIZING GENERATOR TO AN INFINITE BUS

As another application of our model we consider the problem of connecting or "synchronizing" a generator to a particular power system bus (generator bus) and then making the necessary adjustments to deliver a specified complex power. To simplify the discussion we will assume that the MVA rating of our generator is relatively small so that we do not expect to affect significantly the voltage and frequency of the larger system. In this case, as an approximation, we can assume that the (complex) generator bus voltage is independent of how our generator is adjusted (i.e., it is an ideal voltage source). We call such a source an "infinite bus." In mechanics an infinite inertia plays an analogous role and this suggests the terminology. Thus we have the definition.

Definition.   An infinite bus is an ideal voltage source.

Consider the system shown in Fig. 8.12, in which we will synchronize and then deliver complex power to a large system modeled by an infinite bus. Initially, the circuit breaker is open. We wish to close the breaker and deliver power to the (infinite) bus. First, to effect a smooth transition, we must satisfy the following synchronizing

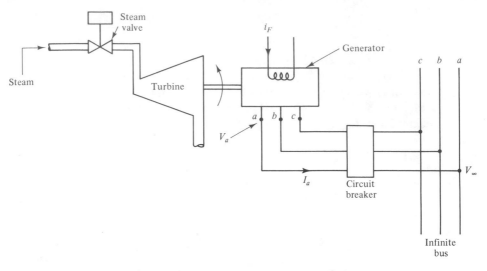

Figure 8.12   Generator to be synchronized.

conditions regarding the voltages V_a and V_∞. We note that before synchronization, $V_a = E_a$, because $I_a = 0$.

1. Frequency is the same.
2. Phase sequence is the same.
3. Phase is the same.
4. $|V_a| = |E_a| = |V_\infty|$.

Note 1: Conditions 3 and 4 may be combined: $V_a = E_a = V_\infty$.

Note 2: The turbine is providing very little power, only enough to overcome rotation losses.

With these conditions satisfied the breaker may be closed. The generator is now synchronized to the power system. It is in a condition known as "floating"; $I_a = 0$ (because $E_a = V_\infty$) and $S_G = 0$. Suppose that we now slowly open the steam valve. The mechanical power driving the generator increases. The coupled generator and turbine rotors tend to accelerate. As we will discuss in Chapter 9, the net effect is for the power angle $\delta_m = \underline{/E_a} - \underline{/V_\infty}$ to increase (V_∞ is the infinite bus voltage and is fixed). Thus $P_G(\delta_m)$ increases. Assuming that a steady state is reached, and neglecting the losses, we get, eventually, a balance between the mechanical power input and electrical power output. If i_F is not changed from its value at synchronization, then $|E_a|$ does not change either. The effect of applying mechanical power, therefore, is to increase $\underline{/E_a}$ while leaving $|E_a|$ unchanged. We get a phasor diagram something like that shown in Fig. 8.13. Here we have assumed a round-rotor model and $r = 0$ for simplicity.

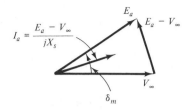

Figure 8.13 Mechanical power changes; i_F fixed.

Continuing with this simplified model, suppose that we consider the effects of changing i_F. If we change i_F, we change $|E_a|$ by the same proportion. If at the same time we keep the mechanical power constant, then $P_G(\delta_m) = (|E_a||V_\infty|/X_s)\sin\delta_m$ remains constant. Thus $|E_a|\sin\delta_m$ remains fixed. This is suggested in Fig. 8.14 by drawing a (dashed) line parallel to V_∞ (i.e., the locus of points E_a satisfying $|E_a|\sin\delta_m = $ constant). For example, $E_a^{(1)}$ might be the value corresponding to i_F at synchronization. $E_a^{(2)}$ corresponds to an increased value of i_F, while $E_a^{(3)}$ corresponds to a decreased value of i_F.

From Fig. 8.14 we can deduce how I_a is affected by changes in i_F. Thus if $|E_a|$ is increased beyond the value $|E_a^{(2)}|$, we would find that $|I_a|$ increases beyond the value $|I_a^{(2)}|$; at the same time the power factor decreases so that $P_G = |V_\infty||I_a|\cos\phi$ remains constant.

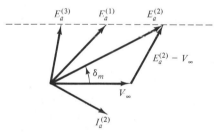

Figure 8.14 i_F changes; mechanical power fixed.

Exercise 5. With constant mechanical power, sketch the locus of I_a as i_F is varied. Show that there is a minimum $|I_a|$.

Note 1: If our generator is salient pole and/or if resistance is considered, we get results that differ only slightly from the above. For a qualitative understanding it is far simpler to use the round-rotor model!

Note 2: Clearly, we can synchronize to a noninfinite bus. The conditions for smooth synchronization remain the same. The discussion of what happens after synchronization is more complicated because the generator bus voltage will vary as a function of the injected bus power S_G.

Example 8.8

A generator is synchronized to an infinite bus. $i_F = 1000$ A (actual) at synchronization. $V_\infty = 1\underline{/0°}$ and $X_s = 1.5$. With i_F unchanged the steam valves at the turbine are adjusted until $P_G = 0.2$.

(a) Find I_a.

(b) With P_G unchanged, i_F is increased to 1600 A (actual). Find I_a.

Solution   (a) At synchronization $E_a = V_a = V_\infty = 1\,\underline{/0°}$. Thus $|E_a| = 1$ corresponds to $i_F = 1000$ A. Then

$$P_G = 0.2 = \frac{|E_a||V_\infty|}{X_s} \sin \delta_m = \frac{1}{1.5} \sin \delta_m$$

Thus $\delta_m = 17.46°$ and $E_a = 1\,\underline{/17.46°}$. Then

$$I_a = \frac{E_a - V_\infty}{jX_s} = \frac{1\,\underline{/17.46°} - 1}{j1.5} = 0.202\,\underline{/8.73°}$$

(b) Increasing i_F by a factor 1.6 also increases $|E_a|$ by the same factor. Then we have

$$P_G = 0.2 = \frac{1.6}{1.5} \sin \delta_m$$

Thus $\delta_m = 10.81°$ and $E_a = 1.6\,\underline{/10.81°}$. Then

$$I_a = \frac{1.6\,\underline{/10.81°} - 1}{j1.5} = 0.430\,\underline{/-62.31°}$$

As a check we note that $0.430 \cos(-62.31°) = 0.200$.

8.8 SYNCHRONOUS CONDENSOR

As a final application of our model we consider a type of synchronous machine called a *synchronous condensor*. Consider the case of a round-rotor machine with $r = 0$. The power delivered by the machine to a bus with voltage V_a is given by (8.35) and is repeated here for convenience.

$$P_G = \operatorname{Re} V_a I_a^* = \frac{|E_a||V_a|}{X_s} \sin \delta_m \tag{8.35}$$

From (8.35) we can see

$$\delta_m > 0 \Rightarrow P_G > 0 \quad \text{(generator)}$$

$$\delta_m < 0 \Rightarrow P_G < 0 \quad \text{(motor)}$$

$$\delta_m = 0 \Rightarrow P_G = 0 \quad \text{(unloaded)}$$

We get similar results if $r \neq 0$ and/or the synchronous machine is salient pole. Consider the case when $\delta_m = 0$ (i.e., the synchronous machine is unloaded). Can such a machine serve a useful purpose?

The answer is yes, if it delivers reactive power. From (8.36), with $\delta_m = 0$, the reactive power is

$$Q_G = \frac{|V_a|(|E_a| - |V_a|)}{X_s} \tag{8.47}$$

Thus if $|E_a|$ is greater than $|V_a|$, the generator supplies reactive power. Since this is also the action of a capacitor bank connected to the bus, a generator specifically built to operate in this mode is called a synchronous condensor. Physically, a synchronous condensor of large MVA rating is large; it looks like a generator without a turbine! In large sizes, synchronous condensors are cheaper than capacitor banks and provide a very convenient and continuous control of reactive power by adjusting the field current i_F.

We note that if the synchronous machine is salient pole, the formula for Q_G, given by (8.46), reduces to (8.47) with X_s replaced by X_d. Finally, we remind the reader that injected reactive power at the receiving end of a long transmission line can be very necessary and very effective in maintaining the receiving-end voltage.

8.9 SUMMARY

For steady-state positive-sequence operation a useful two-reaction synchronous machine model may be derived by considering the various flux components making up the rotating air-gap flux wave. In this way we get an understanding of the physical bases of armature reaction and of direct axis and quadrature axis reactances. With the model we can determine the effects of varying the turbine power and field current on the generator terminal voltage and complex power output.

PROBLEMS

8.1. A generator delivers $P_G = 1.0$ at 0.8 PF lagging to a bus with voltage $|V_a| = 1.0$. Sketch a phasor diagram and calculate $|E_a|$ for
(a) A round rotor: $X_s = 1.0$, $r = 0$.
(b) A salient-pole rotor: $X_d = 1.0$, $X_q = 0.6$, $r = 0$.

8.2. A generator with reactances $X_d = 1.6$ and $X_q = 0.9$ delivers $S_G = 1\ \underline{/45°}$ to a bus with voltage $V_a = 1\ \underline{/0°}$. Find I_a, I_{ad}, I_{aq}, E_a, and the rotor angle θ_0 (i.e., the rotor angle θ at $t = 0$).

8.3. Suppose that in Problem 8.2 there is a symmetric 3ϕ short circuit (fault) at the generator terminals. Find the steady-state values of I_a, I_{ad}, and I_{aq} assuming that E_a does not change from its prefault value.

8.4. Given a generator with $V_a = 1\ \underline{/0°}$, $I_a = 1\ \underline{/60°}$, $X_d = 1.0$, $X_q = 0.6$, and $r = 0.1$, find S_G, I_{ad}, I_{aq}, and E_a.

8.5. A round-rotor generator ($X_s = 1.0$, $r = 0.1$) is synchronized to a bus whose voltage is $1\ \underline{/0°}$. At synchronization $i_F = 1000$ A (actual). The generator is then adjusted until $S_G = 0.8 + j0.6$. ($S_G =$ power supplied to the generator bus.)

(a) Find i_F and the generator efficiency (assuming no generator losses except I^2R losses).

(b) With the same i_F, what is the ultimate (maximum) active power the generator can deliver?

8.6. A round-rotor generator ($X_s = 1.0$, $r = 0$) is synchronized to a bus whose voltage is $1\underline{/0°}$. At synchronization $i_F = 1000$ A (actual). The mechanical power is then increased until $P_G = 0.8$. Now i_F is adjusted.

(a) Plot a curve of $|I_a|$ versus i_F.

(b) Plot a curve of Q_G versus i_F.

(c) When $|I_a|$ is a minimum, what is I_a?

8.7. For a salient-pole generator (with $X_d > X_q$) show that maximum P_G occurs for $\delta < \pi/2$. Assume that $r = 0$.

8.8. Consider a salient-pole generator delivering power through a short transmission line to an infinite bus. $V_\infty = 1\underline{/0°}$, $|E_a| = 1.4$. The active power delivered to the infinite bus is 0.6. We are given the generator reactances $X_d = 1.6$ and $X_q = 1.0$ and the line reactance $X_L = 0.4$. Neglect resistances and find E_a and I_a.

8.9. Refer to Fig. P8.9 and assume that

	G_1	G_2
	$X_d = 1.1$	$X_s = 1.0$
	$X_q = 0.7$	$r = 0.1$

We are given $P_{G2} = 1.0$. Find ϑ_2, Q_{12}, Q_{21}, S_{G1}, S_{G2}, I_{G1}, I_{G2}, E_{a1}, and E_{a2}.

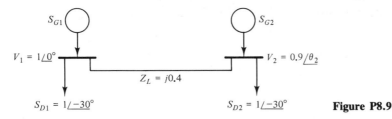

Figure P8.9

Power System Stability

chapter 9

9.0 INTRODUCTION

In Chapter 8 we derived simple round-rotor and salient-pole generator models. Although strictly speaking, these models are limited to (positive sequence) steady-state applications, it turns out they can also be used in cases where the (positive-sequence) generator and bus phasor voltages are "slowly varying." Provided that the variation in amplitude and/or phase is slow enough, we get almost exact results by substituting the variable amplitudes and/or phases in the steady-state expressions; this intuitively reasonable approach is called *pseudo-steady-state analysis*. As applied to (8.35), for example, we would replace the constants P_G, $|E_a|$, $|V_a|$, and δ_m by (slowly varying) functions of time $P_G(t)$, $|E_a(t)|$, $|V_a(t)|$, and $\delta_m(t)$.

We are going to consider the application of pseudo-steady-state analysis to problems involving electromechanical transients. Because of the high moments of inertia of turbine-generator sets, the key requirement of slowly varying phase can be justified. The most important problem we will consider is called the problem of *transient stability*. It concerns the maintenance of synchronism between generators following a severe disturbance.

A typical case is the following. A lightning stroke hits a transmission line, breaking down the air between a pair of conductors, or from line to ground and creating an ionized path. Sixty-hertz line currents also flow through this ionized path, maintaining the ionization even after the lightning stroke energy has been dissipated. This constitutes a short circuit or "fault" as real as if the conductors were in physical contact. To remove the fault, relays detect the short circuit and cause circuit breakers to open at both ends of the line. The line is deenergized and the air deionizes, reestablishing its insulating properties. We say that the fault "clears." The circuit breakers are set to reclose automatically after a preset interval of time, reestablishing the original circuit. This sequence of events (called a *fault sequence*) constitutes a "shock" to the power system and is accompanied by a transient.

The key question of transient stability is this: After a transient period, will the system "lock" back into a steady-state condition, maintaining synchronism? If it does, the system is said to be *transient stable*. If it does not, it is unstable and the system may break up into disconnected subsystems or "islands" which in turn may experience further instability.

In what has been called the "Great Blackout of 1965," improper action of a protective relay initiated a sequence of events that caused a cascade of circuit breaker openings (or trippings) which broke the interconnected Northeast power system into separate parts. Eventually, 30 million people were affected, service was interrupted for up to $17\frac{1}{2}$ hours, and a generator in New York City was severely damaged. Most power interruptions are not as severe. Still, there are about 35 such events a year, on average, in the United States, which are considered serious interruptions of service. We see, then, that the problem of transients and in particular, stability, is an important one.

In general, one must resort to computer simulation to describe these transients quantitatively. In this chapter, however, we will consider a simple case for which there is an elegant and simple solution. The method of solution is called the *equal-area method*. The insights one gets by using this method extend in a qualitatively way to more general cases.

9.1 MODEL

We consider a system initially operating in a steady-state or equilibrium state designated by $\mathbf{x}^0$. In this case we assume that:

1. All generators are rotating at synchronous speed corresponding to 60 Hz.
2. All ac voltages and currents are sinusoids.
3. All field currents i_F are constants.
4. All loads are constants.
5. All mechanical power inputs to the generators are constants.

In particular we note that the system is in the sinusoidal steady state. In the general case we now "disturb" this equilibrium state by a sudden change in input, or load, or structure, or a sequence of such changes, and consider the transient that occurs.

We will be considering a structural change. In Fig. 9.1 we show a single generator connected through a single transmission line to a very large system approx-

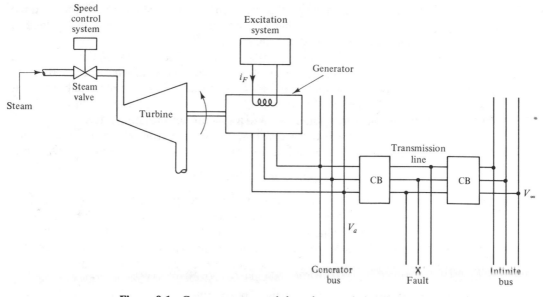

Figure 9.1   Generator connected through transmission line to infinite bus.

imated by an infinite bus. There are circuit breakers (CBs) at each end of the line. The fault location is identified by an ×. For our present purposes the details regarding the fault are unimportant. We simply assume that the fault is severe enough so that relays (not shown) detect and locate the fault and cause the circuit breakers (CBs) to open (to trip). For simplicity we ignore any transformers or assume that they have been "absorbed" into the generator or line models.

Consider the following example of a fault sequence.

Fault Sequence

Stage 1: System is in a prefault steady-state condition, or equilibrium state, $\mathbf{x}^0$.

Stage 2: Fault occurs at $t = 0$, followed by instantaneous openings of CBs. Fault "clears" before circuit breakers attempt to reclose.

Stage 3: Circuit breakers reclose at time $t = T$ and remain closed. Typically, T is less than 1 second.

Question. Does the system return to the equilibrium state $\mathbf{x}^0$?

Definition. System is said to be transient stable (for the given fault sequence) if it returns to the equilibrium state $\mathbf{x}^0$.

Note 1: We will find that there is a T_{critical}. For $T < T_{\text{critical}}$ system is transient stable. For $T \geq T_{\text{critical}}$ it is not.

Note 2: There are many possible modifications of this basic fault sequence. We will consider some simple modifications later in this chapter.

Before we can proceed we need to develop suitable system models for the three stages of the fault sequence. Note that a steady-state model is appropriate for stage 1, but dynamic models are required for stages 2 and 3.

The dynamic models must describe what is observed in practice—that the transients are basically electromechanical in nature, involving what is termed "rotor angle swings." The terminology may be explained as follows. In the steady state, as we discussed in Chapter 8, the rotor angle θ increases uniformly with time (i.e., $\theta = \omega_0 t + \theta_0$); as measured with respect to a synchronous rotating reference frame (at an angle $\omega_0 t$), the rotor angle θ_0 is a constant. During electromechanical transients the rotor angle no longer increases uniformly. We can describe the motion as follows:

$$\theta(t) = \omega_0 t + \theta_0 + \Delta\theta(t) \tag{9.1}$$

In this description there is an increment $\Delta\theta(t)$ superimposed on the uniform rotation. As measured from a synchronous rotating reference frame (at an angle $\omega_0 t$), the rotor angle is $\theta_0 + \Delta\theta(t)$. In one mode of behavior $\Delta\theta(t)$ is observed to oscillate around the mean angle θ_0, and this is described as a "rotor angle swing." The reference, of course, is to the rotor angle measured with respect to the synchronous rotating reference frame.

The very large moments of inertia of the coupled rotors of turbine-generator sets justify the notion that $\Delta\theta(t)$ is "slowly varying" [i.e., that $\Delta\dot\theta(t)$ is relatively small at least in the interval under consideration]. Then, as an approximation,

$$\dot\theta(t) = \omega_0 + \Delta\dot\theta(t) \approx \omega_0 \tag{9.2}$$

We next consider the internal (open-circuit) generator voltage. In Chapter 8 we found the open-circuit voltage associated with uniform rotation. From (8.8) the voltage magnitude is constant (proportional to i_F) and the phase angle is $\theta_0 - \pi/2$. We will make use of something similar in the present case.

Assume for now that during electromechanical transients, the field current i_F remains constant; in Chapter 10 we will reexamine this assumption. With i_F constant, going back to the fundamental relationship given by (8.2), we have the flux linking coil aa',

$$\phi = \phi_{\max} \cos \theta \tag{8.2}$$

Now if $\theta(t) = \omega_0 t + \theta_0 + \Delta\theta(t)$ and ϕ links all N turns of the phase a winding, then

$$\lambda_{aa'}(t) = N\phi_{\max} \cos [\omega_0 t + \theta_0 + \Delta\theta(t)] \tag{9.3}$$

$$e_{aa'}(t) = -\frac{d\lambda_{aa'}}{dt}$$

$$= (\omega_0 + \Delta\dot\theta)N\phi_{\max} \sin [\omega_0 t + \theta_0 + \Delta\theta(t)] \tag{9.4}$$

Using the approximation (9.2), we then have the internal (open-circuit) voltage

$$e_{aa'}(t) = E_{\max} \cos \left[\omega_0 t + \theta_0 + \Delta\theta(t) - \frac{\pi}{2} \right] \tag{9.5}$$

where $E_{max} = \omega_0 N\phi_{max}$ is the same value as calculated in Chapter 8. Defining δ as the phase angle of the phase α internal voltage, we have from (9.5),

$$\delta = \theta_0 + \Delta\theta - \frac{\pi}{2} \tag{9.6}$$

If $\Delta\theta \equiv 0$, (9.5) reduces to (8.6), δ is a constant. If, on the other hand, $\Delta\theta$ varies, then δ also varies and at the same rate. Equation (9.6) gives the vital connection between δ, the (electrical) internal voltage phase angle, and $\theta_0 + \Delta\theta$, the (mechanical) rotor angle (with respect to the synchronous rotating reference frame).

In our modeling of the configuration shown in Fig. 9.1, we will assume a lossless generator ($r = 0$) and a short lossless transmission line ($R_L = 0$) which can be absorbed into the generator model as described at the end of Section 8.5. Thus the "generator" has augmented reactances $\tilde{X}_d = X_d + X_L$, $\tilde{X}_q = X_q + X_L$, and a "terminal" voltage V_∞; to simplify the notation we will assume that $/V_\infty = 0$.

If δ is constant, we can use (8.45) to find the per phase active (average) power delivered by the generator. In (8.45) we would replace $|V_a|$ by $|V_\infty|$, X_d and X_q by $\tilde{X}_d$ and $\tilde{X}_q$, and δ_m by δ (because $/V_\infty = 0$). With these changes we rewrite (8.45) and at the same time we indicate the equivalence, for balanced systems, between instantaneous three-phase and per phase active power in the per unit system. The reader should check that (2.24), after per unit normalization, does in fact give that result. Thus we are led to the following expression for (p.u.) instantaneous generator power for the system in Fig. 9.1:

$$p_{3\phi}(t) = P_G(\delta) = \frac{|E_a||V_\infty|}{\tilde{X}_d} \sin\delta + \frac{|V_\infty|^2}{2}\left(\frac{1}{\tilde{X}_q} - \frac{1}{\tilde{X}_d}\right)\sin 2\delta \tag{9.7}$$

Suppose now that we have a transient where $\Delta\theta(t)$ is slowly varying. From (9.6) this reflects into a slowly varying $\delta(t)$. Nevertheless, in accordance with pseudo-steady-state analysis, as long as $\delta(t)$ varies slowly, we assume that we may still use (9.7).

To simplify (9.7) somewhat, we note that it is in the form

$$p_{3\phi}(t) = a \sin\delta(t) + b \sin 2\delta(t) \tag{9.8}$$

where a and b are constants. We remind the reader that for a round-rotor model, $X_q = X_d$ and therefore $b = 0$.

In the next chapter we will derive a more complete and accurate generator model. At that time we can remove the restriction that the field current i_F is constant. We end up with a more accurate expression, in the form (9.8), but with different constants a and b.

For the rest of this chapter, however, we will continue to use (9.7). Its great simplicity, and the value of its qualitative results, justify its use even if there are some quantitative errors.

Equation (9.7) is an "electrical" equation. We next turn to the "mechanical" equation.

9.2 ENERGY BALANCE

We continue with the modeling of the turbine-generator. Initially, consider the turbine-generator as an ideal lossless converter of mechanical to electrical energy. In describing the conversion process it is convenient to use actual units. Thus we will not be using the per unit system until further notice. Neglecting losses, in the prefault steady state

$$P_M^0 = 3P_G(\delta^0) \qquad (9.9)$$

where $P_G(\delta^0)$ is the actual per phase power and P_M^0 is the mechanical power supplied by the turbine. The superscript 0 designates the prefault equilibrium value. $P_G(\delta^0)$ can be found from (9.7), where we substitute actual values of voltage and reactance.

During a transient, the shaft of the coupled turbine-generator unit accelerates and the kinetic energy stored in the rotating inertia changes. Neglecting all losses and noting that $p_{3\phi}(t) = 3P_G(\delta(t))$, we have

$$P_M^0 = 3P_G(\delta) + \frac{d}{dt}W_{\text{kinetic}} \qquad (9.10)$$

We are assuming that P_M remains constant at its prefault value during the very short interval under consideration. Note that in (9.10) the power angle δ is a variable (i.e., it is no longer fixed at its equilibrium value δ^0).

Next, to improve the realism of the model it is desirable to add a small mechanical power loss term in (9.10). In particular, we include the mechanical friction in the bearings. Then, we get

$$P_M^0 = 3P_G(\delta) + \frac{d}{dt}W_{\text{kinetic}} + P_{\text{friction}} \qquad (9.11)$$

Equation (9.11) is an expression of conservation of instantaneous power; the mechanical power developed by the turbine minus the mechanical friction losses, minus the three phase electrical power out, equals the rate of change of (stored) kinetic energy.

We next consider the term dW_{kinetic}/dt in more detail. We would like to express it in terms of δ. From basic mechanics

$$W_{\text{kinetic}} = \frac{1}{2}J\dot{\theta}^2 \qquad (9.12)$$

where θ is the actual rotor angle and J is the moment of inertia of the coupled turbine and generator rotors. Thus, using (9.1), (9.2), and (9.6),

$$\frac{dW_{\text{kinetic}}}{dt} = J\dot{\theta}\ddot{\theta} \approx J\omega_0\Delta\ddot{\theta} = J\omega_0\ddot{\delta} \qquad (9.13)$$

Note that the constant, $J\omega_0$, is the (actual) angular momentum of the coupled generator and turbine rotors, at synchronous speed. It will be convenient later to normalize (9.13) with respect to $S_B^{3\phi}$, the MVA rating of the generator. In this case instead of $J\omega_0$

we get

$$\frac{J\omega_0}{S_B^{3\phi}} = \frac{\frac{1}{2}J\omega_0^2}{S_B^{3\phi}}\frac{2}{\omega_0} = \frac{1}{\pi f_0}\left(\frac{W_{\text{kinetic}}^0}{S_B^{3\phi}}\right) \tag{9.14}$$

where $f_0 = 60$ Hz. The normalized kinetic energy in (9.14) has physical significance. Let

$$H \triangleq \frac{W_{\text{kinetic}}^0}{S_B^{3\phi}} \qquad \text{megajoules/MVA} \tag{9.15}$$

be the "per unit inertia constant." The units of H are also seconds. Typically, H ranges between 1 and 10 seconds, depending on the size and type of machine. The range of values is small considering the enormous variation in J for different machines.

Consider next the term P_{friction}. Assume linear friction (i.e., friction torque proportional to angular velocity). Then using (9.1), (9.2), and (9.6), we have

$$\begin{aligned}
P_{\text{friction}} &= k\dot{\theta}^2 = k(\omega_0 + \Delta\dot{\theta})^2 \\
&= k\omega_0^2 + 2k\omega_0\Delta\dot{\theta} + k(\Delta\dot{\theta})^2 \\
&\approx k\omega_0^2 + 2k\omega_0\Delta\dot{\theta} \\
&= k\omega_0^2 + 2k\omega_0\dot{\delta}
\end{aligned} \tag{9.16}$$

In accordance with (9.2) we have neglected $(\Delta\dot{\theta})^2$ compared with $2\omega_0\,\Delta\dot{\theta}$. We note that the constant term $k\omega_0^2$ is the power lost in friction when the generator shaft is rotating at synchronous speed. It is convenient in doing the bookkeeping of power balance to associate this term with the turbine. Thus if we subtract $k\omega_0^2$ from P_M^0, we get

$$\tilde{P}_M^0 \triangleq P_M^0 - k\omega_0^2 \tag{9.17}$$

which may be called the "turbine power after synchronous speed losses." The term $k\omega_0^2$ is negligible compared with P_M^0 and for our numerical work it will not be necessary to make a distinction between P_m^0 and $\tilde{P}_M^0$ in what follows.

We now return to (9.11) and normalize with respect to the generator rating. Using (9.13) to (9.17), we get

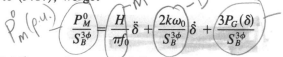

$$\frac{P_M^0}{S_B^{3\phi}} = \frac{H}{\pi f_0}\ddot{\delta} + \frac{2k\omega_0}{S_B^{3\phi}}\dot{\delta} + \frac{3P_G(\delta)}{S_B^{3\phi}} \tag{9.18}$$

Concerning (9.18), we observe the following:

1. We have replaced $\tilde{P}_M^0$ by P_M^0.
2. $P_M^0/S_B^{3\phi} = P_M^0(\text{p.u.})$.
3. It is notationally convenient to define $M \triangleq H/\pi f_0$.
4. We will define a very small constant $D \triangleq 2k\omega_0/S_B^{3\phi}$.
5. $3P_G(\delta)/S_B^{3\phi} = P_G(\delta)/S_B = P_G(\delta)(\text{p.u})$.
6. $P_G(\delta)(\text{p.u.})$ is given by (9.7), with voltages and impedances expressed in p.u.

Finally, we have from (9.18), the following equation involving per unit quantities:

$$M\ddot{\delta} + D\dot{\delta} + P_G(\delta) = P_M^0 \tag{9.19}$$

SWING EQ.

This is a nonlinear differential equation, called the *swing equation* because it describes "swings" in the power angle δ during transients. With this differential equation we can discuss stability in a quantitative way. We note that (9.19) marks our return to the per unit system; to simplify the notation, however, we will suppress the p.u. notation.

First, we note the analogy between (9.19) and a spring-mass system. If we let δ be a displacement, $P_G(\delta)$ a nonlinear spring force, and P_M^0 an applied force, we get the analogy in Fig. 9.2.

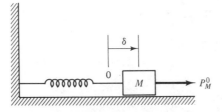

Figure 9.2   Spring-mass analog.

The equilibrium position δ^0 of the mass satisfies $P_G(\delta^0) = P_M^0$. When $P_M^0 = 0$, then $\delta^0 = 0$ (i.e., the spring is in its relaxed position). From the mechanical analogy, and with small D, we expect lightly damped oscillatory responses when the equilibrium state is disturbed. Because the spring is nonlinear, the response will depend critically on the nature of the disturbance.

We can also review the physical connection between the rotor position and δ, which will help us interpret (9.19) more directly, in rotor terms. From (9.6) we see that $\delta = \theta_0 + \Delta\theta(t) - \pi/2$, where $\theta_0 + \Delta\theta(t)$ is the rotor angle with respect to the synchronous rotating reference frame. To find δ we simply subtract 90° from this rotor angle. Thus if we are "riding" the synchronous reference frame we can easily observe δ, as a mechanical angle. Alternating, we can try to visualize the behavior of δ from the outside (i.e., from a stationary reference frame). Noting that $\omega_0 t = 0 \pmod{360°}$ every 1/60 sec, the synchronous reference frame appears stationary if we only consider the instants $t = 0, 1/60, 2/60, \ldots$. Thus δ, at those instants, can be measured with respect to a stationary reference frame. The situation is shown in Fig. 9.3.

We may imagine the rotor to be illuminated by a stroboscope so that we only see the rotor at $t = 0, 1/60, 2/60, \ldots$. During stage 1 of the fault sequence $\delta = \delta^0 =$ constant. (Our strobed rotor stands still.) During stage 2 the circuit breakers are open, so $P_G = 0$, while $P_M = P_M^0$ remains constant. The rotor accelerates and thus δ increases. Finally, during stage 3, the circuit breakers close and remain closed. The graph of δ during the three stages is shown in Fig. 9.4, which shows a stable lightly damped oscillation for $t \geq T$.

We note that unstable behavior is also possible. Curve (b) in Fig. 9.11 illustrates that possibility.

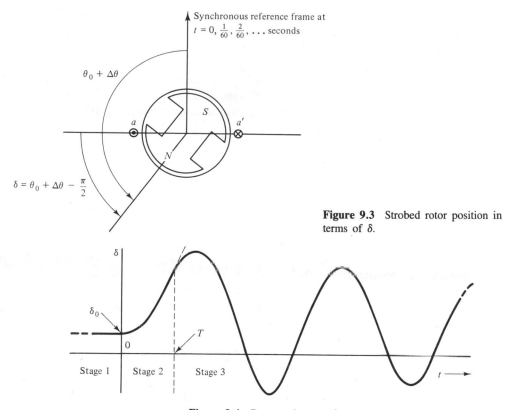

Figure 9.3 Strobed rotor position in terms of δ.

Figure 9.4 Rotor swing transient.

Exercise 1.   Consider the mechanical analog in Fig. 9.2. What is the physical significance of stages 1, 2, and 3 in terms of the mechanical model?

9.3 LINEARIZATION OF SWING EQUATION

In stage 1, $P_M^0 = P_G(\delta^0)$, and the system is in a state of equilibrium with $\delta = \delta^0 =$ constant (and thus $\dot{\delta} = 0$). In stage 2 there is accelerating power which drives the state $(\delta, \dot{\delta})$ away from equilibrium. If the departure from the equilibrium state is "small," we may be able to investigate behavior in stage 3 by the method of linearization.

To linearize (9.19), let $\delta = \delta^0 + \Delta\delta$ with $\Delta\delta$ small, and do the usual two-term Taylor series expansion of $P_G(\delta^0 + \Delta\delta)$. After canceling the constant terms $P_G(\delta^0) = P_M^0$, we get

$$M\Delta\ddot{\delta} + D\,\Delta\dot{\delta} + \mathcal{T}\,\Delta\delta = 0 \tag{9.20}$$

where $\mathcal{T} \triangleq dP_G(\delta^0)/d\delta$ is called a *synchronizing power coefficient* or *stiffness coefficient*. $\mathcal{T}$ is the slope of the power output curve at the operating point δ^0. Its value can be determined by differentiating (9.7). $\mathcal{T}$ is seen to be a maximum at no load ($\delta^0 = 0$). As the load (and δ^0) increase, $\mathcal{T}$ decreases monotonically and, for some value of δ^0 less than or equal to $\pi/2$, goes to zero and then negative.

The consequences of this with respect to system behavior may be seen by solving (9.20) for small arbitrary initial conditions or may be inferred (with less work) by finding the roots of the characteristic polynomial corresponding to (9.20).

The characteristic equation is

$$Ms^2 + Ds + \mathcal{T} = 0 \tag{9.21}$$

with roots (natural frequencies)

$$s_{1,2} = \frac{-D \pm \sqrt{D^2 - 4M\mathcal{T}}}{2M} \tag{9.22}$$

For normal operating conditions $M\mathcal{T} \gg D^2$ and we get $s_{1,2} = \alpha \pm j\omega$, where $\alpha < 0$ and $\omega \approx (\mathcal{T}/M)^{1/2}$. With D very small we get lightly damped oscillatory behavior with the amplitude of the oscillations in $\Delta\delta$, decaying to zero exponentially. Thus $\delta = \delta^0 + \Delta\delta$ returns to δ^0. However, with a negative $\mathcal{T}$, one of the natural frequencies is positive and the linear model predicts "runaway" exponentially increasing behavior for almost all initial conditions.

Example 9.1   Small Power Angle Oscillations

In Fig. 9.1 assume a round-rotor generator delivering power in the steady state to an infinite bus through a transmission line with reactance $X_L = 0.4$. Assume that $|E_a| = 1.8$, $|V_\infty| = 1.0$, $H = 5$ sec, and $X_d = X_q = 1.0$. There is a small disturbance which causes a transient. For example, the breakers open and then quickly close. Neglect damping and find the frequency of the power angle oscillations for $P_G^0 = 0.05, 0.5$, and 1.2.

Solution   Using (9.7), we can find the steady-state power angle δ.

$$P_G(\delta^0) = \frac{1.8 \times 1.0}{1.0 + 0.4} \sin \delta^0 = 1.286 \sin \delta^0$$

Then we can find the synchronizing power coefficient $\mathcal{T}$:

$$\mathcal{T} = \frac{dP_G(\delta^0)}{d\delta} = 1.286 \cos \delta^0$$

Using $M \triangleq H/\pi f_0$, the frequency of oscillations is given by

$$\omega = \left(\frac{\mathcal{T}}{M}\right)^{1/2} = \left(\frac{\mathcal{T}\pi f_0}{H}\right)^{1/2}$$

The results for the three cases are tabulated in Table E9.1. We note that as P_G increases, the synchronizing power coefficient drops, and this decreases the oscillatory frequency. The result is physically reasonable, as can be checked from the spring-mass analogy in Fig. 9.2; the "softer" the spring, the lower the oscillatory frequency we expect.

We observe that the oscillations are fairly low in frequency, in the order of 1 Hz.

TABLE E9.1

P_G^0	δ^0 (deg)	$\mathcal{T}$	ω (rad/sec)	f(Hz)
0.05	2.23°	0.999	6.14	0.977
0.5	22.89°	0.921	5.89	0.938
1.2	68.96°	0.359	3.68	0.586

Thus we expect $\Delta\theta = \delta$ to be small compared with ω_0, as assumed in (9.2). Assume that for $P_G^0 = 0.5$ we have an oscillation in $\Delta\delta$ of amplitude 0.25 rad (14.32°). Then $\delta(t) = \delta^0 + 0.25 \sin(5.89t + \phi)$, and $\Delta\dot{\theta}(t) = \dot{\delta}(t) = 1.47 \cos(5.89t + \phi)$. Thus $|\Delta\dot{\theta}| \leq 1.47$, which is certainly small compared with $\omega_0 = 377$.

9.4 SOLUTION OF NONLINEAR SWING EQUATION

We will be concerned mainly with cases in which the departures from the equilibrium state are large. Thus the method of linearization is inapplicable, and we need to develop an appropriate method. Of course, numerical integration using digital computers is always possible, but we seek a simpler, more portable, method which enhances understanding.

We will start by neglecting the very small mechanical damping. Thus for now assume that $D = 0$. Later we will consider its effect qualitatively. Now consider the behavior in the three stages of the fault sequence. ~per phase power.

Stage 1. $\delta = \delta^0$, $\dot{\delta} = 0$, $P_G(\delta^0) = P_M^0$.

Stage 2.   With the transmission line open, and with $D = 0$, (9.19) reduces to

$$M\ddot{\delta} = P_M^0 \qquad 0 \leq t < T \tag{9.23}$$

with solution, found directly by integration, const?

$$\dot{\delta}(t) = \frac{P_M^0}{M}t + \dot{\delta}(0) = \frac{\pi f_0 P_M^0}{H}t \text{ rad/sec}$$

$$\delta(t) = \frac{P_M^0}{2M}t^2 + \delta^0 = \frac{\pi f_0 P_M^0}{2H}t^2 + \delta^0 \text{ rad} \tag{9.24}$$

This behavior is observed in Fig. 9.4 as a function of t. We will find a state-space description in the δ, $\dot{\delta}$ plane more useful. Thus using (9.24), we note that

$$\delta - \delta^0 = \frac{P_M^0}{2M}t^2 = \frac{M}{2P_M^0}\left(\frac{P_M^0}{M}t\right)^2 = \frac{M}{2P_M^0}\dot{\delta}^2 \tag{9.25}$$

Thus in the state space (or phase plane) we get a parabola in which only the branch in the first quadrant applies to our problem and is shown in Fig. 9.5. In the figure, the arrow shows the direction of the "motion" along the trajectory in the phase plane from the initial state δ^0 to the state δ_T reached after T seconds. We calculate δ_T using (9.24)

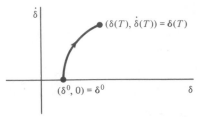

Figure 9.5 Trajectory in phase plane
$(0 \le t \le T)$.

and use it as an "initial state" for stage 3. *Note:* If we had included D, the numerical results would have been indistinguishable from the above for typical (small) values of T.

Stage 3. The transmission line is reconnected at $t = T$. From (9.19), with $D = 0$, we get for $t \ge T$

$$M\ddot{\delta} + P_G(\delta) = P_M^0 \qquad \text{breakers closed.} \qquad (9.26)$$

with "initial" conditions $\delta(T) = \delta_T$ and $\dot{\delta}(T) = \dot{\delta}_T$ determined from (9.24). Defining $P(\delta) = P_G(\delta) - P_M^0$ we get an equivalent, but notationally simpler, version of (9.26):

$$M\ddot{\delta} + P(\delta) = 0 \qquad \text{breakers closed.} \qquad (9.27)$$

We are interested in the properties of the solution of (9.27) relating to transient stability. In this connection "energy" techniques have proved to be useful. Consider the spring-mass analog in Fig. 9.2 as applied to (9.27). Since we have effectively absorbed P_M^0 into the spring force characteristic $P(\delta)$, we now have an isolated system with a mass and a nonlinear spring and, with $D = 0$, no friction. In fact, we now have a conservative system, and know, from basic mechanics, that the sum of potential energy (P.E.) and kinetic energy (K.E.) for such a system is constant. Thus using well-known formulas for K.E. and P.E., we have an expression for the total energy of the system in terms of the state $\boldsymbol{\delta} = (\delta, \dot{\delta})$:

KE + PE

$$V(\boldsymbol{\delta}) = \tfrac{1}{2}M\dot{\delta}^2 + \int_{\delta^0}^{\delta} P(u)\, du \qquad (9.28)$$

We note that at equilibrium (i.e., with $\delta = \delta^0$ and $\dot{\delta} = 0$), both the kinetic energy and the potential energy are zero. We will also use the terminology K.E. and P.E. when discussing our power system model. Although the terminology is very convenient, it can be misleading. Certainly, at equilibrium, with $\dot{\delta} = 0$, the K.E. of the *actual* turbine generator is not zero; it is $J\omega_0^2/2$.

At the beginning of stage 3, $\delta = \delta_T$ and $\dot{\delta} = \dot{\delta}_T$. Thus, using (9.28), there is an initial total energy $V(\boldsymbol{\delta}_T)$ in the system. Thereafter, the motion of δ and $\dot{\delta}$ is constrained by the requirement that the total energy of our conservative system remain constant in value. Thus, for $t \ge T$, we have

$$V(\boldsymbol{\delta}(t)) = \tfrac{1}{2}M\dot{\delta}(t)^2 + \int_{\delta^0}^{\delta(t)} P(u)\, du = \tfrac{1}{2}M\dot{\delta}_T^2 + \int_{\delta^0}^{\delta_T} P(u)\, du = V(\boldsymbol{\delta}_T) \quad (9.29)$$

(handwritten margin notes:) total energy

KE > PE Unstable
KE < PE Stable
(excess absorbed by?
transmission line.)

breakers closed

KE ⟹ rotor speed
PE ⟹ trans line.

We will now show how (9.29) provides useful information about the dynamic behavior of δ and $\dot{\delta}$.

The key factor determining behavior is the potential energy curve. We will use the notation

$$W(\delta) = \int_{\delta^0}^{\delta} P(u) \, du \tag{9.30}$$

for the potential energy. The reader should check that with a linear spring force [i.e., $P(\delta) = k(\delta - \delta^0)$], $W(\delta) = k(\delta - \delta^0)^2/2$. Thus we get a quadratic function with a minimum (zero) at δ^0. With a nonlinear force we get a more interesting "potential well." In Fig. 9.6 we show a representative $P(\delta)$ at the top of the figure, a corresponding $W(\delta)$ below it, and for a given initial condition, a phase plane trajectory at the bottom of the figure. By checking the slopes of the $W(\delta)$ curve against the corresponding values of the $P(\delta)$ curve, the reader may check that the $W(\delta)$ curve has the right shape. In particular, note that W has a local minimum at $\delta = \delta^0$, where $W(\delta^0) = 0$, and has two neighboring local maxima at δ^u and δ^l. Let

$$W_{max} = \min \left(W(\delta^u), W(\delta^l) \right) \tag{9.31}$$

In the case shown in Fig. 9.6, with $P_M^0 > 0$, $W_{max} = W(\delta^u)$.

Let us consider now how we can determine the phase plane trajectory from the potential energy curve. Given δ^0, or equivalently, P_M^0 and $P_G(\delta)$, and the reclosure time T, we can calculate $\boldsymbol{\delta}_T = (\delta_T, \dot{\delta}_T)$ using (9.24). We show the point $\boldsymbol{\delta}_T$ in Fig. 9.6; it is the initial condition for the stage 3 phase plane trajectory. We also calculate the initial store of total energy $V(\boldsymbol{\delta}_T)$ using (9.28). This is shown as a horizontal line above the $W(\delta)$ curve in Fig. 9.6, for the case $V(\boldsymbol{\delta}_T) < W_{max}$. Also, since we know δ_T, we can read off the initial potential energy part of $V(\boldsymbol{\delta}_T)$ directly from the $W(\delta)$ curve; then the initial kinetic energy part is $V(\boldsymbol{\delta}_T) - W(\delta_T)$. These two components are shown in Fig. 9.6.

What happens next? Since $\dot{\delta}_T$ is positive, δ must increase. Say that δ increases to a value $\delta' < \delta_{max}$. From the $W(\delta)$ curve we can measure the new K.E. and therefore the new value of $\dot{\delta}'$. Thus we can find $\dot{\delta}$ as a function of δ and plot a trajectory. Notice what happens when δ reaches δ_{max}. At that point the K.E. $= 0$. This corresponds to $\dot{\delta} = 0$. From the mass-spring analogy we deduce that this corresponds to a maximum extension of the spring and a reversal of phase velocity. Thus δ decreases until at $\delta = \delta^0$ the K.E. is a maximum with $\dot{\delta}$ negative. δ decreases further until it reaches a minimum value δ_{min}. δ then increases until it reaches the value δ_{max} and the process repeats indefinitely with swings between δ_{min} and δ_{max}.

Note that any value of $\delta < \delta_{min}$, or $\delta > \delta_{max}$ is clearly impossible since it would imply a negative K.E. Thus we can think of $W(\delta)$ as a "potential barrier."

Suppose that we do not neglect friction (i.e., $D \neq 0$). In that case it is no longer true that $V(\boldsymbol{\delta}(t)) = V(\boldsymbol{\delta}_T)$. Physically, in terms of the spring-mass analogy, the energy must be slowly draining away, being converted into heat. It is simple to confirm this mathematically. Using (9.28), consider the time rate of change of $V(\boldsymbol{\delta})$. Using the

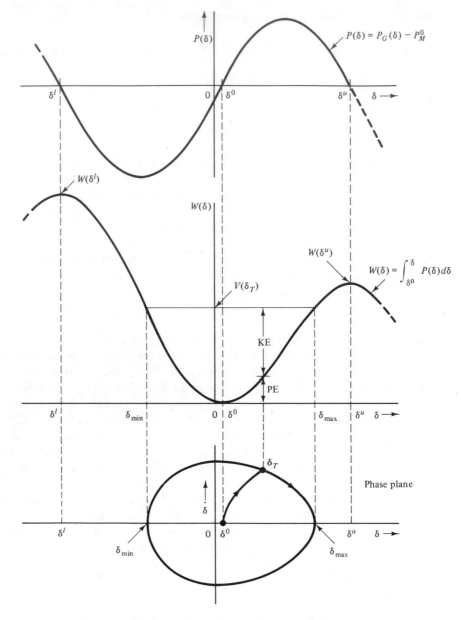

Figure 9.6 Phase plane trajectory from potential energy curve.

chain rule of differentiation, we have

$$
\dot{V}(\boldsymbol{\delta}(t)) = \frac{\partial V}{\partial \dot{\delta}} \frac{d\dot{\delta}}{dt} + \frac{\partial V}{\partial \delta} \frac{d\delta}{dt}
$$

$$
= M\dot{\delta}\ddot{\delta} + P(\delta)\dot{\delta} = [M\ddot{\delta} + P(\delta)]\dot{\delta}
$$

(9.32)

270

With $D = 0$, (9.27) confirms that $\dot{V}(\delta(t)) \equiv 0$; thus $V(\delta(t))$ must be a constant. But when we add the $D\delta$ term to the left side of (9.27), we get $M\ddot{\delta} + P(\delta) = -D\dot{\delta}$. Thus (9.32) reduces to

$$\dot{V}(\delta(t)) = -D\dot{\delta}^2 \le 0 \tag{9.33}$$

As long as the mass is moving the energy is being dissipated and $V(\delta(t))$ is decreasing from swing to swing. In Fig. 9.6, instead of a constant level $V(\delta_T)$ we have a slowly dropping level, and consequently the swings in δ decrease in amplitude. Thus instead of the closed curve in Fig. 9.6, we have a trajectory which spirals inward toward $\delta^0 = (\delta_0, 0)$. This same conclusion is drawn from the linearization analysis. Thus, in the case just described, and including (qualitatively) the effect of damping, we conclude that the system is transient stable; we eventually return to the equilibrium point δ^0.

The case just considered corresponds to a particular choice of reclosure time T. With longer reclosure times, δ_T will be a point farther out on the parabola in Fig. 9.5. This corresponds to a bigger value for $V(\delta_T)$. As we consider increasing values of $V(\delta_T)$, we come to unstable behavior.

In Fig. 9.7 are shown four reclosure times, T_1, T_2, T_3, and T_4. Using (9.24), we calculate the corresponding δ_{T1}, δ_{T2}, δ_{T3}, and δ_{T4}. Using (9.28), we calculate the corresponding V_1, V_2, V_3 and V_4.

By the same reasoning used in describing Fig. 9.5, we get the trajectories shown in the phase plane part of Fig. 9.7. Note that T_3 is the critical reclosure time, T_{critical}. For any $T > T_{\text{critical}}$, $\dot{\delta}(t)$ is always positive; in fact, the trend is for $\dot{\delta}(t)$ to increase with t (on the average); $\delta(t)$ increases monotonically with t. Thus we never return to δ^0 and the system is not transient stable. The phenomenon is called "pull-out" or "loss of synchronism." The generator will have to be resynchronized before it can deliver power to the system again. For those familiar with the terminology, we note that δ^0 is a stable equilibrium point, while $\delta^u = (\delta^u, 0)$ and $\delta^l = (\delta^l, 0)$ are unstable (saddle) points.

We notice that for our one-machine-infinite bus model the question of stability is settled on the basis of the "first swing." In the case we have been considering, if, after reaching a maximum value, δ starts to decrease, stability is assured. It may therefore be seen that the effect of including the small damping term $D\dot{\delta}$ is not critical since it has little effect over the short term. Usually, in deciding on stability the small mechanical damping term $D\dot{\delta}$ is neglected. The results then are slightly conservative (i.e., on the safe side).

We note next that the question of stability reduces to the question: Is $V(\delta_T) < W_{\text{max}}$? This question can be answered without drawing the trajectories in the phase plane. In fact, as we will now show, it is not even necessary to draw the graph of $W(\delta)$. Assume the usual case, with the generator delivering power, that is, assume that $P_M^0 > 0$, in which case $W_{\text{max}} = W(\delta^u)$. From (9.28) and the definition of W_{max}, $V(\delta_T) < W_{\text{max}}$ implies that

$$\tfrac{1}{2}M\dot{\delta}_T^2 + \int_{\delta^0}^{\delta_T} P(u)\, du < \int_{\delta^0}^{\delta^u} P(u)\, du \tag{9.34}$$

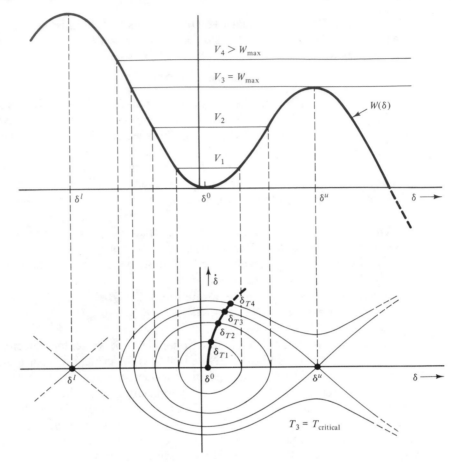

Figure 9.7 Illustrating instability.

or

$$\tfrac{1}{2}M\dot{\delta}_T^2 < \int_{\delta_T}^{\delta^u} P(\delta)\,d\delta \tag{9.35}$$

Next we note from (9.25) that

$$\tfrac{1}{2}M\dot{\delta}_T^2 = P_M^0(\delta_T - \delta^0) \tag{9.36}$$

Substituting (9.36) into (9.35), the condition for stability is

$$P_M^0(\delta_T - \delta^0) < \int_{\delta_T}^{\delta^u} P(\delta)\,d\delta \tag{9.37}$$

Suppose using (9.24) that we calculate $\delta_T = (P_M^0/2M)T^2 + \delta^0$. Knowing δ_T, we can check (9.37) by comparing areas as in the construction shown in Fig. 9.8. The left side of (9.37) is the area A_a and the right side is the area $A_{d\text{max}}$.

Since, as sketched, $A_a < A_{d\text{max}}$, the system is transient stable in this case.

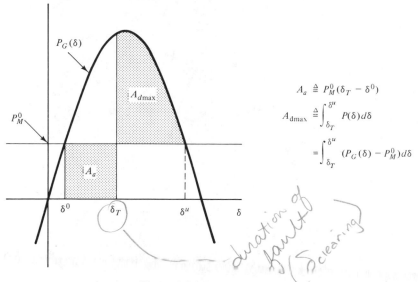

$$A_a \triangleq P_M^0(\delta_T - \delta^0)$$

$$A_{d\max} \triangleq \int_{\delta_T}^{\delta^u} P(\delta)\,d\delta$$

$$= \int_{\delta_T}^{\delta^u} (P_G(\delta) - P_M^0)\,d\delta$$

Figure 9.8 Test for stability.

Note: It is convenient to call A_a the acceleration area because during this stage $\ddot{\delta} > 0$. Similarly, we call $A_{d\max}$ a deceleration area because here $P_G(\delta) > P_M^0$ implies that $\ddot{\delta} < 0$; in terms of the spring-mass analogy the spring restraining force is greater than the applied force P_M^0.

The "equal-area stability criterion" as it applies to this problem can be stated as follows.

Equal-Area Stability Criterion.   If $A_a < A_{d\max}$, the system is transient stable. In words, if the accelerating area is less than the maximum decelerating area, the system is transient stable.

Assuming that the equal-area stability criterion is satisfied, our basic construction can also be used to give information about $\delta_{\max}$ and $\delta_{\min}$. Thus from Fig. 9.6 we see that $\delta_{\max}$ satisfies

$$V(\delta_T) = W(\delta_{\max}) = \int_{\delta^0}^{\delta_{\max}} P(\delta)\,d\delta \tag{9.38}$$

Repeating the calculations (9.34) to (9.37), we get

$$P_M^0(\delta_T - \delta^0) = \int_{\delta_T}^{\delta_{\max}} P(\delta)\,d\delta \tag{9.39}$$

We can draw a figure similar to Fig. 9.8 to show satisfaction of the condition (9.39). For a given δ_T this is shown in Fig. 9.9, where the deceleration area A_d is also defined. Figure 9.9 shows in very graphic terms that $\delta_{\max}$ is the δ for which $A_a = A_d$.

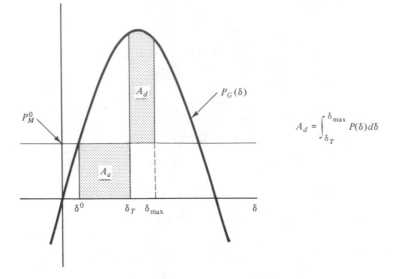

$$A_d = \int_{\delta_T}^{\delta_{max}} P(\delta)\,d\delta$$

Figure 9.9 Calculation of δ_{max}.

Using Fig. 9.9, we can see what happens as the reclosure time T is increased. As T increases, so does δ_T and A_a. For small increases, we can still find an A_d to balance A_a, but we finally reach a value of $T = T_{critical}$ where A_a is just critically balanced by $A_d = A_{dmax}$ and beyond this point we go unstable.

Now consider the calculation of δ_{min}. From Fig. 9.6 we see $W(\delta_{min}) = W(\delta_{max})$; hence

$$\int_{\delta^0}^{\delta_{min}} P(\delta)\,d\delta = \int_{\delta^0}^{\delta_{max}} P(\delta)\,d\delta \qquad (9.40)$$

Equivalently, subtracting the left side from the right,

$$\int_{\delta_{min}}^{\delta_{max}} P(\delta)\,d\delta = 0 \qquad (9.41)$$

Once again we can use a graphical interpretation involving areas (Fig. 9.10).

Once again the accelerating area (in the $-\delta$ direction) and the decelerating area are equal. *Note:* With $D = 0$, δ swings between δ_{min} and δ_{max} forever. With $D > 0$, the swings are attenuated and δ tends to δ^0.

Exercise 2. Consider the case of $P_M^0 < 0$ (i.e., the case of a synchronous motor driving a pump or a fan, for example). Sketch figures analogous to Figs. 9.6, 9.7, 9.9, and 9.10. *Hint:* While the breakers are open, the rotating machines are slowing down. Thus $\dot{\delta}_T$ will be negative.

Example 9.2

Consider the same generator and line as in Example 9.1. Thus $H = 5$ sec and $P_G(\delta) = 1.286 \sin \delta$. Assume that $P_G^0 = 0.5$.

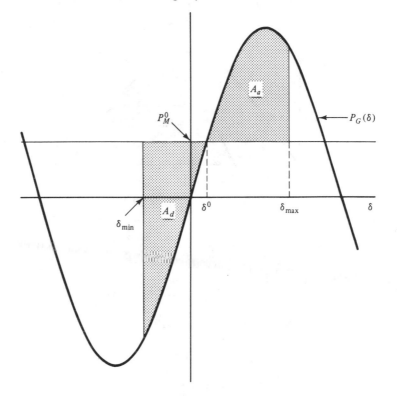

Figure 9.10 Calculation of $\delta_{\min}$.

(a) Find the critical clearing time T_{critical}.

(b) For a clearing time $T = 0.9T_{\text{critical}}$, find $\delta_{\max}$ and $\delta_{\min}$.

Solution   (a) With $P_G(\delta^0) = 1.286 \sin \delta^0 = 0.5$, we find that $\delta^0 = 22.89° = 0.400$ rad. We next find the "critical clearing angle" δ_{Tc} corresponding to T_{critical}, such that $A_a = A_{d\max}$, as suggested in Fig. E9.2.

δ_{Tc} is the largest value δ_T for which we can balance the accelerating area A_a with a deceleration area. This area is $A_{d\max}$. The calculation is simplified if we add the area A_c shown in Fig. E9.2 to both sides of the equation. Then carrying out the indicated operations we get in succession.

$$A_a + A_c = A_{d\max} + A_c$$

$$0.5(\pi - 2\delta^0) = \int_{\delta_{Tc}}^{\pi - \delta^0} (1.286 \sin \delta)\, d\delta$$

$$1.1708 = 1.286(\cos \delta_{Tc} + 0.9211)$$

$$\delta_{Tc} = 90.61° = 1.581 \text{ rad}$$

Note that we do not need to consider the Stage 2 dynamics in calculating δ_{Tc}. Next we calculate T_{critical} using (9.24), i.e., Stage 2 dynamics.

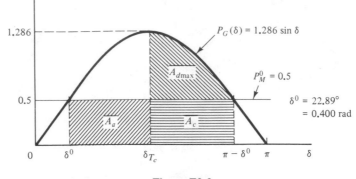

Figure E9.2

$$\delta_{Tc} = 1.581 = \frac{\pi \times 60 \times 0.5}{2 \times 5} T_{critical}^2 + 0.400 \qquad (9.24)$$

Thus $T_{critical} = 0.354$ sec.
(b) With $T = 0.9 T_{critical} = 0.319$ sec, we find, using (9.24)

$$\delta_T = \frac{\pi \times 60 \times 0.5}{2 \times 5}(0.319)^2 + 0.400 = 1.357 \text{ rad} = 77.75°$$

Then $A_a = P_M^0(\delta_T - \delta^0) = 0.5(1.357 \times 0.400) = 0.478$. Thus

$$A_d = 0.478 = \int_{\delta_T}^{\delta_{max}} (1.286 \sin \delta - 0.5) \, d\delta$$

$$= 1.286(\cos \delta_T - \cos \delta_{max}) - 0.5(\delta_{max} - \delta_T)$$

Putting in the value $\delta_T = 77.75° = 1.357$ rad we get

$$1.286 \cos \delta_{max} + 0.5 \delta_{max} = 0.4734$$

Solving for δ_{max} by iteration we get $\delta_{max} = 113.86° = 1.987$ rad. To find δ_{min} we use
(9.41). Then

$$\int_{\delta_{min}}^{\delta_{max}} (1.286 \sin \delta - 0.5) d\delta = 1.286(\cos \delta_{min} - \cos \delta_{max}) - 0.5(\delta_{max} - \delta_{min}) = 0$$

Then we get

$$1.286 \cos \delta_{min} + 0.5 \delta_{min} = 1.286 \cos \delta_{max} + 0.5 \delta_{max} = 0.4734$$

Noting $\delta_{min} < \delta^0$ we solve for δ_{min} by iteration. We get

$$\delta_{min} = -46.74° = -0.816 \text{ rad}$$

It is instructive in the above example to actually plot the swing curves in two
different cases. This is done in Fig. 9.11 assuming $D = 0.1M = 0.1H/\pi f_0 =$
0.00265. We consider two values of T: $T = 0.9 T_{critical} = 0.319$ sec and $T = 0.356$

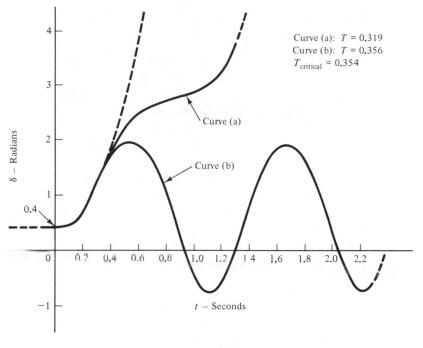

Figure 9.11

sec. The smaller value is the value used in part (b) and corresponds to stable behavior; the larger value, $T = 0.356$, is slightly larger than $T_{critical}$ and corresponds to unstable behavior.

We comment briefly on a few points.

1. The curves were obtained by numerical integration of (9.19) using a fourth-order Runge-Kutta formula, with a step size $h = 0.5$ sec. The initial values of δ_T, $\dot{\delta}_T$ were found from (9.24).

2. The effect of the damping constant D is seen to be small. In curve (a) there is a slight decrease in swing amplitude, cycle to cycle. The maximum δ is 1.977 rad as compared with $\delta_{max} = 1.987$ rad calculated using equal-area methods, neglecting D. Similarly the minimum δ is found to be -0.777 rad compared with $\delta_{min} = -0.816$ rad. Thus the effect of D on the first swing is seen to be relatively minor. In the long term, however, the effect is major, causing the oscillation to disappear and δ to return to its equilibrium value.

3. The Stage 2, quadratic behavior of $\delta(t)$, is also shown in the figure. At $t = T$, the Stage 3 curves peel smoothly off from the quadratic. We show the extrapolation of the Stage 2 quadratic behavior to emphasize the ever increasing rate of buildup of δ_T and $\dot{\delta}_T$ as T increases.

4. It is instructive to compare the curves in Fig. 9.11 (time behavior) with the curves in Fig. 9.7. Thus with $T < T_{critical}$, δ_T and $\dot{\delta}_T$ are small enough to give an initial total energy $V_2 < W_{max}$ and we get the bounded behavior seen in curve (a). With $T > T_{critical}$, δ_T and $\dot{\delta}_T$ have increased sufficiently to give an initial energy typified by $V_4 > W_{max}$ and we get the unbounded behavior seen in curve (b). Note that in this case Fig. 9.7 predicts that, at the beginning, $\dot{\delta}$ will decrease from its initial (positive value) but will not decrease to zero. Thereafter, the tendency is for $\dot{\delta}$ to rapidly increase. This same behavior is observed in curve (b).

5. Figure 9.7 confirms what Fig. 9.11 seems to imply, namely, if on the upswing $\dot{\delta}$ remains positive, synchronism will be lost, while if after the initial upswing, $\dot{\delta}$ goes at all negative, synchronism will be preserved. Thus, at least for the case under consideration, the question of stability is determined on the first swing. We make use of this in the next chapter when we introduce a generator model which is more accurate for the first swing. We also note the reasonability of neglecting (small) D for a first swing determination.

9.5 OTHER APPLICATIONS

The equal-area construction and stability criterion can be used for other problems. In all of these applications we start in an equilibrium state with $P_G(\delta^0) = P_M^0$. Then there is a change to a new $P_G^n(\delta)$, or a new P_M^n, which upsets the balance. In determining behavior during the initial part of the transient, the spring-mass mechanical analogy is very helpful. Thus for example, if $P_M^n > P_G(\delta^0)$, the spring-mass analogy correctly suggests we get acceleration in the positive direction and δ increases from the value δ^0. If $P_M^n < P_G(\delta^0)$, we get acceleration in the negative direction and δ decreases from the value δ^0. The ensuing motion is bounded (δ_{max}, δ_{min}, can be determined) if the accelerating area A_a can be balanced by the decelerating area A_d. In this case, taking damping in account, the rotor angle swings attenuate and δ tends to a new equilibrium value δ^n. On the other hand, if $A_{dmax} < A_a$ synchronism is lost.

Application 1

The system is given in Fig. 9.1. No fault occurs. From an initial steady-state value with $P_M = P_M^0$ we suddenly set $P_M = P_M^n$. Using the techniques previously discussed it is easy to show that the construction in Fig. 9.12 specifies δ_{min} and δ_{max}.

With the new P_M^n the equilibrium δ is δ^n. The way we get to δ^n is as follows. Initially, $\delta = \delta^0$ and $P_M^n > P_G(\delta^0)$, so there is acceleration in the positive direction. The acceleration continues $[P_M^n > P_G(\delta)]$ until we pass δ^n with positive velocity (we overshoot). Then we decelerate $[P_G(\delta) > P_M^n]$ until a maximum δ is reached (i.e., δ_{max}). The acceleration and deceleration areas are equal. The swing between δ_{max} and $\delta_{min} = \delta^0$ would continue indefinitely if there were no damping. With damping the swings attenuate and δ tends to the new equilibrium value δ^n.

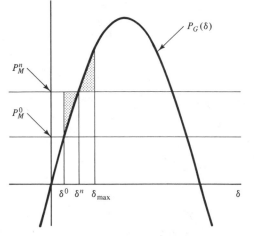

Figure 9.12   Effect of change in P_M.

Exercise 3.   Make up an example (i.e., pick values of P_M^0 and P_M^n) to convince yourself, by using the equal-area method, that it is possible to change from P_M^0 to P_M^n slowly enough (in small steps) to cause δ^0 simply to shift to δ^n in a stable way, while if the change is made suddenly it causes instability.

Application 2

Consider the system of Fig. 9.1 but with two parallel transmission lines. A fault occurs on one of the lines (Fig. 9.13). Assume that the fault is removed instantaneously by opening the breakers of the faulted line. The fault is not transitory and the breakers "lock out" in the open position. We consider $P_G(\delta)$ in the two cases, with the two lines in parallel and with only one line present. In Fig. 9.14 we show the swing accompanying the loss of the line. Synchronism is not lost, and with damping taken into account, δ tends to the new equilibrium value δ^n.

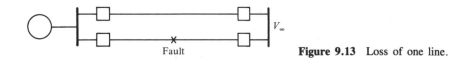

Fault

Figure 9.13   Loss of one line.

Suppose now that the faulted line is repaired and we close the breakers to restore the line to service. For purposes of comparison, suppose that $P_M = P_M^0$ (i.e., the same value as before). Figure 9.15 shows the swing accompanying the regaining of the line. With damping taken into account, δ eventually settles down to its original value δ^0.

Exercise 4.   Provide a sketch, similar to Fig. 9.14, illustrating instability on loss of a line, although the remaining line is capable of carrying the power P_M^0 in the steady state.

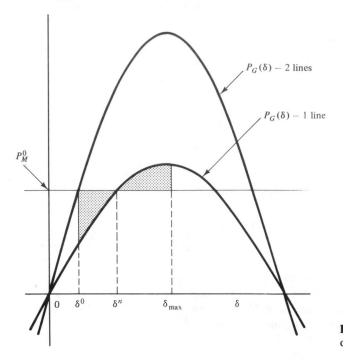

Figure 9.14 Swing accompanying loss of line.

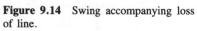

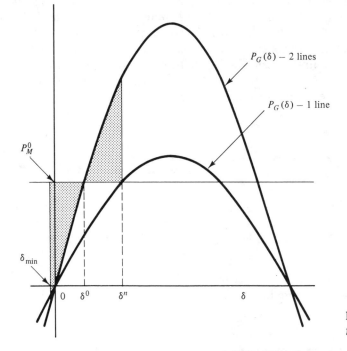

Figure 9.15 Swing accompanying regaining of line.

Application 3

Consider the system of Fig. 9.1 but with two parallel transmission lines and assume that a fault occurs but is not cleared instantaneously by the breakers.

In this case we show three $P_G(\delta)$ curves. In the prefault steady state with two lines in parallel, the power angle is δ^0. When the fault occurs it is not removed instantaneously. With the fault in place, $P_G(\delta)$ is given by the lowest of the three curves in Fig. 9.16. By the time the breakers on the faulted line open, δ has increased to the angle δ_c and an acceleration area A_a has been accumulated. With these breakers open, there is one good line in place, and $P_G(\delta)$ is given by the middle curve. δ increases to a value of δ_{max} corresponding to $A_d = A_a$. Here we are asuming that the fault is not cleared and the breakers on the faulted line remain open. To find δ_{min} we consider the downward swing just as in the case of Fig. 9.10. Taking damping into account, δ tends toward a new equilibrium value corresponding to operating with a single line.

Exercise 5.   Suppose that on the upward swing the fault is cleared and the breakers reclose with $\delta_r < \delta_{max}$. Then for the remainder of the upward swing we have two good lines in place. Redraw Fig. 9.16 to show this case.

Figure 9.16, as drawn, implies that the presence of a faulted line in parallel with a good line reduces the power-transferring capability of the good line. In the next example we would like to offer some supporting evidence.

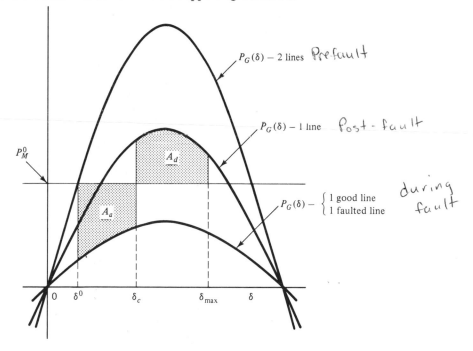

Figure 9.16   Upward swing with fault.

Example 9.3

We wish to obtain an expression for $P_G(\delta)$ with a solid (zero impedance) 3ϕ fault in place, as shown in Fig. 9.13. For simplicity assume a round-rotor generator. We can replace the one-line diagram by the (per phase) circuit diagram shown in Fig. E9.3. In the figure X_{L2} is the total reactance of the faulted line. λ is the fraction of the line to the left of the fault. If $\lambda = 0$, the fault is at the generator bus. $\lambda = 0.5$ puts the fault in the middle of the line, and so on. Find an expression for $P_G(\delta)$ as a function of λ.

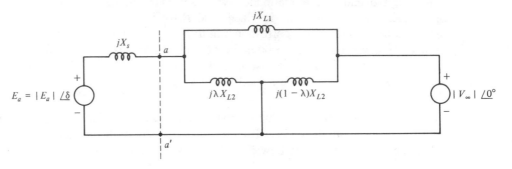

Figure E9.3

Solution   It is simplest to replace the network to the right of the "terminals" $a - a'$ by a Thévenin equivalent circuit. Thus we find, by the voltage-divider law and the rule for finding parallel impedances,

$$V_{\text{Thev}} = \frac{\lambda X_{L2}}{X_{L1} + \lambda X_{L2}} \, |V_\infty| \, \underline{/0^\circ}$$

$$X_{\text{Thev}} = \frac{\lambda X_{L1} X_{L2}}{X_{L1} + \lambda X_{L2}}$$

and then

$$P_G(\delta) = \frac{|E_a| \, V_{\text{Thev}}}{X_s + X_{\text{Thev}}} \sin \delta = \kappa \sin \delta$$

where, after substituting for V_{Thev} and X_{Thev}, we find

$$\kappa \triangleq \frac{|E_a| \, |V_\infty|}{X_s + X_{L1} + (X_s X_{L1}/\lambda X_{L2})}$$

We note that with a fault at the generator bus, $\lambda = 0$ and $\kappa = 0$. Thus the generator delivers zero power and the "runaway" acceleration is maximum. In other cases, as the fault is shifted to the right, away from the generator, κ increases monotonically but never gets as big as $|E_a| \, |V_\infty|/(X_s + X_{L1})$, the gain constant defining power flow in the case of the single line (faulted line removed). Thus we see the importance of removing the faulted line as quickly as possible.

9.6 EXTENSION TO TWO-MACHINE CASE

Suppose that we replace the infinite bus in Fig. 9.1 by a synchronous motor driving a mechanical load. Then we have two synchronous machines, each with its own dynamics, coupled to each other electrically by the transmission line. We can easily extend the results of the previous analysis to this case. Instead of (9.19) we have the pair of equations

$$M_1\ddot{\delta}_1 + D_1\dot{\delta}_1 + P_{G1}(\delta_1 - \delta_2) = P_{M1}^0 \tag{9.42a}$$

$$M_2\ddot{\delta}_2 + D_2\dot{\delta}_2 + P_{G2}(\delta_2 - \delta_1) = P_{M2}^0 \tag{9.42b}$$

where δ_1 and δ_2 are the phase angles of the internal voltages of the two synchronous machines. Assuming a lossless electrical system (i.e., purely reactive line and generator internal impedances), we get

$$P_{G1}(\delta_1 - \delta_2) + P_{G2}(\delta_2 - \delta_1) = 0 \tag{9.43}$$

by considering conservation of power. In this case, in the prefault steady state, with $\delta_1 = \delta_1^0$ and $\delta_2 = \delta_2^0$, we require, and assume, that

$$P_{M1}^0 + P_{M2}^0 = 0 \tag{9.44}$$

Physically, we are assuming that the mechanical power demand at the load is matched by the mechanical power provided by the turbine. Substituting (9.43) and (9.44) in (9.42b), we get

$$M_2\ddot{\delta}_2 + D_2\dot{\delta}_2 - P_{G1}(\delta_1 - \delta_2) = -P_{M1}^0 \tag{9.45}$$

It is convenient now to assume a condition known as *uniform damping*,

$$\frac{D_1}{M_1} = \frac{D_2}{M_2} \tag{9.46}$$

in which case, for stability evaluation, the two equations (9.42a) and (9.45) may be replaced by a single equation in the internodal angle $\delta_{12} = \delta_1 - \delta_2$. We simply divide (9.42a) by M_1 and (9.45) by M_2 and subtract the latter equation from the former. We get

$$\ddot{\delta}_{12} + \frac{D_1}{M_1}\dot{\delta}_{12} + \left(\frac{1}{M_1} + \frac{1}{M_2}\right)P_{G1}(\delta_{12}) = \left(\frac{1}{M_1} + \frac{1}{M_2}\right)P_{M1}^0 \tag{9.47}$$

Equivalently,

$$M_0\ddot{\delta}_{12} + D_0\dot{\delta}_{12} + P_{G1}(\delta_{12}) = P_{M1}^0 \tag{9.48}$$

where $M_0 \triangleq M_1M_2/(M_1 + M_2)$ and $D_0 \triangleq D_1M_2/(M_1 + M_2)$. Equation (9.48) in the variable δ_{12} has the same form as (9.19) in the variable δ. In fact, if we, in effect, convert the second synchronous machine into an infinite bus by letting M_2 tend to infinity, then, with $\delta_2 = \delta_2^0 = 0$, (9.48) reduces to (9.19).

We can apply the equal-area criterion to investigate the stability of (9.48), under various system perturbations, just as we did previously in Sections (9.4) and (9.5). Satisfaction of the stability criterion means that δ_{12} returns to a constant post-fault value. Thus synchronism is maintained. It is to be noted that even in cases where the post-fault δ_{12} returns to its prefault value δ_{12}^0, in general, δ_1 and δ_2 do not return to their prefault values δ_1^0 and δ_2^0. This type of behavior may be more fully understood by considering the spring-mass analogy for (9.42a) and (9.42b); the analogy in this case has two masses coupled by the spring and no wall.

Finally, a brief comment on the assumption of uniform damping. Since the (first swing) stability does not depend significantly on the very small damping constants D_1 and D_2, a convenient choice is justified. In fact, in using the equal-area stability criterion, we neglect D_1 and D_2 entirely. We note that $D_1 = D_2 = 0$ formally satisfies the definition of uniform damping. We note also that if stability is preserved, the eventual, i.e., steady-state value of δ_{12} will also be independent of D_1 and D_2, although this is not the case for δ_1 and δ_2 individually. The main reason we include damping in our present discussion of stability is to preserve the qualitative feature that the oscillations eventually damp out and we return to equilibrium; uniform damping is the simplest assumption that preserves that feature.

Example 9.4

Consider an example similar to Example 9.2. Suppose that the generator is round rotor, with $|E_1| = 1.8$, $H = 5$ sec, and $X_d = X_q = 1.0$. The motor is specified by $|E_2| = 1.215$, $H = 2$ sec, and $X_d = X_q = 1.2$. The line reactance is assumed to be 0.4. Assume that $P_G^0 = 0.5$. Find T_{critical}.

Solution   We first find the expression for $P_{G1}(\delta_{12})$.

$$P_{G1}(\delta_{12}) = \frac{1.8 \times 1.215}{1.0 + 1.2 + 0.4} \sin \delta_{12} = 1.287 \sin \delta_{12}$$

In the prefault steady state, with $P_{G1}(\delta_{12}^0) = 0.5$, we find $\delta_{12}^0 = 22.87° = 0.339$ rad. The data were picked so that these numbers are practically the same as in Example 9.2 and, therefore, we may use a result of that example: namely, the critical clearing angle is $90.61° = 1.581$ rad. We can next find the dynamic behavior during the open-breaker period using (9.24):

$$\delta_1(t) = \frac{\pi \times 60 \times 0.5}{2 \times 5} t^2 + \delta_1^0$$

$$\delta_2(t) = -\frac{\pi \times 60 \times 0.5}{2 \times 2} t^2 + \delta_2^0$$

Therefore,

$$\delta_{12}(t) = \delta_1(t) - \delta_2(t) = 32.99t^2 + \delta_{12}^0$$

Putting in the critical clearing angle and the initial angle, we get

$$1.581 = 32.99 T_{\text{critical}}^2 + 0.399$$

Thus $T_{\text{critical}} = 0.189$ sec.

As compared with the result in Example 9.2, where the critical switching time was 0.354 sec, we have much less time to reclose the breakers if stability is to be maintained. Physically, considering that the generator is speeding up while the motor is slowing down, it is clear why it takes a smaller time to reach the critical clearing angle.

9.7 MULTIMACHINE APPLICATION

Among many possible extensions of the previous results to the general *m*-machine case, we consider next one that is quite similar to the two-machine case. Suppose that we have a power system with a special structure. The generators may be separated into two groups. Within each group the generators are tied together by a "strong" network of transmission lines. On the other hand, the two groups of generators are linked to each other by a comparatively weak set of tie lines. This situation sometimes occurs when two geographically distant metropolitan areas are weakly interconnected.

The situation described above is suggested in Fig. 9.17, where the two groups of generators are shown linked by two tie lines. It will be convenient to define the index set I_A as the set of indices (subscripts) for all the generators, lines, buses, and loads in network A. A similar definition holds for the index set I_B relative to network B. We also define I_C as the set of indices corresponding to tie lines linking the two networks.

It is observed that during electromechanical transients the rotor angles in each subnetwork tend to "swing together" more or less coherently. This phenomenon may be better understood in terms of the spring-mass analogy, extended to *m* masses, by imagining strong coupling (stiff springs) within each group of masses and weak coupling between the two groups. Another way of stating the observed condition is that during transients the voltage angles differences within each network change relatively little compared with the change in the voltage angles across the tie lines between the two networks.

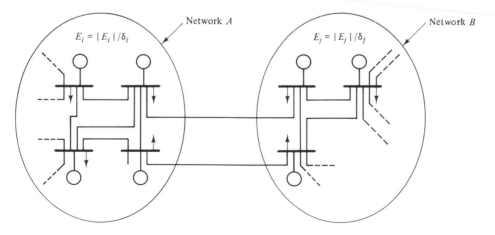

Figure 9.17   *m*-Machine network.

In describing this condition quantitatively, it will be convenient to define the so-called *center of angle* for each of the two networks. This is defined to be a weighted sum of angles as follows:

$$\delta_A = \frac{1}{M_A} \sum_{i \in I_A} M_i \delta_i \tag{9.49}$$

where $M_A \triangleq \sum_{i \in I_A} M_i$ (i.e., we sum all the M_i in network A). Similarly, we define

$$\delta_B = \frac{1}{M_B} \sum_{i \in I_B} M_i \delta_i \tag{9.50}$$

where $M_B \triangleq \sum_{i \in I_B} M_i$. In the spring-mass analogy, δ_A is the center of mass of the masses belonging to network A. δ_B is the center of mass for the masses in network B.

Suppose now that for each of the generators in network A we write the mechanical equation corresponding to (9.19). We get

$$M_i \ddot{\delta}_i + D_i \dot{\delta}_i + P_{Gi} = P_{Mi}^0 \qquad i \in I_A \tag{9.51}$$

As discussed in Section 9.6, to simplify the analysis, we can assume uniform damping. Then if we sum (9.51) for all the generators in network A, we get, using (9.49),

$$M_A \ddot{\delta}_A + D_A \dot{\delta}_A + \sum_{i \in I_A} P_{Gi} = \sum_{i \in I_A} P_{Mi}^0 \tag{9.52}$$

where $D_A = \sum_{i \in I_A} D_i$. Assuming a lossless transmission network, and using the law of conservation of power, we get an expression for P_{AB}, the power transferred from network A to network B:

$$P_{AB} = \sum_{i \in I_A} P_{Gi} - \sum_{i \in I_A} P_{Di}^0 \tag{9.53}$$

In words, the excess of generator power over load in the (lossless) network A is the power transferred to network B. The power is transferred through the lines joining the two networks. Thus

$$P_{AB} = \sum_{k \in I_C} P_k = \sum_{k \in I_C} \frac{|V_i||V_j|}{X_k} \sin \theta_{ij} \tag{9.54}$$

Here P_k is the power from bus i in network A to bus j in network B through line k, which connects the two buses. Equation (9.54) is the familiar expression for active power transferred down the lossless (short) transmission line with reactance X_k.

It is also convenient to introduce notation for the excess of mechanical power over load power for network A. Let

$$P_A^0 \triangleq \sum_{i \in I_A} P_{Mi}^0 - \sum_{i \in I_A} P_{Di}^0 \tag{9.55}$$

Using (9.53) and (9.55) in (9.52), we then get

$$M_A \ddot{\delta}_A + D_A \dot{\delta}_A + P_{AB} = P_A^0 \tag{9.56}$$

Note that under our assumptions the excess power, P_A^0, is a constant just like the mechanical power, P_M^0, in (9.19).

Our objective is to write (9.56) in a form analogous to (9.42a). As the remaining step in this process we need to express P_{AB} in terms of δ_A and δ_B. However, in (9.54), P_{AB} is expressed in terms of the angles $\theta_i - \theta_j$ across the lines joining the two networks. We next show how this conversion may be obtained. In the process we make an approximation consistent with the assumption of coherency previously described. We start with the identity

$$\theta_i - \theta_j = (\theta_i - \delta_A) + (\delta_A - \delta_B) - (\theta_j - \delta_B) \tag{9.57}$$

where $i \epsilon I_A$ and $j \epsilon I_B$.

Consider (9.57) in the prefault steady state. We get

$$\theta_i^0 - \theta_j^0 = (\theta_i^0 - \delta_A^0) + (\delta_A^0 - \delta_B^0) - (\theta_j^0 - \delta_B^0) \tag{9.58}$$

Now think of (9.57) as describing the angles at some time during the transient. If we now subtract (9.58) from (9.57), we can relate the changes in these angle differences.

$$\Delta \theta_{ij} = \Delta(\theta_i - \delta_A) + \Delta \delta_{AB} - \Delta(\theta_j - \delta_B) \tag{9.59}$$

Here $\Delta \theta_{ij} = (\theta_i - \theta_j) - (\theta_i^0 - \theta_j^0)$, $\Delta \delta_{AB} = (\delta_A - \delta_B) - (\delta_A^0 - \delta_B^0)$. We note that $\Delta(\theta_i - \delta_A)$ and $\Delta(\theta_j - \delta_B)$ are the changes, over time, in the angles *within* each network. The notion of coherence is that these changes are small compared to the changes in the angles between the networks. We therefore neglect these changes as compared with $\Delta \delta_{AB}$ and from (9.59), we get

$$\theta_{ij} = \theta_{ij}^0 + \Delta \delta_{AB} \tag{9.60}$$

We can now use (9.60) in (9.54). At the same time let us introduce the notation

$$b_k = \frac{|V_i||V_j|}{X_k} \tag{9.61}$$

Per our previous description, $|V_i|$ and $|V_j|$ are the voltages at the buses terminating line k. If we now assume that these voltage magnitudes are constant during each stage of the fault, the b_k are also constant and (9.54) will be in the required form to apply the equal-area stability criterion. We note that we may assume changes in the $|V_i|$, $|V_j|$, and X_k as we go from one stage of the fault to another but we assume constancy of the b_k within each stage. We then get, using (9.60) and (9.61) in (9.54),

$$P_{AB} = \sum_{k \epsilon I_C} b_k \sin(\theta_{ij}^0 + \Delta \delta_{AB}) \tag{9.62}$$

We can further simplify (9.62) by replacing the sum of sinusoids by a single sinusoid. Using complex arithmetic in the usual way, we have

$$P_{AB} = \sum_{k \in I_C} b_k \, \text{Im} \; e^{j(\theta_{ij}^0 + \Delta\delta_{AB})}$$

$$= \text{Im} \left(\sum_{k \in I_C} b_k e^{j\theta_{ij}^0} e^{j\Delta\delta_{AB}} \right)$$

$$= \text{Im} \; b_s e^{j\theta_s^0} e^{j\Delta\delta_{AB}} \tag{9.63}$$

$$= b_s \sin \left(\theta_s^0 + \Delta\delta_{AB} \right)$$

where

$$b_s e^{j\theta_s^0} = \sum_{k \in I_C} b_k e^{j\theta_{ij}^0} \tag{9.64}$$

That is, b_s and θ_s^0 are the magnitude and phase of the sum of the complex numbers which individually have magnitude b_k and phase θ_{ij}^0.

Using (9.63) in (9.56), and defining $\Delta\delta_A = \delta_A - \delta_A^0$, we get

$$M_A \, \Delta\ddot{\delta}_A + D_A \, \Delta\dot{\delta}_A + b_s \sin \left(\theta_s^0 + \Delta\delta_{AB} \right) = P_A^0 \tag{9.65}$$

which is analogous to (9.42). Similarly for network B, carrying out the steps which led to (9.56), we get

$$M_B \, \Delta\ddot{\delta}_B + D_B \, \Delta\dot{\delta}_B + P_{BA} = P_B^0 \tag{9.66}$$

where $\Delta\delta_B = \delta_B - \delta_B^0$. Because the lines are assumed lossless, $P_{BA} = -P_{AB}$. Assuming power balance in the prefault steady state, and with a lossless transmission system and lossless generators, we have $P_B^0 = -P_A^0$. Thus we may replace (9.66) with

$$M_B \, \Delta\ddot{\delta}_B + D_B \, \Delta\dot{\delta}_B - b_s \sin \left(\theta_s^0 + \Delta\delta_{AB} \right) = -P_A^0 \tag{9.67}$$

which is analogous to (9.45). Our assumption of uniform damping for all the generators implies that $D_A/M_A = D_B/M_B$. We get, in a parallel development to the two-machine case,

$$M_s \, \Delta\ddot{\delta}_{AB} + D_s \, \Delta\dot{\delta}_{AB} + b_s \sin \left(\theta_s^0 + \Delta\delta_{AB} \right) = P_A^0 \tag{9.68}$$

where $M_s \triangleq M_A M_B/(M_A + M_B)$, and $D_s \triangleq D_A M_B/(M_A + M_B)$. Equation (9.68) is analogous to (9.48) and may be used similarly in the investigation of stability of two coherent groupings of generators.

Example 9.5

Consider a case of the type described in application 2 in Section 9.5. In the prefault steady state we have three connecting lines between networks A and B. Assume that there is a fault on one line, removed instantaneously by the opening of its breakers, and the faulted line is then locked out, leaving two lines in service. The data for the three lines in the prefault steady state are tabulated in Table E9.5. We also show the prefault transmitted power calculated using $P_k = b_k \sin \theta_{ij}^0$. Note that $P_A^0 = \sum_{k=1}^{3} P_k = 3.738$. Assume coherent groupings of generators in networks A and B and use the analysis method of Section 9.7 to answer the following questions.

Does the equal-area stability criterion predict stability if

(a) Line 3 is lost?
(b) Line 1 is lost?

Assume that the b_k do not change when a line is lost.

TABLE E9.5

	Line 1	Line 2	Line 3
θ_{ij}^0	45°	30°	17°
b_k	2.5	1.6	4.0
P_k	1.768	0.800	1.170

Solution   We assume that just after a line is lost the θ_{ij} do not change appreciably from their prefault values θ_{ij}^0. This is because the centers of angles δ_A and δ_B cannot change instantaneously and the angular differences $\Delta(\theta_i - \delta_A)$ and $\Delta(\theta_j - \delta_B)$ are assumed to be small. We next find an expression for P_{AB}.
   (a) With line 3 out we find from (9.64),

$$b_s \underline{/\theta_s^0} = 2.5 \underline{/45°} + 1.6 \underline{/30°} = 4.067 \underline{/39.16°}$$

Then using (9.63),

$$P_{AB} = 4.067 \sin (39.16° + \Delta\delta_{AB})$$

Next we sketch P_A^0 and P_{AB}. We can think of network A as a single generator with "mechanical" power P_A^0 and electrical output P_{AB}. Initially, the power supplied down the remaining lines is $1.768 + 0.800 = 2.568$. This is less than $P_A^0 = 3.603$, so, initially, δ_A increases. At the same time network B is a net consumer of power, so δ_B decreases. Hence δ_{AB} initially increases, as does the angle $\theta_s \triangleq \theta_s^0 + \Delta\delta_{AB}$. A power balance is reached when $\theta_s = \theta_s^n = 66.8° = 1.166$ rad. The acceleration and maximum deceleration areas are shown in Fig. E9.5(a). We can calculate the areas.

$$A_a = \int_{\theta_s^0}^{\theta_s^n} (3.738 - 4.067 \sin u) \, du = 1.804 - 1.551 = 0.2526$$

$$A_{dmax} = 2 \int_{\theta_s^n}^{\pi/2} (4.067 \sin u - 3.738) \, du = 0.1781$$

Since $A_a > A_{dmax}$, the criterion predicts instability. In this case networks A and B "pull apart" (i.e., lose synchronism and form separate "islands" each with generation and load imbalances). In network A there would be an excess of generation and in network B there would be a deficiency of generation. Corrective action would need to be taken to prevent further deterioration.
   (b) With line 1 out we find, from (9.64),

$$b_s \underline{/\theta_s^0} = 1.6 \underline{/30°} + 4.0 \underline{/17°} = 5.571 \underline{/20.7°}$$

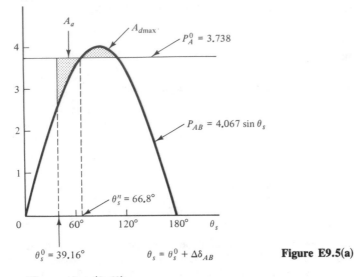

Thus, using (9.63),

$$P_{AB} = 5.571 \sin (20.7° + \Delta\delta_{AB})$$

The sketch of P_A^0 and P_{AB} is shown in Fig. E9.5(b). Clearly, $A_a < A_{dmax}$ and the equal-area criterion predicts stability. In the sketch we show the balancing deceleration area $A_d = A_a$.

What makes this case so different from part (a) is the presence of a relatively strong line ($b_3 = 4$) compared to the other two lines, which is initially lightly loaded. Thus the line is capable of making a strong contribution to the deceleration area.

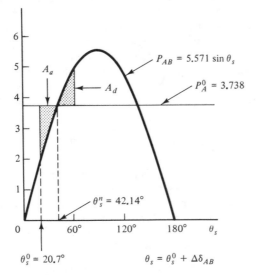

Figure E9.5(b)

Exercise 6.   Suppose for the case under consideration in this section (i.e., two coherent groups of generators) that we lose all the lines connecting a hypothetical network A to network B at $t = 0$, and then all the breakers reclose after T seconds. All other factors being equal (or disregarded), what is the most favorable grouping of the M_i to minimize $\Delta \delta_{AB}$ after a fixed time T? More specifically, if M_T is the sum of all the M_i and $M_A = \lambda M_T$, $M_B = (1 - \lambda)M_T$, how would you pick λ? λ is a number between 0 and 1.

In concluding this section we make a few general comments.

Although the assumption of coherency is not expected to be fully realized in practice, we can obtain useful qualitative results, nevertheless, which help in thinking about stability problems in the system; these results can then be checked by more precise models and/or by simulation. Thus we can consider the effects of increasing (or decreasing) moments of inertia. We can consider the effects of readjusting generation versus load distributions (this will change P_A^0 and the θ_{ij}^0 across the lines joining the two networks A and B), the effect of losing particular lines between the two networks A and B (thereby changing b_s and θ_s^0), and the effect of low voltages during fault-on periods.

Using even our very simple system models and stability analysis techniques we can appreciate certain transient stability enhancement techniques which are used in practice. The basic objective, whether we have two generators or two coherent groups of generators, is to reduce the acceleration area and increase the maximum available deceleration area. The importance of rapid removal of faulty lines to allow the remaining parallel line(s) to deliver more power is seen in Fig. 9.16. This figure also illustrates the importance of maintaining voltage during fault-on conditions. During a fault, voltages tend to drop throughout the system; this is one of the reasons for the low $P_G(\delta)$ during the fault. Fast-acting voltage control systems at generator buses counter this drop by increasing generator excitations. One of the effects is to increase the transferred power during the fault. This then tends to decrease the acceleration area. In other cases, when the fault is temporary, the importance of rapid reclosing of circuit breakers may be inferred from the very simple case considered in Fig. 9.8.

Fast-closing turbine steam valves actuated during faults can help reduce the mechanical input P_M and thereby both decrease the accelerating area and increase the decelerating area. It may also be possible to use up some of the excess accelerating power by momentarily activating resistance loads (braking resistances) when fault conditions are detected. Switchable series capacitors to cancel some of the series line inductance and thereby increase the transferred power is another possibility.

In the design phase the importance of paralleling lines to strengthen weak transmission links is clear. It would also be desirable to increase the inertia constants, H, and decrease the internal impedances of the generating units. Unfortunately, current manufacturing practice runs contrary to this need.

We note that there are many natural extensions of the technique we have been describing to more general systems with more general assumptions. One powerful technique, the so-called *second method of Liapunov,* is a generalization of the energy technique we used in Section 9.4. In fact, (9.28) is an example of a so-called *Liapunov function,* and its use is demonstrated in the material through (9.33). With a more general Liapunov function we can undertake the investigation of the stability of multimachine systems without the coherency restriction.

Another extension concerns the development of a more accurate generator model. This will be undertaken in Chapter 10. This more accurate model is well suited for investigations of first swing behavior.

9.8 SUMMARY

In this chapter we consider the problem of power system stability, in particular the problem of transient stability. For this purpose a dynamic model of the turbine-generator unit is required. Since it is observed in practice that the rotor angle (measured with respect to a synchronous rotating reference frame) undergoes relatively slow variations during these transients, we can derive a very simple generator model. In this model the internal voltage magnitude is assumed to be constant while the phase angle δ is variable. Using the method of pseudo-steady-state analysis, the power output of the generator is found by using the steady-state formula derived in Chapter 8; in this formula we allow δ to be slowly varying. For the case of a turbine-generator unit connected through a lossless transmission link to an infinite bus, the end result is (9.19), a second-order nonlinear differential equation in the power angle δ. We note that the same equation describes an analogous spring-mass system with a nonlinear spring characteristic.

For the transient stability problem, linearization is not usually suitable; we can use an energy method, however, as suggested by the spring-mass analogy. We find that the sum of kinetic energy and potential energy decreases with time and use this property to infer the system behavior.

Once we understand the principles involved, we may simplify the determination of stability by using the equal-area criterion. If the system is stable, we can also use the equal-area construction to determine the extent of the swings in the power angle. We note that the question of stability is settled in the first swing; if the power angle is initially increasing but reaches a maximum value and then starts to decrease, stability is assured.

The analysis can also be extended to the case of two machines connected through a lossless transmission line. Another extension to the case of two groupings of generators is possible if the machines are closely coupled within each group and weakly coupled between groups.

Although the equal-area criterion is based on a model too simple to give accurate quantitative results, its simplicity enhances understanding and helps explain the measures taken in industry practice to improve stability.

PROBLEMS

9.1. Given a turbine-generator unit rated 100 MVA with a per unit inertia constant, $H = 5$ sec.
   (a) Calculate the kinetic energy stored at synchronous speed (3600 rpm).
   (b) Compare this figure with the kinetic energy of a 10-ton truck going at 60 mph.
   (c) Suppose that the generator is delivering 100 MW and then, at $t = 0$, the line circuit breakers open. Calculate the shaft acceleration in rad/sec^2.
   (d) At this rate, how long does it take for δ to increase from δ^0 to $\delta^0 + 2\pi$ radians?

9.2. In the model used to study the stability of the single generator connected to an infinite bus, we assumed that $P_M = P_M^0$ was constant. Consider steam governor action such that the mechanical power input to the generator is given by (Fig. P9.2)

$$P_M = P_M^0 - k(\omega - \omega_0)$$

mechanical power supplied by turbine.

where k is positive and ω is the angular velocity of the (two-pole) turbogenerator shaft. *(pre-fault equil')* Show how this feedback, in effect, changes the mechanical friction constant D. Is this a stabilizing factor? What happens if the constant k is negative?

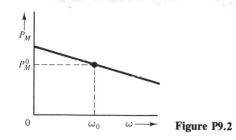

Figure P9.2

9.3. Consider the spring-mass system shown in Fig. P9.3 and described by the differential equation

$$m\frac{d^2x}{dt^2} + a\frac{dx}{dt} + \sin x = f(t)$$

   (a) For $t < 0$, $f(t) = 0.5$. Find the steady state x^0.
   (b) For $0 \le t$, $f(t) = 0.5 + \Delta f(t)$, where $\Delta f(t)$ is small. Find a linear differential equation which describes $\Delta x = x - x^0$ in terms of $\Delta f(t)$.
   (c) Find the natural frequencies of the linearized system found in part (b) with $m = 1.0$ and $a = 0.01$. Suppose that $\Delta f(t) = 0.1u(t)$, where $u(t)$ is the unit step function. Can you roughly sketch $x(t)$ for $t \ge 0$, showing the initial value, final value, frequency, and decay of the lightly damped oscillatory component?

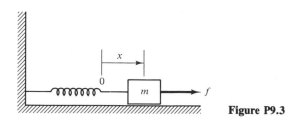

Figure P9.3

9.4. Consider a round-rotor generator delivering a steady-state power $P_G = 0.5$ to an infinite bus through a transmission line with reactance $X_L = 0.4$. Assume that $|E_a| = 1.8$, $V_\infty = 1\ \underline{/0°}$, $H = 5$ sec, and $X_d = X_q = 1.0$. Neglect resistance.

    (a) Find the two possible steady-state (equilibrium) values of power angle $\delta = \underline{/E_a} - \underline{/V_\infty}$ which lie in the interval $[0, 2\pi]$.

    (b) Which of these two equilibrium values are we likely to observe in practice? *Hint:* Linearize around the equilibrium value and consider the natural frequencies found from the linear differential equation in $\Delta\delta$.

9.5. Suppose that in Problem 9.4 at $t = 0$, the line circuit breakers open. $P_M^0 = 0.5$ remains constant. Calculate δ and $\dot{\delta}$ at the end of 1 sec under two assumptions and compare:

    (a) $D = 0$ and **(b)** $D = 0.001$.

9.6. A round-rotor generator is delivering a steady-state power, $P_G = 0.4$, to an infinite bus through a transmission line with reactance $X_L = 0.3$. Assume that $|E_a| = 2.0$, $V_\infty = 1\ \underline{/0°}$, $H = 0.4$ sec, and $X_d = X_q = 1.2$. At $t = 0$ the circuit breakers open and reclose T seconds later.

    (a) Find the critical clearing angle δ_{Tc}.

    (b) Find the corresponding critical clearing time, $T_{critical}$.

    (c) Using $T = 0.9 T_{critical}$, find δ_{max} and δ_{min}.

9.7. Refer to Fig. P9.7 and assume that

$$M = 1.0$$

$$P_M = 0.5$$

$$|E_a| = 1.5$$

$$X_d = X_q = 1.0$$

    (a) At $t = 0$ the circuit breakers open and reclose T seconds later. Find $T_{critical}$.

    (b) Assume that $T = 0.7\ T_{critical}$. Find the bounds on the swing of δ (i.e., find δ_{max} and δ_{min}).

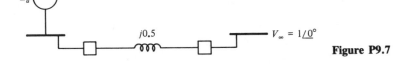

$V_\infty = 1\underline{/0°}$

Figure P9.7

9.8. Refer to Fig. P9.8 and assume that

$$P_M = 1.0$$

$$|E_a| = 1.5$$

$$X_d = X_q = 0.9$$

    (a) At $t = 0$ the circuit breakers open and remain open. Determine if the transient is stable.

    (b) Repeat if $X_d = 1.0$, $X_q = 0.6$.

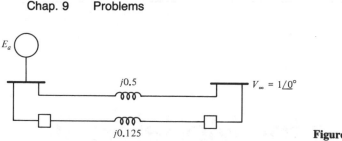

Figure P9.8

9.9. Refer to Fig. P9.9 and assume that

$$P_M = 1.0$$

$$|E_a| = 1.8$$

$$X_d = X_q = 0.9$$

$$M = 1.0$$

The system is operating in the steady state. At $t = 0$ a solid 3ϕ symmetric fault occurs as shown, and the fault is not cleared until t_1, when $\delta(t_1) = \pi/2$. At t_1 the breakers open and remain open until a time t_2 when $\delta(t_2) = 2\pi/3$. By t_2 the fault has deenergized, and at t_2 the breakers close and remain closed. Is the system transient stable with respect to the fault sequence described above?

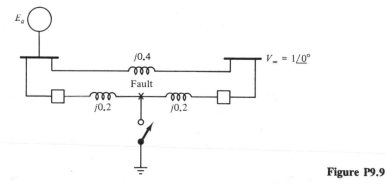

Figure P9.9

9.10. Jack and Jill are asked to check for possible transient stability problems for a remote salient-pole generator delivering power through a transmission line to the main part of the system, which is modeled by an infinite bus. They collect the following data (converted to per unit): $P_G = 0.8$, $V_\infty = 1 \underline{/0°}$, $X_d + X_L = 1.0$, $X_q + X_L = 0.6$, and power factor $= 0.8$. When they are back in the office they notice that they did not check on whether the power factor was positive or negative. Jack says: "Let's check the leading power factor possibility. If that one gives enough margin of stability, the lagging case will certainly be okay." Do you agree?

Generator Modeling II
(Circuit Viewpoint)

chapter 10

10.0 INTRODUCTION

The generator model developed in Chapter 8 is a steady-state model. More precisely, it is a positive-sequence steady-state model, since we assumed that the generator emfs and stator currents were both positive sequence. Although the model was used to good effect for a transient analysis in Chapter 9, in many cases a more accurate model is essential. For some transients, for example, those accompanying short circuits at the generator terminals, the steady-state model gives completely erroneous results. The steady-state model is inapplicable even in certain cases of steady-state operation, for example, if the (positive sequence) generator is subjected to negative-sequence currents; this is a case we will consider in Chapter 13.

In this chapter we develop a more general model of the generator. The model reduces to the steady-state model derived in Chapter 8 as a special case. A circuit theory viewpoint is used in the development.

10.1 ENERGY CONVERSION

We start by considering the general case of a set of coupled coils in which one or more of the coils is mounted on a shaft and can rotate. Thus there is one mechanical degree of freedom. The other coils are fixed. The situation is shown schematically in Fig. 10.1.

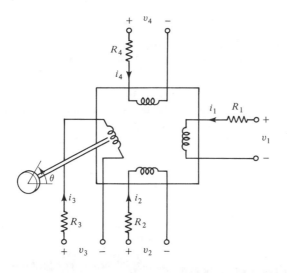

Figure 10.1 Coupled coils.

In Fig. 10.1 we see that some of the mutual inductances may vary with θ. In Section 10.2 we will be interested in the specific coupling configuration of a three-phase machine, but for now we proceed in general terms.

Assume associated reference directions for the v, i, λ. For example, pick a reference direction for i_1, find the associated reference direction for v_1 as in circuit theory, and also find the associated reference direction for flux by the right-hand rule; the flux linkages and the flux have the same sign. Assume that for any fixed θ there is a linear relationship between λ and i. Then we get the relationship

$$\boldsymbol{\lambda} = \mathbf{L}(\theta)\mathbf{i} \tag{10.1}$$

where, in the case of Fig. 10.1, $\mathbf{i}$ and $\boldsymbol{\lambda}$ are 4-vectors, and $\mathbf{L}$ is a symmetric 4×4 matrix. The fact that the matrix is symmetric is known from circuit theory. Application of KVL to the circuit in Fig. 10.1 and use of (10.1) give

$$\boldsymbol{v} = \mathbf{R}\mathbf{i} + \frac{d\boldsymbol{\lambda}}{dt}$$

$$= \mathbf{R}\mathbf{i} + \mathbf{L}(\theta)\frac{d\mathbf{i}}{dt} + \frac{d\mathbf{L}(\theta)}{dt}\mathbf{i} \tag{10.2}$$

where $\boldsymbol{v} = \mathrm{col}\,\{v_1, v_2, v_3, v_4\}$ and $\mathbf{R} = \mathrm{diag}\,\{R_i\}$. If θ is constant with time (i.e., the coils are fixed), then $\mathbf{L}(\theta)$ is a constant matrix; the flux changes are then due only to "transformer action." On the other hand, if $\mathbf{i}$ does not change with time (i.e., $\mathbf{i}$ is dc), the flux changes are due entirely to the motion (or speed) of the coils relative to each other. For this reason the terminology "transformer voltage" and "speed voltage" are used to designate the second and third terms on the right side of (10.2).

We are interested in the case where θ can change (i.e., the shaft can rotate). To investigate the behavior of θ, we need a mechanical equation. Then, assuming a rigid shaft with total coupled inertia J and linear friction, the equation will be

$$T = T_M + T_E = J\ddot{\theta} + D\dot{\theta} \tag{10.3}$$

where T is total applied torque in the positive θ direction. T is the sum of an externally applied mechanical torque T_M and an electrical torque T_E due to interactions of magnetic fields.

We next derive an expression for T_E by use of the principle of conservation of instantaneous power. It is helpful to take a "motor" point of view, regarding the device in Fig. 10.1 as one that converts electrical power to mechanical power. Let p be the instantaneous electrical power into the coupled coils. Using (10.2),

$$p = \mathbf{i}^T \boldsymbol{\nu} = \mathbf{i}^T \mathbf{R} \mathbf{i} + \mathbf{i}^T \mathbf{L}(\theta) \frac{d\mathbf{i}}{dt} + \mathbf{i}^T \frac{d\mathbf{L}(\theta)}{dt} \mathbf{i} \tag{10.4}$$

where $\mathbf{i}^T$ is the transpose of $\mathbf{i}$ (i.e., $\mathbf{i}^T$ is a row vector). In the present case, $\mathbf{i}^T \boldsymbol{\nu} = i_1 v_1 + i_2 v_2 + i_3 v_3 + i_4 v_4$. We also note that since $\mathbf{R}$ is a diagonal matrix, $\mathbf{i}^T \mathbf{R} \mathbf{i} = R_1 i_1^2 + R_2 i_2^2 + R_3 i_3^2 + R_4 i_4^2$ is the instantaneous power dissipated in the resistances.

From elementary circuit theory we also have an expression for the instantaneous stored magnetic energy of coupled coils.

$$W_{\text{mag}} = \tfrac{1}{2} \mathbf{i}^T \mathbf{L}(\theta) \mathbf{i} \tag{10.5}$$

By differentiating we get the power supplied to the magnetic field.

$$\begin{aligned}
\frac{dW_{\text{mag}}}{dt} &= \frac{1}{2} \mathbf{i}^T \mathbf{L}(\theta) \frac{d\mathbf{i}}{dt} + \frac{1}{2} \frac{d\mathbf{i}^T}{dt} \mathbf{L}(\theta) \mathbf{i} + \frac{1}{2} \mathbf{i}^T \frac{d\mathbf{L}(\theta)}{dt} \mathbf{i} \\
&= \mathbf{i}^T \mathbf{L}(\theta) \frac{d\mathbf{i}}{dt} + \frac{1}{2} \mathbf{i}^T \frac{d\mathbf{L}(\theta)}{dt} \mathbf{i}
\end{aligned} \tag{10.6}$$

In deriving (10.6) we have used the fact that the transpose of a scalar is the same scalar [each product of terms in (10.6) is a scalar], that the transpose of a product of matrices is the product of the transposed matrices in reverse order, and that $\mathbf{L}(\theta)$ is a symmetric matrix. Thus

$$\frac{1}{2} \frac{d\mathbf{i}^T}{dt} \mathbf{L}(\theta) \mathbf{i} = \frac{1}{2} \mathbf{i}^T \mathbf{L}(\theta) \frac{d\mathbf{i}}{dt}$$

and the two terms may be combined. Using (10.6) in (10.4), we have

$$p = \mathbf{i}^T \mathbf{R} \mathbf{i} + \frac{dW_{\text{mag}}}{dt} + \frac{1}{2} \mathbf{i}^T \frac{d\mathbf{L}(\theta)}{dt} \mathbf{i} \tag{10.7}$$

Considering all the mechanisms allowed for in our model, we obtain, by conservation of instantaneous power,

$$\begin{pmatrix} \text{electrical} \\ \text{power in} \end{pmatrix} = \begin{pmatrix} \text{power dissipated} \\ \text{in resistances} \end{pmatrix} + \begin{pmatrix} \text{power supplied to} \\ \text{magnetic fields} \end{pmatrix} + \begin{pmatrix} \text{converted into} \\ \text{mechanical power} \end{pmatrix}$$

Comparing this with (10.7), we can identify the right-hand term in (10.7) as the electrical power converted into the mechanical form. Finally, using the chain rule of differentiation, we get $d\mathbf{L}(\theta)/dt = (d\mathbf{L}(\theta)/d\theta)(d\theta/dt)$, where $d\theta/dt$ is the (scalar) angular velocity. Thus the electrical power converted into the mechanical form is

$$P_E = \frac{d\theta}{dt}\frac{1}{2}\mathbf{i}^T\frac{d\mathbf{L}(\theta)}{d\theta}\mathbf{i} = \omega T_E \tag{10.8}$$

where $\omega = d\theta/dt$ is the shaft angular velocity and we can identify

$$T_E = \frac{1}{2}\mathbf{i}^T\frac{d\mathbf{L}(\theta)}{d\theta}\mathbf{i} \tag{10.9}$$

as the mechanical torque on the shaft in the direction of increasing θ due to the interaction of the magnetic fields set up by the electrical currents. Substituting (10.9) in (10.3), we get

$$J\ddot{\theta} + D\dot{\theta} - \frac{1}{2}\mathbf{i}^T\frac{d\mathbf{L}(\theta)}{d\theta}\mathbf{i} = T_M \tag{10.10}$$

We can call (10.10) the *mechanical equation*. With (10.2) and (10.10) we get the interaction of the mechanical and electrical variables.

In the derivation of (10.9) we assumed that only rotational motion was possible. Suppose that we had considered a different case, the case of a single translational degree of freedom. Suppose that $\mathbf{L}$ depends on x, for example. The reader should check that we still get (10.9) with θ replaced by x, and torque replaced by force.

Example 10.1

Calculate average torque of a "reluctance" motor (Fig. E10.1). (This is a very simple synchronous motor used in clocks.) Assume that $i(t) = \sqrt{2}|I|\cos\omega_0 t$. As an approximation assume that $L(\theta) = L_s + L_m\cos 2\theta$, $L_s > L_m > 0$. Also assume that the rotor moment of inertia is "reasonably high."

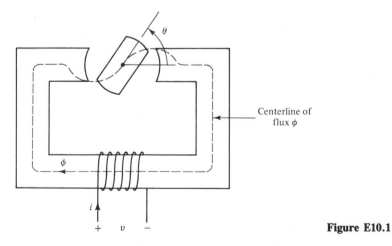

Figure E10.1

Solution We note that in this case Fig. 10.1 reduces to a single coil whose self-inductance is $L(\theta)$. We can first check the physical basis for the assumed form of $L(\theta)$.

When $\theta = n\pi$, with n an integer, the average air-gap distance is a minimum (reluctance also is a minimum), so $L(\theta)$ must be a maximum. (See Appendix 1.) When $\theta = n\pi + \pi/2$, $L(\theta)$ must be a minimum (but still positive). We see that $L(\theta)$ goes through two cycles of variation when θ increases from 0 to 2π. Thus the assumed value $L(\theta) = L_s + L_m \cos 2\theta$, $L_s > L_m > 0$, seems to be a reasonable simplification. Then using (10.9),

$$T_E = -\tfrac{1}{2}i^2(t)2L_m \sin 2\theta = -L_m i^2(t) \sin 2\theta$$

Using (10.10), we get

$$J\ddot{\theta} + D\dot{\theta} + L_m i^2 \sin 2\theta = 0$$

Note: With $i = $ constant, this is the equation of a pendulum, not a motor! Alternatively, we may note that as θ increases uniformly (as it would for a motor with a very large moment of inertia), the average torque is zero. This is not consistent with motor behavior; in order to get motor action we need a nonzero (positive) average torque.

In our case, $i(t) = \sqrt{2}|I| \cos \omega_0 t$, and we will get a nonzero average torque. The mechanical equation is nonlinear and we will not attempt an exact solution. Instead, as an approximation, we will calculate the average value of T_E assuming a uniform rate of rotation in the steady state. Assume, then, that in the steady state, $\theta(t) = \omega_0 t + \delta$, where ω_0 corresponds to synchronous speed and δ is a constant. Then

$$\begin{aligned}
T_E &= -2L_m|I|^2 \cos^2 \omega_0 t \sin 2(\omega_0 t + \delta) \\
&= -L_m|I|^2(1 + \cos 2\omega_0 t) \sin 2(\omega_0 t + \delta) \\
&= -\tfrac{1}{2}L_m|I|^2[\sin 2\delta + 2 \sin (2\omega_0 t + 2\delta) + \sin (4\omega_0 t + 2\delta)]
\end{aligned}$$

From this, averaging over time, we get

$$T_{Eav} = -\tfrac{1}{2}L_m|I|^2 \sin 2\delta$$

and thus with $-\pi/2 < \delta < 0$ we develop a positive average torque which is consistent with the rotation.

As may be seen from the preceding, there are sinusoidal torque "pulsations" (of frequency $2\omega_0$ and $4\omega_0$) superimposed on the average torque. However, we assumed a "reasonably high" moment of inertia J and thus we do not expect these pulsations to significantly alter the assumed uniform rotation rate. Thus the analysis is consistent, and we have found at least an approximate value for the average torque.

We turn next to an example that will be helpful in understanding the calculation of inductance parameters for synchronous machines.

Example 10.2

Consider the machine sketched in Fig. E10.2(a). Sketch $L_{11}(\theta)$ and $L_{12}(\theta)$ as a function of θ. Neglect all but the dc and the lowest frequency components (in the Fourier series) of the periodic variations.

Solution As seen in the figure, the reference directions for currents, voltages, and fluxes (flux linkages) are picked in associated reference directions. $L_{11}(\theta)$ and $L_{12}(\theta)$ are the inductance parameters relating λ_1 to i_1 and i_2 as follows.

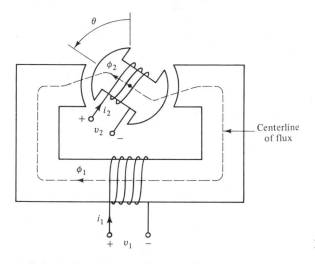

Figure E10.2(a)

$$\lambda_1 = L_{11}(\theta)i_1 + L_{12}(\theta)i_2$$

λ_1 is the flux linkages of coil 1; λ_1 has the same sign as ϕ_1. To find L_{11} we set $i_2 = 0$. Then

$$L_{11}(\theta) = \frac{\lambda_1}{i_1} = \frac{N_1\phi_1}{i_1}$$

We get a maximum (air-gap) reluctance when $\phi = n\pi$, and a minimum when $\theta = n\pi + \pi/2$. Thus L_{11} will depend on θ somewhat as sketched in Fig. E10.2(b). Thus, discarding higher harmonics, L_{11} is in the form $L_{11}(\theta) = L_s - L_m \cos 2\theta$. To find L_{12} we set $i_1 = 0$. Then

$$L_{12}(\theta) = \frac{\lambda_1}{i_2} = \frac{N_1\phi_1}{i_2}$$

ϕ_1 is the flux linking coil 1 due to i_2. L_{12} is sketched in Fig. E10.2(c).

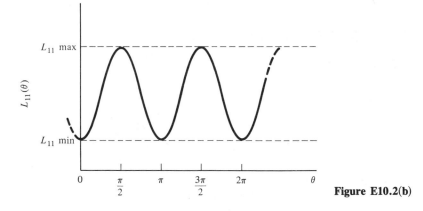

Figure E10.2(b)

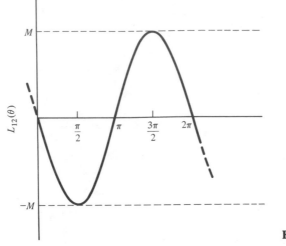

Figure E10.2(c)

We can verify the general shape as follows. For convenience assume that $i_2 > 0$, so ϕ_2 is in its positive reference direction. When $\theta = 0$, none of ϕ_2 links coil 1; $L_{12}(0) = 0$. When $\theta = \pi/2$, ϕ_2 reaches its maximum value (minimum reluctance) and most of it threads coil 1 from left to right. Thus ϕ_1 and L_{12} reach minimum (negative) values. When $\theta = \pi$, $L_{12} = 0$ again. When $\theta = 3\pi/2$, ϕ_2 again reaches a maximum value and most of it threads coil 1 from right to left (i.e., ϕ_1 is in its positive reference direction). Thus L_{12} reaches a maximum. Neglecting harmonics, we get L_{12} in the form $L_{12} = -M \sin \theta$. The reader should compare L_{11} and L_{12} and note that they exhibit very different behavior as a function of θ.

Exercise 1.   Sketch L_{22} and L_{21} by repeating the steps in Example 10.2. As a check, note the result from circuit theory: $L_{21}(\theta) = L_{12}(\theta)$.

Some readers may wish to examine additional examples in Appendix 2. These are derivations of force equations for solenoids or relays and are needed for later use in Chapter 14.

10.2 APPLICATION TO SYNCHRONOUS MACHINE

We will now develop (10.2) and (10.10) for a three-phase machine. In the process we introduce the Park (Blondel) transformation.

Assume that the machine consists of three phase-windings mounted on the stator or stationary part of the machine, a field winding mounted on the rotating part (the rotor), and two additional fictitious windings on the rotor which model the short-circuited paths of the damper windings and/or solid-iron rotor. These two windings are assumed aligned with the direct axis and quadrature axis of the rotor.

The six windings, represented by one-turn coils, are shown schematically in Fig. 10.2. Compared with Fig. 8.1, we note the addition of the two fictitious coils and the introduction of the direct axis and quadrature axis reference directions.

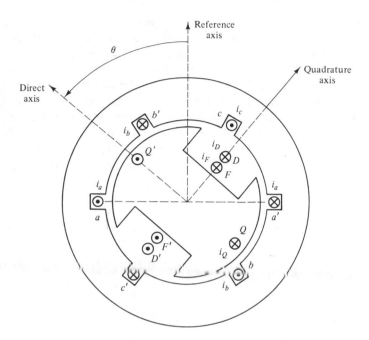

Figure 10.2 Machine schematic.

The six currents are i_a, i_b, i_c, i_F, i_D and i_Q. Lowercase letters indicate stator quantities. Uppercase letters indicate rotor quantities. i_F is the field current and is subject to control through an "excitation" control system. i_D and i_Q are the currents in the fictitious rotor coils.

Using the circuits convention on associated reference directions, we get the usual relationship between voltages, currents, and flux linkages.

$$
\begin{bmatrix} v_{a'a} \\ v_{b'b} \\ v_{c'c} \\ v_{FF'} \\ v_{DD'} \\ v_{QQ'} \end{bmatrix} = \begin{bmatrix} r & & & & & \\ & r & & & 0 & \\ & & r & & & \\ & & & r_F & & \\ & 0 & & & r_D & \\ & & & & & r_Q \end{bmatrix} \begin{bmatrix} i_a \\ i_b \\ i_c \\ i_F \\ i_D \\ i_Q \end{bmatrix} + \frac{d}{dt} \begin{bmatrix} \lambda_{aa'} \\ \lambda_{bb'} \\ \lambda_{cc'} \\ \lambda_{FF'} \\ \lambda_{DD'} \\ \lambda_{QQ'} \end{bmatrix} \qquad (10.11)
$$

$$
= \mathbf{Ri} + \frac{d\boldsymbol{\lambda}}{dt}
$$

The matrix and vector quantities are defined by comparing the two lines above. For example, $\mathbf{R} = \text{diag}\ \{r,\ r,\ r,\ r_F,\ r_D,\ r_Q\}$, where r is the resistance of each phase winding, r_F is the resistance of the field winding, and r_D and r_Q are the resistances of the fictitious rotor coils modeling the short-circuited paths. The flux linkage terms are identified by the subscripts; they agree in sign with their corresponding fluxes, which in turn agree, by the right-hand rule, with their associated currents. For example, $\lambda_{aa'}$

is positive if the flux links the coil aa' in the upward direction. $\lambda_{FF'}$ is positive if the flux links the coil FF' in the direct axis direction. Note that although we have defined **R**, **i**, and **λ**, we have not yet defined a voltage vector **v**.

We wish first to introduce the generator convention on reference directions for the stator coil voltages; we remind the reader that the voltages in (10.11) follow the circuits convention on reference directions. Earlier, in Section 8.3, with the same objective in mind, we introduced the (open circuit) voltage $e_{aa'}$ rather than $e_{a'a}$. Similarly, we now adopt the generator convention for stator voltages and deal with $v_{aa'}$, $v_{bb'}$, and $v_{cc'}$ instead of $v_{a'a}$, $v_{b'b}$, and $v_{c'c}$. The notation for the remaining voltages is not changed. As in Section 8.3 we also simplify by using a single-subscript notation to indicate the voltage of the unprimed subscript with respect to the primed subscript. For example, $v_a \triangleq v_{aa'} = -v_{a'a}$, $v_F \triangleq v_{FF'}$, and so on. Now we define a voltage vector **v** as follows:

$$\mathbf{v} \triangleq \begin{bmatrix} v_a \\ v_b \\ v_c \\ -v_F \\ -v_D \\ -v_Q \end{bmatrix}$$

Using this definition in (10.11), we get

$$\mathbf{v} = -\mathbf{R}\mathbf{i} - \frac{d\boldsymbol{\lambda}}{dt} \tag{10.12}$$

It is to be noted that v_D and v_Q above are actually zero because in our model the fictitious rotor coils are short circuited. It is convenient to call (10.12) the *electrical equation*.

We next relate **λ** to **i** through $\boldsymbol{\lambda} = \mathbf{L}(\theta)\mathbf{i}$ assuming a linear relationship. $\mathbf{L}(\theta)$ is a 6×6 matrix relating six flux linkages to the six currents. Just as in Example 10.2, we can find the element L_{ij} by finding the flux linkages of the ith coil after setting all currents equal to zero except the current in the jth coil. We will also simplify, just as we did in Example 10.2, by considering only dc and the lowest harmonic terms.

We now consider the calculation of some of the terms. It is simplest in the following to use actual rather than per unit quantities. Thus the inductances are in henrys. In all the definitions of inductance to follow, to simplify the notation, we will leave out the proviso "all other currents equal zero." We will start by specifying the stator self-inductances.

Self-inductances of stator coils

$$L_{aa} = \frac{\lambda_{aa'}}{i_a} = L_s + L_m \cos 2\theta \qquad L_s > L_m \geq 0$$

$$L_{bb} = \frac{\lambda_{bb'}}{i_b} = L_s + L_m \cos 2\left(\theta - \frac{2\pi}{3}\right)$$

$$L_{cc} = \frac{\lambda_{cc'}}{i_c} = L_s + L_m \cos 2\left(\theta + \frac{2\pi}{3}\right)$$

The form of L_{aa} may be explained as follows. Referring to Fig. 10.2, the mmf due to the current in coil aa' is effective in the vertical direction. The resulting flux, with centerline in the vertical direction, is maximum when $\theta = 0$ or π, and minimum when $\theta = \pi/2$ or $3\pi/2$. The variation of $L_{aa}(\theta)$ is π-periodic; by taking only dc and fundamental terms we get the result claimed. Note that with a round rotor, $L_{aa}(\theta) = L_s$ is constant.

L_{bb} may be similarly explained. Here we note that the mmf due to the current in coil bb' is effective for producing flux along an axis at an angle of $2\pi/3$ rather than 0. Thus $\theta = 2\pi/3$ results in a maximum flux rather than $\theta = 0$. This explains the shift in argument in L_{bb} compared with L_{aa}. In L_{bb} we replace θ by $\theta - 2\pi/3$, so that the maximum occurs when $\theta = 2\pi/3$ rather than $\theta = 0$. In the case of L_{cc}, similar considerations lead to a replacement of θ by $\theta + 2\pi/3$, or equivalently, $\theta - 4\pi/3$.

We note that these replacements or shifts will occur frequently as the theory is developed, and occasionally it is convenient to introduce a *shift operator* $\mathcal{T}$ to simplify the notation. Let $\mathcal{T}L_{aa}$ indicate L_{aa} with θ shifted (delayed) by $2\pi/3$ (i.e., θ replaced by $\theta - 2\pi/3$). Then we have $L_{bb} = \mathcal{T}L_{aa}$ and $L_{cc} = \mathcal{T}L_{bb} = \mathcal{T}^2 L_{aa}$.

We turn next to some mutual inductances.

Mutual inductances between stator coils

$$L_{ab} = \frac{\lambda_{aa'}}{i_b} = -\left[M_s + L_m \cos 2\left(\theta + \frac{\pi}{6} \right) \right] \qquad M_s > L_m \geq 0$$

$$L_{bc} = \frac{\lambda_{bb'}}{i_c} = -\left[M_s + L_m \cos 2\left(\theta - \frac{\pi}{2} \right) \right]$$

$$L_{ca} = \frac{\lambda_{cc'}}{i_a} = -\left[M_s + L_m \cos 2\left(\theta + \frac{5\pi}{6} \right) \right]$$

Intuitively, we may justify the first of these results as follows. Consider a round rotor first, in which case $L_m = 0$. Noting the reference directions for $\phi_{aa'}$ and i_b, we see that L_{ab} must be negative as shown. Now consider the salient-pole case. L_{ab} will still be negative but now depends on θ. We want to make it plausible that $|L_{ab}|$ is maximum for $\theta = -\pi/6$. If $\theta = -\pi/3$ the mmf due to i_b is most effective in generating flux, but the flux is misdirected as far as linkages of coil aa' are concerned. On the other hand, if $\theta = 0$, the flux is most effectively directed but there is less of it. A quantitative analysis indicates that $-\pi/6$, the average value of $-\pi/3$ and 0, is the compromise which gives the "best" result. We note that $L_{bc} = \mathcal{T}L_{ab}$ and $L_{ca} = \mathcal{T}L_{bc} = \mathcal{T}^2 L_{ab}$, where $\mathcal{T}$ is the shift operator; examination of the coil geometry in Fig. 10.2 provides the physical basis for this result.

Other inductances.   We will list and indicate the derivation of only a few more of the remaining terms.

$$L_{aF} = \frac{\lambda_{aa'}}{i_F} = M_F \cos \theta \qquad M_F > 0$$

$$L_{FF} = \frac{\lambda_{FF'}}{i_F} = L_F \qquad\qquad L_F > 0$$

$$L_{FQ} = \frac{\lambda_{FF'}}{i_Q} = 0$$

The first result is implied by a sinusoidal distribution of field flux; this condition is an objective of generator design and was discussed in Section 8.2; the expression for (the open circuit) $\lambda_{aa'}$, given in (8.3a), is in the form $L_{aF} i_F$. The second result may be inferred by noting that the magnetic circuit "seen" from the rotor is the same for all θ. The last result follows because the two coils involved are perpendicular.

We will not continue further with this list. We note that in the 6×6 inductance matrix, because of symmetry, there are 21 elements to be determined, of which we have listed and attempted to justify 9. Completing the process, we finally get the symmetric matrix

$$\mathbf{L}(\theta) = \begin{bmatrix} \mathbf{L}_{11}(\theta) & \mathbf{L}_{12}(\theta) \\ \mathbf{L}_{21}(\theta) & \mathbf{L}_{22}(\theta) \end{bmatrix} \tag{10.13}$$

whose submatrices are as follows:

$$\mathbf{L}_{11} = \begin{bmatrix} L_s + L_m \cos 2\theta & -M_s - L_m \cos 2\left(\theta + \frac{\pi}{6}\right) & -M_s - L_m \cos 2\left(\theta + \frac{5\pi}{6}\right) \\ -M_s - L_m \cos 2\left(\theta + \frac{\pi}{6}\right) & L_s + L_m \cos 2\left(\theta - \frac{2\pi}{3}\right) & -M_s - L_m \cos 2\left(\theta - \frac{\pi}{2}\right) \\ -M_s - L_m \cos 2\left(\theta + \frac{5\pi}{6}\right) & -M_s - L_m \cos 2\left(\theta - \frac{\pi}{2}\right) & L_s + L_m \cos 2\left(\theta + \frac{2\pi}{3}\right) \end{bmatrix}$$

$$\mathbf{L}_{12} = \mathbf{L}_{21}^T = \begin{bmatrix} M_F \cos \theta & M_D \cos \theta & M_Q \sin \theta \\ M_F \cos \left(\theta - \frac{2\pi}{3}\right) & M_D \cos \left(\theta - \frac{2\pi}{3}\right) & M_Q \sin \left(\theta - \frac{2\pi}{3}\right) \\ M_F \cos \left(\theta + \frac{2\pi}{3}\right) & M_D \cos \left(\theta + \frac{2\pi}{3}\right) & M_Q \sin \left(\theta + \frac{2\pi}{3}\right) \end{bmatrix}$$

$$\mathbf{L}_{22} = \begin{bmatrix} L_F & M_R & 0 \\ M_R & L_D & 0 \\ 0 & 0 & L_Q \end{bmatrix} \tag{10.14}$$

All constants above are *nonnegative*. We can think of their values as being determined experimentally. Note that $\mathbf{L}_{11}$ relates stator flux linkages to stator currents, $\mathbf{L}_{22}$ relates rotor flux linkages to rotor currents, while $\mathbf{L}_{12}$ gives the flux linkages of the stator coils in terms of the currents in the rotor coils.

It may be seen that $\mathbf{L}(\theta)$ depends on θ in a complicated way. The generator voltage-current relation reflects that complication. From (10.12), using $\boldsymbol{\lambda} = \mathbf{L}(\theta)\mathbf{i}$, we get

$$\mathbf{v} = -\mathbf{R}\mathbf{i} - \frac{d\mathbf{L}(\theta)}{dt}\mathbf{i} - \mathbf{L}(\theta)\frac{d\mathbf{i}}{dt} \tag{10.15}$$

Hence even in the simplest case with uniform shaft rotation, where $\theta = \omega_0 t + \theta_0$, the form of (10.14) indicates that we get linear differential equations with periodically time-varying coefficients. It is hard to understand the general nature of system behavior in this case.

10.3 THE PARK TRANSFORMATION

To simplify the equations and in some important cases obtain linear time-invariant equations, we use a *Park transformation* (also called a Blondel transformation, or 0*dq* transformation, or transformation to "rotor coordinates") of the stator *abc* quantities. We transform *abc* voltages, currents, and flux linkages. The transformation, in the case of currents, yields

$$\begin{bmatrix} i_0 \\ i_d \\ i_q \end{bmatrix} = \sqrt{\frac{2}{3}} \begin{bmatrix} \dfrac{1}{\sqrt{2}} & \dfrac{1}{\sqrt{2}} & \dfrac{1}{\sqrt{2}} \\ \cos\theta & \cos\left(\theta - \dfrac{2\pi}{3}\right) & \cos\left(\theta + \dfrac{2\pi}{3}\right) \\ \sin\theta & \sin\left(\theta - \dfrac{2\pi}{3}\right) & \sin\left(\theta + \dfrac{2\pi}{3}\right) \end{bmatrix} \begin{bmatrix} i_a \\ i_b \\ i_c \end{bmatrix} \tag{10.16}$$

or, using matrix notation,

$$\mathbf{i}_{0dq} = \mathbf{P}\mathbf{i}_{abc} \tag{10.17}$$

which defines the $\mathbf{P}$ matrix and the *abc* and 0*dq* vectors. Similarly, for voltages and flux linkages, we get

$$\mathbf{v}_{0dq} = \mathbf{P}\mathbf{v}_{abc} \tag{10.18}$$

$$\boldsymbol{\lambda}_{0dq} = \mathbf{P}\boldsymbol{\lambda}_{abc} \tag{10.19}$$

The new 0*dq* variables are also called Park's variables, or, for reasons we will indicate in the next paragraph, rotor-based variables. Note that the transformation $\mathbf{P}$ depends on θ.

Physically, i_d and i_q have an interpretation as currents in a fictitious pair of rotating windings "fixed in the rotor." i_d flows in a fictitious coil whose axis is aligned with the direct or *d* axis; i_q flows in a fictitious coil whose axis is aligned with the quadrature or q axis. These currents produce the same flux components as do the actual *abc* currents.

Recalling from Chapter 8 that in the steady state with balanced (positive sequence) *abc* currents we got a synchronously rotating flux (i.e., constant when viewed from the rotor), it is understandable that we can generate this *constant* rotating flux with *constant* currents in the fictitious rotating coils. Hence the inductance matrix relating these transformed quantities should be a constant matrix.

In fact, we now show this mathematically. Note first that $\mathbf{P}$ is nonsingular; in fact, $\mathbf{P}^{-1} = \mathbf{P}^T$.

$$\mathbf{P}^{-1} = \mathbf{P}^T = \sqrt{\frac{2}{3}} \begin{bmatrix} \dfrac{1}{\sqrt{2}} & \cos\theta & \sin\theta \\[2mm] \dfrac{1}{\sqrt{2}} & \cos\left(\theta - \dfrac{2\pi}{3}\right) & \sin\left(\theta - \dfrac{2\pi}{3}\right) \\[2mm] \dfrac{1}{\sqrt{2}} & \cos\left(\theta + \dfrac{2\pi}{3}\right) & \sin\left(\theta + \dfrac{2\pi}{3}\right) \end{bmatrix} \quad (10.20)$$

The reader may check this by multiplying $\mathbf{P}$ into $\mathbf{P}^T$ to get the identity matrix. Since the rows of $\mathbf{P}$ are the same as the columns of $\mathbf{P}^T$, we find that the columns of $\mathbf{P}^T$ are orthonormal (i.e., if $\mathbf{u}_i$ designates a column of $\mathbf{P}$, then $\mathbf{u}_i^T\mathbf{u}_j = 0$, $i \neq j$, $\mathbf{u}_i^T\mathbf{u}_i = 1$).

We need to consider all six components of each current, voltage, or flux linkage vector. While we wish to transform the stator-based (*abc*) variables into rotor-based (*0dq*) variables, we wish to leave the original rotor quantities unaffected. With this objective we define the 6-vector $\mathbf{i}_B$ and the 6×6 matrix $\mathbf{B}$ as follows:

$$\mathbf{i}_B \triangleq \begin{bmatrix} i_0 \\ i_d \\ i_q \\ \hline i_F \\ i_D \\ i_Q \end{bmatrix} = \left[\begin{array}{c|c} \mathbf{P} & \mathbf{0} \\ \hline \mathbf{0} & \mathbf{1} \end{array}\right] \begin{bmatrix} i_a \\ i_b \\ i_c \\ \hline i_F \\ i_D \\ i_Q \end{bmatrix} = \mathbf{B}\mathbf{i} \quad (10.21)$$

where $\mathbf{1}$ is the 3×3 identity matrix and $\mathbf{0}$ is the 3×3 zero matrix. Similarly, we define

$$\mathbf{v}_B = \mathbf{B}\mathbf{v} \quad (10.22)$$

$$\boldsymbol{\lambda}_B = \mathbf{B}\boldsymbol{\lambda} \quad (10.23)$$

The reader can easily check that

$$\mathbf{B}^{-1} = \mathbf{B}^T = \begin{bmatrix} \mathbf{P}^T & \mathbf{0} \\ \mathbf{0} & \mathbf{1} \end{bmatrix} \quad (10.24)$$

Now we wish to find the relationship between the rotor-based variables $\boldsymbol{\lambda}_B$ and $\mathbf{i}_B$. Starting with (10.1) and using (10.21) and (10.23), we get, successively,

$$\lambda = \mathbf{Li}$$

$$\mathbf{B}^{-1}\lambda_B = \mathbf{LB}^{-1}\mathbf{i}_B \tag{10.25}$$

$$\lambda_B = \mathbf{BLB}^{-1}\mathbf{i}_B = \mathbf{L}_B\mathbf{i}_B$$

where $\mathbf{L}_B \triangleq \mathbf{BLB}^{-1}$. Using the rule for finding the transpose of a product of matrices (product of transposes in reverse order), we find that $\mathbf{L}_B^T = (\mathbf{BLB}^T)^T = \mathbf{BL}^T\mathbf{B}^T = \mathbf{BLB}^T = \mathbf{L}_B$. Thus $\mathbf{L}_B$ is symmetric because $\mathbf{L}$ is symmetric. By premultiplying by $\mathbf{B}^{-1}$ and postmultiplying by $\mathbf{B}$ we also find the relationship $\mathbf{L} = \mathbf{B}^{-1}\mathbf{L}_B\mathbf{B}$. Next, we can calculate $\mathbf{L}_B$ using (10.13) and (10.24):

$$
\mathbf{L}_B = \begin{bmatrix} \mathbf{P} & 0 \\ 0 & 1 \end{bmatrix}\begin{bmatrix} \mathbf{L}_{11} & \mathbf{L}_{12} \\ \mathbf{L}_{21} & \mathbf{L}_{22} \end{bmatrix}\begin{bmatrix} \mathbf{P}^T & 0 \\ 0 & 1 \end{bmatrix}
$$

$$
= \begin{bmatrix} \mathbf{PL}_{11}\mathbf{P}^T & \mathbf{PL}_{12} \\ \mathbf{L}_{21}\mathbf{P}^T & \mathbf{L}_{22} \end{bmatrix} \tag{10.26}
$$

Consider now the terms in (10.26). As expected, $\mathbf{L}_{22}$ is unchanged. By a tedious but straightforward calculation, we can show that

$$
\mathbf{PL}_{11}\mathbf{P}^T = \begin{bmatrix} L_0 & 0 & 0 \\ 0 & L_d & 0 \\ 0 & 0 & L_q \end{bmatrix} \tag{10.27}
$$

where

$$L_0 \triangleq L_s - 2M_s$$

$$L_d \triangleq L_s + M_s + \tfrac{3}{2}L_m$$

$$L_q \triangleq L_s + M_s - \tfrac{3}{2}L_m$$

With a knowledge of linear algebra we can show that this simplification occurs because the columns of $\mathbf{P}^T$ are (orthonormal) eigenvectors of $\mathbf{L}_{11}$ and hence the similarity transformation yields a diagonal matrix of eigenvalues. *Note:* L_d and L_q are the inductances associated with X_d and X_q in the two-reaction model derived in Chapter 8.

Exercise 2.   Show that the first column of $\mathbf{P}^T$ [i.e., $(1/\sqrt{3})(1, 1, 1)^T$] is an eigenvector of $\mathbf{L}_{11}$ corresponding to the eigenvalue $L_0 = L_s - 2M_s$.

Consider next the off-diagonal submatrix $\mathbf{L}_{21}\mathbf{P}^T = (\mathbf{PL}_{12})^T$. Using (10.14),

$$
\mathbf{L}_{21}\mathbf{P}^T = \begin{bmatrix} M_F\cos\theta & M_F\cos\left(\theta - \frac{2\pi}{3}\right) & M_F\cos\left(\theta + \frac{2\pi}{3}\right) \\ M_D\cos\theta & M_D\cos\left(\theta - \frac{2\pi}{3}\right) & M_D\cos\left(\theta + \frac{2\pi}{3}\right) \\ M_Q\sin\theta & M_Q\sin\left(\theta - \frac{2\pi}{3}\right) & M_Q\sin\left(\theta + \frac{2\pi}{3}\right) \end{bmatrix}\sqrt{\frac{2}{3}}\begin{bmatrix} \frac{1}{\sqrt{2}} & \cos\theta & \sin\theta \\ \frac{1}{\sqrt{2}} & \cos\left(\theta - \frac{2\pi}{3}\right) & \sin\left(\theta - \frac{2\pi}{3}\right) \\ \frac{1}{\sqrt{2}} & \cos\left(\theta + \frac{2\pi}{3}\right) & \sin\left(\theta + \frac{2\pi}{3}\right) \end{bmatrix} \tag{10.28}
$$

Here we recognize that the first two rows of $\mathbf{L}_{21}$ are proportional to the second column of $\mathbf{P}^T$, which we previously indicated was orthogonal to the remaining two columns of $\mathbf{P}^T$. The third row of $\mathbf{L}_{21}$ is proportional to the third column of $\mathbf{P}^T$. Thus, multiplying rows into columns, it is easy to evaluate the product of the two matrices. We get

$$
\mathbf{L}_{21}\mathbf{P}^T = \begin{bmatrix} 0 & \sqrt{\dfrac{3}{2}}M_F & 0 \\[2ex] 0 & \sqrt{\dfrac{3}{2}}M_D & 0 \\[2ex] 0 & 0 & \sqrt{\dfrac{3}{2}}M_Q \end{bmatrix} \tag{10.29}
$$

To simplify the notation, let $k \triangleq \sqrt{3/2}$; then, from (10.14), (10.26), (10.27), and (10.29), we get

$$
\mathbf{L}_B = \left[\begin{array}{ccc:ccc} L_0 & 0 & 0 & 0 & 0 & 0 \\ 0 & L_d & 0 & kM_F & kM_D & 0 \\ 0 & 0 & L_q & 0 & 0 & kM_Q \\ \hdashline 0 & kM_F & 0 & L_F & M_R & 0 \\ 0 & kM_D & 0 & M_R & L_D & 0 \\ 0 & 0 & kM_Q & 0 & 0 & L_Q \end{array} \right] \tag{10.30}
$$

Note that the matrix $\mathbf{L}_B$ is simple, sparse, symmetric, and *constant*. The reader should compare it with the more complicated $\mathbf{L}$ in (10.13) and (10.14).

10.4 PARK'S VOLTAGE EQUATION

We next derive the voltage-current relations using Park's variables. Starting with (10.12), which is repeated below,

$$
\mathbf{v} = -\mathbf{R}\mathbf{i} - \frac{d\boldsymbol{\lambda}}{dt} \tag{10.12}
$$

and using (10.21) to (10.23), we find that

$$
\mathbf{B}^{-1}\mathbf{v}_B = -\mathbf{R}\mathbf{B}^{-1}\mathbf{i}_B - \frac{d}{dt}(\mathbf{B}^{-1}\boldsymbol{\lambda}_B)
$$

Premultiplying on the left by **B**, we get

$$v_B = -BRB^{-1}i_B - B\frac{d}{dt}(B^{-1}\lambda_B) \tag{10.31}$$

Using the identity

$$BRB^{-1} = \begin{bmatrix} P & 0 \\ 0 & 1 \end{bmatrix}\begin{bmatrix} r & & & 0 \\ & r & & \\ & & r & \\ 0 & & & \begin{matrix} r_F & & \\ & r_D & \\ & & r_Q \end{matrix} \end{bmatrix}\begin{bmatrix} P^{-1} & 0 \\ 0 & 1 \end{bmatrix} = R \tag{10.32}$$

we can simplify (10.31) by replacing BRB^{-1} by **R**. Continuing, and using the rule for differentiating a product, we get

$$v_B = -Ri_B - B\frac{dB^{-1}}{dt}\lambda_B - \frac{d\lambda_B}{dt} \tag{10.33}$$

Thus, just as in (10.2), we get "speed voltages" and "transformer voltages."

Next, we wish to obtain a more explicit expression for the matrix $B(dB^{-1}/dt)$ in (10.33). We first calculate $B(dB^{-1}/d\theta)$. Using(10.21) and (10.24), we get

$$B\frac{dB^{-1}}{d\theta} = \begin{bmatrix} P & 0 \\ 0 & 1 \end{bmatrix}\begin{bmatrix} \dfrac{dP^{-1}}{d\theta} & 0 \\ 0 & 0 \end{bmatrix} = \begin{bmatrix} P\dfrac{dP^{-1}}{d\theta} & 0 \\ 0 & 0 \end{bmatrix} \tag{10.34}$$

where

$$P\frac{dP^{-1}}{d\theta} = \sqrt{\frac{2}{3}}\begin{bmatrix} \dfrac{1}{\sqrt{2}} & \dfrac{1}{\sqrt{2}} & \dfrac{1}{\sqrt{2}} \\ \cos\theta & \mathcal{T}\cos\theta & \mathcal{T}^2\cos\theta \\ \sin\theta & \mathcal{T}\sin\theta & \mathcal{T}^2\sin\theta \end{bmatrix}$$

$$\times\sqrt{\frac{2}{3}}\begin{bmatrix} 0 & -\sin\theta & \cos\theta \\ 0 & -\mathcal{T}\sin\theta & \mathcal{T}\cos\theta \\ 0 & -\mathcal{T}^2\sin\theta & \mathcal{T}^2\cos\theta \end{bmatrix} \tag{10.35}$$

To simplify the notation, we have introduced the shift operator. Using the orthogonality properties of the rows and columns in (10.35), we get

$$P\frac{dP^{-1}}{d\theta} = \frac{2}{3}\begin{bmatrix} 0 & 0 & 0 \\ 0 & 0 & \frac{3}{2} \\ 0 & -\frac{3}{2} & 0 \end{bmatrix} = \begin{bmatrix} 0 & 0 & 0 \\ 0 & 0 & 1 \\ 0 & -1 & 0 \end{bmatrix} \tag{10.36}$$

Then, using (10.36) in (10.34), we get

$$
\mathbf{B}\,\frac{d\mathbf{B}^{-1}}{d\theta} =
\left[
\begin{array}{ccc:c}
0 & 0 & 0 & \\
0 & 0 & 1 & \mathbf{0} \\
0 & -1 & 0 & \\
\hdashline
& \mathbf{0} & & \mathbf{0}
\end{array}
\right]
\tag{10.37}
$$

a 6×6 matrix with only two nonzero elements! Finally, noting that $\mathbf{B}(d\mathbf{B}^{-1}/dt) = \mathbf{B}(d\mathbf{B}^{-1}/d\theta)(d\theta/dt)$ and substituting (10.37) in (10.33), we get the *electrical equation*:

$$
\mathbf{v}_B = -\mathbf{R}\mathbf{i}_B - \dot{\theta}
\begin{bmatrix}
0 \\
\lambda_q \\
-\lambda_d \\
0 \\
0 \\
0
\end{bmatrix}
- \frac{d\boldsymbol{\lambda}_B}{dt}
\tag{10.38}
$$

Note 1: If the shaft rotation is uniform (i.e., $\dot{\theta} = d\theta/dt = \text{constant}$),(10.38) is *linear* and *time invariant*! Very often as a good approximation we can assume that $\dot{\theta} = \text{constant}$.

Note 2: Although, superficially, (10.38) looks very much like (10.12), it is important to note that (10.38) is basically much simpler. In (10.38), $\boldsymbol{\lambda}_B = \mathbf{L}_B\mathbf{i}_B$, where $\mathbf{L}_B$ is the *constant* matrix in (10.30), while in (10.12), $\boldsymbol{\lambda} = \mathbf{L}\mathbf{i}$, where $\mathbf{L} = \mathbf{L}(\theta)$ is the very complicated matrix in (10.13) and (10.14).

Equation (10.38) is the *electrical equation* in terms of the transformed variables $\mathbf{i}_B$, $\mathbf{v}_B$, and $\boldsymbol{\lambda}_B$. To study dynamic behavior we need a companion *mechanical equation*. Thus, we next find the mechanical torque T_E, in terms of $\mathbf{L}_B$ and $\mathbf{i}_B$ and $\boldsymbol{\lambda}_B$, for use in the mechanical equation (10.10).

10.5 PARK'S MECHANICAL EQUATION

We start by deriving an expression for the electrical torque in terms of Park's variables. The original expression (10.9) repeated below is

$$
T_E = \frac{1}{2}\mathbf{i}^T\,\frac{d\mathbf{L}(\theta)}{d\theta}\,\mathbf{i}
\tag{10.9}
$$

We wish to replace $\mathbf{L}$ and $\mathbf{i}$ with $\mathbf{L}_B$ and $\mathbf{i}_B$. We can use $\mathbf{i} = \mathbf{B}^{-1}\mathbf{i}_B = \mathbf{B}^T\mathbf{i}_B$ and $\mathbf{L} = \mathbf{B}^{-1}\mathbf{L}_B\mathbf{B}$. Thus

$$
T_E = \frac{1}{2}\,\mathbf{i}_B^T\mathbf{B}\,\frac{d}{d\theta}\,(\mathbf{B}^{-1}\mathbf{L}_B\mathbf{B})\mathbf{B}^T\mathbf{i}_B
\tag{10.39}
$$

We note again that $\mathbf{L}_B$ is not a function of θ, but $\mathbf{B}$ and $\mathbf{B}^{-1}$ are. Differentiating with respect to θ, we get

$$T_E = \frac{1}{2} \mathbf{i}_B^T \mathbf{L}_B \frac{d\mathbf{B}}{d\theta} \mathbf{B}^T \mathbf{i}_B + \frac{1}{2} \mathbf{i}_B^T \mathbf{B} \frac{d\mathbf{B}^{-1}}{d\theta} \mathbf{L}_B \mathbf{i}_B \tag{10.40}$$

where we have freely interchanged $\mathbf{B}^T$ and $\mathbf{B}^{-1}$. We now will show the two products of terms in (10.40) are equal. The product of terms on the left, a scalar, is replaced by its transpose, the same scalar. Then we get

$$T_E = \frac{1}{2} \mathbf{i}_B^T \mathbf{B} \left(\frac{d\mathbf{B}}{d\theta}\right)^T \mathbf{L}_B \mathbf{i}_B + \frac{1}{2} \mathbf{i}_B^T \mathbf{B} \frac{d\mathbf{B}^T}{d\theta} \mathbf{L}_B \mathbf{i}_B \tag{10.41}$$

Since the operations of differentiation and transposition commute, $(d\mathbf{B}/d\theta)^T = d\mathbf{B}^T/d\theta$, and the two products are equal. Thus taking twice the second (right-hand) term, we get

$$T_E = \mathbf{i}_B^T \mathbf{B} \frac{d\mathbf{B}^{-1}}{d\theta} \mathbf{L}_B \mathbf{i}_B \tag{10.42}$$

Finally, using (10.37) and $\boldsymbol{\lambda}_B = \mathbf{L}_B \mathbf{i}_B$, we get

$$T_E = \mathbf{i}_B^T \left[\begin{array}{ccc|c}
0 & 0 & 0 & \\
0 & 0 & 1 & \mathbf{0} \\
0 & -1 & 0 & \\
\hline
& \mathbf{0} & & \mathbf{0}
\end{array}\right] \boldsymbol{\lambda}_B$$

$$= [i_0, i_d, i_q, i_F, i_D, i_Q] \begin{bmatrix} 0 \\ \lambda_q \\ -\lambda_d \\ 0 \\ 0 \\ 0 \end{bmatrix} = i_d \lambda_q - i_q \lambda_d \tag{10.43}$$

which is a very compact expression for the torque. Using (10.30) to relate flux linkages to currents, we may obtain an expression entirely in terms of currents. Note that λ_q is related to i_q and i_Q, while λ_d depends on i_d, i_F, and i_D. Note that i_0 does not enter into the torque equation at all.

Substituting (10.43) in (10.3) we get a *mechanical equation*

$$J\ddot{\theta} + D\dot{\theta} + i_q \lambda_d - i_d \lambda_q = T_M \tag{10.44}$$

in addition to the *electrical equation* (10.38). Note that both equations are in terms of the Park variables.

10.6 CIRCUIT MODEL

A close examination of $\mathbf{L}_B$, (10.30), which relates $\boldsymbol{\lambda}_B$ to $\mathbf{i}_B$, reveals a useful property. λ_0 depends only on i_0; λ_d, λ_F, and λ_D depend only on i_d, i_F, and i_D; λ_q and λ_Q depend only on i_q and i_Q. The decoupling of the direct axis and quadrature axis circuits may be understood by noting the orientations of the physical and/or fictitious coils.

We can take advantage of this decoupling by grouping the scalar equations that make up the components of vector equation (10.38). The groupings are into zero sequence, direct axis, and quadrature axis equations as follows:

Zero sequence:

$$v_0 = -ri_0 - \frac{d\lambda_0}{dt} \tag{10.45}$$

Direct axis:

$$v_d = -ri_d - \dot{\theta}\lambda_q - \frac{d\lambda_d}{dt}$$

$$v_F = r_F i_F + \frac{d\lambda_F}{dt} \tag{10.46}$$

$$v_D = r_D i_D + \frac{d\lambda_D}{dt} = 0$$

Quadrature axis:

$$v_q = -ri_q + \dot{\theta}\lambda_d - \frac{d\lambda_q}{dt}$$

$$v_Q = r_Q i_Q + \frac{d\lambda_Q}{dt} = 0 \tag{10.47}$$

Note that except for the $\dot{\theta}\lambda_q$ and $\dot{\theta}\lambda_d$ terms, the three groups of equations are completely decoupled.

We also have the completely decoupled equations relating flux linkages to currents as follows:

Zero sequence:

$$\lambda_0 = L_0 i_0 \tag{10.48}$$

Direct axis:

$$\begin{bmatrix} \lambda_d \\ \lambda_F \\ \lambda_D \end{bmatrix} = \begin{bmatrix} L_d & kM_F & kM_D \\ kM_F & L_F & M_R \\ kM_D & M_R & L_D \end{bmatrix} \begin{bmatrix} i_d \\ i_F \\ i_D \end{bmatrix} \tag{10.49}$$

Quadrature axis:

$$\begin{bmatrix} \lambda_q \\ \lambda_Q \end{bmatrix} = \begin{bmatrix} L_q & kM_Q \\ kM_Q & L_Q \end{bmatrix} \begin{bmatrix} i_q \\ i_Q \end{bmatrix} \tag{10.50}$$

Since the matrix (10.30) is symmetric, we can find a circuit model for the equations above and this is shown in Fig. 10.3.

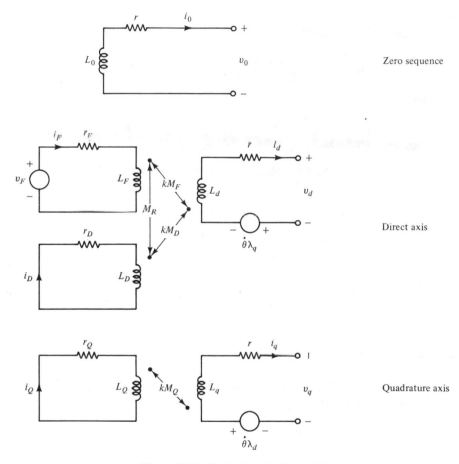

Figure 10.3 Equivalent-circuit model.

By writing KVL for each circuit, the circuit model above may be checked. The advantage of a circuit diagram is that we can use our experience with other circuits to understand, at least qualitatively, the behavior of the circuit. For example, the time constants of the "damper" circuits (in which i_D and i_Q flow) are very low, even taking the coupling into account. In this case, in many transients, i_D and i_Q decay very rapidly to zero, and, except for a very short initial interval in the transient, we expect little error if we neglect i_D and i_Q (i.e., set them equal to zero). This discussion will be clearer in the context of specific applications.

We should note that a normalization of rotor-based quantities is possible (an extension of the per unit system) such that simpler equivalent circuits are obtained for the direct axis and quadrature axis circuits. This is analogous to the elimination of the ideal transformers in the transformer per phase equivalent circuits in Chapter 5.

10.7 INSTANTANEOUS POWER OUTPUT

Before going to some applications we consider an additional useful result. We calculate the *instantaneous* three-phase power output of the generator in terms of $0dq$ variables. We start with

$$p_{3\phi}(t) = i_a v_a + i_b v_b + i_c v_c$$
$$= \mathbf{i}_{abc}^T \mathbf{v}_{abc} \tag{10.51}$$

where we are using the notation introduced in (10.17). Thus, using $\mathbf{P}^T = \mathbf{P}^{-1}$,

$$p_{3\phi}(t) = (\mathbf{P}^{-1}\mathbf{i}_{0dq})^T \mathbf{P}^{-1}\mathbf{v}_{0dq}$$
$$= \mathbf{i}_{0dq}^T \mathbf{P}\mathbf{P}^{-1}\mathbf{v}_{0dq}$$
$$= \mathbf{i}_{0dq}^T \mathbf{v}_{0dq}$$
$$= i_0 v_0 + i_d v_d + i_q v_q \tag{10.52}$$

The important result is that we can calculate instantaneous three-phase power using $0dq$ variables, without converting back to *abc* variables.

10.8 APPLICATIONS

We next turn to three examples which illustrate the use of the Park transformation and equations. The general procedure follows:

General procedure

1. The problem as usually stated includes references to *abc* (stator) voltage and/or current variables. We transform these variables to $0dq$ variables using (10.17) and/or (10.18).

2. The problem is then solved using the mechanical equation (10.44) and/or the electrical equation (10.38) together with (10.30). Frequently, it is simpler to use the electrical equations (10.45) to (10.47) instead, together with (10.48) to

(10.50). *Note:* If $\dot{\theta}$ = constant, the equations are linear and time invariant and may be solved conveniently using the Laplace transform.

3. We can then transform back to *abc* variables using the inverse Park transformation (10.20).

The reader may recognize the parallel between these steps and those in any transformation technique. We now consider the first of our three examples.

Example 10.3   Voltage Buildup

We are given the system schematic shown in Fig. E10.3. Assume

1. The generator is rotating at synchronous speed with

$$\theta = \omega_0 t + \frac{\pi}{2} + \delta \qquad \delta = \text{constant}$$

2. The stator is open circuited (i.e., $i_a = i_b = i_c \equiv 0$).
3. The initial field current $i_F(0) = 0$.
4. At $t = 0$, a step function of constant voltage v_F^0 is applied.

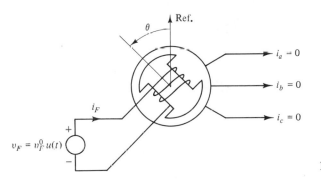

Figure E10.3

Neglect the damper circuits (i.e., set $i_D = i_Q = 0$) and derive an expression for the open-circuit terminal voltages v_a, v_b, and v_c for $t \geq 0$.
Note the following:

1. The assumptions imply that the generator is initially at "rest" electrically (i.e., in equilibrium), with all currents, fluxes, and voltages equal to zero. We wish to see how the generator terminal voltages come to life upon application of the field voltage.
2. Inclusion of the $\pi/2$ in the expression for θ seems awkward at first but leads to simpler expressions for voltage in terms of δ. Henceforth we will routinely include the $\pi/2$.
3. We expect the errors caused by neglecting the damper circuits to be confined to a very short initial interval.

Solution   From (10.17) $\mathbf{i}_{abc} \equiv \mathbf{0}$ implies that $\mathbf{i}_{0dq} = \mathbf{0}$. Since we are neglecting the damper circuits, $i_D = i_Q = 0$. Thus the only nonzero current is i_F. From (10.48) to (10.50), we find that

$$\lambda_0 = 0$$

$$\lambda_d = kM_F i_F$$

$$\lambda_F = L_F i_F$$

$$\lambda_q = 0$$

Since we are neglecting the damper circuits, we do not need to calculate λ_D and λ_Q. Using (10.45) to (10.50), we get, for $t \geq 0$,

$$v_0 = 0$$

$$v_d = -\frac{d\lambda_d}{dt} = -kM_F \frac{di_F}{dt}$$

$$v_F^0 = r_F i_F + L_F \frac{di_F}{dt}$$

$$v_q = \omega_0 \lambda_d = \omega_0 kM_F i_F$$

Solving the third equation with $i_F(0) = 0$, we get

$$i_F(t) = \frac{v_F^0}{r_F}(1 - e^{-(r_F/L_F)t}) \qquad t \geq 0$$

With no armature reaction ($\mathbf{i}_{abc} \equiv \mathbf{0}$) this is exactly what we expect! Using this result in the fourth equation, we get

$$v_q(t) = \frac{\omega_0 kM_F v_F^0}{r_F}(1 - e^{-(r_F/L_F)t})$$

and from the second equation, we get

$$v_d(t) = -\frac{kM_F v_F^0}{L_F}e^{-(r_F/L_F)t}$$

Going back to abc variables, $\mathbf{v}_{abc} = \mathbf{P}^{-1}\mathbf{v}_{0dq}$; thus, using (10.20),

$$v_a(t) = \frac{1}{k}\left(\frac{1}{\sqrt{2}}v_0 + v_d \cos\theta + v_q \sin\theta\right)$$

$$= -\frac{M_F v_F^0}{L_F}e^{-(r_F/L_F)t}\cos\left(\omega_0 t + \frac{\pi}{2} + \delta\right)$$

$$+ \frac{\omega_0 M_F v_F^0}{r_F}(1 - e^{-(r_F/L_F)t})\sin\left(\omega_0 t + \frac{\pi}{2} + \delta\right)$$

From (10.20) we can see that the only change required for the calculation of v_b and v_c is the appropriate shift in argument of the sine and cosine by $2\pi/3$. In fact, $v_b(t)$ and $v_c(t)$ are identical to $v_a(t)$ except for a phase shift of $-120°$ and $-240°$, respectively.

As a practical matter, $\omega_0 L_F \gg r_F$, so that very early in the transient the second term dominates. Then

$$v_a(t) \approx \frac{\omega_0 M_F v_F^0}{r_F}(1 - e^{-(r_F/L_F)t}) \cos(\omega_0 t + \delta)$$

Thus the "envelope" of voltage "builds up" to its steady-state value with the buildup having a time constant $T'_{do} \triangleq L_F/r_F$. T'_{do} is called the (direct axis) transient open-circuit time constant. "Open circuit" here refers to the stator circuit. The value of T'_{do} is typically in the range 2 to 9 sec.

In the steady state we get the (open-circuit) voltage

$$v_a(t) = \frac{\omega_0 M_F v_F^0}{r_F} \cos(\omega_0 t + \delta) = \sqrt{2}\,|E_a| \cos(\omega_0 t + \delta)$$

which matches the result found earlier in (8.6). We note that the open-circuit voltage magnitude is $|E_a| = \omega_0 M_F v_F^0/\sqrt{2}\,r_F$, a result we will be using in the next example. Finally, we note that the phase of the (steady-state) open-circuit voltage is δ; the reason we chose to include the $\pi/2$ in $\theta = \omega_0 t + \pi/2 + \delta$ was to obtain this simple result.

Example 10.4   Symmetrical Short Circuit

Suppose that in Example 10.3, after the voltage has reached its steady-state value, we have a "solid" three-phase short circuit or "fault" at the stator terminals of the machine. This is the simplest case of a short circuit we can consider. It is important to determine the short-circuit current in order to pick circuit breakers and relay settings to protect the generator. We note that initially these currents can be as much as 10 times the steady-state short-circuit currents! An introduction to the subject of generator protection, the location of circuit breakers and relays, is considered in Chapter 14.

Since this example yields results of great practical interest, we will consider it in more detail than necessary simply to illustrate the use of Park's transformations. Consider the system shown in Fig. E10.4(a). Assume that:

1. Generator is open circuited for $t < 0$ and short circuited for $t \geq 0$.
2. $\theta = \omega_0 t + (\pi/2) + \delta$, $\delta = $ constant.
3. $v_F = v_F^0 = $ constant. i_F is in the steady state at $t = 0$ [i.e., $i_F(0) = v_F^0/r_F$].
4. Neglect the effect of damper windings.

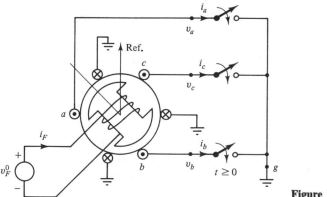

Figure E10.4(a)

Find i_a, i_b, i_c, and i_F for $t \geq 0$, concentrating on the behavior at the beginning of the transient.

Note: At the beginning of our analysis we will include the damper circuits. We will drop them later.

Solution.   We make the following observations.

1. For $t < 0$, $\mathbf{i}_{abc} = \mathbf{0} \Rightarrow \mathbf{i}_{0dq} = \mathbf{0}$.
2. For $t \geq 0$, $\mathbf{v}_{abc} = \mathbf{0} \Rightarrow \mathbf{v}_{0dq} = \mathbf{0}$.

From what we know about inductive circuits with (finite) voltage sources, the currents will not change instantaneously. Thus we may assume for the circuit established at $t = 0$,

3. $\mathbf{i}_{0dq}(0) = \mathbf{0}$.

For $t \geq 0$ we have the short circuit, so from the observations above and (10.45) to (10.47),

$$0 = -ri_0 - L_0 \frac{di_0}{dt} \qquad \text{zero sequence}$$

$$\left. \begin{aligned} 0 &= -ri_d - \omega_0\lambda_q - \frac{d\lambda_d}{dt} \\[2mm] v_F^0 &= r_F i_F + \frac{d\lambda_F}{dt} \\[2mm] 0 &= r_D i_D + \frac{d\lambda_D}{dt} \end{aligned} \right\} \text{direct axis}$$

$$\left. \begin{aligned} 0 &= -ri_q + \omega_0\lambda_d - \frac{d\lambda_q}{dt} \\[2mm] 0 &= r_Q i_Q + \frac{d\lambda_Q}{dt} \end{aligned} \right\} \text{quadrature axis}$$

The zero-sequence equation describes an *RL* circuit without a source and with a zero initial condition. Thus $i_0(t) \equiv 0$ and need not be considered further.

In the remaining equations we may express flux linkages in terms of currents; we get five linear constant coefficient differential equations in the current variables i_d, i_F, i_D, i_q, and i_Q. In principle the solution is straightforward but requires numerical data. Later we will simplify by neglecting the damper circuits. First, however, let us digress to obtain the steady-state solution.

Steady-State Solution. The only source is the constant voltage, v_F^0, and with resistance, all variables tend to constant values in the steady state. Thus $d\lambda_d/dt \to 0$, . . . , $d\lambda_Q/dt \to 0$, and then, from the equations, it may be seen that $i_D \to 0$, $i_Q \to 0$, and $i_F \to v_F^0/r_F$. We get the following steady-state relations:

$$ri_d + \omega_0 L_q i_q = 0$$

$$ri_q - \omega_0 \left(L_d i_d + \frac{kM_F v_F^0}{r_F} \right) = 0$$

The second of these equations may be simplified by noting that the prefault open-circuit voltage magnitude (at $t = 0$) is $|E_a(0)| = \omega_0 M_F v_F^0 / \sqrt{2}\, r_F$; this was shown at the end of Example 10.3. Then we can write

$$\begin{bmatrix} r & \omega_0 L_q \\ -\omega_0 L_d & r \end{bmatrix} \begin{bmatrix} i_d \\ i_q \end{bmatrix} = \begin{bmatrix} 0 \\ 1 \end{bmatrix} \sqrt{2}\, k\, |E_a(0)|$$

Solving for i_d and i_q and using $\mathbf{P}^{-1}$, we get the steady state $i_a(t)$.

$$i_a(t) = \frac{\sqrt{2}\, |E_a(0)|}{r^2 + \omega_0^2 L_d L_q} \left[r \sin \left(\omega_0 t + \frac{\pi}{2} + \delta \right) - \omega_0 L_q \cos \left(\omega_0 t + \frac{\pi}{2} + \delta \right) \right]$$

We get the same result by using the two-reaction model of Chapter 8. Finally, neglecting r, we get the simple result,

$$i_a(t) = -\frac{\sqrt{2}\, |E_a(0)|}{\omega_0 L_d} \cos \left(\omega_0 t + \frac{\pi}{2} + \delta \right) = \frac{\sqrt{2}\, |E_a(0)|}{X_d} \sin \left(\omega_0 t + \delta \right)$$

This matches the result obtained in Example 8.7.

We return now to a consideration of the transient.

Solution for Transient. To avoid the necessity of using numerical data, we will simplify the system of differential equations by neglecting the damper circuits. In this case we assume that $i_D = i_Q = 0$ and eliminate two differential equations. We note that neglecting the damper circuit is equivalent to ignoring the very first part of the transient (one or two cycles in duration). We are then left with three linear constant coefficient differential equations. Using (10.49) and (10.50), we can express the flux linkages directly in terms of currents. Changing sign where convenient, we get

$$0 = ri_d + \omega_0 L_q i_q + L_d \frac{di_d}{dt} + kM_F \frac{di_F}{dt}$$

$$v_F^0 = r_F i_F + kM_F \frac{di_d}{dt} + L_F \frac{di_F}{dt} \tag{10.53}$$

$$0 = ri_q - \omega_0 L_d i_d - \omega_0 kM_F i_F + L_q \frac{di_q}{dt}$$

The most convenient way to get a solution (in closed form) is to use the Laplace transform. We remind the reader that the Laplace transform of dx/dt is $sX(s) - x(0^-)$, where $X(s)$ is the Laplace transform of $x(t)$. We use this result in (10.53), noting the initial conditions, $i_d(0^-) = i_q(0^-) = 0$ and $i_F(0^-) = i_F^0 = v_F^0 / r_F$. Introducing matrix notation, we get

$$\begin{bmatrix} r + sL_d & skM_F & \omega_0 L_q \\ skM_F & r_F + sL_F & 0 \\ -\omega_0 L_d & -\omega_0 kM_F & r + sL_q \end{bmatrix} \begin{bmatrix} \hat{i}_d \\ \hat{i}_F \\ \hat{i}_q \end{bmatrix} = \begin{bmatrix} kM_F \\ L_F + r_F/s \\ 0 \end{bmatrix} i_F^0 \tag{10.54}$$

Here we are using the notation $\hat{x}$ for the Laplace transform of $x(t)$ to save the notation I_d and I_q for a different use.

Now we can solve for $\hat{i}_d$, $\hat{i}_F$, and $\hat{i}_q$ using Cramer's rule. For example, in the case of $\hat{i}_d$ we get

$$\hat{i}_d = \frac{\begin{vmatrix} kM_F & skM_F & \omega_0 L_q \\ \dfrac{r_F}{s} + L_F & r_F + sL_F & 0 \\ 0 & -\omega_0 kM_F & r + sL_q \end{vmatrix}}{\Delta(s)} i_F^0$$

where the characteristic polynomial $\Delta(s)$ is the determinant of the matrix in (10.54).

The location of the zeros of $\Delta(s)$ are important in specifying the transient behavior of i_d, i_F, and i_q. By a routine but tedious calculation, we get a characteristic polynomial in the form

$$\Delta(s) = as^3 + bs^2 + cs + d$$

where

$$a = L_d' L_F L_q$$
$$b = (L_d' + L_q)L_F r + L_d L_q r_F$$
$$c = \omega_0^2 L_d' L_F L_q + L_F r^2 + (L_d + L_q)rr_F$$
$$d = (\omega_0^2 L_d L_q + r^2)r_F$$

and $L_d' \triangleq L_d - (kM_F)^2/L_F$. L_d' is called the *direct axis transient inductance;* its physical significance will be discussed later. L_d' can be shown to be always positive. A typical value is $L_d' \approx 0.2L_d$. Thus all the coefficients of the characteristic polynomial $\Delta(s)$ are positive.

In fact, we can show more. We can show that all the zeros of $\Delta(s)$ lie in the open left half-plane. The reader who knows the Routh–Hurwitz test can easily verify that result. In practice because the resistance values r and r_F are relatively small, the zeros of $\Delta(s)$ are not very different from the values found by setting them equal to zero. In this case we get the very simple result

$$\Delta(s) = L_d' L_F L_q s (s^2 + \omega_0^2)$$

and thus the zeros of $\Delta(s)$ are on the $j\omega$ axis at $s_1 = 0$, $s_2 = j\omega_0$, and $s_3 = -j\omega_0$. The effect of including the resistance is to move the zeros of $\Delta(s)$ slightly into the left half-plane.

As an approximation we will assume that $r = r_F = 0$. This will not significantly affect behavior for the first few cycles. The main effect is to eliminate damping in the solution of the differential equation; later we will reintroduce the damping (in a qualitative way) to explain the transition to the steady-state condition.

Now we return to the solution for $\hat{i}_d$. Evaluating the numerator determinant (with $r = r_F = 0$), we get $-\omega_0^2 kM_F L_F L_q$. Thus

$$\hat{i}_d = -\frac{kM_F i_F^0}{L_d'} \frac{\omega_0^2}{s(s^2 + \omega_0^2)}$$

Now, after partial fraction expansion, we can write

$$\frac{\omega_0^2}{s(s^2 + \omega_0^2)} = \frac{1}{s} - \frac{s}{s^2 + \omega_0^2}$$

Taking inverse Laplace transforms,

$$i_d(t) = \frac{kM_F i_F^0}{L_d'} (\cos \omega_0 t - 1)$$

Defining $X_d' = \omega_0 L_d'$, called the *direct axis transient reactance*, and using quantities previously defined [i.e., $\sqrt{2} |E_a(0)| = \omega_0 M_F v_F^0/r_F = \omega_0 M_F i_F^0$ and $k = \sqrt{3/2}$], we get $i_d(t)$ in the simplest terms.

$$i_d(t) = \frac{\sqrt{3} |E_a(0)|}{X_d'} (\cos \omega_0 t - 1)$$

where $|E_a(0)|$ is the prefault open-circuit voltage we calculated in Example 10.3.

Returning to (10.54), by similar calculations the reader may verify the following results:

$$i_q(t) = \frac{\sqrt{3} |E_a(0)|}{X_q} \sin \omega_0 t$$

$$i_F(t) = \left[1 + \frac{X_d - X_d'}{X_d'} (1 - \cos \omega_0 t) \right] i_F^0$$

$$= \left[\frac{X_d}{X_d'} - \left[\frac{X_d}{X_d'} - 1 \right] \cos \omega_0 t \right] i_F^0$$

While i_d and i_q are intermediate results (we still need to calculate $\mathbf{i}_{abc}$), the calculation of the (approximate) i_F is complete and we can discuss the result. Because X_d/X_d' is large, typically in the range 4 to 8, we get a large average (dc) component and a large sinusoidal component. This is shown in Fig. E10.4(b) on the left.

Figure E10.4(b) on the right indicates an experimentally determined field transient. From this we can infer the effect of including resistance in the model; we observe the

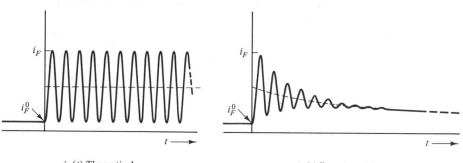

$i_F(t)$ Theoretical
(neglecting resistance)

$i_F(t)$ Experimental

Figure E10.4(b)   Field current during transient.

damping and the return of $i_F(t)$ to the steady-state value i_F^0. Note that i_F is far from constant during this transient, with or without resistance. Note also that for the initial part of the transient (say the first cycle) the effect of neglecting resistance does not lead to serious errors; the two waveforms in Fig. E10.4(b) are reasonably close.

Now we use the inverse Park transformation to find i_a, i_b, and i_c.

$$\mathbf{i}_{abc} = \mathbf{P}^{-1}\mathbf{i}_{0dq}$$

Then, noting that $i_0(t) \equiv 0$, we get

$$i_a(t) = \frac{\sqrt{2}}{\sqrt{3}}\left[\cos\left(\omega_0 t + \frac{\pi}{2} + \delta\right)\frac{\sqrt{3}\,|E_a(0)|}{X_d'}(\cos\omega_0 t - 1)\right.$$
$$\left. + \sin\left(\omega_0 t + \frac{\pi}{2} + \delta\right)\frac{\sqrt{3}\,|E_a(0)|}{X_q}\sin\omega_0 t\right]$$

Using trigonometric identities for the various products above, we separate the terms of frequency 0, ω_0, and $2\omega_0$.

$$i_a(t) = -\frac{\sqrt{2}\,|E_a(0)|}{X_d'}\cos\left(\omega_0 t + \frac{\pi}{2} + \delta\right)$$
$$+ \frac{\sqrt{2}\,|E_a(0)|}{X_d'}\left[\frac{1}{2}\cos\left(2\omega_0 t + \frac{\pi}{2} + \delta\right) + \frac{1}{2}\cos\left(-\frac{\pi}{2} - \delta\right)\right]$$
$$+ \frac{\sqrt{2}\,|E_a(0)|}{X_q}\left[\frac{1}{2}\cos\left(-\frac{\pi}{2} - \delta\right) - \frac{1}{2}\cos\left(2\omega_0 t + \frac{\pi}{2} + \delta\right)\right]$$

$$= \frac{\sqrt{2}\,|E_a(0)|}{X_d'}\sin(\omega_0 t + \delta) \tag{10.55a}$$

$$- \frac{|E_a(0)|}{\sqrt{2}}\left[\frac{1}{X_d'} + \frac{1}{X_q}\right]\sin\delta \tag{10.55b}$$

$$- \frac{|E_a(0)|}{\sqrt{2}}\left[\frac{1}{X_d'} - \frac{1}{X_q}\right]\sin(2\omega_0 t + \delta) \tag{10.55c}$$

The first term, (10.55a), is the fundamental or 60-Hz component. The second term, (10.55b), is the *dc* or *offset* or *unidirectional* component. The third term, (10.55c), is the second harmonic or 120-Hz component. It should be recalled that in the derivation of these results, resistance was neglected. Thus these results are only valid during the initial portion of the transient before the damping due to resistance has had a significant effect.

The reader should check that i_b and i_c can be found from (10.55) by replacing δ by $\delta - 2\pi/3$ and $\delta + 2\pi/3$, respectively. We note that while the change in δ only changes the phase of the ω_0 and $2\omega_0$ terms in (10.55a) and (10.55c), it changes the *magnitude* of the unidirectional term in (10.55b).

This completes the solution of Example 10.4. We now make some additional observations about the solutions just obtained.

1. We previously digressed to calculate the steady-state short-circuit current neglecting resistance. The result was

$$i_a(t) = \frac{\sqrt{2}\,|E_a(0)|}{X_d} \sin(\omega_0 t + \delta)$$

Now we note from (10.55a) that the initial behavior of the 60-Hz component is given by the above with X_d replaced by X_d'. When (small) resistance is included in the analysis, we get a smooth transition from the initial to the steady-state behavior. The approximate transient is given by

$$i_a'(t) = \sqrt{2}\,|E_a(0)| \left[\frac{1}{X_d} + \left(\frac{1}{X_d'} - \frac{1}{X_d} \right) e^{-t/T_d'} \right] \sin(\omega_0 t + \delta)$$

where i_a' designates the 60-Hz component. The time constant describing the transition is T_d', called the *direct axis transient short circuit time constant*, with numerical values of 0.5 to 3 sec for large machines.

2. When the resistance is included, the unidirectional term in (10.55b) is no longer constant but decays rapidly (exponentially) to zero (8 to 10 cycles). Similarly, the phase b and phase c terms decay exponentially. Fast-acting circuit breakers can operate in 1 to 2 cycles and thus must "interrupt" currents which include these unidirectional components.

3. To compare the initial values of these unidirectional components for the different phases, we may use Fig. 10.4. We draw three "phasors" of magnitude $[|E_a(0)|/\sqrt{2}](1/X_d' + 1/X_q)$ and phase angles δ, $\delta - 120°$, and $\delta - 240°$, corresponding to the unidirectional components of phases a, b, and c respectively. Then we take the negative of the vertical components of these "phasors" as shown in Fig. 10.4. Thus for this particular δ, the biggest unidirectional component (it is positive) occurs in phase b.

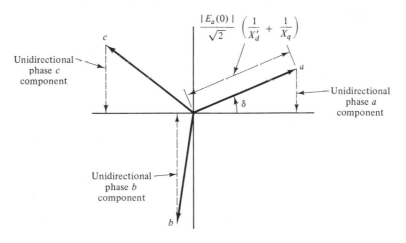

Figure 10.4 Unidirectional components.

In the derivation we have assumed that the short circuit occurs at $t = 0$. Physically, applying the short circuit at $t = 0$ is equivalent to applying the short circuit when the rotor angle $\theta = (\pi/2) + \delta$. Thus from Fig. 10.4 we can estimate that the short occurred when the rotor direct axis was at an angle of about 110°. Using this point of view, one can determine the effect of a short circuit occurring at different times. For example, if $\theta = 90°$ when the short occurs ($\delta = 0$), the unidirectional component in i_a is zero. If $\theta = 0°$ or $180°$ (maximum flux linkages of coil aa') when the short occurs, the phase a unidirectional component has a maximum magnitude.

4. The 120-Hz term (10.55c) also decays to zero when resistance is included. When this relatively small term is added to the 60-Hz term in (10.55a), we get the nonsinusoidal wave shape observable in practice. The effect, however, is considered insignificant and the 120-Hz component is usually ignored when calculating short-circuit currents.

5. Suppose that a short occurs for $\theta = 0$ (i.e., when $\delta = 90°$). Then the phase a unidirectional component is zero, the phase b component is positive, and the phase c component (equal in magnitude) is negative. Each of the three components decays rapidly to zero. A very crude sketch of the three-phase currents including the unidirectional offsets is shown in Fig. 10.5.

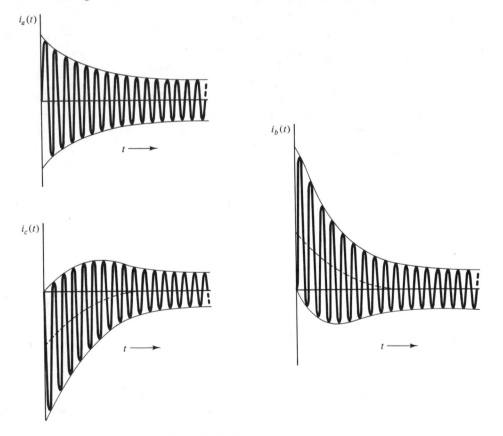

Figure 10.5 Short-circuit currents.

6. If the damper windings are included in the analysis, we get a larger 60-Hz component during the first few cycles. Thus we get, approximately, for the 60-Hz component.

$$i_a'' = \sqrt{2}\,|E_a(0)|\left[\frac{1}{X_d} + \left(\frac{1}{X_d'} - \frac{1}{X_d}\right)e^{-t/T_d'} + \left(\frac{1}{X_d''} - \frac{1}{X_d'}\right)e^{-t/T_d''}\right]\sin(\omega_0 t + \delta)$$

X_d'' is called the *direct axis subtransient reactance* with a typical value $X_d'' = 0.1X_d$ (for large turbine-generators). T_d'' is called the *direct axis subtransient short circuit time constant* with a typical value of less than 0.035 sec (two cycles). By comparison, T_d' is typically in the order of 1 sec. Within the brackets, from right to left, we find terms identified with the subtransient component (with time constant T_d''), the transient component (with T_d'), and the steady-state component. It is instructive to sketch the envelope of this waveform as in Fig. 10.6. Because $T_d'' \ll T_d'$, the transient and subtransient components show up as distinct components and are easily identified.

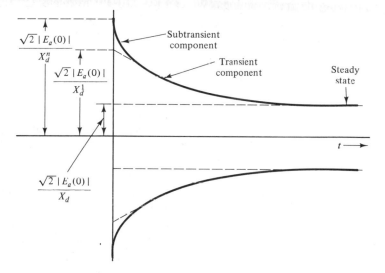

Figure 10.6   Short-circuit current components.

Before continuing with another example, it may be of interest to try to understand physically why the initial 60-Hz short-circuit currents are so much bigger than the steady-state values. A highly simplified answer follows. For simplicity we will neglect the effect of the damper winding circuits, and also assume that the small resistances r and r_F are negligible.

In the steady state, before the short circuit occurs, there is a constant field current i_F^0 and corresponding constant flux linkages $\lambda_F^0 = L_F i_F^0$ of the field winding. The stator currents are zero and do not contribute to the field flux linkages. This is no longer true once stator currents begin to flow during the short circuit. The short-circuit currents lag the internal voltages by 90°; thus, as was shown in Chapter 8, they tend to set up a demagnetizing armature reaction flux in the field winding, which opposes the flux due to the field current. Equivalently, in the notation of this chapter, i_d is negative. Thus λ_F now has two components, one due to i_F and the other due to the stator currents.

Now with $r_F = 0$ we get a remarkable result: No matter what the variation of i_F and the stator currents, λ_F remains constant. For example, in Example 10.4 just concluded, although $i_d(t)$ and $i_F(t)$ were each highly variable [see i_F in Fig. E10.4(b)], a calculation of $\lambda_F(t)$ shows it to be constant! We note that the constancy of λ_F follows from a particular application of the so-called principle of conservation of flux linkages.

With λ_F constant, and in the face of the demagnetizing component due to the stator currents, i_F must increase. In fact, in Fig. E10.4(b) we did observe an increase (on average) of i_F during the transient. This increase in i_F may be viewed as the reason for the larger short-circuit current.

Suppose that r_F is not equal to zero but is small in value. Then λ_F changes slowly from its initial value. For an initial interval we can assume that it is almost constant.

We can carry the consequences of constant (or almost constant) λ_F a step further. Neglecting damper windings, we can use (10.49) to relate λ_d, λ_F, i_d, and i_F more succinctly, by using the "hybrid" form. Picking i_d and λ_F as independent variables, we solve for λ_d and i_F and after carrying out simple algebra, we get

$$\begin{bmatrix} \lambda_d \\ i_F \end{bmatrix} = \begin{bmatrix} L'_d & kM_F/L_F \\ -kM_F/L_F & 1/L_F \end{bmatrix} \begin{bmatrix} i_d \\ \lambda_F \end{bmatrix} \tag{10.56}$$

where $L'_d = L_d - (kM_F)^2/L_F$ is the previously defined direct axis transient reactance. With λ_F constant, we get $d\lambda_d/dt = L'_d di_d/dt$ [i.e., L'_d (rather than L_d) appears to be the "appropriate" inductance parameter when there is a constraint: $\lambda_F =$ constant]. This may help explain the frequent occurrence of L'_d in the short-circuit derivations.

Although we will not pursue this development in detail, it is interesting to consider the extension of these results to the case where the damping circuits are included in the analysis. Once again assuming field and damper resistances negligible and using the principle of conservation of flux linkages, we find λ_F, λ_D, and λ_Q all constant. Consider the implications with respect to the direct axis equations. Writing hybrid equations in the form (10.56), instead of (10.49), we find λ_d, i_F, and i_D in terms of i_d, λ_F, and λ_D. We find, after simple matrix algebra, that

$$\lambda_d = L''_d i_d + \gamma_2 \lambda_F + \gamma_2 \lambda_D \tag{10.57}$$

where $L''_d = L_d - k^2(M_F^2 L_D + L_D M_D^2 - 2M_D M_R M_F)/(L_F L_D - M_R^2)$ and γ_1 and γ_2 are constants. We mentioned $X''_d \triangleq \omega_0 L''_d$ earlier in Note 6 following the example. Since only i_d is variable on the right side of the equation, L''_d is the appropriate inductance parameter describing the initial interval of the short circuit. We note that taking the constancy of λ_Q into account there is also a new L''_q to be introduced. Of course, with resistance in the damper circuit, λ_D and λ_Q can be assumed to be constant in only a *very* short interval. In fact, they quickly decay from their initial values, and there is a smooth transition to a transient dominated by the effects of field flux decay.

Exercise 3.   Verify the expression for L_d'' by writing (10.49) in the suggested hybrid form and doing the matrix algebra. *Hint:* Partition (10.49) with λ_d and i_d as 1-vector partitions and solve for λ_d in terms of i_d and $\lambda_2 = [\lambda_F, \lambda_D]^T$.

We turn now to a somewhat different kind of example. In the two examples considered so far, we have assumed "synchronous operation" (i.e., a rotor angle θ which increases at a uniform rate ω_0). We next wish to consider what happens when the rotation rate is not necessarily ω_0.

Example 10.5   Nonsynchronous Operation

Suppose that we connect a generator to an infinite bus which imposes a set of positive sequence (balanced) terminal voltages with $v_a(t) = \sqrt{2}|V| \cos(\omega_0 t + \underline{/V})$. The rotation of the generator rotor is described by $\theta = \omega_1 t + (\pi/2) + \delta$. In general, $\omega_1 \neq \omega_0$. In fact, ω_1 may even be negative, in which case the rotor is turning clockwise instead of counterclockwise! Find v_0, v_d, and v_q.

Solution   Using the Park transformation as specified by (10.16),

$$
\begin{bmatrix} v_0 \\ v_d \\ v_q \end{bmatrix} = \frac{2|V|}{\sqrt{3}} \begin{bmatrix} \frac{1}{\sqrt{2}} & \frac{1}{\sqrt{2}} & \frac{1}{\sqrt{2}} \\ \cos\theta & \mathcal{T}\cos\theta & \mathcal{T}^2\cos\theta \\ \sin\theta & \mathcal{T}\sin\theta & \mathcal{T}^2\sin\theta \end{bmatrix} \begin{bmatrix} \cos\psi \\ \cos\left(\psi - \frac{2\pi}{3}\right) \\ \cos\left(\psi - \frac{4\pi}{3}\right) \end{bmatrix}
$$

where $\psi \triangleq \omega_0 t + \underline{/V}$. Thus we have

$$v_0 = 0$$

$$
\begin{aligned}
v_d &= \frac{2|V|}{\sqrt{3}} \left[\cos\theta\cos\psi + \mathcal{T}\cos\theta\cos\left(\frac{\psi - 2\pi}{3}\right) + \mathcal{T}^2\cos\theta\cos\left(\frac{\psi - 4\pi}{3}\right) \right] \\
&= \frac{2|V|}{\sqrt{3}} \left[\frac{1}{2}\cos(\theta - \psi) + \frac{1}{2}\cos(\theta + \psi) \right.\\
&\quad + \frac{1}{2}\cos(\theta - \psi) + \frac{1}{2}\cos\left(\frac{\theta + \psi - 4\pi}{3}\right) \\
&\quad \left. + \frac{1}{2}\cos(\theta - \psi) + \frac{1}{2}\cos\left(\frac{\theta + \psi + 4\pi}{4}\right) \right] \\
&= \sqrt{3}|V|\cos(\theta - \psi) = \sqrt{3}|V|\cos(\psi - \theta)
\end{aligned}
$$

Substituting for ψ and θ, we get

$$
\begin{aligned}
v_d &= \sqrt{3}|V|\cos\left[(\omega_0 - \omega_1)t + \underline{/V} - \frac{\pi}{2} - \delta \right] \\
&= \sqrt{3}|V|\sin[(\omega_0 - \omega_1)t + \underline{/V} - \delta]
\end{aligned} \tag{10.58}
$$

By a similar development, we get

$$v_q = -\sqrt{3}\,|V|\,\sin\left[(\omega_0 - \omega_1)t + \underline{/V} - \frac{\pi}{2} - \delta\right]$$

$$= \sqrt{3}\,|V|\,\cos[(\omega_0 - \omega_1)t + \underline{/V} - \delta]$$

(10.59)

This completes the example. We would now like to discuss the result briefly.

1. We invite the reader to check the derivation of (10.58) and (10.59) without assuming that $|V|$, $\underline{/V}$, and δ are constant. The reader will find that (10.58) and (10.59) hold even if $|V|$, $\underline{/V}$, and δ are time varying. We still require the positive-sequence relation between the voltages in phases a, b, and c (i.e., whatever the voltage variation in phase a, that in phases b and c must be identical except for lags of 120° and 240°, respectively). We will still use the term "positive sequence" to describe this condition.

2. The result in (10.58) and (10.59) applies equally well to the calculation of i_d and i_q or λ_d and λ_q with obvious substitutions in the formulas.

3. Returning to the case of constant V_a and δ, notice the interesting fact that we get sinusoids of frequency $\omega_0 - \omega_1$. We claim that from a physical point of view it makes sense. We may see this more clearly by considering current instead of voltage. Replacing voltage sources by current sources, we get currents i_d and i_q of frequency $\omega_0 - \omega_1$. This is the frequency of the currents in the fictitious coils on the rotor that generate the armature reaction flux in the air gap. Since the sinusoidal currents in the fictitious coils are appropriately shifted in space and time (90° in each case), they set up a rotating flux in the air gap. Relative to the rotor, this rotating field has an angular velocity $\omega_0 - \omega_1$, but since the rotor angular velocity is ω_1, the rotating flux in the air gap (relative to a stationary reference frame) is $(\omega_0 - \omega_1) + \omega_1 = \omega_0$, which checks (i.e., it is the frequency of the current sources).

4. Note that at synchronous speed ($\omega_1 = \omega_0$) v_d and v_q (or i_d and i_q) are at zero frequency (dc). At standstill ($\omega_1 = 0$) the frequency of v_d and v_q is ω_0. For those familiar with induction motor operation, it is helpful to note that the frequency of v_d and v_q is the slip frequency.

5. We note the possibility of calculating the quasi-steady-state torque of a synchronous motor during "startup" (i.e., during the time the motor is being brought up to speed). We will not attempt the calculation since it is lengthy and not central to our subject. In principle, however, we may proceed as follows. Suppose that the motor is turning at an angular velocity ω_1. Then v_d and v_q are sinusoids of frequency $\omega_0 - \omega_1$ and may be represented by phasors, for quasi-steady-state analysis. Using our synchronous machine model (10.45) to (10.50), we can then solve for the currents and flux linkages using impedances (calculated for the frequency $\omega_0 - \omega_1$). The average steady-state torque may then be calcu-

lated using (10.43). The analysis is simplified if we assume that the field circuit is open circuited ($i_F = 0$) during startup. It is necessary, however, to retain the damper circuits in the analysis.

This completes the discussion of Example 10.5. We conclude this section by suggesting that the three examples just considered demonstrate the great power and flexibility of the synchronous machine model presented in (10.45) to (10.50). Adding the physical insight we get from the simpler but more restricted model of Chapter 8, we can develop a good grasp of synchronous machine behavior.

10.9 SYNCHRONOUS OPERATION

The results of Example 10.5 can be used to obtain a very nice result of use in the extremely important case of synchronous operation. Suppose that the generator terminal voltages are a set of positive-sequence voltages with

$$v_a(t) = \sqrt{2}|V| \cos{(\omega_0 t + \underline{/V})}$$

and the generator rotor is turning at synchronous speed [i.e., $\theta = \omega_0 t + (\pi/2) + \delta$]. For now we will assume that $|V|$, $\underline{/V}$, and δ are constants. Since $\omega_1 = \omega_0$, (10.58) and (10.59) in Example 10.5 reduce to

$$v_q = \sqrt{3}|V| \cos{(\underline{/V} - \delta)}$$
$$v_d = \sqrt{3}|V| \sin{(\underline{/V} - \delta)}$$

Note that v_d and v_q are now constants (dc). We next consider a simple relation between the phasor V_a, the angle δ, and v_d and v_q. Consider the following complex quantity.

$$v_q + jv_d = \sqrt{3}|V|[\cos{(\underline{/V} - \delta)} + j \sin{(\underline{/V} - \delta)}]$$
$$= \sqrt{3}|V| e^{j(\underline{/V}-\delta)}$$
$$= \sqrt{3} V_a e^{-j\delta}$$

where $V_a = |V| e^{j\underline{/V}}$ is the (complex) voltage phasor. Rearranging, we get

$$V_a = \left(\frac{v_q}{\sqrt{3}} + j\frac{v_d}{\sqrt{3}}\right)e^{j\delta}$$

Let $V_q \triangleq v_q/\sqrt{3}$ and $V_d \triangleq v_d/\sqrt{3}$; then

$$V_a = (V_q + jV_d)e^{j\delta} \tag{10.60}$$

Note that V_q and V_d are purely real. Equation (10.60) gives the simple connection between V_a, δ, and V_q and V_d. Suppose that we are given V_a and δ. By an algebraic or geometric calculation, we then find V_q and V_d. Finally, we find v_q and v_d by the formulas $v_q = \sqrt{3} V_q$ and $v_d = \sqrt{3} V_d$. Consider how very much simpler this is than using the Park transformation, (10.16), to find these same quantities. We caution the reader, however, that (10.60) applies only to synchronous (positive sequence) operation.

In solving (10.60) geometrically, we can use Fig. 10.7. The figure is drawn for V_q and V_d positive. It is useful to think of $e^{j\delta}$ and $je^{j\delta}$ as unit vectors of a new coordinate system at an angle δ with respect to the real axis and imaginary axis; we label the new axes q axis and d axis. Thus the task of finding V_q and V_d is simply that of resolving V_a into components with respect to the new coordinate system.

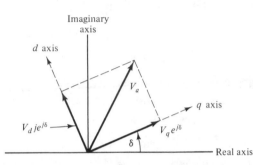

Figure 10.7 Geometry of V_a $= V_q + jV_d e^{j\delta}$

Example 10.6

Find v_q and v_d corresponding to synchronous operation with $V_a = \sqrt{2}\,\underline{/75°}$ and $\delta = 30°$.

Solution In the complex plane we sketch $V_a = \sqrt{2}\,\underline{/75°}$ and the new coordinate frame (the q axis and d axis) at an angle of 30°. From the sketch V_a is at an angle of $75° - 30° = 45°$ with respect to the q axis. Then resolving into components we get $V_q = V_d = 1$. Then $v_d = v_q = \sqrt{3}$. We can also solve the problem algebraically using (10.60).

Example 10.7

For synchronous operation, $\delta = 90°$, $v_q = \sqrt{3}$, and $v_d = -\sqrt{3}$. Find V_a.

Solution First we find $V_q = 1$ and $V_d = -1$. Using (10.60), we find that

$$V_a = (1 - j1)e^{j\pi/2} = \sqrt{2}\,\underline{/45°}$$

We can also solve the problem geometrically using Fig. 10.7.

Note: What has been done for V_a holds as well for I_a or Λ_a.

Example 10.8

Synchronous operation again. $\delta = 0°$ and $I_a = -j1$. Find i_d and i_q.

Solution $\delta = 0°$, so the "new" and "old" coordinate systems are aligned. $I_q = 0$ and $I_d = -1$. Then $i_q = 0$ and $i_d = -\sqrt{3}$.

Example 10.9

A synchronous machine is in steady-state operation. The rotor angle $\theta(t) = \omega_0 t + \pi/3$. $v_a(t) = \sqrt{2}\,100\cos(\omega_0 t + \pi/6)$ volts and $i_a(t) = -\sqrt{2}\,100\cos\omega_0 t$ amperes. Find $i_0, i_d, i_q, v_0, v_d, v_q,$ and $p_{3\phi}(t)$. Is the machine a generator or a motor?

Solution $\theta(t) = \omega_0 t + \pi/2 + \delta = \omega_0 t + \pi/3$ implies that $\delta = -\pi/6 = -30°$. Unless stated to the contrary, we assume synchronous operation. Thus the voltages and currents are in positive-sequence sets with $V_a = 100 \underline{/30°}$ and $I_a = -100$. Since $\delta_m = \delta - \underline{/V_a} = -30° -30° = -60°$ is negative, we have a motor. Since the currents and voltages are in balanced sets, $i_0 = v_0 = 0$. We next resolve V_a and I_a into components along the q axis and d axis, as in Fig. 10.7. This is shown in Fig. E10.9. We find $V_q = 50$, $V_d = 86.6$, $I_q = -86.6$, and $I_d = -50$. Then $v_q = \sqrt{3} \times 50 = 86.6$, $v_d = 150$, $i_q = -150$, and $i_d = -86.6$. The units are volts and amperes.
We can calculate $p_{3\phi}(t)$ using (10.52):

$$p_{3\phi}(t) = i_0 v_0 + i_d v_d + i_q v_q$$

$$= 0 + (-86.6)(15) + (-150)(86.6) = -25{,}981 \text{ W}$$

which checks that the synchronous machine is a motor.
Alternatively, we can calculate $p_{3\phi}(t)$ using V_a and I_a. We get

$$p_{3\phi}(t) = 3P = 3|V_a||I_a|\cos\phi$$

$$= 3 \times 100 \times 100 \times \cos(30° - 180°)$$

$$= -25{,}981 \text{ W}$$

which checks.

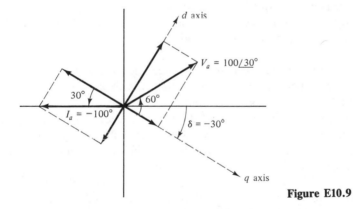

Figure E10.9

While sketches give the most insight into the relations between the phasors, δ, and direct and quadrature components, it is frequently more convenient in calculations to use algebra relating real numbers. Thus we note, taking V_a as an example, that $V_a = \text{Re } V_a + j \text{ Im } V_a$. Then, using (10.60), $V_a = \text{Re } V_a + j \text{ Im } V_a = (V_q + jV_d)(\cos\delta + j\sin\delta)$. Thus, equating real parts and equating imaginary parts, we get

$$\begin{bmatrix} \text{Re } V_a \\ \text{Im } V_a \end{bmatrix} = \begin{bmatrix} \cos\delta & -\sin\delta \\ \sin\delta & \cos\delta \end{bmatrix} \begin{bmatrix} V_q \\ V_d \end{bmatrix}$$

Also, taking the inverse,

$$\begin{bmatrix} V_q \\ V_d \end{bmatrix} = \begin{bmatrix} \cos\delta & \sin\delta \\ -\sin\delta & \cos\delta \end{bmatrix} \begin{bmatrix} \text{Re } V_a \\ \text{Im } V_a \end{bmatrix}$$

We note that the terminology *machine reference frame* for the *d-q* axes, and the *network, common,* or *system reference frame* for the real and imaginary axes, are frequently used in the literature. The two reference frames and the relationship between components in the two reference frames are shown in Fig. 10.8. We note again that a given phasor V_a distributes itself into very different components V_q and V_d, depending on the angle δ of the machine reference frame.

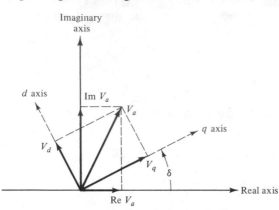

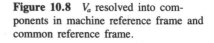

Figure 10.8 V_a resolved into components in machine reference frame and common reference frame.

We note that the relationships we have derived also apply to the case where V_a and δ are (slowly varying) functions of time. In this case (10.60) and Fig. 10.8 describe a dynamic situation in which δ and/or V_a can vary. In general, then, V_d and V_q will also be varying.

Finally, let us consider the case of m machines connected by a transmission system. In this case there are m internal voltages with angles $\delta_1, \delta_2, \ldots, \delta_m$ to consider; until now we have considered only a single angle δ. To help fix ideas, we will very briefly consider the solution of the Park equations in the m-machine case. As in Chapter 6, we assume n buses linked by a transmission system. To simplify, assume that there are no loads at generator buses. The system interconnection is then given by the mixed circuit/block diagram shown in Fig. 10.9.

We assume positive-sequence voltages and currents with the amplitude and phase of I_i and V_i changing slowly enough to use quasi-steady-state analysis. In this case we can use the relation $\mathbf{I}_{\text{bus}} = \mathbf{Y}_{\text{bus}} \mathbf{V}_{\text{bus}}$, where $\mathbf{Y}_{\text{bus}}$ is a constant matrix relating the slowly varying $\mathbf{I}_{\text{bus}}$ and $\mathbf{V}_{\text{bus}}$.

For each generator model we can use (10.44) to (10.50), which involve the Park voltages and currents. These variables after scaling by the factor $\sqrt{3}$ give the V_{di}, V_{qi}, I_{di}, and I_{qi}. These variables, in machine or d-q reference frames, must be converted into a common reference frame by axis transformations. After axis transformations we get the phasors I_i and V_i. This collection of voltage and current phasors must satisfy the constraint $\mathbf{I}_{\text{bus}} = \mathbf{Y}_{\text{bus}} \mathbf{V}_{\text{bus}}$. We will not be able to pursue the interesting problem of the numerical integration of the equations beyond these "setup" considerations.

We return now to the case of a single machine and consider the Park equations for steady-state analysis. We show that we get the same model as in Chapter 8.

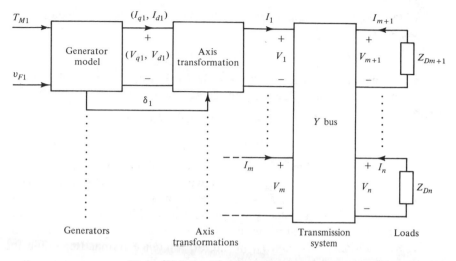

Figure 10.9 m-Machine interconnection.

10.10 STEADY-STATE MODEL

Assume synchronous, positive-sequence, steady-state operation. In that case, $v_0 = 0$ and, as we have seen in Section 10.9, v_d and v_q are constants. With v_F also constant, all the left sides of (10.45) to (10.47) are constants. Then in the steady state all the currents and flux linkages are constants. In particular, $i_0 = 0$ and the rotor damper currents $i_D = i_Q = 0$. We note also that $\dot{\theta} = \omega_0$. Then (10.45) to (10.50) reduce to

$$v_d = -ri_d - \omega_0\lambda_q \tag{10.61}$$

$$v_q = -ri_q + \omega_0\lambda_d \tag{10.62}$$

$$v_F = r_F i_F \tag{10.63}$$

where

$$\lambda_d = L_d i_d + kM_F i_F \tag{10.64}$$

$$\lambda_F = kM_F i_d + L_F i_F \tag{10.65}$$

$$\lambda_q = L_q i_q \tag{10.66}$$

We want to show that the first two equations (10.61) and (10.62) are equivalent to the generator model in Chapter 8. Using (10.64) and (10.66) to replace λ_q and λ_d in (10.61) and (10.62) and using the definitions $v_d = \sqrt{3}\,V_d$, $v_q = \sqrt{3}\,V_q$, $i_d = \sqrt{3}\,I_d$, and $i_q = \sqrt{3}\,I_q$, we can form the following single complex equation:

$$(V_q + jV_d)e^{j\delta} = -r(I_q + jI_d)e^{j\delta} + \omega_0 L_d I_d e^{j\delta} - j\omega_0 L_q I_q e^{j\delta} + \frac{1}{\sqrt{3}}\sqrt{\frac{3}{2}}\,\omega_0 M_F i_F e^{j\delta}$$

Using (10.60) as it applies to voltage and current, and introducing $jX_d = j\omega_0 L_d$ and $jX_q = j\omega_0 L_q$, we get

$$V_a = -rI_a - jX_d jI_d e^{j\delta} - jx_q I_q e^{j\delta} + E_a$$

where we have defined $\sqrt{2}E_a = \omega_0 M_F i_F e^{j\delta}$. Since $V_a = E_a$ when $I_a = 0$, E_a is the open-circuit voltage. Rearranging the equation, we get

$$
\begin{aligned}
E_a &= V_a + rI_a + jX_d jI_d e^{j\delta} + jX_q I_q e^{j\delta} \\
&= V_a + rI_a + jX_d I_{ad} + jX_q I_{aq}
\end{aligned}
\tag{10.67}
$$

where $I_{ad} \triangleq jI_d e^{j\delta}$ and $I_{aq} \triangleq I_q e^{j\delta}$. The relations in (10.67), and the fact that $I_a = (I_q + jI_d)e^{j\delta} = I_{aq} + I_{ad}$, are shown in Fig. 10.10. In the case illustrated in the figure, I_q is positive and I_d is negative. Comparing with Fig. 8.8, or (9.30), we see that we get the same steady-state model as in Chapter 8. I_{aq} and I_{ad} are the same complex quantities that we used in Chapter 8. The quantities I_q and I_d are simply the representations of I_{aq} and I_{ad} in the d-q reference frame. Regarding the use of the quantities I_{ad} and I_{aq} versus I_d and I_q, we note that we get more compact expressions using the complex quantities I_{ad} and I_{aq}, but in analytical work it is frequently easier to use the real quantities I_d and I_q. We will take (10.67) to mean the equation in either form.

It is clearly much simpler to use (10.67) or the equivalent phasor diagram, Fig. 10.10, than to use the general three-stage process involving the Park transformations and variables. We next consider a dynamic model which has similar advantages.

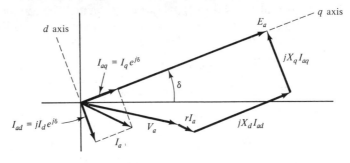

Figure 10.10   Phasor diagram.

10.11 SIMPLIFIED DYNAMIC MODEL

Frequently, it is possible to simplify the differential equations (10.44) to (10.47). This is particularly true if we are considering transients in which the terminal voltages and currents are sinusoids with "slowly varying amplitude and phase." One such case is that of electromechanical transients considered in Chapter 9, but the applicability is much more general.

We make the following assumptions regarding the generator, which are frequently justified in practice.

1. Positive-sequence, synchronous operation.
2. $\theta = \omega_0 t + (\pi/2) + \delta$, with $|\dot{\delta}| \ll \omega_0$.
3. $\dot{\lambda}_d$ and $\dot{\lambda}_q$ are small compared with $\omega_0 \lambda_q$ and $\omega_0 \lambda_d$.
4. The effect of damper circuits is negligible (i.e., we can set $i_D = i_Q = 0$).

In assumption 1 we are referring to positive-sequence operation in the broad sense; that is, we will allow the waveform of phase a to change but whatever that waveform, the waveform in phases b and c are identical except for a phase shift $-120°$ and $-240°$, respectively. In this case, as mentioned earlier, we can use (10.60) to relate time-varying quantities. In some respect, assumptions 2 and 3 approximate steady-state behavior; $\dot{\delta}$, $\dot{\lambda}_d$, and $\dot{\lambda}_q$ are not zero, but they are relatively small. Assumption 4 is consistent with assumptions 2 and 3 and is reasonable provided that we neglect a very short interval at the very beginning of the transient.

Assumption 1 implies that $v_0 = i_0 \equiv 0$; using assumption 3, the remaining electrical equations may be approximated by

$$v_d = -ri_d - \dot{\theta}\lambda_q \tag{10.68}$$

$$v_q = -ri_q + \dot{\theta}\lambda_d \tag{10.69}$$

$$v_F = r_F i_F + \frac{d\lambda_F}{dt} \tag{10.70}$$

where the flux linkages are related to currents by (10.64) to (10.66). Note the very great simplification in these equations compared with (10.45) to (10.47)!

We note that by assumption 2, $\dot{\theta} \approx \omega_0$ and thus, if we compare these equations with the steady-state equations, (10.61) to (10.63), we may take the first two equations in each set to be the same. Therefore, by the same development using (10.60), as it applies to voltage and current, we are led to (10.67) and the associated phasor diagram in Fig. 10.10.

The quantities in (10.67) and the associated Fig. 10.10 are not necessarily constant. For example, suppose that $i_a(t) = \sqrt{2}\,|I_a|\cos(\omega_0 t + \underline{/I_a})$ is varying in amplitude and phase. Then $I_a = |I_a|\,\underline{/I_a}$ is a phasor that is also varying in time. In Fig. 10.10 we would see I_a moving around the complex plane. Thus even with δ constant, we would expect I_q and I_d (the projections of I_a on the q axis and d axis) to vary in time. With I_a fixed and δ variable, the projections I_q and I_d vary again. In the general case I_a, V_a, E_a, and δ all can vary, but they must vary in such a way that the algebraic constraint specified by (10.67) or Fig. 10.10 is satisfied.

We note that, so far, we have been able to preserve the simplicity of the steady-state model. In particular, we do not need to calculate the rotor-based terms i_F, i_q, i_d, v_q, and v_d, but can work with the stator-based terms E_a, I_a, and V_a directly.

Turning to the differential equation (10.70), we see that this involves a rotor-based quantity, the state λ_F. We would like to express λ_F in stator terms. Starting with (10.65), repeated here for convenience,

$$\lambda_F = kM_F i_d + L_F i_F \tag{10.65}$$

we next define a "stator voltage" $E_a' \triangleq (\omega_0 M_F/\sqrt{2} L_F) e^{j\delta} \lambda_F$. The magnitude of E_a' is proportional to λ_F just as the magnitude of E_a is proportional to i_F. Using (10.65) in the definition, we get

$$E_a' = \frac{\omega_0 M_F}{\sqrt{2} L_F} e^{j\delta} \lambda_F = \frac{\omega_0 k M_F^2}{\sqrt{2} L_F} e^{j\delta} i_d + \frac{\omega_0 M_F}{\sqrt{2}} e^{j\delta} i_F \tag{10.71}$$

We can very effectively simplify and reduce (10.71) by using a number of previously defined quantities: $i_d = \sqrt{3} I_d$, $k = \sqrt{3/2}$, $E_a = \omega_0 M_F e^{j\delta} i_F/\sqrt{2}$, and $L_d' = L_d - (kM_F)^2/L_F$. We get

$$
\begin{aligned}
E_a' &= \omega_0(L_d - L_d')I_d e^{j\delta} + E_a \\
&= j(X_d' - X_d)jI_d e^{j\delta} + E_a
\end{aligned}
\tag{10.72}
$$

Using (10.67) to replace E_a in (10.72), we get

$$
\begin{aligned}
E_a' &= V_a + rI_a + jX_d'jI_d e^{j\delta} + jX_q I_q e^{j\delta} \\
&= V_a + rI_a + jX_d'I_{ad} + jX_q I_{aq}
\end{aligned}
\tag{10.73}
$$

which is in the same form as (10.67) except that X_d' replaces X_d and E_a' replaces E_a. Furthermore, E_a' and E_a both have the same phase δ.

It is thus a simple matter to draw the phasor diagram that is associated with (10.73). Instead, we will draw a more useful phasor diagram with both E_a and E_a' shown. Thus we have Fig. 10.11. The voltage E_a' is not as easy to describe physically as is E_a. While, physically, E_a is the open-circuit or Thévenin equivalent voltage, E_a' can only be described as an "internal" voltage whose magnitude is proportional to λ_F. Nevertheless, with familiarity, E_a' will become as physically meaningful as E_a. We note that Fig. 10.11 shows very clearly the relationships between E_a, E_a', V_a, and I_a, and thereby the relationships between i_F, λ_F, δ, V_a, and I_a.

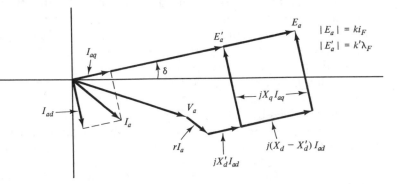

Figure 10.11 Phasor diagram showing E_a' and E_a.

We next consider the differential equation, (10.70), and rewrite to introduce the notation $|E_a|$ and $|E'_a|$. Multiply each term in (10.70) by $\omega_0 M_F/\sqrt{2}\, r_F$. We get

$$\frac{\omega_0 M_F}{\sqrt{2}\, r_F} v_F = \frac{\omega_0 M_F}{\sqrt{2}\, r_F} r_F i_F + \frac{\omega_0 M_F}{\sqrt{2}\, r_F} \frac{d\lambda_F}{dt} \tag{10.74}$$

Next we define

$$E_{fd} = \frac{\omega_0 M_F}{\sqrt{2}\, r_F} v_F$$

Thus E_{fd} is proportional to v_F. Then using the definitions of $|E_a|$, $|E'_a|$, and T'_{do}, (10.74) reduces to

$$E_{fd} = |E_a| + \frac{L_F}{r_F} \frac{d|E'_a|}{dt} = |E_a| + T'_{do} \frac{d|E'_a|}{dt} \tag{10.75}$$

E_{fd} may be viewed as v_F in equivalent stator terms; one unit of E_{fd} is as "effective" as one unit of $|E_a|$. In using the per unit system we can pick the same voltage base for E_{fd} and $|E_a|$ and relate the bases of v_F and E_{fd} (just as we do for transformers), so that v_F and E_{fd} have the same per unit value. In that case no distinction need be made between v_F and E_{fd}. However, in what follows, we prefer to use the E_{fd} notation.

Finally, to complete the simplified synchronous machine model, we consider the mechanical equation, repeated here for convenience.

$$J\ddot{\theta} + D\dot{\theta} + i_q \lambda_d - i_d \lambda_q = T_M \tag{10.44}$$

Multiplying by $\dot{\theta}$, we get

$$\frac{d}{dt}\left(\frac{1}{2} J\dot{\theta}\right)^2 + D\dot{\theta}^2 + \dot{\theta}(i_q \lambda_d - i_d \lambda_q) = \dot{\theta} T_M \tag{10.76}$$

or, using the notation of Chapter 9,

$$\frac{d}{dt} W_{\text{kinetic}} + P_{\text{friction}} + \dot{\theta}(i_q \lambda_q - i_d \lambda_q) = P_M \tag{10.77}$$

We next examine the electrical terms in (10.77). Using (10.68) and (10.69), we find that

$$
\begin{aligned}
\dot{\theta}(i_q \lambda_d - i_d \lambda_q) &= i_q v_q + i_d v_d + r(i_q^2 + i_d^2) \\
&= 3(I_q V_q + I_d V_d) + 3r(I_q^2 + I_d^2) \\
&= 3\,\text{Re}\, V_a I_a^* + 3r|I_a|^2 \\
&= 3P_G + 3r|I_a|^2
\end{aligned} \tag{10.78}
$$

In going from the second to the third line, we have used (10.60) as it applies to both I_a and V_a. The right side of (10.78) may be seen to be the instantaneous three-phase generated electrical power "before" resistance losses in the generator. These $I^2 R$ losses are very small and as an approximation we are going to neglect them. In this case (10.77) reduces to (9.11) and, as discussed in Section 9.2, we end up with

$$M\ddot{\delta} + D\dot{\delta} + P_G = P_M(\text{p.u.}) \tag{10.79}$$

Equation (10.79) is the desired simplified mechanical equation. Equation (10.75) is the simplified electrical equation subject to the algebraic constraints (10.67) and (10.73). We summarize these results below and at the same time return to the use of the per unit system of units.

Summary: Simplified dynamic model (per unit). With the assumptions stated at the beginning of this section, we get two differential equations:

$$M\ddot{\delta} + D\dot{\delta} + P_G = P_M \tag{10.79}$$

$$T'_{do}\frac{d|E'_a|}{dt} + |E_a| = E_{fd} \tag{10.75}$$

and two algebraic equations

$$E_a = V_a + rI_a + jX_dI_{ad} + jX_qI_{aq} \tag{10.67}$$

$$E'_a = V_a + rI_a + jX'_dI_{ad} + jX_qI_{aq} \tag{10.73}$$

In (10.67) and (10.73) it may be convenient to substitute $I_{ad} = jI_de^{j\delta}$ and $I_{aq} = I_qe^{j\delta}$. In (10.79), in general, $P_G = \text{Re}\,V_aI_a^*$. If $r = 0$ (or as an approximation), we may use the following expressions instead:

$$P_G = \frac{|E_a||V_a|}{X_d} \sin \delta_m + \frac{|V_a|^2}{2}\left(\frac{1}{X_q} - \frac{1}{X_d}\right) \sin 2\delta_m \tag{10.80}$$

or

$$P_G = \frac{|E'_a||V_a|}{X'_d} \sin \delta_m + \frac{|V_a|^2}{2}\left(\frac{1}{X_q} - \frac{1}{X'_d}\right) \sin 2\delta_m \tag{10.81}$$

where $\delta_m \triangleq \underline{/E_a} - \underline{/V_a} = \underline{/E'_a} - \underline{/V_a} = \delta - \underline{/V_a}$.

Note: Equation (10.80) is the same as (8.45). In the derivation, in Section 8.6, I_{ad} and I_{aq} were expressed in terms of E_a and V_a. If, instead, I_{ad} and I_{aq} are expressed in terms of E'_a and V_a, using (10.73), then by exactly the same algebraic steps we get (10.81). There is no need to carry out the algebra; noting the similarity of (10.67) and (10.73), we need only change E_a to E'_a and X_d to X'_d to go from (10.80) to (10.81). Note that even in the case of a round-rotor machine ($X_q = X_d$), we get a "saliency" effect in (10.81). We should also point out that since $X'_d < X_q$, then $1/X_q - 1/X'_d$ is negative! Thus the shapes of the P_G versus δ_m curves are quite different, when plotted for *constant* $|E_a|$, $|V_a|$ or *constant* $|E'_a|$, $|V_a|$. However, if we introduce the constraint between $|E_a|$ and $|E'_a|$, the two formulas must give identical results (for the same δ_m) since they describe the power output of the same generator under the same conditions!

Exercise 4.   Can you prove the result just stated?

We conclude this section with some examples using the simplified model.

Example 10.10   Voltage Buildup

Solve the problem in Example 10.3 using the simplified synchronous machine model.

Solution   Since $I_a = 0$, from (10.67) and (10.73) we see that $E_a'(t) = E_a(t) = V_a(t)$. Also, $|E_a'(0)| = 0$, because we start from rest with zero field flux. Using (10.75), in which we can replace E_a with E_a',

$$T_{do}' \frac{d|E_a'|}{dt} + |E_a'| = E_{fd}^0 \qquad |E_a'(0)| = 0$$

where E_{fd}^0 is constant. Solving, we get

$$|E_a'(t)| = (1 - e^{-t/T_{do}'})\, E_{fd}^0 \qquad t \geq 0$$

Since $|V_a(t)| = |E_a'(t)|$, while $\underline{/V_a(t)} = \underline{/E_a'(t)} = \delta$, we get

$$V_a(t) = (1 - e^{-t/T_{do}'})\, E_{fd}^0 e^{j\delta}$$

Here we get a phasor with a magnitude that varies with time. The corresponding instantaneous voltage is

$$v_a(t) = \sqrt{2}\, E_{fd}^0 (1 - e^{-t/T_{do}'}) \cos(\omega_0 t + \delta)$$

which, noting the definition of E_{fd} just after (10.74), is the same as the end result in Example 10.3. This calculation is much simpler.

Example 10.11   Symmetrical Short Circuit

For the problem in Example 10.4, neglect the stator resistance r and the damper windings, and find the short-circuit current.

Solution   Consider first the prefault steady-state condition. $|E_a'|$ is constant. Thus from (10.75) (with $d|E_a'|/dt = 0$), $|E_a(0)| = E_{fd}^0$. Also, with $I_a = 0$, we have from (10.67) and (10.73), $E_a'(0) = V_a(0) = E_a(0)$. Putting these two results together we have the initial condition $|E_a'(0)| = |E_a(0)| = E_{fd}^0$ to use in solving (10.75) in the "fault on" period, $t \geq 0$.

For $t \geq 0$ we have the fault with $V_a \equiv 0$. Thus, with $r = 0$, we get from (10.67) and (10.73),

$$E_a = jX_d I_{ad} + jX_q I_{aq}$$

$$E_a' = jX_d' I_{ad} + jX_q I_{aq}$$

From the associated phasor diagram geometry, $I_{aq} = 0$, and also, therefore, $I_a = I_{ad}$. Thus we can simplify. We get

$$E_a = jX_d I_a$$

$$E_a' = jX_d' I_a$$

We digress from the calculation of the transient to consider the "initial" sinusoidal current. Since λ_F cannot change instantaneously when the fault occurs, neither can $|E_a'|$. Thus using $E_a' = jX_d' I_a$, we can calculate the current during the initial part of the transient:

$$i_a(t) \approx \frac{\sqrt{2}\,|E_a'(0)|}{X_d'} \sin(\omega_0 t + \delta)$$

Noting that $E_a'(0) = E(0)$, this is the same as the 60-Hz component in (10.55a).

Returning now to the solution for the transient, we express E_a in terms of E_a' and get

$$E_a = \frac{X_d}{X_d'} E_a'$$

We next use this result in (10.75). We get

$$T_{do}' \frac{d|E_a'|}{dt} + \frac{X_d}{X_d'} |E_a'| = E_{fd}^0 \qquad |E_a'(0)| = E_{fd}^0$$

We can write in time-constant form. Defining $T_d' = (X_d'/X_d) T_{do}' = (L_d'/L_d) T_{do}'$, we get

$$T_d' \frac{d|E_a'|}{dt} + |E_a'| = \frac{X_d'}{X_d} E_{fd}^0 \qquad |E_a'(0)| = E_{fd}^0$$

with solution

$$|E_a'(t)| = \frac{X_d'}{X_d} E_{fd}^0 + \left(1 - \frac{X_d'}{X_d}\right) E_{fd}^0 e^{-t/T_d'}$$

We get $E_a' = |E_a'| \underline{/\delta}$ as a time-varying phasor. Since $E_a' = jX_d'I_a$, we can solve for $i_a(t)$.

$$i_a(t) = \sqrt{2} E_{fd}^0 \left[\frac{1}{X_d} + \left(\frac{1}{X_d'} - \frac{1}{X_d}\right) e^{-t/T_d'}\right] \sin(\omega_0 t + \delta)$$

Noting that $E_{fd}^0 = |E_a(0)|$, this result is the same as that given in Note 1 following Example 10.4. Furthermore, in the present case ($r = 0$) we get an explicit formula for T_d'.

In summary, we get the same 60-Hz component as in Example 10.4, and with much less work, but lose the unidirectional and 120-Hz components.

Note: Suppose that in Example 10.11 instead of an open circuit at $t = 0$, there is a nonzero initial stator current $I_a(0)$. This time $E_a'(0) \neq E_a(0)$, but we can calculate it using (10.67) and (10.73) and then proceed with the calculation in Example 10.11.

In the examples considered so far in this section, it was reasonable to assume that δ remained fixed at its initial angle. Thus the mechanical equation was not involved. We turn next to an example involving the mechanical equation.

Example 10.12   Equal-Area Criterion

Consider the circuit shown in Fig. E10.12(a). Assume that

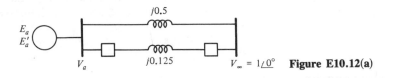

$V_\infty = 1\underline{/0°}$ **Figure E10.12(a)**

$$E_{fd}^0 = 1.5$$

$$X_d = X_q = 0.9$$

$$X_d' = 0.3$$

Neglect all resistances. The system is in the steady state. Then, at $t = 0$, the circuit breakers shown open and remain open. The mechanical power remains fixed at its initial value P_M^0. Determine if transient stable for $P_M^0 = 1.0$, using two different modeling assumptions:

    (a) $|E_a| = ki_F =$ constant for the transient
    (b) $|E_a'| = k'\lambda_F =$ constant for the transient

Solution   In either case we can absorb the line reactances into the generator model just as in (8.33). Before the fault, with both lines in service, we have $j0.5$ in parallel with $j0.125$ to give an equivalent impedance of $j0.1$. Thus for $t < 0$ we have $\tilde{X}_d \triangleq X_d + X_L = 1.0$, $\tilde{X}_q = 1.0$, and $\tilde{X}_d' = 0.4$.

    Prefault Steady State. We have $P_G(\delta^0) = P_M^0 = 1.0$, $|E_a(0)| = E_{fd}^0 = 1.5$. Using the round-rotor formula,

$$P_G(\delta^0) = \frac{1.5 \times 1.0}{1.0} \sin \delta^0 = 1.0$$

we get $\delta^0 = 41.81°$. To find $|E_a'(0)|$ we use a version of Fig. 10.11 as shown in Fig. E10.12(b). From the figure we get for the prefault quantities, $|E_a| = 1.5 = |V_\infty| \cos 41.81° + \tilde{X}_d |I_{ad}| \Rightarrow |I_{ad}| = 0.7546$. Then

$$|E_a'| = |V_\infty| \cos 41.81° + \tilde{X}_d' |I_{ad}| = 1.047$$

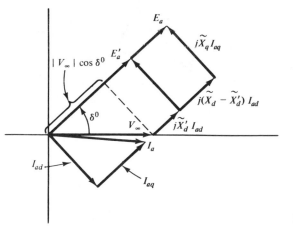

Figure E10.12(b)

    Transient. For $t \geq 0$ we have $\tilde{X}_d = \tilde{X}_q = 0.9 + 0.5 = 1.4$ and $\tilde{X}_d' = 0.8$. We now consider the use of the two alternative models.

    (a) This is the model we used in Chapter 9. Assuming that $|E_a|$ is constant, we get

$$P_G(\delta) = \frac{1.5 \times 1.0}{1.4} \sin \delta = 1.07 \sin \delta$$

(b) Assuming that $|E'_a|$ is constant, we get, by using (10.81),

$$P_G(\delta) = \frac{1.047 \times 1.0}{0.8} \sin \delta + \frac{1}{2}\left(\frac{1}{1.4} - \frac{1}{0.8}\right) \sin 2\delta$$

$$= 1.31 \sin \delta - 0.27 \sin 2\delta$$

With E'_a as the internal voltage instead of E_a, we get the same swing equation (9.19), but with a different expression for $P_G(\delta)$.

In Fig. E10.12(c) we next plot the power versus power angle curves in (a) and (b) to use the equal-area criterion. Initially, in either case, with $\delta^0 = 41.81°$, loss of a line means less electrical power output. Thus, since P_M^0 does not change, there is initial acceleration. Using the equal-area criterion, we compare the accelerating area with the maximum available decelerating area A_{dmax}. In this case the constant $|E_a|$ model predicts

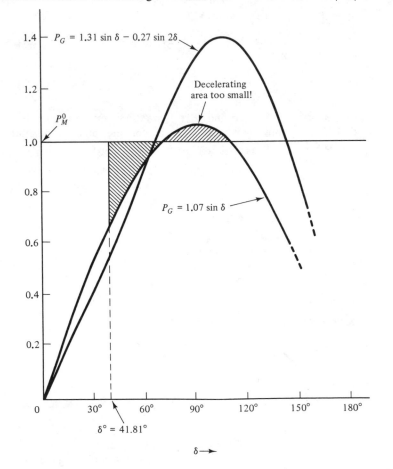

Figure E10.12(c)   Equal area construction, $P_M^0 = 1.0$.

instability while the constant $|E'_a|$ model predicts stability. Which do we believe? As we saw in Chapter 9, for our system the "first swing" is the critical one because if synchronism is not lost in the first swing, it will not be lost thereafter. In this case, since $|E'_a|$ is approximately constant for an initial interval, we are inclined to trust the constant $|E'_a|$ model. It is to be noted that the constant $|E_a|$ model usually gives results on the "safe" side (i.e., it gives a "conservative" design). The constant $|E_a|$ model also has the advantage that if stability is maintained, the new equilibrium angle δ'' may immediately be read off the P_G curve for constant $|E_a|$. For getting a rough idea of transient behavior the simpler, constant $|E_a|$ model is attractive, but as this example shows, the quantitative results are questionable.

We conclude this section with another example.

Example 10.13

Consider the circuit shown in Fig. E10.13(a). Assume that

$$X_d = 1.0$$

$$X_q = 0.6$$

$$X'_d = 0.2$$

$$T'_{do} = 4 \text{ sec}$$

Neglect resistances. The generator has just been synchronized to the infinite bus. Then:

(a) P_M is slowly increased until $P_G = 0.5$. E_{fd} is not changed. In the (new) steady state, find V_a, I_a, E_a, and E'_a.

(b) Now at $t = 0$, E_{fd} is suddenly doubled. Assume that $\theta = \omega_0 t + (\pi/2) + \delta$, δ = constants and find $|E'(t)|$ and $|E(t)|$.

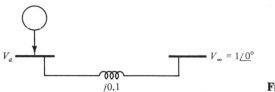

$$V_\infty = 1\underline{/0°}$$

$$j0.1$$

Figure E10.13(a)

Solution   At synchronization

$$I_a = 0 \Rightarrow E'_a = E_a = V_a = V_\infty = 1 \underline{/0°}$$

Also from (10.75) we get, in the presynchronism steady state, $E_{fd} = |E_a| = 1.0$.

(a) At synchronization $\delta = 0$, but as the generator delivers power, δ increases. Assuming that the power is increased slowly, we can assume a quasi-steady-state transition with $i_F = v_F^0/r_F$ = constant. Thus when $P_G = 0.5$, $|E_a| = 1.0$. Considering the infinite bus to be the machine "terminal," we have

$$0.5 = \frac{1.0 \times 1.0}{1.1} \sin \delta + \frac{1}{2}\left(\frac{1}{0.7} - \frac{1}{1.1}\right) \sin 2\delta$$

$$= 0.909 \sin \delta + 0.260 \sin 2\delta$$

Solving iteratively, we find that $\delta = 21.0°$. Thus $E_a = 1.0 \underline{/21.0°}$. To find I_a we can use Fig. 10.11, as in Example 10.12, or we can proceed analytically. Let us use algebra this time. Recalling that the machine "terminal" is V_∞ and using (10.67), we have

$$E_a = 1.0e^{j\delta} = V_\infty + j\tilde{X}_d jI_d e^{j\delta} + j\tilde{X}_q I_q e^{j\delta}$$

where we are using I_d and I_q. Next multiply by $e^{-j\delta}$ and introduce numerical values.

$$1.0 = 1 \underline{/-21.0°} + 1.1I_d + j0.7I_q$$

I_d and I_q are real numbers, so it is easy to separate real and imaginary parts. Taking real parts first and then imaginary parts, we get

$$1.0 = 0.934 - 1.1I_d \Rightarrow I_d = -0.0604$$
$$0 = -0.358 + 0.7I_q \Rightarrow I_q = 0.5120$$

Thus

$$I_a = (I_q + jI_d)e^{j\delta} = 0.5155 \underline{/14.27°}$$
$$V_a = V_\infty + j0.1I_a = 1.0 + 0.0515 \underline{/104.27°}$$
$$= 0.9886 \underline{/2.897°}$$
$$E'_a = E_a - j(X_d - X'_d)jI_d e^{j\delta}$$
$$= 1 \underline{/21.0°} + 0.8(-0.0604) \underline{/21.0°}$$
$$= 0.952 \underline{/21.0°}$$

Note that all four quantities V_a, I_a, E_a, and E'_a have changed from their values at synchronization. Only $|E_a|$ has not changed (because we have not yet changed E_{fd}).

(b) In this part we are assuming that $\delta = 21.0° =$ constant. We have $|E'_a(0)| = 0.952$, $E_{fd} = 2$, so, using (10.75),

$$4\frac{d|E'_a|}{dt} + |E_a| = 2 \qquad |E'_a(0)| = 0.952 \tag{10.82}$$

Next we express $|E_a|$ in terms of $|E'_a|$, δ, and $|V_\infty|$. From Fig. E10.13(b) [or using (10.67) and (10.73)] we see that

$$\frac{|E_a| - |V_\infty| \cos \delta}{|E'_a| - |V_\infty| \cos \delta} = \frac{\tilde{X}_d |I_{ad}|}{\tilde{X}'_d |I_{ad}|} = \frac{\tilde{X}_d}{\tilde{X}'_d}$$

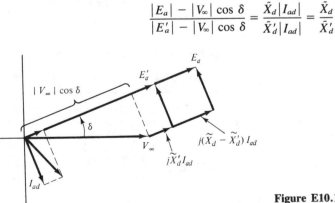

Figure E10.13(b)

From this we can solve for $|E_a|$ in terms of $|E'_a|$ and δ.

$$|E_a| = \frac{\tilde{X}_d}{\tilde{X}'_d}|E'_a| + |V_\infty|\frac{\tilde{X}'_d - \tilde{X}_d}{\tilde{X}'_d}\cos\delta \qquad (10.83)$$

$$= 3.6667\,|E'_a| - 2.4895$$

Substituting in (10.82), we get

$$4\frac{d|E'_a|}{dt} + 3.6667|E'_a| = 4.4895$$

or

$$1.0909\frac{d|E'_a|}{dt} + |E'_a| = 1.2244$$

with solution

$$|E'_a(t)| = 0.9520 + 0.2724(1 - e^{-t/1.0909})$$

$$= 1.2244 - 0.2724e^{-t/1.0909}$$

Now we go back and find $|E_a(t)|$ from (10.83):

$$|E_a(t)| = 2 - e^{-t/1.0909}$$

Note that the steady-state value of $|E_a|$ checks with the value we get from (10.75).

Exercise 5.   Suppose that we want to find $v_a(t)$ during the transient above. Indicate how you would proceed.

10.12 GENERATOR CONNECTED TO INFINITE BUS (LINEAR MODEL)

In Example 10.13(b) we assumed that $\delta = $ constant. If we had not, δ would appear in (10.83) in a nonlinear fashion (cos δ). In Example 10.12 we assumed $|E'_a|$ or $|E_a|$ to be constant, and used a graphical technique (equal-area construction) to investigate the behavior of δ. If $|E'_a|$ or $|E_a|$ were not constant, the technique breaks down.

In general, since the equations are nonlinear, we need to introduce numerical values and obtain computer solutions using standard numerical integration algorithms. In many cases, however, linearization of the nonlinear equations is appropriate, and in such cases we can gain valuable insights into general behavior over a variety of operating conditions by the powerful techniques of linear systems analysis. As an example we consider the special case of a generator connected to an infinite bus through a transmission line (Fig. 10.12). As in Example 10.13 we will absorb the transmission line into the generator model, with V_∞ the "terminal" voltage. Assume that both the stator and line resistances are zero. As in Example 10.13, it is most convenient to express $|E_a|$ in terms of the state variables $|E'_a|$ and δ by using (10.83). As a preliminary, define

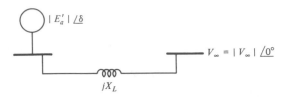

$$V_\infty = |V_\infty| \underline{/0^\circ}$$

Figure 10.12 System to be linearized.

$$K_3 = \frac{\tilde{X}_d'}{\tilde{X}_d} = \frac{X_d' + X_L}{X_d + X_L}$$

Note: Since $X_L > 0$ and $X_d' < X_d$, we have $0 < K_3 < 1$.

Substituting in (10.83), we get

$$|E_a| = \frac{1}{K_3}|E_a'| + \left(1 - \frac{1}{K_3}\right)|V_\infty|\cos\delta$$

Substituting in (10.75) and multiplying by K_3, we get

$$K_3 T_{do}' \frac{d|E_a'|}{dt} + |E_a'| = K_3 E_{fd} + (1 - K_3)|V_\infty|\cos\delta \qquad (10.84)$$

Also, from (10.79), we have

$$M\ddot{\delta} + D\dot{\delta} + P_G = P_M \qquad (10.79)$$

where from (10.81),

$$P_G = P_G(\delta, |E_a'|) = \frac{|E_a'||V_\infty|}{\tilde{X}_d'}\sin\delta + \frac{|V_\infty|^2}{2}\left(\frac{1}{\tilde{X}_q} - \frac{1}{\tilde{X}_d'}\right)\sin 2\delta \qquad (10.85)$$

Now assume steady-state operating conditions, in which case $\delta = \delta^0$,

$$|E_a'| = |E_a'|^0, \quad P_G = P_G^0 = P_G(\delta^0, |E_a'|^0), \quad E_{fd} = E_{fd}^0, \text{ and so on.}$$

We call the collection of these constant values the *operating point.* Now consider small perturbations around the operating point. We let $\delta = \delta^0 + \Delta\delta, |E'| = |E'|^0 + \Delta|E'|$, and so on.

In the usual way, the increments $\Delta\delta$, $\Delta|E'|$, ΔE_{fd}, and ΔP_M are related by linear differential and/or algebraic equations. Thus instead of (10.84) and (10.79), we get

$$K_3 T_{do}' \frac{d\Delta|E_a'|}{dt} + \Delta|E_a'| = K_3\,\Delta E_{fd} - [(1 - K_3)|V_\infty|\sin\delta^0]\Delta\delta \qquad (10.86)$$

$$M\,\Delta\ddot{\delta} + D\,\Delta\dot{\delta} + \frac{\partial P_G(\delta^0, |E_a'|^0)}{\partial\delta}\Delta\delta + \frac{\partial P_G(\delta^0, |E_a'|^0)}{\partial|E_a'|}\Delta|E_a'| = \Delta P_M \qquad (10.87)$$

We define $K_4 = [(1/K_3) - 1]|V_\infty|\sin\delta^0$, $K_2 = \partial P_G^0/\partial|E_a'|$, and $T = \partial P_G^0/\partial\delta$. We have previously seen T, the synchronizing coefficient, in Chapter 9. If we plot P_G versus δ, with $|E_a'| = |E_a'|^0$, then T is the slope of P_G at $\delta = \delta^0$. We can also find analytic expressions for T and K_2, by differentiating (10.85).

Suppose that we are at some operating point, and at $t = 0$ there is an increment

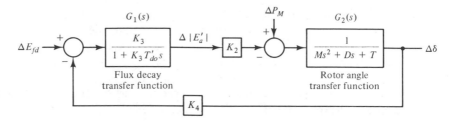

Figure 10.13 Block diagram of linearized system.

in E_{fd} and/or P_M (i.e., we have a ΔE_{fd} and/or ΔP_M). We can study the transient by using the Laplace transform. Noting that $\Delta |E'(0^-)| = 0$, $\Delta \delta(0^-) = 0$, we Laplace transform (10.86) and (10.87) and get

$$(K_3 T'_{do} s + 1)\Delta |\hat{E}'_a| = K_3 \Delta \hat{E}_{fd} - K_3 K_4 \Delta \hat{\delta} \qquad (10.88)$$

$$(Ms^2 + Ds + T) \Delta \hat{\delta} = \Delta \hat{P}_M - K_2 \Delta |\hat{E}'_a| \qquad (10.89)$$

where the circumflex indicates the Laplace transform of the increment. We express these relations between inputs ΔE_{fd} and ΔP_M and the (output) variables $\Delta |E'_a|$ and $\Delta \delta$ in Fig. 10.13. The reader should check that the block diagram conforms to (10.88) and (10.89). We comment briefly on the block diagram. The feedback loop arises because of the "natural" coupling between $\Delta |E'_a|$ and $\Delta \delta$; we have not introduced feedback externally. Later when we consider the operation of the excitation system and the speed governor, we will show how ΔE_{fd} and ΔP_M depend on the output variables through external feedback loops. From Fig. 10.13 we can easily find the transfer function from any input to any output by using simple rules. For example, we can write the transfer function from ΔP_M to $\Delta \delta$ by inspection:

$$\frac{\Delta \hat{\delta}}{\Delta \hat{P}_M} = \frac{G_2(s)}{1 - K_2 K_4 G_1(s) G_2(s)} \qquad (10.90)$$

where $G_1(s)$ and $G_2(s)$ are defined in Fig. 10.13. Similarly, we get, by inspection,

$$\frac{\Delta |\hat{E}'_a|}{\Delta \hat{E}_{fd}} = \frac{G_1(s)}{1 - K_2 K_4 G_1(s) G_2(s)} \qquad (10.91)$$

The transfer functions from ΔP_M to $\Delta |E'_a|$, and from ΔE_{fd} to $\Delta \delta$, may also be written by inspection. We call all these transfer functions *closed-loop* transfer functions. Equations (10.90) and (10.91) are examples of closed-loop transfer functions expressed in terms of the *open-loop* transfer functions $G_1(s)$ and $G_2(s)$.

From the closed-loop transfer functions we can find the responses to a particular input. Usually, we are more interested in general properties, such as stability and damping of oscillatory modes. The stability and damping properties depend on the location of the poles of the closed-loop transfer functions. These poles are the same for all four closed-loop transfer functions. For our purposes it is most convenient to find, or estimate, the location of these poles by the root-locus method. The method, which is based on the relation of closed-loop poles to open-loop poles and zeros and the loop gain, is described in Appendix 4. One obtains the loci of closed-loop poles as the loop gain varies from zero to infinity. Since various gains in our model, for example K_2 and K_4, depend on the operating point, the root-locus method is particularly appropriate to obtain the range of possible behavior.

We will next apply the method to investigate the general stability and damping properties of the (linearized) generator model of Fig. 10.13. We note that this is a case of "positive" feedback. We see this as follows: The loop gain, as defined in Appendix 4, is $K = K_2 K_4 / T'_{do} M$. T'_{do} and M are positive constants. Using (10.95) to evaluate K_2 and K_4 and noting that $K_3 \triangleq \tilde{X}'_d / \tilde{X}_d < 1$, we get

$$K_2 K_4 = \frac{|V_\infty|}{\tilde{X}'_d} \sin \delta^0 \left(\frac{1}{K_3} - 1 \right) |V_\infty| \sin \delta^0$$

$$= \frac{(X_d - X'_d)|V_\infty|^2 \sin^2 \delta^0}{X'^2_d} \geq 0$$

Thus $K \geq 0$. On the other hand, if we look at the two summing points in Fig. 10.13, we subtract twice, so in effect we have *positive* feedback. Consider then, in Fig. 10.14, a positive feedback root locus plotted versus $K_2 K_4$. The branches of the root locus start at the open-loop poles s_1, s_s, and s_3. s_1 and s_2 are the open-loop poles corresponding to the lightly damped rotor angle modes; s_3 is the pole associated with the "flux decay" effects.

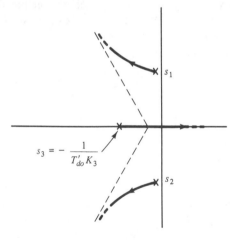

Figure 10.14 Root locus diagram.

From this general sketch one can see the beneficial effects of the (natural) loop on the damping of the rotor angle modes. With the loop open, for example with $K_4 = 0$, we are stuck with the lightly damped poles of $G_2(s)$. But with positive feedback the lightly damped poles move deeper into the left half-plane. From a physical point of view, the changes in field flux linkages (proportional to $\Delta|E'_a|$) that accompany the changes in δ are conductive to rotor angle damping. Note, in particular, the effect of T'_{do}. If $T'_{do} = 0, G_1(s) = K_3$ (i.e., we lose the real-axis pole and the root-locus branches become parallel to the $j\omega$ axis); we completely lose the beneficial damping. Thus the "delays" associated with the buildup and decay of flux in the field winding are required for good damping!

On the other hand, suppose that $T'_{do} \to \infty$. Then

$$G_2(s) = \frac{K_3/T'_{do}}{1/T'_{do} + K_3 s} \to \frac{0}{K_3 s}$$

which implies zero loop gain, and therefore once again we lose the beneficial damping. Thus we can pose a "machine design" problem. Find T'_{do} for the best damping. Note that since the K_i depend on system operating conditions, the design must be a compromise.

We note also that the main source of rotor angle damping is due to T'_{do}. By comparison the shaft friction damping, $D\dot{\delta}$, is almost negligible! Finally, we note that while damping of the rotor angle poles improves as the loop gain increases, the real-axis pole (associated with T'_{do}) becomes less damped. For small loop gains this may not be a problem but must definitely be considered when the gains are large.

In Chapter 11 we will consider what happens when we connect a voltage control system which supplies the input ΔE_{fd} in Fig. 10.13. We will see that care must be taken to avoid losing the beneficial natural damping.

We now leave the topic of generator modeling. However, we will need some additional results, derived in Appendix 5, when we consider nonsymmetric short circuits in Chapter 13.

10.13 SUMMARY

A synchronous machine may be modeled by three stator coils and three rotor coils. The inductance parameters depend on the rotor angle θ. Using actual flux linkages, phase voltages, and currents, we get (10.10) and (10.12) as the mechanical and electrical equations.

The equations may be simplified by introducing the Park transformation of abc variables into $0dq$ variables. In terms of the new variables, we get new mechanical and electrical equations, (10.44) and (10.38). The new equations are much simpler. In particular, we note that if the shaft angular velocity is constant, the electrical equation is linear and time invariant. In applications it is convenient to separate the electrical equations into zero-sequence, direct axis, and quadrature axis sets. This is done in (10.45) to (10.47). Correspondingly, the equations relating flux linkages to currents are separated in the same way in (10.48) to (10.50).

The Park equations may be used to investigate behavior under very general conditions. In the case of steady-state (positive sequence) operation, they reduce to (10.67), with associated phasor diagram in Fig. 10.10; this model was derived on a different basis in Chapter 8.

Another simplified version of the Park equations may be obtained under assumptions of "slowly varying amplitude and phase" (i.e., a relatively small departure from steady-state operating conditions); the more precise conditions are given at the beginning of Section 10.10. Under these conditions we obtain the simplified dynamic model summarized in Section 10.11. This consists of a scalar second-order mechanical differential equation, a first-order electrical differential equation, and two algebraic equations. The variables in these equations are abc variables rather than $0dq$ variables; no Park transformations are required.

For a single synchronous machine connected through a transmission link to an infinite bus, application of the simplified dynamic model results, after linearization, in the block diagram in Fig. 10.13. A root-locus analysis of the system shows the beneficial effects of the field flux changes accompanying rotor angle swings. This provides a strong natural damping of the rotor angle modes.

PROBLEMS

10.1. We are given a solenoid with two coils whose inductance parameters depend on the displacement x of a movable iron core. Find the force on the movable core (in the direction of increasing x) if $i_1 = 10$, $i_2 = -5$, $L_{11} = 0.01/(200x + 1)$, $L_{22} = 0.01/(3 - 200x)$, and $L_{12} = 0.002$. Assume that the units are meters, amps, and henrys, in which case the force will be in newtons. Graph the force for $0 \leq x \leq 0.01$ m.

10.2. A generator is rated 10,000 kVA, 13.8 kV line-line. $X_d'' = 0.06$ p.u., $X_d' = 0.15$ p.u., and $X_d = X_q = 1.0$ p.u. The generator is at no-load and rated voltage when a 3ϕ short occurs. Find the following currents in p.u.
   (a) The steady-state short-circuit current.
   (b) The initial 60 Hz component of short-circuit current.
   (c) The initial 60 Hz component of short-circuit current if damper windings are neglected.
   (d) In case (c), what is the maximum possible unidirectional component (magnitude)?
   (e) Convert the per unit value found in case (b) into actual amperes.

10.3. A generator is operating under open-circuit steady-state conditions with $V_a = E_a = 1\underline{/0°}$ (Fig. P10.3). Use Park's equations to consider the effect of switching on a 3ϕ Y-connected set of resistors at $t = 0$. Assume that:

   1. $\theta(t) = \omega_0 t + \pi/2$.
   2. $i_F = i_F^0 = $ constant (current source).
   3. Neglect damper windings.
   4. $r = 0.1$, $R = 0.9$, $X_d = \omega_0 L_d = 1.0$, and $X_q = \omega_0 L_q = 0.6$.

   (a) Show that $\mathbf{v}_{0dq} = \mathbf{R}i_{0dq}$, where $\mathbf{R} = \text{diag }\{R\}$.
   (b) Find differential equations for i_d and i_q for $t \geq 0$.
   (c) Write Laplace transforms for i_d and i_q.
   (d) Find i_d and i_q as $t \to \infty$.
   (e) Find I_a and V_a as $t \to \infty$ (i.e., steady-state behavior).
   (f) Repeat part (e) using the E_a, X_d, X_q model of Chapter 8, and compare.

Note: The final value theorem of the Laplace transform: *If there is a steady state value,* then $\lim_{t \to \infty} f(t) = \lim_{s \to 0} sF(s)$, where $F(s)$ is the Laplace transform of $f(t)$.

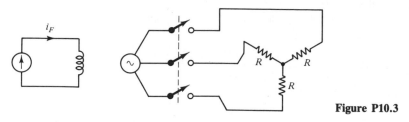

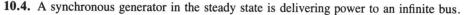

Figure P10.3

10.4. A synchronous generator in the steady state is delivering power to an infinite bus.

$$\theta(t) = \omega_0 t + \frac{\pi}{2} + \delta, \quad \delta = \frac{\pi}{4}$$

$$\lambda_q = \lambda_d = \frac{1}{\sqrt{2}\omega_n}, \quad i_q = \frac{1}{\sqrt{2}}, \quad i_d = -\frac{1}{\sqrt{2}}$$

$$r = 0, \ X_d = \omega_0 L_d = 1, \ \omega_0 k M_F = 1, \ T'_{do} = 1 \text{ sec}$$

(a) Find the torque T_E.
(b) Find $v_a(t)$, $i_a(t)$, and i_F.
(c) At $t = 0$, the generator is suddenly disconnected from the infinite bus. Assume that $v_F = $ constant, $i_D = i_Q = 0$. Sketch $i_F(t)$. *Hint:* To find $i_F(0^+)$, consider the consequences of the fact that $\lambda_d(0^+) = \lambda_d(0^-)$.

10.5. Neglecting damper circuits and the field resistance r_F, the direct axis equivalent circuit (Fig. 10.4) reduces to that shown in Fig. P10.5. Find the Thévenin equivalent circuit for the network in the box and show that the Thévenin equivalent inductance is

$$L'_d = L_d - \frac{(kM_F)^2}{L_F}$$

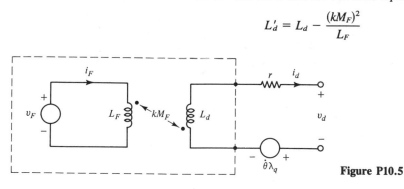

Figure P10.5

10.6. The shaft of a synchronous machine (motor) is clamped or "blocked" and is thus not free to turn. A set of (positive-sequence) voltages is applied to the motor terminals. Assume that:

1. $V_a = |V_a|\underline{/0°}$.
2. $\theta(t) = \pi/2$.
3. $i_F = 0$ (open-circuited field).
4. Neglect damper windings.

    (a) Find i_d and i_q in the steady state.
    (b) Find an expression for the average torque T_{Eav}.
    (c) Suppose that $r = 0$. What is T_{Eav}?
    (d) Do you think that assumption 4 is a realistic assumption?

10.7. Suppose that $\theta(t) = \omega_0 t + \pi/2$ and

$$i_a(t) = \sqrt{2}\,\cos \omega_0 t$$

$$i_b(t) = \sqrt{2}\,\sin \omega_0 t$$

$$i_c(t) = -\sqrt{2}\,\cos \omega_0 t$$

Note that the currents are not a balanced set. Find the frequency components in $i_0(t)$, in $i_d(t)$, and in $i_q(t)$. Consider dc to be zero frequency.

10.8. A synchronous machine is in the (positive sequence) sinusoidal steady state. $\theta = \omega_0 t + (\pi/2) + \delta$ and $\delta = 30°$. At its terminals $V_a = 1\underline{/0°}$ and $I_a = 1\underline{/-30°}$.
    (a) Find $i_0, i_d, i_q, v_0, v_d,$ and v_q.
        In parts (b), (c), and (d), neglect r.
    (b) Find $\lambda_0, \lambda_d,$ and λ_q.
    (c) Find the torque T_E.
    (d) Find the 3ϕ power output $p_{3\phi}(t)$ in terms of $0dq$ components.

10.9. Repeat Problem 10.8 with $\delta = 20°$. Compare with the results in Problem 10.8.

10.10. The terminal conditions of a synchronous generator are $V_a = 1\underline{/0°}$ and $I_a = 1\underline{/90°}$. The generator parameters are $X_d = 1.0$, $X'_d = 0.2$, $X_q = 0.6$, and $r = 0$. Find $\delta, I_d, I_q, i_d, i_q, I_{ad}, I_{aq}, |E_a|, |E'_a|,$ and P_G.

10.11. Repeat Problem 10.10 with $I_a = 1\underline{/-30°}$.

10.12. Refer to Fig. P10.12 and assume that

$$\theta(t) = \omega_0 t + \frac{\pi}{2} + \delta, \quad \delta = \text{constant}$$

$$X_d = X_q = 0.9$$

$$X'_d = 0.2$$

$$X_L = 0.1$$

$$T'_{do} = 2.0 \text{ sec}$$

The generator is operating in steady state with $P_G = 0.5$ and $|E_a| = 1.5$. At $t = 0$ there is a fault and the circuit breakers open. Use the *simplified dynamic model* and find an expression for $|V_a(t)| = |E'_a(t)|$, for $t \geq 0$. Assume that $\delta = $ constant at its prefault value.

V_a jX_L $V_\infty = 1\underline{/0°}$

                                                         Figure P10.12

10.13. Repeat Example 10.12 but with $E^0_{fd} = 1.6$, $X_d = X_q = 1.1$, and $X'_d = 0.4$.

10.14. Repeat Example 10.13 but with $X_d = 1.6$, $X_q = 0.9$, $X'_d = 0.3$.

10.15. Refer to Fig. P10.15 and assume that

$$X_d = 1.15$$
$$X_q = 0.6$$
$$X'_d = 0.15$$
$$X_L = 0.2$$
$$r = 0$$
$$T'_{do} = 2 \text{ sec}$$

The generator is in steady state with $E_{fd} = 1$ and $E'_a = 1\underline{/15°}$. Then at $t = 0$, E_{fd} is changed to a new constant value: $E_{fd} = 2$. Assume that the rotation is still uniform. Find $E'_a(t)$ for $t \geq 0$. Find $v_a(t)$ for $t \geq 0$.

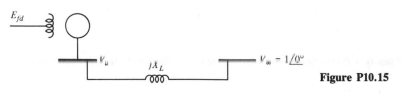

Figure P10.15

Voltage Control System

chapter 11

11.0 INTRODUCTION

The generator model derived in Chapter 10 shows two inputs, field voltage and mechanical power. In this chapter we consider the first of these inputs.

The field voltage (and current) may be supplied by a dc generator driven by a motor or by the shaft of the turbine-generator. Rectifier and thyristor units converting ac to dc in various ways are also common. In any case, the units are called *exciters,* and the complete voltage control system, including an error detector and various feedback loops, is usually called an *excitation system* or *automatic voltage regulator* (AVR).

For a large synchronous generator the exciter may be required to supply field currents as large as $i_F \approx 6500$ A at $v_F = 500$ V, so the exciter may be a fairly substantial machine.

An example of an excitation system is shown in Fig. 11.1. It typifies a variety of available excitation systems. A simplified description of the system operation follows. The generator output voltage $|V_a|$ is compared with a reference voltage and any error is amplified (transistor and magnetic amplifiers) and fed to the field of a special "high-gain" dc generator called an *amplidyne*. The amplidyne provides incremental changes to the exciter field in a so-called "boost-buck" scheme. The exciter output provides the rest of the exciter field excitation in a "self-exciting" mode. Thus,

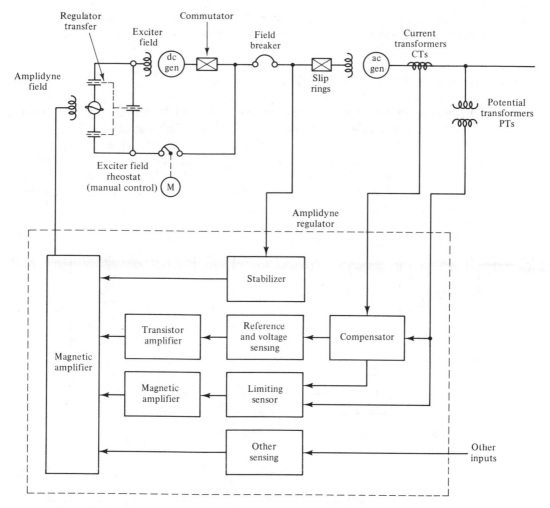

Figure 11.1 Excitation system. Excitation control system with dc generator-commutator exciter. (Adapted from IEEE *Transactions on Power Apparatus and Systems,* Vol. PAS-88, "Proposed Excitation System Definitions for Synchronous Machines," IEEE Committee Report, © 1969 IEEE.)

even if the amplidyne output is zero (i.e., no correction in $|V_a|$ is needed), the exciter supplies the required level of voltage v_F to the synchronous generator field.

The system is a feedback system which turns out to be unstable unless the stabilizing loop is provided. The purpose of the limiter is to prevent undesirable levels of excitation. The purpose of the feedback of current $|I_a|$ is to introduce a voltage drop proportional to $|I_a|$. It turns out this is required to ensure that reactive power is properly divided between parallel generators. The purpose of the "other inputs" will be discussed in a later section. Finally, we note that if the amplidyne "regulator" is

out of service, one can use the regulator transfer switches to remove the amplidyne
from the circuit and put the exciter field on manual control.

11.1 EXCITER SYSTEM BLOCK DIAGRAM

A much-simplified version of the system we have been describing is shown in Fig.
11.2. The basic operation can be described as follows. The control input is V_{ref}. If V_{ref}
is raised (while the generator terminal voltage $|V_a|$ initially remains the same), v_e goes
up, v_R increases, v_F increases, and $|V_a|$ tends to increase. A new equilibrium is
reached with a higher $|V_a|$. If $|V_a|$ changes because of a change in load, we also get
a "correcting" action. Note there is no boost-buck feature here; all the exciter field
excitation comes from the amplifier.

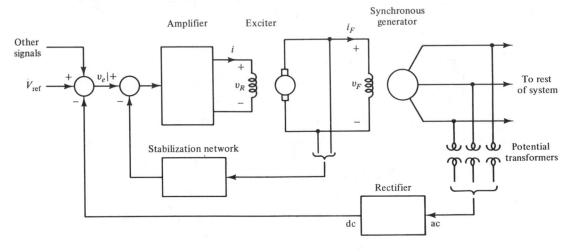

Figure 11.2 Simplified excitation system.

For more quantitative results we need a model. Figure 11.3 shows a typical result
of such modeling. All the variables shown are expressed in per unit. In this case the
gain of the "measurement" block is unity in the steady state (i.e., a 1% change in $|V_a|$
leads to a 1% change at the summing point). T_R is very small. It is usually neglected.
The amplifier gain K_A is typically in the range 25 to 400. T_A is typically in the range
0.02 to 0.4 sec. In modeling the amplifier for large-signal behavior, the limits need
to be included. Thus we have $V_{Rmin} \leq v_R \leq V_{Rmax}$. Away from the limits, the gain in
the linear zone is 1. As we will see soon, some form of stabilization is needed. A
typical stabilizer is shown. Typical parameter values are K_F in the range 0.02 to 0.1,
and T_F in the range 0.35 to 2.2 sec.

We next turn to the modeling of the exciter. We will consider the case of the dc
generator shown in Fig. 11.2. A simplified description follows. The open-circuit
voltage of a dc generator is proportional to the product of speed times air-gap flux per

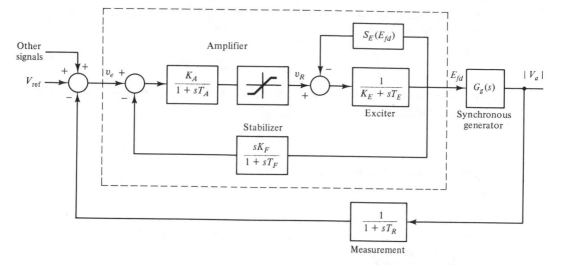

Figure 11.3 Block diagram of excitation system and generator.

pole. The flux is a nonlinear function of the dc generator field current because of saturation effects in the magnetic circuit. Thus we get the familiar "saturation curve" of dc generator open-circuit voltage versus field current. When the loading of the dc generator by the ac generator field winding is considered, we get a different curve, the "load-saturation" curve. We assume that this curve is available and show the plot of exciter output voltage v_F in terms of exciter field current i in Fig. 11.4. It will be simplest in the following development to work with actual voltages, currents, and flux linkages, not per unit values.

The "air-gap" line is tangent to the lower, approximately linear portion of the

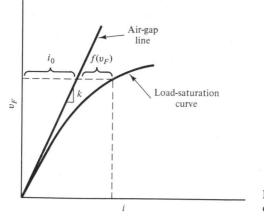

Figure 11.4 Exciter load-saturation curve.

open-circuit saturation curve. $f(v_F)$ measures the departure of the load saturation curve from that simple ideal. In terms of $f(v_F)$ we have

$$i = i_0 + f(v_F) = \frac{1}{k} v_F + f(v_F) \tag{11.1}$$

where k is the slope of the air-gap line. We next consider the field circuit of the dc generator and get

$$v_R = Ri + \frac{d\lambda}{dt} \tag{11.2}$$

where v_R is the output of the amplifier and R, i, and λ are the resistance, current, and flux linkages of the dc generator field winding connected across the amplifier terminals.

We next assume that the dc generator output voltage v_F is proportional to the dc generator air-gap flux, which in turn is proportional to λ. Making this reasonable assumption, we have a relation of the form

$$v_F = \beta\lambda \tag{11.3}$$

Substituting (11.1) and (11.3) in (11.2), we get

$$v_R = R\left(\frac{1}{k} v_F + f(v_F)\right) + \frac{1}{\beta} \frac{dv_F}{dt} \tag{11.4}$$

Multiplying by k/R, we get

$$\frac{k}{R} v_R = v_F + v_F S(v_F) + T_E \frac{dv_F}{dt} \tag{11.5}$$

where $T_E \triangleq k/R\beta$ and $S(v_F) \triangleq kf(v_F)/v_F$. The multiplicative factor $S(v_F)$ accounts for the departure of the load-saturation curve from the air-gap line.

We would now like to replace v_F in (11.5) with a scaled version, E_{fd}, defined just after (10.74). At the same time we can convert (11.5) to the per unit system. The base for v_R can be chosen (relative to that for E_{fd}) to eliminate unnecessary constants. Finally (without explicit p.u. notation), we get the per unit version of (11.5):

$$v_R = E_{fd} + E_{fd} S_E(E_{fd}) + T_E \frac{dE_{fd}}{dt} \tag{11.6}$$

where $S_E(E_{fd})$ is called the (per unit) *saturation function*. Equation (11.6) conforms to the exciter portion of the block diagram in Fig. 11.3 with $K_E = 1.0$. $K_E = 1.0$ is appropriate for our exciter model with a separately excited shunt field. Modeling a self-excited shunt field we would get approximately $K_E = -0.1$. Typical values of T_E are in the range 0.5 to 1.0 sec. Over the allowable values of E_{fd}, typical values of S_E are under 1.3. We turn next to the modeling of the generator with input E_{fd} and output $|V_a|$.

11.2 GENERATOR MODELS

We will consider a few cases.

Generator model, case I: open circuit.   In this case $I_a = 0$; thus we have $E'_a = E_a = V_a$. In using (10.75), repeated here for convenience:

$$T'_{do} \frac{d\,|E'_a|}{dt} + |E_a| = E_{fd} \qquad (10.75)$$

we can replace $|E_a|$ by $|E'_a|$. By taking the Laplace transform, we get the transfer function

$$G_g(s) = \frac{|\hat{V}_a|}{\hat{E}_{fd}} = \frac{|\hat{E}'_a|}{\hat{E}_{fd}} = \frac{1}{1 + sT'_{do}} \qquad (11.7)$$

Generator model, case II: impedance load.   We assume an isolated generator supplying an impedance load Z. In this case $V_a = ZI_a$. Also, assume uniform rotation, $\theta = \omega_0 t + \pi/2 + \delta$, $\delta = $ constant. Again we use (10.75). The problem of relating $|V_a|$ to E_{fd} will be solved in three steps. We will first express $|E_a|$ in terms of $|E'_a|$ and the generator and load impedances. Then from (10.75) we find the transfer function from E_{fd} to $|E'_a|$. Finally, we find $|V_a|$ in terms of $|E'_a|$.

Expressing $|E_a|$ in terms of $|E'_a|$ is similar to the calculation in Example 10.13 leading to (10.83), except that now we must include resistance. We start with (10.73), repeated here for convenience.

$$E'_a = V_a + rI_a + jX'_d jI_d e^{j\delta} + jX_q I_q e^{j\delta} \qquad (10.73)$$

With an impedance load, $Z = R + jX$, and

$$V_a + rI_a = (Z + r)I_a = (r + R + jX)(I_q + jI_d)e^{j\delta} \qquad (11.8)$$

Thus, substituting (11.8) in (10.73) and multiplying by $e^{-j\delta}$, we get

$$|E'_a| = (r + R)I_q - (X'_d + X)I_d + j[(r + R)I_d + (X_q + X)I_q] \qquad (11.9)$$

Equating real and imaginary parts, we get

$$\begin{bmatrix} -(X'_d + X) & r + R \\ r + R & X_q + X \end{bmatrix} \begin{bmatrix} I_d \\ I_q \end{bmatrix} = \begin{bmatrix} |E'_a| \\ 0 \end{bmatrix} \qquad (11.10)$$

Solving for I_d and I_q,

$$I_d = K_d |E'_a| \qquad \text{where } K_d \triangleq \frac{-(X + X_q)}{(r + R)^2 + (X + X'_d)(X + X_q)} \qquad (11.11)$$

$$I_q = K_q |E'_a| \qquad \text{where } K_q \triangleq \frac{(r + R)}{(r + R)^2 + (X + X'_d)(X + X_q)} \qquad (11.12)$$

Continuing with the analysis, we use another equation from Chapter 10:

$$E_a = E'_a + j(X_d - X'_d)jI_d e^{j\delta} \tag{10.72}$$

Substituting I_d from (11.11), we get

$$|E_a| = |E'_a| - (X_d - X'_d)K_d|E'_a|$$

$$= \left[1 - (X_d - X'_d)K_d\right]|E'_a| = \frac{1}{\sigma}|E'_a| \tag{11.13}$$

where

$$\sigma \triangleq \frac{1}{1 - (X_d - X'_d)K_d} = \frac{(r + R)^2 + (X + X'_d)(X + X_q)}{(r + R)^2 + (X + X_d)(X + X_q)} \tag{11.14}$$

Note: For inductive loads $X \geq 0$, which implies that $\sigma < 1$.

We next substitute (11.13) into (10.75) (the first equation in this section) and get

$$\sigma T'_{do} \frac{d|E'_a|}{dt} + |E'_a| = \sigma E_{fd} \tag{11.15}$$

Taking the Laplace transform of (11.15), the transfer function from E_{fd} to $|E'_a|$ is

$$\frac{|\hat{E}'_a|}{\hat{E}_{fd}} = \frac{\sigma}{1 + s\sigma T'_{do}} \tag{11.16}$$

Finally, we express $|V_a|$ in terms of $|E'_a|$.

$$V_a = ZI_a = Z(I_q + jI_d)e^{j\delta}$$

$$= Z(K_q + jK_d)|E'_a|e^{j\delta} \tag{11.17}$$

Let $K \triangleq K_q + jK_d$; then

$$V_a = ZKE'_a \tag{11.18}$$

$$|V_a| = |Z||K||E'_a| = k_v|E'_a| \tag{11.19}$$

where $k_v \triangleq |Z||K|$. Finally, using (11.16) and (10.19), we get the desired transfer function.

$$G_g(s) = \frac{k_v\sigma}{1 + s\sigma T'_{do}} \tag{11.20}$$

Note the differences in the transfer functions of the loaded and unloaded generators.

Example 11.1 Transfer Function of Loaded Generator

Given $r = 0$, $R = 1$, $X = 0$, $X_d = 1$, $X'_d = 0.2$, $X_q = 0.6$, and $T'_{do} = 2$, find $G_g(s)$.

Solution Using the equations in the preceding section, we find successively

$$K_d = \frac{-X_q}{R^2 + X'_d X_q} = -0.54$$

$$K_q = 0.89$$

$$k_v = 1 \times 1.04 = 1.04$$

$$\sigma = 0.7$$

$$G_g(s) = \frac{0.73}{1 + 1.4s}$$

Note: For the unloaded generator we would get

$$G_g(s) = \frac{1}{1 + 2s}$$

Exercise 1.   Assuming that r is sufficiently small, can you find a value of Z such that σ is negative? Suppose that σ is negative. Suppose that E_{fd} is an arbitrarily small step function. Show that $|V_a|$ builds up "spontaneously" (self-excitation) Can you explain the result physically? *Hint:* Assume that the load is a capacitor.

Note: We have not yet considered the important case of a generator connected through a transmission line to an infinite bus. We defer consideration of this model until Section 11.5.

11.3 STABILITY OF EXCITATION SYSTEM

Now that we have a model for $G_g(s)$ with an impedance load, we can look at a possible stability problem for the feedback loop In Fig. 11.3. To proceed with linear analysis we first need to linearize the excitation system around an operating point. Let us assume that the amplifier is in the linear range and S_E is small enough to be negligible. Suppose that the stabilizer loop in Fig. 11.3 is absent. We then have a feedback loop with no zeros and four real poles. Feedback control theory tells us that to have a good regulator of voltage magnitude, we need a high dc loop gain. In this case the steady-state error will be small (provided that the system is stable!).

Unfortunately, even moderate loop gains make the system unstable, as may be seen by using a root locus. We consider an example.

Example 11.2   Stability of Basic Excitation System

In Fig. 11.3 assume that $T_A = 0.05$, $T_R = 0.06$, $K_E = 1$, and $T_E = 0.8$. The generator is open circuited with $T'_{do} = 5$. All time constants are given in seconds. Assume that there is no stabilizer. Neglect limiting and the S_E function and draw the root locus.

Solution   The open-loop poles are at -20, -16.67, -1.25, and -0.2. There are no zeros. Using the rules for plotting (negative feedback) root loci supplemented by a few calculated points, we get the root locus shown in Fig. E11.2. We note that the root-locus gain $K = K_A/(T_A T_E T'_{do} T_R) = 416.7K_A$.

For K_A slightly greater than 11, the closed-loop poles move into the right half-plane. We would like to be able to maintain stability with a much larger gain; gains of 400 are common in practice.

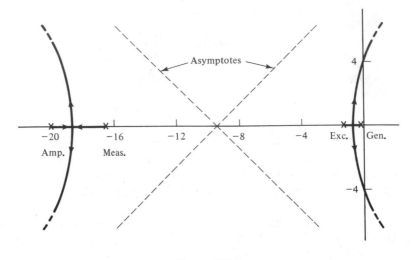

Figure E11.2

The problem of raising the gain while maintaining stability is a classic one in feedback control theory and an effective solution is to use "rate" feedback sta-bilization, as shown by the inner loop in Fig. 11.3. Note that E_{fd}, which is proportional to v_F, is a readily available signal, as may be seen in Figs. 11.1 and 11.2. Note that the stabilizer does not affect the steady-state (dc) behavior, because of the s in the numerator of the stabilizer transfer function.

11.4 VOLTAGE REGULATION

Next, assuming stability, we would like to show the effect of feedback in maintaining voltage under varying conditions of load. Suppose that we compare the volt-ampere ("regulation") curves, or V-I curves, at the terminals of an isolated generator with and without a voltage control system.

We can consider an example.

Example 11.3   Steady-State V-I Curves

Find steady-state V-I curves, for:

(a) A generator without a voltage control system.
(b) A generator with a voltage control system.

Assume that:

1. The field current is adjusted so that in both cases, (a) and (b), the open-circuit voltage $|V_a| = 1$.
2. $Z = R$, where R is varied from ∞ to 0.
3. $r = 0$, $X_q = X_d$.
4. Nonlinear limiting and effects due to S_E can be ignored.

Note: In both cases $|V_a| = 1$ when the generator is open circuited (i.e., $R = \infty$) and $|V_a| = 0$ when the generator is short circuited (i.e., $R = 0$).

Solution   In both cases the generator model is

$$G_g(s) = \frac{k_v \sigma}{1 + s\sigma T'_{do}}$$

Since only steady-state behavior is being considered, $|V_a|$ and E_{fd} are constant values related by

$$|V_a| = G_g(0)E_{fd} = k_v \sigma E_{fd}$$

From the definitions of k_v and σ and the assumption $X_d = X_q$,

$$k_v \sigma = \frac{R}{\sqrt{R^2 + X_q^2}}$$

Thus

$$|V_a| = \frac{R}{\sqrt{R^2 + X_q^2}} E_{fd}$$

Alternatively, we may derive this same result from the steady-state (round-rotor) synchronous machine model of Chapter 8.

(a) In this case we are assuming that E_{fd} remains fixed at the value giving $|V_a| = 1$ when $R = \infty$. This implies $E_{fd} = 1$. Then as R is varied from ∞ to 0,

$$|V_a| = \frac{R}{\sqrt{R^2 + X_q^2}}$$

Also,

$$|I_a| = \frac{|V_a|}{R} = \frac{1}{\sqrt{R^2 + X_q^2}}$$

Thus both $|V_a|$ and $|I_a|$ may be described in terms of the parameter R. We can relate $|V_a|$ to $|I_a|$ more directly by noting that $|V_a|^2 + X_q^2|I_a|^2 = 1$. Thus the *V-I* curves lie on an ellipse in the $|V_a|$, $|I_a|$ plane, as suggested by Fig. E11.3(a).

(b) Consider the case with feedback. In the steady state, and neglecting nonlinear effects, Fig. 11.3 reduces to the block diagram shown in Fig. E11.3(b). Physically, it is clear from the nature of negative feedback that as $|V_a|$ drops, E_{fd} will rise, countering the tendency of $|V_a|$ to drop. More precisely, using the relationships implied by the block diagram,

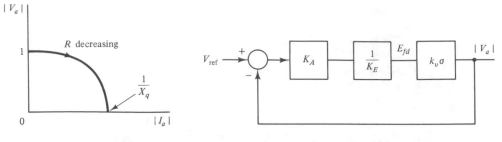

Figure E11.3(a)　　　　　　　　　　　　　　Figure E11.3(b)

$$|V_a| = \frac{K_A k_v \sigma}{K_E + K_A k_v \sigma} V_{\text{ref}} = \frac{K_A R}{\sqrt{R^2 + X_q^2}\, K_E + K_A R} V_{\text{ref}}$$

Since we assume that $|V_a| = 1$ when $R = \infty$, we must choose $V_{\text{ref}} = (K_E + K_A)/K_A$. Thus under more general conditions of loading,

$$|V_a| = \frac{(K_E + K_A)R}{\sqrt{R^2 + X_q^2}\, K_E + K_A R}$$

We do not attempt to get a closed-form solution for $|V_a|$ versus $|I_a|$ as in case (a). However, we note that

$$R \to \infty \Rightarrow |V_a| \to 1, \qquad |I_a| \to 0$$

$$R \to 0 \Rightarrow |V_a| \to 0, \qquad |I_a| \to \frac{K_E + K_A}{X_q K_E}$$

Since the $|I_a|$-axis intercept is now increased by the factor $(K_E + K_A)/K_E$ as compared with case (a), we can assess the improvement. A numerical result will be interesting. Suppose that $K_E = 1$, $K_A = 25$, and $X_q = 0.6$; then we can draw the V-I curves for case (b) and compare with case (a) [Fig. E11.3(c)]. The case (a) curve and part

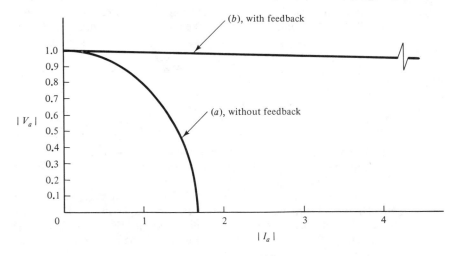

Figure E11.3(c)

of the case (b) curve are shown below. The $|I_a|$ intercept for case (b) is at $|I_a| = 43.3$. Thus even with a moderate loop gain, we get a tremendous improvement in voltage regulation.

Exercise 2.   Show that the (feedback) *V-I* curves in Example 11.3 are approximately a straight line for small R, with slope $X_q K_E / K_A$.

11.5 GENERATOR CONNECTED TO INFINITE BUS

We now return to the interesting and important case of a generator connected through a transmission line to an infinite bus and consider the effects of adding a voltage control system. The system we are considering is shown in Fig. 11.5.

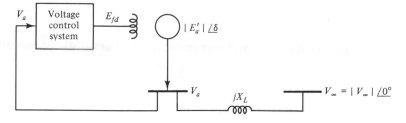

Figure 11.5 System under consideration.

In Section 10.12 we considered a linearized generator model for this case. It led to the block diagram in Fig. 10.14. The input was ΔE_{fd}, which may now be obtained from the excitation system block diagram in Fig. 11.3. Considering a linearized version of Fig. 11.3, we have $\Delta |V_a|$ as an input and ΔE_{fd} as an output of the excitation system. To close the loop through the generator, we need $\Delta |V_a|$ as an output from the generator model. In looking at Fig. 10.14, we do not see the variable $\Delta |V_a|$. In fact, as we will show,

$$\Delta |V_a| = K_5 \Delta\delta + K_6 \Delta |E_a'| \tag{11.21}$$

where $\Delta\delta$ and $\Delta |E_a'|$ may be seen to be available as "outputs" of the generator model and K_5 and K_6 are parameters depending on the system reactances and operating condition. Thus we are able to add the voltage regulator loop to Fig. 10.14.

We start by showing how (11.21) applies to the case under consideration, that of a lossless generator connected to an infinite bus through a lossless transmission line.

Following a familiar technique, we absorb the line series reactance into the generator model and consider $V_\infty = |V_\infty| \underline{/0^\circ}$ as the terminal voltage. Using (10.73) as it applies in the present context, we get

$$E_a' = V_\infty + j\tilde{X}_d' jI_d e^{j\delta} + j\tilde{X}_q I_q e^{j\delta} \tag{10.73}$$

After multiplication by $e^{-j\delta}$, we get

$$|E_a'| = |V_\infty| \cos\delta - j|V_\infty| \sin\delta - \tilde{X}_d' I_d + j\tilde{X}_q I_q \tag{11.22}$$

from which, by equating real and imaginary parts, we get

$$I_d = \frac{|V_\infty| \cos \delta - |E_a'|}{\tilde{X}_d'} \tag{11.23}$$

$$I_q = \frac{|V_\infty| \sin \delta}{\tilde{X}_q} \tag{11.24}$$

Now we are in a position to calculate $|V_a|$. As we note from Fig. 11.5,

$$V_a = V_\infty + jX_L I_a \tag{11.25}$$

Introducing d and q variables, we get $V_a = (V_q + jV_d)e^{j\delta}$ and $I_a = (I_q + jI_d)e^{j\delta}$. Substituting in (11.25), multiplying by $e^{j\delta}$, and noting that $V_\infty = |V_\infty| \underline{/0^\circ}$, we get

$$V_q + jV_d = |V_\infty| \cos \delta - j|V_\infty| \sin \delta + jX_L(I_q + jI_d) \tag{11.26}$$

Now using (11.23) and (11.24) in (11.26) and separating the real and imaginary parts, we can find V_q and V_d:

$$V_q = |V_\infty| \cos \delta - X_L I_d = \frac{X_L}{\tilde{X}_d'}|E_a'| + \frac{X_d'}{\tilde{X}_d'}|V_\infty| \cos \delta \tag{11.27}$$

$$V_d = -\frac{X_q}{\tilde{X}_q}|V_\infty| \sin \delta \tag{11.28}$$

Next we find $|V_a|^2$.

$$|V_a|^2 = V_a V_a^* = (V_q + jV_d)e^{j\delta}(V_q - jV_d)e^{-j\delta}$$

Thus

$$|V_a| = (V_q^2 + V_d^2)^{1/2} \tag{11.29}$$

We note that (11.29) together with (11.27) and (11.28) gives us the dependence of $|V_a|$ on $|E_a'|$ and δ in terms of the system parameters X_d', X_q, and X_L. The dependence is seen to be nonlinear, and we next consider a linearization. We assume steady-state operation with $|E_a'| = |E_a'|^0$, $\delta = \delta^0$, $|V_a| = |V_a|^0$ and then allow small perturbations of these quantities. Relating the perturbations, we get the linear equation

$$\Delta|V_a| = \frac{\partial|V_a|^0}{\partial \delta} \Delta\delta + \frac{\partial|V_a|^0}{\partial|E_a'|} \Delta|E_a'| \tag{11.30}$$

The superscript 0 in (11.30) refers to the evaluation of the partial derivatives at the steady-state operating condition. Comparing (11.30) an (11.21), we see that

$$K_5 \triangleq \frac{\partial|V_a|^0}{\partial \delta} \qquad K_6 \triangleq \frac{\partial|V_a|^0}{\partial|E_a'|}$$

We next calculate K_5. Using (11.29) and the chain rule of differentiation,

$$\frac{\partial|V_a|}{\partial \delta} = \frac{1}{2}(V_q^2 + V_d^2)^{-1/2} 2\left(V_q \frac{\partial V_q}{\partial \delta} + V_d \frac{\partial V_d}{\partial \delta}\right) \tag{11.31}$$

Evaluating by use of (11.27) and (11.28),

$$K_5 = \frac{\partial |V_a|^0}{\partial \delta} = -|V_\infty| \left(\frac{X_d' V_q^0}{\tilde{X}_d' |V_a|^0} \sin \delta^0 + \frac{X_q V_d^0}{\tilde{X}_q |V_a|^0} \cos \delta^0 \right) \qquad (11.32)$$

In calculating K_5, we must first find the steady-state operating condition (i.e., we must find δ^0, $|E_a'|^0$). Then we can find V_q^0 and V_d^0 using (11.27) and (11.28) and $|V_a|^0$ from (11.29). Then we use (11.32). It is important to note that K_5 can be positive or negative depending on the steady-state operating condition. This possibility may be inferred as follows. From (11.27) it may be seen that V_q^0 is positive under normal operating conditions. From (11.28), for a generator, V_d^0 is negative. Thus in (11.32), for a generator, the two terms on the right side of the equation are opposite in sign. Typical behavior shows K_5 positive for light loads (small P_G) and negative for heavy loads (large P_G). The fact that K_5 may vary in sign leads to some stability problems to be discussed in the next section.

We next calculate K_6. Using (11.29), (11.27), and (11.28) again, we get

$$\frac{\partial V_a}{\partial |E_a'|} = \tfrac{1}{2}(V_q^2 + V_d^2)^{-1/2} \, 2 \left(V_q \frac{\partial V_q}{\partial |E_a'|} + V_d \frac{\partial V_d}{\partial |E_a'|} \right)$$

$$K_6 = \frac{\partial |V_a|^0}{\partial |E_a'|} = \frac{X_L}{\tilde{X}_d'} \frac{V_q^0}{|V_a|^0} |E_a'|^0 \qquad (11.33)$$

Under normal operating conditions this constant is positive.

Using Fig. 11.3, we now complete the block diagram shown in Fig. 10.14 to show the effect of adding a voltage control system. This leads to the block diagram shown in Fig. 11.6. In Fig. 11.6, $G_e(s)$ is the transfer function corresponding to the (linearized) "feedforward" part of the excitation system shown within the dashed outline in Fig. 11.3.

The new incremental system inputs are ΔV_{ref} and ΔP_M. We may wish to investigate steady-state and/or transient behavior in response to various inputs. As an

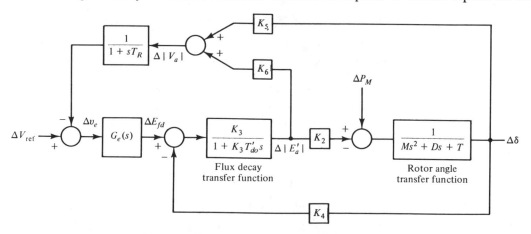

Figure 11.6 Voltage control system added.

example we wish to use the model to show what happens to the "natural" damping (discussed in Section 10.12) when a high-gain fast-acting voltage control system is used. In the analysis in Section 10.12 we treated E_{fd} as an (independent) input variable. Now with the voltage control loop added, ΔE_{fd} is a function of $\Delta|E_a'|$ and $\Delta\delta$. What happens to the damping in this case?

To simplify matters, assume that the voltage control system is very fast compared with the dynamics of the rest of the system. Thus we assume, in Fig. 11.6, that any change $\Delta|V_a|$ causes an essentially instantaneous change $\Delta|v_e|$ and any change Δv_e causes an essentially instantaneous change in ΔE_{fd}. The model implications of these assumptions are the following. In the measurement block we set the time constant $T_R = 0$; equivalently, we replace the measurement block by its dc gain, 1. Similarly, we replace $G_e(s)$ by its dc gain, $K_e \triangleq G_e(0)$. We note, from Fig. 11.3, that K_e is positive. Thus if Δv_e is a (positive) unit step function, ΔE_{fd} is a positive step function of amplitude K_e. Making these changes in Fig. 11.6, we get Fig. 11.7. We note that the approximation is reasonable for a fast-acting solid-state (thyristor) type of exciter.

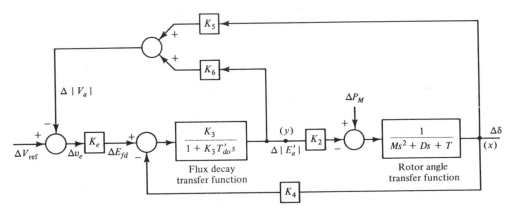

Figure 11.7 Simplification with fast-acting exciter.

To investigate the effect of the fast-acting exciter on the damping of the rotor angle poles, we will use the root-locus method. To do this we need to reduce Fig. 11.7 to a single loop. Suppose that we redraw Fig. 11.7 appropriate for finding the closed-loop transfer function from ΔP_M to $\Delta\delta$. We save that (feedforward) portion of Fig. 11.7 between the points labeled (y) and (x). With $\Delta V_{\text{ref}} = 0$, we can replace the rest (all the feedback paths) by a single transfer function $G_E(s)$ from (x) to (y).

We now calculate $G_E(s) \triangleq |\hat{E}_a'|/\Delta\hat{\delta}$. From Fig. 11.7,

$$\Delta|\hat{E}_a'| = \frac{K_3}{1 + K_3 T_{do}'}[-K_4\Delta\hat{\delta} - K_e(K_5\,\Delta\hat{\delta} + K_6\,\Delta|\hat{E}_a'|)] \qquad (11.34)$$

Collecting terms in $\Delta|\hat{E}_a'|$ on the left, we get

$$\left(1 + \frac{K_3 K_e K_6}{1 + K_3 T_{do}'s}\right)\Delta|\hat{E}_a'| = \frac{-K_3(K_4 + K_e K_5)}{1 + K_3 T_{do}'s}\Delta\hat{\delta} \qquad (11.35)$$

Multiplying both sides of the equation by $1 + K_3 T'_{do} s$. we then find that

$$G_E(s) = \frac{\Delta |\hat{E}'_a|}{\Delta \hat{\delta}} = \frac{-K_3(K_4 + K_e K_5)}{1 + K_3 K_e K_6 + K_3 T'_{do} s} \tag{11.36}$$

With this simplification we get the single-loop feedback system shown in Fig. 11.8. The summing point at the left is included to facilitate a comparison with Fig. 10.14. It is *not* the ΔV_{ref} input.

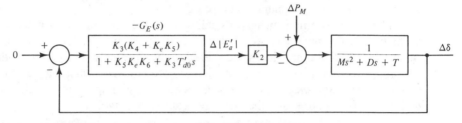

Figure 11.8   Single-loop system.

If we now compare Fig. 11.8 with Fig. 10.14, we observe that except for different constants, we have the same kind of system. There are still three open-loop poles. The open-loop rotor angle poles are the same; since K_3, K_e, and K_6 are all positive, the real-axis pole associated with T'_{do} is farther in the left half-plane. Since K_4 is also positive, then if K_5 is positive, the overall situation is not qualitatively different from the case considered in Section 10.12. Thus we can expect to preserve the natural damping.

But consider the common case when K_5 is negative. This is the troublesome case and occurs when the generator is supplying a large active power over a long transmission line. In this case $K_4 + K_e K_5$ is frequently negative and in effect the feedback changes from positive to negative! Thus we get the root locus shown in Fig. 11.9; we can think of K_2 as the gain parameter which varies from zero to infinity.

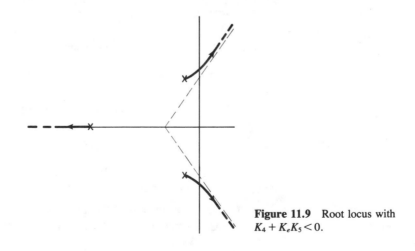

Figure 11.9   Root locus with $K_4 + K_e K_5 < 0$.

Since the rotor angle poles are so close to the $j\omega$ axis, we can expect the system to be unstable with typical loop gains. Thus we have not only lost the natural (positive) damping, but the system may now even be unstable! The situation is changed in detail but not in its general features if a more detailed voltage control system model is considered.

In solving the problem we would like to retain the fast-acting exciter. The importance of maintaining voltage during faults to improve transient stability was discussed in Chapter 9. Nor is reversing the feedback the solution (simply done with an inverter); $K_4 + K_e K_5$ can also be positive!

The most common means of dealing with the problem is to make the system insensitive to the parameter K_5 with an auxiliary loop. The loop provides a signal usually derived from machine speed (i.e., $\Delta\hat{\delta}$), but it can also be derived from terminal voltage, power, and so on, or a combination thereof. The "shaping" of this signal is done in a *power system stabilizer* (PSS). As a simple example consider the case shown in Fig. 11.10. Assume that the stabilizer gain is positive.

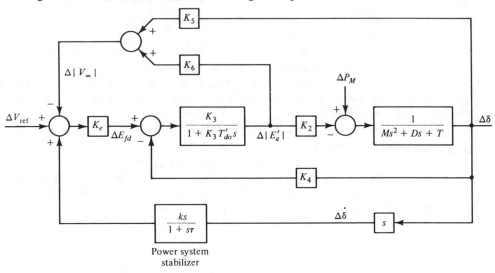

Figure 11.10 Power system stabilizer added.

We again calculate the gain $G_E(s) = \Delta|\hat{E}'_a|/\Delta\hat{\delta}$.

$$\Delta|\hat{E}'_a| = \frac{K_3}{1 + K_3 T'_{do}s}\left[-K_4\,\Delta\hat{\delta} - K_e\left(K_5\Delta\hat{\delta} + K_6\,\Delta|\hat{E}'_a| - \frac{ks^2}{1 + s\tau}\,\Delta\hat{\delta}\right)\right] \quad (11.37)$$

Collecting terms in $\Delta|E'_a|$ on the left, after some algebra we get

$$G_E(s) = \frac{\Delta|\hat{E}'_a|}{\Delta\hat{\delta}} = K_3\,\frac{kK_es^2 - \tau(K_4 + K_eK_5)s - (K_4 + K_eK_5)}{(1 + K_3K_eK_6 + K_3T'_{do}s)(1 + s\tau)} \quad (11.38)$$

Compared with (11.36) we have an additional pole (at $s = -1/\tau$) and two zeros. The location of the zeros depends on K_5. If $K_4 + K_eK_5$ is positive, we get real-axis zeros. If $K_4 + K_eK_5$ is negative, we can have either real or complex zeros. Typically, the zeros are quite close to the origin in either case.

Thus the effect of a negative $K_4 + K_eK_5$, which previously reversed the sign of the feedback, now only shifts the location of a pair of zeros in an unimportant way. Note also that the sign of $G_E(s)$ in pole-zero form is now positive; thus we plot a negative feedback root locus. The results of such a plot are shown in Fig. 11.11 for values of $M = 3/377$, $D = 0$, $T = 1.01$, $K_3 = 0.36$, $K_4 = 1.47$, $K_5 = -0.097$, $K_6 = 0.417$, $K_e = 25$, $k = 10/377$, and $\tau = 0.05$. The root-locus gain is $K = 276.5K_2$. As the gain K is increased, the rotor angle poles move farther into the left half-plane. Thus the damping is restored. We show the closed-loop pole location by a square symbol for a value $K_2 = 1.15$. The system is very well damped.

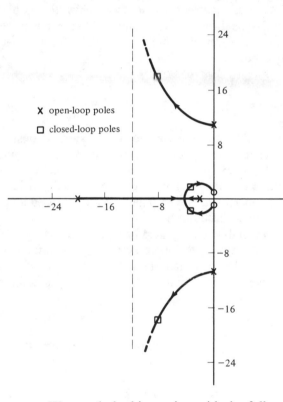

Figure 11.11 Root locus with power system stabilizer added.

We conclude this section with the following observations:

1. There are better power system stabilizers than the one shown in Fig. 11.10. The basic idea, however, is nicely illustrated with this very simple type.
2. Earlier, in Figs. 11.1 to 11.3, we showed *other inputs* as inputs to the excitation system. Now we can give the PSS output as an example of one of these other inputs.
3. We have been considering the stabilization of a single generator. Suppose that

we have many interconnected generators, each with its own voltage control system. In this case, in the PSS realization, we can decide to use only signals available locally at each generator (decentralized control), or we can consider the use of shared system measurements (centralized control). If at the ith generator we only use $\dot{\delta}_i$, this is a case of decentralized control; if we use additional $\dot{\delta}_j$s derived from other generators, this is a case of centralized control.

4. The problem addressed in this section belongs to the larger class of *dynamic stability* problems. One manifestation of this kind of instability is the spontaneous buildup of oscillations under certain operating conditions. There may be a variety of causes (not just the voltage control system) and solutions (not just a PSS). The identification of the source of the instability can be a very challenging assignment. Once identified, finding a *robust* solution which will work under all operating conditions may also be difficult. Fortunately, unlike in the case of the transient stability problem, the powerful techniques of linear (feedback) system theory can be effectively used.

11.6 SUMMARY

Figure 11.3 shows a simple block diagram of a voltage control system consisting of a generator and an excitation system. The input to the generator is E_{fd}. The output of the generator, $|V_a|$, is fed back to the exciter, which generates E_{fd}. The model facilitates the study of steady-state and transient behavior. For an isolated generator, the voltage regulation is improved by increasing the loop gain. With increased loop gain, however, care must be taken to avoid instability.

Stability must also be considered when a generator with a fast-acting exciter is connected to a larger system. Modeling the larger system by an infinite bus, a root-locus analysis shows the possibility of instability under certain operating conditions. Stability is restored, for general conditions of operation, by use of a power system stabilizer.

PROBLEMS

11.1. Consider an isolated generator with an inductive load $Z = R + jX$. Show that $G_g(0)$, which is the dc gain of the transfer function from E_{fd} to $|V_a|$, is always less than 1.

11.2. Refer to Fig. P11.2 and assume that

$$Z = 0.05 - j0.8$$

$$r = 0.05$$

$$X_d = 1.0, X'_d = 0.2, X_q = 0.6$$

$$T'_{do} = 4.0 \text{ sec}$$

Initially, $|V_a| = 0$. At $t = 0$, $E_{fd} = 1$ is applied. Find $|V_a|$ as a function of time.

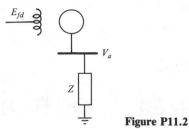

Figure P11.2

11.3 For an isolated generator with a local (impedance) load, $Z = 1/j\omega C = -j0.5$, $X_d = X_q = 1$, $X'_d = 0.2$, $T'_{do} = 1$, and $E_{fd} = \epsilon$, $\epsilon > 0$ arbitrarily small. The system is initially at rest with $|V_a(0)| = 0$. Find an expression for $|V_a|(t)$ for $t \geq 0$.

11.4. Assume the simplified linearized voltage control system model shown in Fig. P11.4 and an isolated generator supplying a local load. We are given the values

$$T'_{do} = 4 \text{ sec}$$
$$X_d = 1$$
$$X_q = 0.6$$
$$X'_d = 0.2$$
$$r = 0$$
$$Z = -j0.3$$

Is there a positive K for which the feedback system is stable? If so, find such a K.

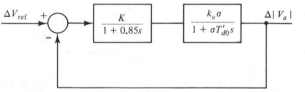

Figure P11.4

11.5. Consider the system in Problem 11.4 with the following change: $Z = R$. Assume that $K > 0$ is large enough for a damped oscillatory response. Do we get more damping with R large or R small? *Hint:* Look at the real part of the (complex) closed-loop poles.

11.6. For a generator with a local (impedance) load, plot the steady-state voltage regulation curves ($|V_a|$ versus $|I_a|$) for **(a)** no feedback and **(b)** feedback. Assume that:

1. $Z = R, 0 \leq R \leq \infty$
2. $X_d = X_q = 1$
3. $K_A/K_E = 3$
4. Open circuit $|V_A| = 1.0$

Note: In (a) pick E_{fd} so that assumption 4 is true; in (b) pick V_{ref} so that assumption 4 is true.

11.7. Repeat Problem 11.6 except assume that $Z = jX, 0 \leq X \leq \infty$.

Power Control System

chapter 12 _____

12.0 INTRODUCTION

The power delivered by a generator is controlled by controlling the mechanical power of a "prime mover" such as a steam turbine, or water turbine, or gas turbine, or diesel engine. In the case of a steam or water turbine the mechanical power is controlled by opening or closing valves regulating steam or water flow.

Since the load of a power system is always changing, at least one generator in a power system must respond to the changing load in order to maintain the power balance. If the load increases, the generated power must increase. Thus steam and/or water valves must open wider. If the load decreases, generation must also decrease, and this requires valve openings to be made smaller. The way we sense the power imbalance is through its effect on generator speeds and/or frequency. Thus if there is excess generation, the generator sets will tend to speed up and the frequency will rise. If there is a deficiency of generation, the generator speeds and frequency will drop. These deviations from nominal speed and/or frequency are used as control signals to cause appropriate valve action automatically.

In addition to its control function, in maintaining the power balance, we note that the maintenance of system frequency, close to the nominal value, is also important in its own right. The speed and output of fans and pumps depend on frequency. The speed and accuracy of (synchronous) electric clocks depend on frequency. In addition, there is the danger that steam turbine blades may resonate at frequencies off of synchronism. Sustained operation of these frequencies could lead to metal fatigue and failure of the unit. To protect the turbine, there are underfrequency relays which trip the unit on low frequency, and there are also relays which trip the unit on overspeed.

12.1 POWER CONTROL SYSTEM MODELING

We consider next a device called a *speed governor* which senses a speed deviation, or a power change command, and converts it into appropriate valve action. We will model its dynamics, and later, join it with turbine and generator dynamics to describe the system that controls power output and frequency.

    We will describe the basic operation of the classic Watt centrifugal governor. Since the governor itself does not generate enough force to operate the steam or water valves, we also consider the operation of a single-stage hydraulic servomotor interposed between the governor and the valve. A schematic is shown in Fig. 12.1.

    We are looking at a vertical section. The horizontal rotating shaft on the lower left may be viewed as an extension of the turbine-generator shaft and has a fixed axis.

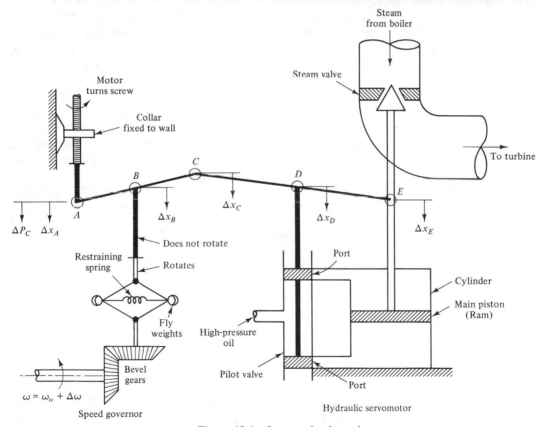

Figure 12.1 Servo-assisted speed governor.

The vertical shaft above the fly weight mechanism also rotates between fixed bearings. Although its axis is fixed, it can move up and down, transferring its vertical motion to the pivot point labeled B. Similarly, the motor-driven screw is free to rotate on a vertical axis and transfers its vertical motion to the pivot point A. The pivot points A, B, and C are joined by a rigid bar or linkage. Similarly, C, D, and E are joined by a rigid bar. We assume that the various points of connection between linkages and vertical attaching members are slotted or otherwise arranged so that free vertical motion of the pivot points is possible.

The operation is as follows. If the generator speed increases above synchronous speed, the fly weights will be flung farther apart (by centrifugal force), lowering point B. Assume that point A is fixed so that point C goes down. Initially, point E is fixed, so point D goes down. This uncovers the lower and upper ports of the "pilot" valve, admitting oil under high pressure under the main piston and at the same time providing for release of the oil above the piston. Thus the piston moves up, tending to close the steam valve. This action, by reducing the mechanical power, will tend to slow down the generator. However, mainly because of the high moment of inertia of the turbine-generator set, this process takes a fair amount of time, and there is the danger of overreacting (i.e., of closing the steam valve too far). This then could cause the speed in time to drop far below synchronous speed. A continuation of the "corrective" process could therefore lead to ever larger swings in speed (i.e., to instability). To prevent this, as point E moves up, it raises point D, tending to close the pivot valve ports. This "anticipates" the correct, but delayed, action to come.

We can similarly discuss the action of the "command" input P_C. This input permits the operator to move the steam valve by using the servomotor. For example, to shift generation from one unit to another, the operator lowers point A on one unit (by rotating the "speeder" motor in one direction) and raises it on the other. When point A goes down, point C goes up, since initially ω does not change and hence point B remains fixed. Then point D goes up, allowing oil to enter the top of the cylinder and escape from the bottom. Thus the steam valve opens wider and the generator output goes up. By rotating the speeder motor of the other unit in the opposite direction, point A of that unit goes up and the generator output goes down.

We note that in the steady state, with ω constant, and all the pivot points stationary, the pilot valve ports must be covered (i.e., as shown in Fig. 12.1). If this were not the case, the main piston would be in motion. We note that with the pilot valve ports closed, the steam valve is locked in position and able to withstand the considerable forces developed by the high-pressure steam.

We next wish to develop a linearized model of the speed governor operating near a specified operating point. Suppose at this specified operating point that all the linkages are in the positions shown in Fig. 12.1. We next consider the relationships between the increments ΔP_C. $\Delta \omega$, Δx_A, . . . , Δx_E; these increments are shown in the figure. Since points A, B, and C are on the same straight line, the position of point C is determined by points A and B. Thus for small deviations there are positive constants (depending on the operating point) such that the linearized relationships are

$$\Delta x_C = k_B \Delta x_B - k_A \Delta x_A$$

$$= k_1 \Delta \omega - k_2 \Delta P_C \tag{12.1}$$

The geometric factor k_B is absorbed into k_1 together with the sensitivity of the fly weight mechanism (i.e., $\partial x_B / \partial \omega$). The geometric factor k_A is absorbed in the constant k_2 together with the scale factor $\partial x_A / \partial P_C$.

Since points C, D, and E lie on the same straight line, the position of point D depends on that of points C and E. Linearizing the geometry, we get a relationship in the form

$$\Delta x_D = k_3 \, \Delta x_C + k_4 \, \Delta x_E \tag{12.2}$$

where k_3 and k_4 are positive.

Next, we turn to the servomotor dynamics, i.e., to the relation between input Δx_D and output Δx_E. It is simplest to assume that the linkage between D and E has been disconnected, say, at point D. Suppose that we depress D a small amount. Then with the ports (partially) opened oil flows at a certain rate and the main piston rises at a certain rate. For simplicity assume that the oil flow *rates* through the ports of the pilot valve are proportional to Δx_D. Then Δx_E depends on Δx_D as follows:

$$\frac{d \, \Delta x_E}{dt} = -k_5 \, \Delta x_D \tag{12.3}$$

where the (positive) constant k_5 depends on the oil pressure and the geometry of the servomotor.

Assume that the transient starts at $t = 0$ with the system at the operating point. In that case $\Delta x_E(0^-) = 0$. Taking Laplace transforms, we get

$$\Delta \hat{x}_E = -\frac{k_5}{s} \, \Delta \hat{x}_D \tag{12.4}$$

where we are using the circumflex to indicate the Laplace transform. Equations (12.1), (12.2), and (12.4) may now be combined. The reader may check that we get the block diagram shown in Fig. 12.2.

We can simplify the description. Using the relationships specified by the block diagram, we obtain

$$\Delta \hat{x}_E = k_3 \frac{-k_5}{s + k_4 k_5} (-k_2 \Delta \hat{P}_C + k_1 \Delta \hat{\omega})$$

$$= \frac{k_2 k_3 k_5}{s + k_4 k_5} \left(\Delta \hat{P}_C - \frac{k_1}{k_2} \Delta \hat{\omega} \right) \tag{12.5}$$

$$= \frac{K_G}{1 + T_G s} \left(\Delta \hat{P}_C - \frac{1}{R} \Delta \hat{\omega} \right)$$

where $K_G = k_2 k_3 / k_4$, $T_G = 1 / k_4 k_5$, and $R = k_2 / k_1$. The transfer function in (12.5) is

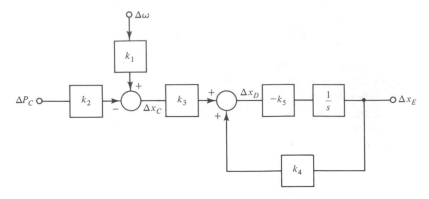

Figure 12.2   Speed governor block diagram.

in time-constant form with K_G the gain; T_G the time constant, typically around 0.1 sec; and R the *regulation constant* or setting. The reason for the terminology "regulation constant" will be seen later in the application.

Example 12.1

In (12.5) assume that at $t = 0$ an increment $\Delta P_C = 1.0$ is applied. Assume that the (servo-assisted) speed governor is on test, operating open loop, so that as an independent input, we may assume that $\Delta\omega = 0$. Find the increment in steam valve opening $\Delta x_E(t)$ and sketch the response.

Solution   Using the Laplace transform, we have

$$\Delta \hat{x}_E = \frac{K_G}{s(1 + T_G s)} = \frac{K_G/T_G}{s(s + 1/T_G)}$$

Using the partial fraction expansion, we get, by evaluating the residues,

$$\Delta \hat{x}_E = \frac{K_G}{s} - \frac{K_G}{s + 1/T_G}$$

Taking the inverse Laplace transform,

$$\Delta x_E(t) = K_G(1 - e^{-t/T_G}) \qquad t \geq 0$$

This is sketched as shown in Fig. E12.1. The curve has the familiar features of a single time-constant response with time constant T_G.

We have found the response for a unit step change in P_C. The response scales for other step changes, provided that the changes remain small enough to justify the use of the linear model.

We next propose a highly simplified model for the turbine which relates changes in mechanical power output to changes in the steam valve position. We will not attempt a detailed physical description but rather attempt to justify the model based on intuition. Imagine that we conduct an experiment. Suppose that there is a small step function increase in x_E (i.e., Δx_E is a small positive step function). In the steady state

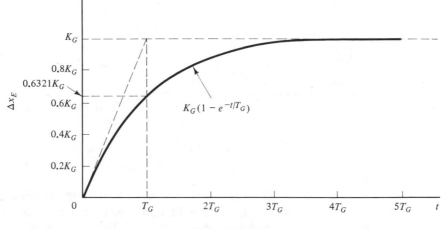

Figure E12.1

ΔP_M will also be (a positive) constant but the change does not occur instantaneously. It takes time before the increased flow of steam penetrates into all the blades of the turbine. The detailed behavior is complicated, but if the turbine is of the nonreheating (or nonextracting) type, ΔP_M increases approximately like the step response of a single time-constant circuit. Thus we may approximate the transfer function using experimental data. In particular, we can plot the step response and read off the gain and time constants for the turbine. In Fig. 12.3 we show such a turbine model in cascade with the speed governor model of (12.5). The time constant T_T is in the range 0.2 to 2.0 sec.

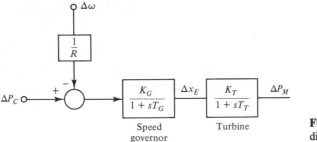

Figure 12.3 Turbine-generator block diagram.

Suppose that the command increment is ΔP_C. Then in the steady state we get $\Delta P_M = K_G K_T \, \Delta P_C$. It makes sense to pick the scale factor $\partial x_A / \partial P_C$ so that $\Delta P_M = \Delta P_C$. This is equivalent to picking $K_G K_T = 1$. Then (suppressing Δx_E) we get the simpler description shown in Fig. 12.4.

This completes the derivation of an incremental or small-signal model for the turbine and (servo-assisted) governor. Frequently, the model is adapted for large-signal use by adding a "saturation"-type nonlinear element which introduces the obvious fact that the steam valve must operate between certain limits. The valve cannot be more open than fully open, nor more closed than fully closed. Other than

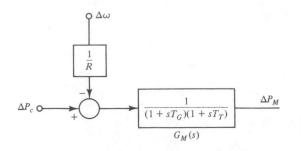

Figure 12.4 Turbine-governor block diagram.

this, the linearization derived in this section may be used as an approximation to relate the (large-signal) dependence of P_M on P_C; we still make the reasonable assumption that $\Delta \omega$ is small.

The model in Fig. 12.4 may also be modified to account for reheat cycles in the turbine and more accurate representation of fluid dynamics in the steam inlet pipes or, in the case of hydraulic turbines, in the penstock. For our purposes, however, the simple model in Fig. 12.4 will suffice.

We conclude this section by sketching some "static" speed-power curves for the turbine-governor. These curves relate P_M and ω with P_C as a parameter. The curves also relate the quantities if they are slowly varying. From the block diagram, Fig. 12.4, we get the static, algebraic relation

$$\Delta P_M = \Delta P_C - \frac{1}{R} \Delta \omega \tag{12.6}$$

from which the local shape of the speed-power curves may be inferred. Assuming that the local behavior can be extrapolated to a larger domain, we sketch two static speed-power curves in Fig. 12.5. Suppose that we adjust P_M by using the speeder motor in Fig. 12.1 until $P_M = P_C^{(1)}$, the desired command power, at synchronous speed ω_0. Thereafter, with fixed $P_C^{(1)}$ (i.e., $\Delta P_C = 0$) (12.6) predicts the straight-line relationship shown, with $-R$ as slope. Suppose now that we wish to get more power, $P_C^{(2)}$, at synchronous speed. We must then make the adjustment with the speeder motor. This then puts us on the speed-power curve to the right.

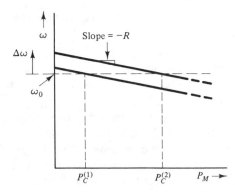

Figure 12.5 Static speed-power curves.

We note that the term "regulation" as applied in the present context refers to the variation of ω with P_M. The less the "droop," the better the regulation. The choice of the symbol R, for regulation, may then be understood.

We note that the units of R depend on those chosen for P_M and ω. We often will use P_M in per unit and ω in radians per second. Then the units of R are radians per second/p.u. power. Frequently, the units of R are stated in Hz/MW, or, as in the next example, they may be stated in per unit.

Example 12.2

A standard figure for R is 0.05 p.u. (or 5%). This relates fractional changes in ω to fractional (i.e., per unit) changes in P_M. Thus we have $\Delta\omega/\omega_0 = -0.05\,\Delta P_M$, where ΔP_M is in per unit.

(a) Suppose that the frequency changes from 60 Hz to 59 Hz. Find the increase in P_M.

(b) What change in frequency would cause P_M to change from 0 to 1 (i.e., no load to full load)?

Solution   (a) We have $\Delta\omega/\omega_0 = \Delta f/f_0 = -1/60 = -0.05\Delta P_M$. Thus $\Delta P_M = 0.333$ p.u. For example, if the turbine-generator unit is a 100-MW unit, the power output would be increased by 33.3 MW.

(b) A 5% drop in frequency, from 60 Hz to 57 Hz, changes the output from 0 to 1.

Note: If all the turbines on our system have the same R in p.u., based on their own ratings, they operate nicely in parallel in the sense that as the frequency drops, each generating unit picks up a share of the total power increase proportional to its own rating.

12.2 APPLICATION TO SINGLE MACHINE–INFINITE BUS SYSTEM

As a first application of the model, we can add the power or frequency control system to "close the loop" in Fig. 11.6. In that figure we showed ΔP_M as an independent input, but in fact, with a speed governor, ΔP_M depends on $\Delta\omega$ and ΔP_C. Since ω is the angular velocity of the turbine-generator shaft, we have

$$\omega = \dot{\theta} = \omega_0 + \dot{\delta} \Rightarrow \Delta\omega = \Delta\dot{\delta} \qquad (12.7)$$

$\Delta\delta$ is available as an output in Fig. 11.6 and we can now add the turbine-generator loop. At the same time we include a PSS loop as in Fig. 11.10. This is done in Fig. 12.6.

Figure 12.6 clearly shows the mathematic relationships between incremental system variables and inputs. It is important also to keep in mind the physical arrangement of the control elements. This is shown schematically in Fig. 12.7.

It may be seen that the "primary" inputs to the generating unit are P_C and V_{ref}. We note that, in turn, P_C and V_{ref} are supplied from a higher level of control. For example, P_C is calculated to minimize fuel costs. In Chapter 7 we considered the

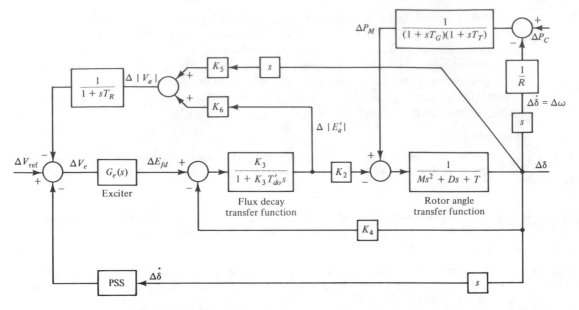

Figure 12.6 Block diagram of generator with voltage and power control loops.

optimal P_{Gi} without specifying the necessary control action. Now we can see how the optimal allocation of P_{Gi} (calculated by computer "in real time") can be physically implemented.

Finally, it should be mentioned that in addition to the power and voltage control system, there is a boiler control system which we have neglected. Clearly, as the generator load increases, it is necessary promptly to increase the rate of steam generation. For short-duration transients, however, the assumption of constant steam pressure (and temperature) is a good one and we can neglect the modeling and dynamics of the boiler and its control system.

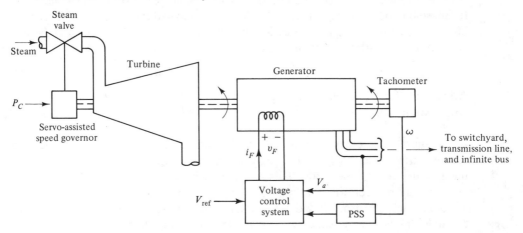

Figure 12.7 Physical arrangement of control elements (schematic).

12.3 SIMPLIFIED ANALYSIS OF POWER CONTROL SYSTEM

Even with simplified modeling of the turbine, governor, and voltage control system, Fig. 12.6 is fairly complicated. And we remind the reader that this is only the single machine–infinite bus case. The multimachine case that we usually need to consider would be even more complicated. Although the complications are not necessarily serious if we intend to rely on computer simulation, they are a definite disadvantage if we want to understand system behavior in general terms.

Fortunately, the system lends itself to simplifications. There are two inputs, ΔV_{ref} and ΔP_C. We can take $\Delta|V_a|$ and/or $\Delta\delta$ to be the outputs. Suppose that we are interested in the step response to ΔV_{ref} acting alone (i.e., with $\Delta P_C = 0$). It is observed that the transient is usually of such a short duration that P_M does not change appreciably during the interesting part of the transient; the power control loop is too slow to affect the transient significantly. In that case, as an approximation, we can set $\Delta P_M = 0$ and ignore the power control loop. We then have the situation considered in Section 11.5, leading to models such as in Fig. 11.10 with $\Delta P_M = 0$.

On the other hand, suppose that we are interested in the step response to ΔP_C acting alone, with $\Delta V_{ref} = 0$. Once again there is the possibility of a modeling simplification. We can leave out the various loops associated with the voltage control system. Thus in Fig. 12.6 we are left with a single loop through the rotor angle transfer function and the power control system. In this simplified model the effect of the voltage control system may be accounted for by modifying the damping constant D and the stiffness constant T in the rotor angle transfer function. The justification of the simplified model is quite involved and will not be undertaken here. With this simplification, redrawing Fig. 12.6, we get Fig. 12.8.

Note that we are led to two very different models starting from the same Fig. 12.6. If we are interested in the response to ΔV_{ref}, we neglect the power control loop. If we are interested in the response to ΔP_C, we suppress the detailed dynamics of the voltage control loop.

The development outlined here, for the single machine–infinite bus case, extends also to more general interconnections. Henceforth we will assume that in modeling the power control system for each generator, the voltage control system need not be considered in detail. With reference to the simplified dynamic model in Section 10.11, we assume that the electrical equation (10.75) is dropped while the mechanical

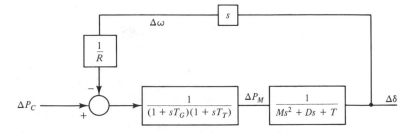

Figure 12.8   Simplified power control system.

equation (10.70) suffices. We will not attempt to justify these modifications. We note that, in general, the rigorous justification of this process of "reduced order modeling" (i.e., the "elimination" of some of the system differential equations) is studied by methods such as the method of "singular perturbations."

Returning now to Fig. 12.8, we note that it is easy to investigate the stability of the power control loop, and we do so in the following example.

Example 12.3

We wish to estimate the effect of the regulation setting R on dynamic behavior. Assume the following parameters in Fig. 12.8: $T_G = 0.1$, $T_T = 0.5$, $M = H/\pi f° = 5/60\pi = 0.0265$, $T = 2.0$, and $D = 0.1$.

Solution   We will use the root-locus method. Starting with the time-constant form and changing into pole-zero form, we get

$$KG(s)H(s) = \frac{1}{R} \frac{s}{(1 + 0.1s)(1 + s)(0.0265s^2 + 0.1s + 2.0)}$$

$$= \frac{377}{R} \frac{s}{(s + 10)(s + 1)(s^2 + 3.77s + 75.5)}$$

$$= \frac{377}{R} \frac{s}{(s + 10)(s + 1)(s - 1.88 + j8.48)(s - 1.88 - j8.48)}$$

The root-locus diagram is shown in Fig. E12.3. For a gain $K = 377/R = 1001$, two root-locus branches cross the $j\omega$ axis; the two other corresponding root-locus locations shown by a square are also indicated. For $K < 1001$ the system is stable. In terms of R, the stability condition is $R > 0.377$.

The units of R are in radians per second/p.u. power. If we consider the open-loop turbine-generator, like that shown in Fig. 12.4, this means that a drop of only 0.377 rad/sec, or only 0.06 Hz, drives the generating unit from no load to full load. This much sensitivity is high destabilizing! The more common figure of $R = 0.05$ p.u. translates to a value of $0.05\omega_0 = 18.85$ rad/sec/p.u. power, a figure large enough to be well within the stability boundary. Thus for the system under consideration with a reasonable value of R, stability does not appear to be a problem.

Exercise 1.   Assume that the system in Fig. 12.8 (i.e., generator connected to an infinite bus) is stable. Suppose that we change P_C. Show that in the steady state the frequency does not change at all.

12.4 POWER CONTROL, MULTIGENERATOR CASE

We wish to apply the power control model derived in Section 12.1 to the case of a number of generating units connected together by a transmission system, with local loads at each generator bus. For example, we can consider the three-bus system in Fig. 12.9. We are interested in active power flows and assume that we can use the active power model discussed in Section 6.9 with all voltages set equal to their nominal values. Not shown, but to be considered later, are the turbine-governor systems

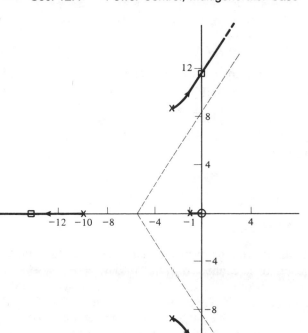

Figure E12.3

similar to Fig. 12.4, which can be used to control the power supplied by each generating unit. The objective of the analysis is to obtain an incremental model relating power commands, electrical and mechanical power outputs, and power angles for the different generators. We note that the analysis may be easily extended to include pure load buses in the model; however, we will be assuming that every bus is a generator bus.

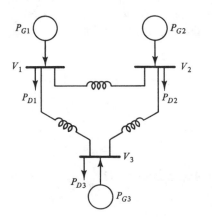

Figure 12.9 Power flow for three generators.

At each bus

$$P_{Gi} = P_{Di} + P_i \tag{12.8}$$

where P_i is the bus power (i.e., the total active power from the ith bus into the transmission system). Assume an operating point

$$P_{Gi}^0 = P_{Di}^0 + P_i^0 \tag{12.9}$$

and then with $P_{Gi} = P_{Gi}^0 + \Delta P_{Gi}$, $P_{Di} = P_{Di}^0 + \Delta P_{Di}$, and $P_i = P_i^0 + \Delta P_i$, we get the relationship between increments:

$$\Delta P_{Gi} = \Delta P_{Di} + \Delta P_i \tag{12.10}$$

For each generator we use the linearized version of (10.79). Thus we consider

$$M_i \, \Delta \ddot{\delta}_i + D_i \, \Delta \dot{\delta}_i + \Delta P_{Gi} = \Delta P_{Mi} \tag{12.11}$$

where δ_i is the angle of the internal voltage of the ith generator. We will number the generators so that the ith generator is connected to the ith bus.

We next consider the behavior of P_{Di}. In general, P_{Di} depends on both the voltage magnitude and frequency. In our active power model we assume that the voltage magnitudes are fixed at their nominal values and we therefore need not consider the voltage magnitude dependence. Concerning the frequency dependence, we assume that P_{Di} increases with frequency. This assumption is in accord with observed behavior. We would also like to be able to introduce changes in load as external (disturbance) inputs. Thus we can imagine the sudden increase in load due to the switching of an electrical device which "draws" a specified power. In accordance with this description, for small changes in frequency, we assume that

$$P_{Di} = P_{Di}^0 + \frac{\partial P_{Di}(\omega^0)}{\partial \omega_i} \Delta \omega_i + \Delta P_{Li} \tag{12.12}$$

where the first two terms describe the linearization of the frequency dependence of the load and ΔP_{Li} is the change of load *input*, by switching, referred to above. Note that the operating-point frequency is ω^0, which is expected to be close to but not necessarily equal to the nominal system frequency $\omega_0 = 2\pi 60$. In defining the frequency at each bus, we note that the instantaneous (phase a) voltage is

$$v_i(t) = \sqrt{2}|V_i| \cos [\omega^0 t + \theta_i^0 + \Delta \theta_i(t)] \tag{12.13}$$

Thus by taking the time derivative of the argument of the cosine function, we get

$$\omega_i = \omega^0 + \Delta \dot{\theta}_i \tag{12.14}$$

and thus

$$\Delta \omega_i = \omega_i - \omega^0 = \Delta \dot{\theta}_i \tag{12.15}$$

Defining $D_{Li} = \partial P_{Di}(\omega^0)/\partial \omega_i$, and using (12.15), we can replace (12.12) with

$$\Delta P_{Di} = D_{Li} \, \Delta \dot{\theta}_i + \Delta P_{Li} \tag{12.16}$$

Next we use (12.16) and (12.10) in (12.11). We get

$$M_i \, \Delta \ddot{\delta}_i + D_i \, \Delta \dot{\delta}_i + D_{Li} \, \Delta \dot{\theta}_i + \Delta P_{Li} + \Delta P_i = \Delta P_{Mi} \qquad (12.17)$$

We next calculate ΔP_i in (12.17). Assume that the line admittance parameters are purely imaginary, and using (6.4) we get (in the general n-bus case)

$$P_i = \sum_{k=1}^{n} |V_i||V_k| b_{ik} \sin (\theta_i - \theta_k) \qquad (12.18)$$

where $\theta_i = \underline{/V_i}$. In our model the $|V_i|$ are constant. In this case the $|V_i||V_k| b_{ik}$ are constants.

Next we linearize (12.18) around the operating point. Thus we replace P_i by $P_i^0 + \Delta P_i$ and θ_i by $\theta_i^0 + \Delta \theta_i$. For variety, instead of using Taylor series, we can proceed as follows, using trigonometric identities.

$$P_i = P_i^0 + \Delta P_i = \sum_{k=1}^{n} |V_i||V_k| b_{ik} \sin (\theta_i^0 + \Delta \theta_i - \theta_k^0 - \Delta \theta_k)$$

$$= \sum_{k=1}^{n} |V_i||V_k| b_{ik} [\sin (\theta_i^0 - \theta_k^0) \cos (\Delta \theta_i - \Delta \theta_k) \qquad (12.19)$$
$$+ \cos (\theta_i^0 - \theta_k^0) \sin (\Delta \theta_i - \Delta \theta_k)]$$

This result is exact. As we let the increments go to zero, we get

$$\Delta P_i = \sum_{k=1}^{n} [|V_i||V_k| b_{ik} \cos (\theta_i^0 - \theta_k^0)](\Delta \theta_i - \Delta \theta_k) \qquad (12.20)$$

which is the same as the linearization result found by using Taylor series. Defining $T_{ik} = |V_i||V_k| b_{ik} \cos (\theta_i^0 - \theta_k^0)$, we get

$$\Delta P_i = \sum_{k=1}^{n} T_{ik}(\Delta \theta_i - \Delta \theta_k) \qquad (12.21)$$

The constants T_{ik} are called *stiffness* or *synchronizing power coefficients*. The larger the T_{ik} the greater the exchange of power for a given change in bus phase angle. We are now in a position to substitute (12.21) into (12.17).

First, however, we will make a further approximation. We assume coherency between the internal and terminal voltage phase angles of each generator so that these angles tend to "swing together." Stated differently, we assume that the increments $\Delta \delta_i$ and $\Delta \theta_i$ are equal. The assumption that $\Delta \delta_i = \Delta \theta_i$ significantly simplifies the analysis and gives results which are qualitatively correct.

Making the assumption, and substituting (12.21) into (12.19), we get for $i = 1$, 2, . . . , n,

$$M_i \, \Delta \ddot{\delta}_i + \tilde{D}_i \, \Delta \dot{\delta}_i + \Delta P_i = \Delta P_{Mi} - \Delta P_{Li} \qquad (12.22)$$

where $\tilde{D}_i + D_i + D_{Li}$ and $\Delta P_i = \Sigma_{k=1}^{n} T_{ik}(\Delta \delta_i - \Delta \delta_k)$. From the way D_{Li} adds to D_i, we expect the (positive) load dependence on frequency to contribute to system damping.

The relations above are shown in block diagram form in Fig. 12.10. The reader is invited to check that with $K_{Pi} \triangleq 1/\bar{D}_i$ and $T_{Pi} \triangleq M_i/\bar{D}_i$, Fig. 12.10 represents (12.22) in block diagram form. In Fig. 12.10 we have $\Delta\omega_i$ as an output and ΔP_{Mi} as an input and can close the power control loop by introducing the turbine-governor block diagram shown in Fig. 12.4. Thus for generating unit 1, for example, we get Fig. 12.11. All the other generating units, $i = 2, 3, \ldots, n$, have similar block diagrams. In the figure, in accordance with our coherency assumption, we have replaced the dependence of ΔP_i on the $\Delta\theta_{ik}$ by a dependence on the $\Delta\delta_{ik}$.

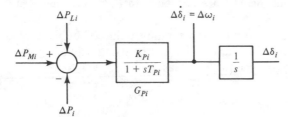

Figure 12.10   Generator block diagram.

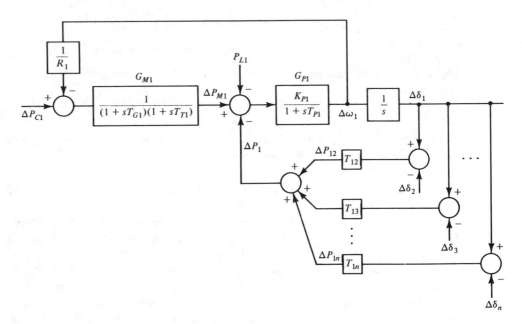

Figure 12.11   Power control of generating unit 1.

12.5 SPECIAL CASE: TWO GENERATING UNITS

In this case we have two synchronizing coefficients to consider, T_{12} and T_{21}.

$$T_{12} = |V_1||V_2|b_{12} \cos(\theta_1^0 - \theta_2^0)$$

$$T_{21} = |V_2||V_1|b_{21} \cos(\theta_2^0 - \theta_1^0)$$

Since $T_{21} = T_{12}$, the block diagram for the two units may be combined and simplified as shown in Fig. 12.12.

We can describe the operation of the system in qualitative terms as follows. Starting in the steady state ($\Delta\omega_1 = \Delta\omega_2 = 0$), suppose that additional load is switched onto bus 2 (i.e., ΔP_{L2} is positive). Initially, the mechanical power input to generator 2 does not change; the power imbalance is supplied from the rotating kinetic energy. Thus ω_2 begins to drop (i.e., $\Delta\omega_2$ goes negative). As a consequence, $\Delta\delta_2$ decreases and the phase angle $\delta_1 - \delta_2$ across the connecting transmission line increases. Thus the transferred power P_{12} increases. The sending-end power P_{12} *and* the receiving-end power $-P_{21}$ both increase. Unit 1 sees an increased power demand and ω_1 begins to drop. Each governor senses a frequency (speed) drop and acts to increase the mechanical power output from its turbine. In the steady state there is a new (lower) system frequency and an increased P_{12}. The frequency may then be restored and P_{12} adjusted, if desired, by operator action (i.e., by adjustment of P_{C1} and P_{C2}).

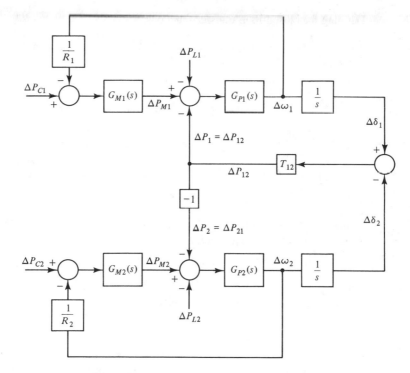

Figure 12.12 Power control of two generating units.

We next consider an example in which we calculate the steady-state change in frequency by automatic (governor) control and see the benefit of the interconnection in quantitative terms. We will consider the more general case when both P_{L1} and P_{L2} change.

Example 12.4

Consider two turbine-generator units connected by a tie line (i.e., a transmission line) operating in the steady state (with $\omega_0 = 2\pi 60$). Suppose that $\Delta P_{L1} = a_1 u(t)$ and $\Delta P_{L2} = a_2 u(t)$, where $u(t)$ is the unit step function. Calculate and compare the steady-state change in frequency (speed) in two cases:

(a) $T_{12} = 0$ (tie line open)
(b) $T_{12} > 0$ (tie line in operation)

Solution   From Fig. 12.12 using the block diagram relations, we get the following equations:

$$\Delta \hat{\omega}_1 = \frac{G_{P1}(s)}{1 + G_{P1}(s)\,G_{M1}(s)/R_1}\left[-\frac{a_1}{s} - \frac{T_{12}}{s}(\Delta \hat{\omega}_1 - \Delta \hat{\omega}_2)\right]$$

$$\Delta \hat{\omega}_2 = \frac{G_{P2}(s)}{1 + G_{P2}(s)\,G_{M2}(s)/R_2}\left[-\frac{a_2}{s} + \frac{T_{12}}{s}(\Delta \hat{\omega}_1 - \Delta \hat{\omega}_2)\right]$$

Defining the closed-loop transfer functions,

$$H_i(s) \triangleq \frac{G_{Pi}(s)}{1 + G_{Pi}(s)\,G_{Mi}(s)/R_i} \qquad i = 1, 2$$

and introducing them into the equations, we get

$$\Delta \hat{\omega}_1 = \frac{H_1(s)}{s}[-a_1 - T_{12}(\Delta \hat{\omega}_1 - \Delta \hat{\omega}_2)]$$

$$\Delta \hat{\omega}_2 = \frac{H_2(s)}{s}[-a_2 + T_{12}(\Delta \hat{\omega}_1 - \Delta \hat{\omega}_2)]$$

(a) $T_{12} = 0$. Using the final value theorem of the Laplace transform [i.e.,

$$\lim_{t \to \infty} f(t) = \lim_{s \to 0} sF(s)$$

where $F(s)$ is the Laplace transform of $f(t)$], we get $\Delta\omega_{1ss} = -H_1(0)a_1$ and $\Delta\omega_{2ss} = -H_2(0)a_2$. Evaluating $H_1(0)$ and $H_2(0)$, we get

$$\Delta \omega_{1ss} = \frac{-K_{P1}a_1}{1 + K_{P1}/R_1} = \frac{-a_1}{D_1 + D_{L1} + 1/R_1} = -\frac{a_1}{\beta_1} \qquad (12.23)$$

where $\beta_1 \triangleq D_1 + D_{L1} + 1/R_1$. Similarly,

$$\Delta \omega_{2ss} = -\frac{a_2}{\beta_2} \qquad (12.24)$$

where $\beta_2 \triangleq D_2 + D_{L2} + 1/R_2$. Equations (12.23) and (12.24) imply (static) frequency "droop" characteristics as shown in Fig. E12.4 in the case of (12.23).

(b) $T_{12} > 0$. We need to solve for $\Delta \hat{\omega}_1$ and $\Delta \hat{\omega}_2$ before using the final value theorem. In matrix form,

$$\begin{bmatrix} 1 + \dfrac{H_1}{s}T_{12} & -\dfrac{H_1}{s}T_{12} \\ -\dfrac{H_2}{s}T_{12} & 1 + \dfrac{H_2}{s}T_{12} \end{bmatrix}\begin{bmatrix} \Delta \hat{\omega}_1 \\ \Delta \hat{\omega}_2 \end{bmatrix} = \begin{bmatrix} -\dfrac{H_1}{s}a_1 \\ -\dfrac{H_2}{s}a_2 \end{bmatrix}$$

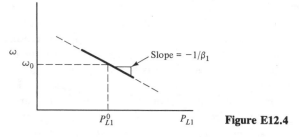

Figure E12.4

Solving for $\Delta\hat{\omega}_1$ using Cramer's rule, we get

$$\Delta\hat{\omega}_1 = \frac{1}{s}\frac{H_1a_1(s + H_2T_{12}) + H_2a_2H_1T_{12}}{s + (H_1 + H_2)T_{12}}$$

Then, using the final value theorem, we get

$$\Delta\omega_{1ss} = -\frac{(a_1 + a_2)H_1(0)H_2(0)}{H_1(0) + H_2(0)} = -\frac{a_1 + a_2}{1/H_1(0) + 1/H_2(0)} = -\frac{a_1 + a_2}{\beta_1 + \beta_2} \tag{12.25}$$

where the β_i are the frequency droop characteristics for the individual units. Based on physical reasoning or direct calculation,

$$\Delta\omega_{2ss} = \Delta\omega_{1ss}$$

Exercise 2.   Do you think the formula (12.25) generalizes to n machines?

The advantages of the interconnection may be seen by comparing (12.25) to (12.23) and (12.24). If we connect to a system with a large β, we benefit. The benefit is mutual if there is "diversity" in the load. For example, if $a_1 > 0$ and $a_2 < 0$, there is a smaller change in frequency than for isolated units. Of course, there are other advantages to the connection as well. In Chapter 7 we saw the possibilities of reducing production costs by adjusting the generation of interconnected units. There is also more flexibility in operating an interconnected system in terms of having enough available generating capacity to accommodate scheduled and/or emergency shutdowns with minimum loss of load.

Exercise 3.   Suppose that in Fig. 12.12, generating unit 2 has a very, very large moment of inertia. Approximate it by an infinite bus (i.e., $\Delta\delta_2 = 0$) and show that the block diagram for generating unit 1 reduces to the same form as the block diagram as Fig. 12.8.

12.6 DIVISION OF POWER SYSTEM INTO CONTROL AREAS

For an interconnected system with a large number of generating units, in the order of, say, 1000, modeling the power control system of each individual generator as suggested by Fig. 12.11 is far too complex to be useful. Simpler models of lower dimension give more understandable results. Fortunately, in many cases, we can identify groups of generators which are closely coupled internally but relatively weakly coupled between groups. We considered a similar case of coherent generators earlier, in Section 9.7, and can use the same aggregation techniques again in obtaining a simplified approximate model. The accuracy of the model and its usefulness are enhanced when the boundaries of the coherent groups of generators coincide with those of an operating company or utility. In this case, under a single operating jurisdiction, the generators are frequently controlled "as a single unit" and in such cases the groups are called control areas. The model using control areas is very useful in describing the exchange of power between neighboring power companies.

The result of the process of aggregation is a system model with block diagram just like Fig. 12.11, or Fig. 12.12 in the case of a two-area system, with the blocks representing control areas instead of individual machines. For example, in Fig. 12.12, ΔP_{C1} is to be interpreted as the increase in command power from the "area energy control" (dispatch) center. ΔP_{12} is the increase in power transferred from area 1 to area 2 (through all the connecting tie lines), in response to an increase in the center of angle difference $\Delta \delta_1 - \Delta \delta_2$. We can think of T_{12} as determined either analytically, by (9.63) for example, or experimentally.

We can also determine other elements in the model experimentally. For example, in the case of a two-area system, suppose that area 1 is supplying a small (positive) power P_{12} to area 2 through a tie line (or tie lines). Suppose that these connections are removed. The ensuing transient will be followed by a new steady state, assuming each "island" is stable. Since $\Delta P_{12} = -P_{12}$, in accordance with the results of Example 12.4, $\Delta \omega_{1ss} = P_{12}/\beta_1$ and $\Delta \omega_{2ss} = -P_{12}/\beta_2$. Thus by measuring the change in frequency we find the area frequency characteristics β_1 and β_2. With these and other measurements and calculations, we may specify all the elements in Fig. 12.12.

Suppose that we have developed the control area model. Let us now consider some applications related to so-called pool operation. A power pool is an interconnection of the power systems of individual utilities. Each company operates independently within its own jurisdiction, but there are contractual agreements about intercompany exchanges of power through the tie lines and other agreements about operating procedures to maintain system frequency. There are also agreements about operating procedures to be followed in the event of major faults or emergencies which we will not consider.

The basic principle of pool operation is that in the normal steady state,

1. Scheduled interchanges of tie-line power are maintained
2. Each area absorbs its own load changes

We note that these objectives cannot be satisfied during transients, nor, to obtain the benefits of pool operation, would this be desirable. Consider, for example, the case of a sudden step increase in P_{L2}, in Fig. 12.12. If the tie-line power is not allowed to change, then with respect to incremental changes the two systems are effectively isolated. From the results in Example 12.4, $\Delta \omega_2 = -\Delta P_{L2}/\beta_2$ and $\Delta \omega_1 = 0$. If we allow the tie-line power to change naturally, we get the better result, $\Delta \omega_1 = \Delta \omega_2 =$

$-\Delta P_{L2}/(\beta_1 + \beta_2)$. The tie-line power will also have increased. These changes take place automatically in only a few seconds. We may then think of restoring the system frequency and the scheduled tie-line power by adjustment of P_{C1} and P_{C2}; in doing so, we are guided by the basic principle of pool operation.

In practice the adjustment of P_{C1} and P_{C2} is done automatically by "higher-level" control loops which drive so-called area control errors (ACEs) to zero; the associated control scheme is called tie-line bias control. As applied to the two-area system, the two ACEs are

$$\text{ACE}_1 = \Delta P_{12} + B_1 \Delta \omega_1 \qquad (12.26)$$

$$\text{ACE}_2 = \Delta P_{21} + B_2 \Delta \omega_2 \qquad (12.27)$$

where ΔP_{12} and ΔP_{21} are departures from scheduled interchanges. The constants B_1 and B_2 are called *frequency bias settings* and are positive. In our example, with ΔP_{L2} positive, P_{12} tends to increase above its scheduled value. Thus ΔP_{12} is positive while ΔP_{21} is negative. With frequency dropping, both $\Delta \omega_1$ and $\Delta \omega_2$ are negative and thus ACE_2 is the sum of two negative quantities. ACE_1, on the other hand, is the sum of a positive and negative quantity with some tendency for cancellation. Since, in accordance with the basic principle of pool operation, we want P_{C2} to increase while P_{C1} should not change, use of the area control errors as "actuating" signals to effect change in P_{C1} and P_{C2} seems highly effective in this case.

A representation of an implementation of tie-line bias control in the case of two areas is shown in Fig. 12.13. It consists of Fig. 12.12 augmented by additional loops showing the generation of the ACE_i and their use in changing the area command powers. The K_i are positive constants; the associated negative sign is needed since we want a negative ACE_i to increase P_{Ci}. The transfer function is that of an integrator. From a hardware point of view we can understand the presence of the integrator by considering (suitably adjusted versions of) the ACE voltages distributed to the (dc) speeder motors of the individual generating units. These motors turn at a rate $\dot{\theta}$ proportional to the ACE voltage and continue to turn until they are driven to zero. This hardware implies an integrator in the block diagram.

From a control theory point of view the integrator has the merit that it drives the ACE_i to zero in the steady state. Of course, we are implicitly assuming that the system is stable, so the steady state is achievable. Thus in the steady state $\text{ACE}_1 = \text{ACE}_2 = 0$, and also, $\Delta \omega_1 = \Delta \omega_2$. Applying (12.26) and (12.27) in this case, we find $\Delta \omega_1 = \Delta \omega_2 = \Delta P_{12} = \Delta P_{21} = 0$. Thus there is a return to nominal frequency and scheduled interchanges. Furthermore, since both areas are in energy balance (with scheduled interchanges), any load changes must have been absorbed within each area. Thus the requirements for proper pool operation are met.

We can easily extend tie-line bias control to the n-area case. The definition of the area control error becomes

$$\text{ACE}_i = \sum_{j=1}^{n} \Delta P_{ij} + B_i \, \Delta \omega_i \qquad (12.28)$$

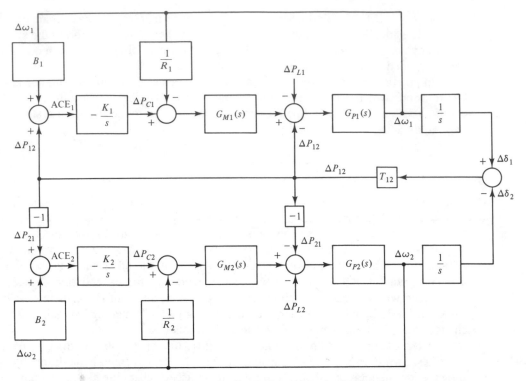

Figure 12.13   Tie-line bias control added.

where we sum over all the deviations from schedule interchanges from area i to obtain the net interchange error. The ACE_i are then used to adjust the area command powers just as in the two-area case. We note that our model predicts that in the steady state the system frequency returns to its nominal value and the interchange powers return to their scheduled values modulo posssible circulating power exchanges not sensed by the control algorithm. The reader can gain an appreciation for the problem by considering a three-area system with an arbitrary circulating component of power not sensed by (12.28), $i = 1, 2, 3$. We will not consider the auxiliary control action needed to eliminate this problem.

Finally, we wish to touch upon how economic dispatch fits into the picture. Usually, this is the responsibility of individual power companies rather than the pool. When the area generation must be changed, the individual generations are assigned, taking economic operation into account. Thus if the area generation is to be raised, a higher area λ is required, and by techniques similar to those in Chapter 7, the new generation assignments are calculated, in a company energy control center, and communicated to the speeder motors of individual units. Economic considerations also enter when scheduling future tie-line transfers between areas. For example, advantage may be taken of attractive rates in contracting for the purchase of excess hydropower from a neighboring power company, seasonally, during the spring runoff, or all through the year.

12.7 SUMMARY

In this chapter we considered the modeling of the servo-assisted speed governor and turbine. This led to the block diagram shown in Fig. 12.4. This may be used to complete the block diagram of the single machine–infinite bus system previously described and is shown in Fig. 12.6. The model now includes both voltage and power control loops. In analyzing the power control function, we frequently use the much simpler block diagram in Fig. 12.8, in which the loops associated with the voltage control system are suppressed.

Similar simplifications are made in dealing with power control in the multi-machine case. Figure 12.11 shows the resulting block diagram in the case of generating unit 1. The utility of the model is illustrated by its use in calculating frequency changes following increases in load as a function of the frequency droop characteristics of the individual generating units.

We may also use the model to describe pool operation in terms of the behavior of individual control areas. As an application, in the two-area case, it is shown how tie-line bias control is successful in meeting the basic objectives of pool operation.

PROBLEMS

12.1. A turbine generator set is operating open loop (i.e., the generator speed ω is not fed back to the governor). The turbine output is 500 MW at 3600 rpm. System parameters are $R = 0.01$ rad/MW-sec, $T_G = 0.001$ sec, and $T_T = 0.1$ sec. Suppose now that the output increases to 600 MW. Find the new turbine speed.

12.2. For the isolated generating station with local load shown in Fig. P12.2, it is observed that $\Delta P_L = 0.1$ brings about $\Delta \omega = -0.2$ in the steady state.
(a) Find $1/R$.
(b) Specify ΔP_C to bring $\Delta \omega$ back to zero (i.e., back to the frequency $\omega = \omega_0$).

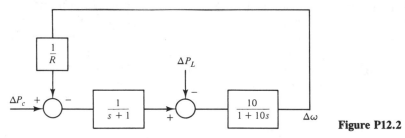

Figure P12.2

12.3. Assume an isolated generating station with a local load and with $T_G = 0$, $T_T = 1$, $T_P = 20$, $\bar{D} = 10^{-2}$, $R = 2.5$, $\Delta P_L = u(t)$, and $\Delta P_C = 0$. These constants are appropriate for calculating $\Delta \omega$ in rad/sec. Determine $\Delta \omega(t)$.

12.4. Suppose that the generator in Problem 12.3 is connected to a large system through a tie line with $T_{12} = 10$. Assume that the large system is approximated by an infinite bus. Making reasonable approximations, solve for $\Delta \omega(t)$ and compare the results with those in Problem 12.3.

Unbalanced System Operation

chapter 13

13.0 INTRODUCTION

Normally, a power system operates under balanced conditions. Efforts are made to ensure this desirable state of affairs. Unfortunately, under abnormal (i.e., fault) conditions, the system may become unbalanced.

Some typical nonsymmetric transmission-line faults are shown in Fig. 13.1. We show the faults with zero impedance, but in general, nonzero impedances must also be considered. In addition, there are faults on generators and other equipment.

Line faults are the most common because lines are exposed to the elements. Usually, faults are triggered by lightning strokes, which may cause the insulators to flash over. Occasionally, high winds topple towers. Wind and ice loading may cause insulator strings to fail mechanically. A tree may fall across a line. Fog or salt spray on dirty insulators may provide conduction paths associated with insulation failure.

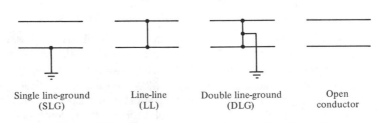

| Single line-ground (SLG) | Line-line (LL) | Double line-ground (DLG) | Open conductor |

Figure 13.1 Some nonsymmetric transmission line faults.

The most common type of fault by far is the single line-to-ground (SLG) fault, followed in frequency of occurrence by line-to-line (LL) faults, double line-to-ground (DLG) faults, and balanced three-phase (3ϕ) faults.

Much less common are faults on cables, circuit breakers, generators, motors, and transformers. A generator, motor, or transformer may fail if thermal ratings have been exceeded for long periods of time; the insulation may then have deteriorated sufficiently so that mechanical vibration and expansion and contraction due to thermal cycling may bring about a total failure of the insulation at some point.

It is important to study the system under fault conditions, in order to provide for system protection. We need to confirm that the ratings of circuit breakers and the settings of the relays which trip them are correct. In addition, in the design phase, the transformer connections and the grounding scheme are specified with concern for these same abnormal conditions.

Initially, we are going to consider only steady-state behavior. This will help fix the fundamental ideas. Later we will modify the analysis to give an estimate of the more extreme conditions during the initial few cycles of the transient.

For the steady-state analysis of unbalanced systems, we are going to use the method of "symmetrical components" due to Fortesque. As will be seen, it not only simplifies the problem numerically, but more important, its use definitely improves our understanding of system behavior during fault conditions.

We will next describe symmetrical components and illustrate their use.

13.1 SYMMETRICAL COMPONENTS

Suppose that we are given an arbitrary set of three phasors, say I_a, I_b, and I_c. We emphasize that the set is perfectly arbitrary. We claim that we can represent I_a, I_b, and I_c in terms of nine symmetrical components—I_a^0, I_a^+, I_a^-, I_b^0, I_b^+, I_b^-, I_c^0, I_c^+, I_c^-—as follows:

$$I_a = I_a^0 + I_a^+ + I_a^-$$

$$I_b = I_b^0 + I_b^+ + I_b^-$$

$$I_c = I_c^0 + I_c^+ + I_c^- \tag{13.1}$$

where I_a^0, I_b^0, I_c^0 are a zero-sequence set, I_a^+, I_b^+, I_c^+ are a positive-sequence set, and I_a^-, I_b^-, I_c^- are a negative-sequence set.

By now we are familiar with (balanced) positive (abc)-sequence and (balanced) negative (acb)-sequence sets. A zero-sequence set I_a^0, I_b^0, and I_c^0 has the following property:

$$I_a^0 = I_b^0 = I_c^0$$

Thus, like positive- and negative-sequence sets, all the magnitudes are the same. But here the *phases* of the a, b, and c components are also identical. Although I_a^0, I_b^0, and I_c^0 are not a balanced set, the set has a useful symmetry.

We can rewrite (13.1) as a matrix equation:

$$\begin{bmatrix} I_a \\ I_b \\ I_c \end{bmatrix} = \begin{bmatrix} I_a^0 \\ I_b^0 \\ I_c^0 \end{bmatrix} + \begin{bmatrix} I_a^+ \\ I_b^+ \\ I_c^+ \end{bmatrix} + \begin{bmatrix} I_a^- \\ I_b^- \\ I_c^- \end{bmatrix} \tag{13.2}$$

defining $\mathbf{I}$ as the (current) vector with components I_a, I_b, and I_c, and $\mathbf{I}^0$, $\mathbf{I}^+$, and $\mathbf{I}^-$ as the zero-, positive-, and negative-sequence (current) vectors whose components are, respectively, the zero-sequence set, the positive-sequence set, and the negative-sequence set, we can write (13.2) using vector notation:

$$\mathbf{I} = \mathbf{I}^0 + \mathbf{I}^+ + \mathbf{I}^- \tag{13.3}$$

We now will show how to find the nine symmetrical components. As a notational convenience we first introduce the complex number $\alpha = e^{j2\pi/3} = 1\underline{/120°}$. Just as in the case of the complex number $j = e^{j\pi/2}$, it is helpful to think of α as an operator. Multiplication of a complex number I by α leaves the magnitude unchanged and increases the angle by $120°$ (i.e., it rotates I by a positive angle of $120°$).

It is clear that only three of the nine symmetrical components may be chosen independently. The common convention is that I_a^+, I_a^-, and I_a^0 are chosen as the independent (lead) variables and the remainder are then expressed in terms of the lead variables. Applied to (13.2), we get

$$\begin{bmatrix} I_a \\ I_b \\ I_c \end{bmatrix} = I_a^0 \begin{bmatrix} 1 \\ 1 \\ 1 \end{bmatrix} + I_a^+ \begin{bmatrix} 1 \\ \alpha^2 \\ \alpha \end{bmatrix} + I_a^- \begin{bmatrix} 1 \\ \alpha \\ \alpha^2 \end{bmatrix} \tag{13.4}$$

which is equivalent to

$$\begin{bmatrix} I_a \\ I_b \\ I_c \end{bmatrix} = \begin{bmatrix} 1 & 1 & 1 \\ 1 & \alpha^2 & \alpha \\ 1 & \alpha & \alpha^2 \end{bmatrix} \begin{bmatrix} I_a^0 \\ I_a^+ \\ I_a^- \end{bmatrix} \tag{13.5}$$

Define the symmetrical components (current) vector $\mathbf{I}_s$ with components I_a^0, I_a^+, and I_a^-. Also define

$$\mathbf{A} = \begin{bmatrix} 1 & 1 & 1 \\ 1 & \alpha^2 & \alpha \\ 1 & \alpha & \alpha^2 \end{bmatrix} \tag{13.6}$$

$\mathbf{A}$ is called the *symmetrical components transformation matrix*. Now we can use matrix notation in (13.5).

$$\mathbf{I} = \mathbf{A}\mathbf{I}_s \tag{13.7}$$

where $\mathbf{I}$ is the phase (current) vector and $\mathbf{I}_s$ is the symmetrical components (current) vector.

The reader can confirm that $\det \mathbf{A} = 3(\alpha - \alpha^2)$. Since $\det \mathbf{A} \neq 0$, the inverse exists. In fact,

$$\mathbf{A}^{-1} = \frac{1}{3}\begin{bmatrix} 1 & 1 & 1 \\ 1 & \alpha & \alpha^2 \\ 1 & \alpha^2 & \alpha \end{bmatrix} \tag{13.8}$$

Thus we have a formula for calculating $\mathbf{I}_s$ from $\mathbf{I}$:

$$\mathbf{I}_s = \mathbf{A}^{-1}\mathbf{I} \tag{13.9}$$

or more explicitly, in terms of components,

$$\begin{bmatrix} I_a^0 \\ I_a^+ \\ I_a^- \end{bmatrix} = \frac{1}{3}\begin{bmatrix} 1 & 1 & 1 \\ 1 & \alpha & \alpha^2 \\ 1 & \alpha^2 & \alpha \end{bmatrix}\begin{bmatrix} I_a \\ I_b \\ I_c \end{bmatrix}$$

Once we know $\mathbf{I}_s$ (which gives us the leading symmetrical components I_a^0, I_a^+, I_a^-), we can calculate the remaining six components by using the properties of the positive-, negative-, and zero-sequence sets which they represent.

Note the following points.

1. Usually, we leave out the subscript a when writing the components of $\mathbf{I}_s$. The subscript a is understood, just as in per phase analysis, where we write, say, V_1 and V_2, reserving the subscript to designate bus number.
2. In all the algebra that we need to do, we use the identities $\alpha^2 = \alpha*$, $\alpha^3 = 1$, $1 + \alpha + \alpha^2 = 0$, where $\alpha* = e^{-j2\pi/3}$ is the complex conjugate of α.
3. In many texts the notation 0, 1, 2 is used instead of 0, +, −.
4. Of course, any set of three phasors (not just I_a, I_b, I_c) can be represented by symmetrical components and we will take (13.1) to (13.9) to apply in general.
5. In (10.16), together with i_d and i_q we defined the "zero-sequence" variable i_0. In the examples the zero-sequence variables always turned out to be zero because we assumed balanced conditions.

We next consider an example.

Example 13.1.   Finding Symmetrical Components

Given $I_a = 1\underline{/0°}$, $I_b = 1\underline{/-90°}$, and $I_c = 2\underline{/135°}$, find I_a^0, I_a^+, and I_a^-.

Solution   Using (13.9), we have

$$\begin{bmatrix} I_a^0 \\ I_a^+ \\ I_a^- \end{bmatrix} = \frac{1}{3}\begin{bmatrix} 1 & 1 & 1 \\ 1 & \alpha & \alpha^2 \\ 1 & \alpha^2 & \alpha \end{bmatrix}\begin{bmatrix} 1\underline{/0°} \\ 1\underline{/-90°} \\ 2\underline{/135°} \end{bmatrix}$$

Carrying out the complex arithmetic, we get

$$I_a^0 = \frac{1}{3}(1\underline{/0°} + 1\underline{/-90°} + 2\underline{/135°}) = 0.195\underline{/135°}$$

$$I_a^+ = \frac{1}{3}(1\underline{/0°} + 1\underline{/30°} + 2\underline{/375°}) = 1.311\underline{/15°}$$

$$I_a^- = \frac{1}{3}(1\underline{/0°} + 1\underline{/150°} + 2\underline{/255°}) = 0.494\underline{/-105°}$$

As a check we can use the calculated values and find that $I_a = I_a^0 + I_a^+ + I_a^- = 1\underline{/0°}$.

Exercise 1. Using the results of Example 13.1, sketch in the complex plane and label the three symmetrical components that make up $\mathbf{I}^0$. Repeat for $\mathbf{I}^+$ and $\mathbf{I}^-$. Using all nine symmetrical components, as in (13.1), show graphically that the (appropriate) symmetrical components add up to give I_a, I_b, and I_c.

Exercise 2. Suppose that we have a set of unbalanced line-line voltages, V_{ab}, V_{bc}, and V_{ca}. Show that the corresponding zero-sequence voltage is always zero.

The results of the next example will be used in explaining how symmetrical components are used in fault analysis.

Example 13.2. Finding Symmetrical Components of Single Line-to-Ground (SLG) Fault Currents

Suppose that we are given an SLG fault current I^f on phase a. By that we mean the phase a fault current $I_{af} = I^f$ and the phase b and phase c fault currents, I_{bf} and I_{cf}, respectively, are zero. The situation is shown in Fig. E13.2(a). Even though the currents $I_{bf} = I_{cf} = 0$, it is useful to define a hypothetical three-phase "stub" with currents I_{af}, I_{bf}, and I_{cf}. We then define a *vector* fault current

$$\mathbf{I}^f \triangleq \begin{bmatrix} I_{af} \\ I_{bf} \\ I_{cf} \end{bmatrix} = \begin{bmatrix} I^f \\ 0 \\ 0 \end{bmatrix}$$

With this general introduction we next wish to find the symmetrical components of $\mathbf{I}^f$, and interpret the result.

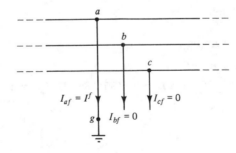

Figure E13.2(a)

Solution From (13.9),

$$\begin{bmatrix} I_{af}^0 \\ I_{af}^+ \\ I_{af}^- \end{bmatrix} = \frac{1}{3} \begin{bmatrix} 1 & 1 & 1 \\ 1 & \alpha & \alpha^2 \\ 1 & \alpha^2 & \alpha \end{bmatrix} \begin{bmatrix} I^f \\ 0 \\ 0 \end{bmatrix} = \frac{I^f}{3} \begin{bmatrix} 1 \\ 1 \\ 1 \end{bmatrix}$$

Thus $I_{af}^+ = I_{af}^- = I_{af}^0 = I^f/3$. With these three leading symmetrical components we can easily find the remaining six.

By way of interpretation we can consider (13.4) as it applies in this example. We get

$$\begin{bmatrix} I^f \\ 0 \\ 0 \end{bmatrix} = \frac{I^f}{3}\begin{bmatrix} 1 \\ 1 \\ 1 \end{bmatrix} + \frac{I^f}{3}\begin{bmatrix} 1 \\ \alpha^2 \\ \alpha \end{bmatrix} + \frac{I^f}{3}\begin{bmatrix} 1 \\ \alpha \\ \alpha^2 \end{bmatrix}$$

Thus the (actual) fault current vector $\mathbf{I}^f$ may be seen to be the sum of a zero-sequence fault vector $\mathbf{I}_f^0$, a positive-sequence fault vector $\mathbf{I}_f^+$, and a negative-sequence fault vector $\mathbf{I}_f^-$.

We can check the result graphically. For simplicity we pick I^f real and show all nine symmetrical components [Fig. E13.2(b)]. We note that

$$I_{af} = I_{af}^+ + I_{af}^- + I_{af}^0 = I^f$$

$$I_{bf} = I_{bf}^+ + I_{bf}^- + I_{bf}^0 = 0$$

$$I_{cf} = I_{cf}^+ + I_{cf}^- + I_{cf}^0 = 0$$

which checks.

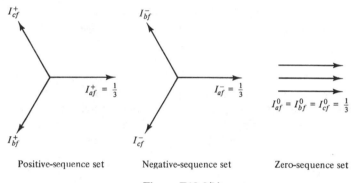

Positive-sequence set   Negative-sequence set   Zero-sequence set

Figure E13.2(b)

Exercise 3.   Convince yourself by doing the algebra that if $\{I_a, I_b, I_c\}$ is a positive-sequence set, then $I_a^+ = I_a$, $I_a^- = I_a^0 = 0$. Consider what happens for negative-sequence and zero-sequence sets.

13.2 USE OF SYMMETRICAL COMPONENTS FOR FAULT ANALYSIS

We consider next a very simple problem which illustrates the whole idea behind the use of symmetrical components in finding voltages and currents for faulted systems. Consider the circuit in Fig. 13.2 consisting of ideal voltage sources and impedances. Later, once we understand the basic idea, we will introduce a realistic power system model. Suppose that the problem is to find I^f, $V_{a'g}$, $V_{b'g}$, and $V_{c'g}$.

We note the following from Fig. 13.2.

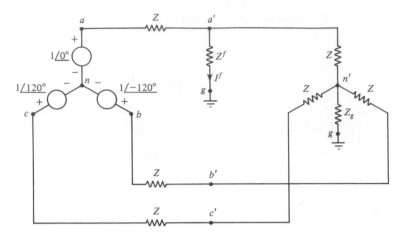

Figure 13.2 System with single line-ground fault.

1. $V_{a'g} = Z^f I_f$
2. Voltage sources are a positive-sequence set.
3. Except for the fault, the network is symmetrical.

It is clear that we cannot solve the problem by per phase analysis because the network does not have the required three-phase symmetry. We can solve the network by general circuit analysis techniques but we wish to introduce an alternative which makes full use of the symmetry of the unfaulted network. We will try to develop the basic idea by presenting the following procedure. This procedure is for purposes of explanation; later we can use a less unwieldy procedure.

Procedure

Step 1: Replace the fault (i.e., the branch with current I^f) by a current source with current I^f.

Step 2: Introduce a three-phase stub, as in Example 13.2, with $I_{af} = I^f$ and $I_{bf} = I_{cf} = 0$, and replace by an equivalent sum of symmetrical components. These may be grouped into a zero-sequence set, a positive-sequence set, and a negative-sequence set.

Step 3: Find $V_{a'g}$ in terms of I^f by the method of superposition (i.e., add the responses to the positive-sequence sources, negative-sequence sources, and zero-sequence sources). This gives one relationship between $V_{a'g}$ and I^f.

Step 4: As a second relationship, use the constraint imposed by the fault (i.e., use $V_{a'g} = Z^f I^f$). With the two equations we can solve for I^f.

Step 5: With I^f known we can solve for $V_{a'g}$, $V_{b'g}$, and $V_{c'g}$.

We next consider the steps in more detail.

Step 1.   For the rest of the network it does not matter how I^f is generated. All the network "knows" is that a current I^f is being extracted from the node labeled a'. Thus if we replace the branch in which the current I^f flows by a current source of value I^f, nothing will change in the solution of the rest of the network. In circuit theory textbooks, a more general version is offered under the title of the substitution theorem. Note that the replacement of I^f at this stage is at the conceptual level; we do not yet know its numerical value. Note also that the process works in reverse; we can replace a current source by a branch, provided that the branch impedance is chosen so that the current does not change.

Step 2.   Here we use (13.1):

$$I_{af} = I_{af}^0 + I_{af}^+ + I_{af}^-$$

$$I_{bf} = I_{bf}^0 + I_{bf}^+ + I_{bf}^-$$

$$I_{cf} = I_{cf}^0 + I_{cf}^+ + I_{cf}^-$$

where from Example 13.2, $I_{af}^0 = I_{af}^+ = I_{af}^- = \frac{1}{3}I^f$. We have thus replaced the fault by the stub with current sources I^f, 0, 0, and each of these in turn by the appropriate sum of three symmetrical components. This is shown in Fig. 13.3. So far it looks like we have complicated the problem considerably; we have nine nonzero current sources instead of only one nonzero current source!

Step 3.   The system is linear. We can use superposition to find the responses to the 12 sources. We do this by grouping the sources into a positive-sequence set, negative-sequence set, and zero-sequence set. Then we find the responses to each set individually, and, finally, add. In particular, in this manner we find the response $V_{a'g}$.

First consider the response to the positive-sequence set. In accordance with the superposition principle, we set all other sources equal to zero. Thus Fig. 13.3 reduces to Fig. 13.4.

In this figure note that all the sources are in positive-sequence sets, and that the network is symmetrical. In this (balanced) case we know that all the responses will occur in positive-sequence sets; no zero- or negative-sequence voltages or currents will be present in Fig. 13.4. Anticipating a similar decoupling when we consider the responses to the negative- and zero-sequence sources, we use the same superscript + notation to indicate any response (e.g., $V_{a'g}^+$) to the positive-sequence sources as well as the positive-sequence component in the symmetrical components representation of $V_{a'g}$.

Since the network is balanced, we can use per phase analysis to find the a phase variables. Thus we are led to consider Fig. 13.5. We note that under balanced conditions the current in Z_g is zero and hence n, n', and g are all at the same potential, as shown in Fig. 13.5.

Solving for $V_{a'g}^+$, we get

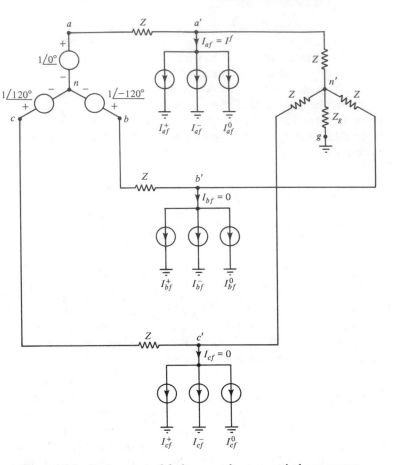

Figure 13.3 Replacement of fault current by symmetrical components.

$$V_{a'g}^+ = \tfrac{1}{2} - \tfrac{1}{2}ZI_{af}^+ \tag{13.10}$$

(easiest to use superposition again). As usual for per phase analysis, there is no need to repeat the calculation if we wish to find $V_{b'g}^+$ and $V_{c'g}^+$. As implied by the notation, $V_{a'g}^+$, $V_{b'g}^+$, and $V_{c'g}^+$ form a positive sequence set.

Consider next the negative-sequence set of sources. Setting positive- and zero-sequence sources to zero, Fig. 13.3 reduces to Fig. 13.6.

Once again the network is balanced. The responses will occur in negative-sequence sets. The a phase variables may be found by using per phase analysis. Thus we are led to consider Fig. 13.7. Solving for $V_{a'g}^-$, we get

$$V_{a'g}^- = -\tfrac{1}{2}ZI_{af}^- \tag{13.11}$$

Without detailed further calculation, we obtain the negative-sequence set $V_{a'g}^-$, $V_{b'g}^-$, and $V_{c'g}^-$.

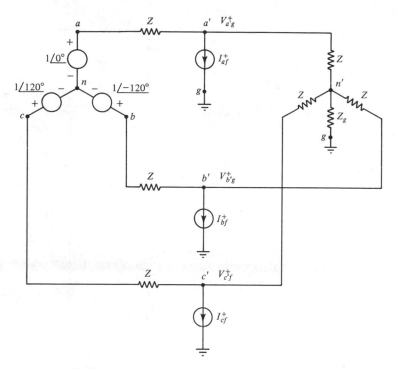

Figure 13.4   Positive-sequence network (three-phase).

Finally, we consider the zero-sequence set of sources acting alone. It is easy to prove and intuitively clear that zero-sequence sources applied to a symmetric network must produce only zero-sequence responses. In particular, in the network in Fig. 13.8, $I_a^0 = I_b^0 = I_c^0$. On the other hand, applying KCL at node n, $I_a^0 + I_b^0 + I_c^0 = 0$. Thus $I_a^0 = I_b^0 = I_c^0 = 0$ (i.e., looking from the fault to the left, we see an open circuit). It is also true that $I_{a'}^0 = I_{b'}^0 = I_{c'}^0$, so looking to the right, we get

$$V_{a'g}^0 = ZI_{a'}^0 + 3Z_g I_{a'}^0 \tag{13.12}$$

with $V_{a'g}^0 = V_{b'g}^0 = V_{c'g}^0$. Equation (13.12) implies that we can calculate $V_{a'g}^0$ from the single-phase zero-sequence network shown in Fig. 13.9. Since the current from n' to g is $I_{a'}^0$ instead of $3I_{a'}^0$, it is necessary to substitute $3Z_g$ for Z_g in order to preserve the

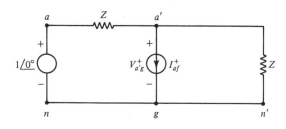

Figure 13.5   Positive-sequence network.

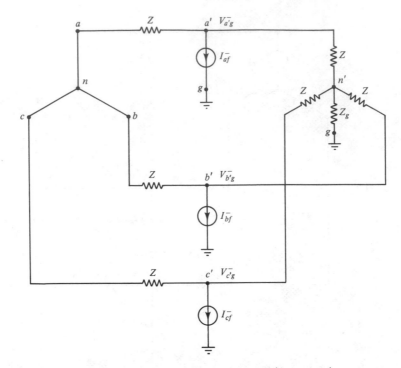

Figure 13.6   Negative-sequence network (three-phase).

equivalence. We also note, in Figs. 13.8 and 13.9, that n, n', and g are not at the same potential.

We now have the individual responses $V_{a'g}^+$, $V_{a'g}^-$, and $V_{a'g}^0$. By the principle of superposition, $V_{a'g}$, the response to all the sources taken together, is

$$V_{a'g} = V_{a'g}^+ + V_{a'g}^- + V_{a'g}^0 \tag{13.13}$$

$$= \tfrac{1}{2} - \tfrac{1}{2} Z I_{af}^+ - \tfrac{1}{2} Z I_{af}^- - Z I_{af}^0 - 3 Z_g I_{af}^0$$

$$= \tfrac{1}{2} - \tfrac{1}{2} Z \frac{I^f}{3} - \tfrac{1}{2} Z \frac{I^f}{3} - Z \frac{I^f}{3} - 3 Z_g \frac{I^f}{3}$$

$$= \tfrac{1}{2} - \tfrac{2}{3} Z I^f - Z_g I^f \tag{13.14}$$

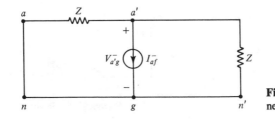

Figure 13.7   Negative-sequence network.

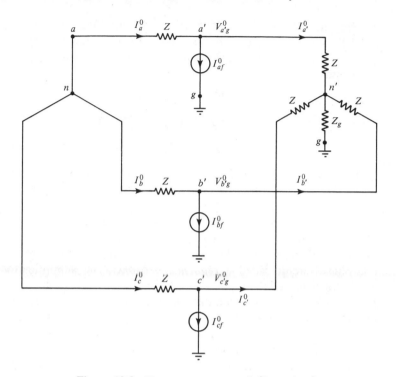

Figure 13.8   Zero-sequence network (three-phase).

In going from (13.13) to (13.14), we have used (13.10) to (13.12) and the result from Example 13.2 that $I_{af}^+ = I_{af}^- = I_{af}^0 = I^f/3$.

Step 4.   Equation (13.14) gives $V_{a'g}$ in terms of I^f. But at the fault point we also have $V_{a'g} = Z^f I^f$. Solving for I^f, we get

$$I^f = \frac{\frac{1}{2}}{Z^f + \frac{2}{3}Z + Z_g} \tag{13.15}$$

Note: $V_{a'g} = Z^f I^f$ may now be found immediately.

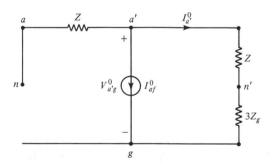

Figure 13.9   Zero-sequence network.

Step 5. With I^f determined we can now find $V_{a'g}^{+}$, $V_{a'g}^{-}$, and $V_{a'g}^{0}$ using (13.10) to (13.12), or the sequence networks in Figs. 13.5, 13.7, and 13.9. Once we have $V_{a'g}^{+}$, $V_{a'g}^{-}$, $V_{a'g}^{0}$ we can specify all the remaining six symmetrical components of $V_{a'g}$. Then we can find

$$V_{a'g} = V_{a'g}^{0} + V_{a'g}^{+} + V_{a'g}^{-}$$

$$V_{b'g} = V_{b'g}^{0} + V_{b'g}^{+} + V_{b'g}^{-}$$

$$V_{c'g} = V_{c'g}^{0} + V_{c'g}^{+} + V_{c'g}^{-}$$

Alternatively, we can use (13.5) directly (as it applies to voltage). *Note:* We have included $V_{a'g}$ in the listing above for completeness; it is easier to use $V_{a'g} = Z^f I_f$.

From this example we can see an advantage of the symmetrical components method. We replace the given faulted (nonsymmetrical) network with three symmetrical three-phase networks, each of which may be analyzed by considering a single-phase circuit called a *sequence network*. The simplifications of per phase analysis for balanced three-phase circuits are preserved. In this connection we point out that once we are convinced that the method of analysis is legitimate, we can skip the preliminary stages and go directly to the sequence networks (i.e., go directly from Fig. 13.2 to Figs. 13.5, 13.7, and 13.9).

There is an additional reason for our interest in the symmetrical components method. Our circuits example can be solved by traditional methods of circuit analysis, but as will be seen in later sections, many of the impedances that arise in modeling elements of power systems are sequence dependent. For example, the Thévenin equivalent generator impedances in the positive-, negative-, and zero-sequence networks are all different. Thus the traditional methods of circuit analysis are not even available as options.

We next consider a development that makes the use of the sequence networks even more attractive.

13.3 SEQUENCE NETWORK CONNECTIONS FOR SINGLE LINE-GROUND FAULT

If in the problem of the preceding section, we consider the consequences of the relations

$$V_{a'g} = V_{a'g}^{+} + V_{a'g}^{-} + V_{a'g}^{0} \qquad (13.13)$$

and

$$I_{af}^{+} = I_{af}^{-} = I_{af}^{0} = \frac{I^f}{3} \qquad (13.16)$$

it is clear that $V_{a'g}$ is the voltage across the series connection of the positive-, negative-, and zero-sequence networks when $I^f/3$ flows in the series connection. In addition, the terminal constraint

$$V_{a'g} = Z^f I^f = 3Z^f \frac{I^f}{3} \tag{13.17}$$

can be introduced by placing an impedance $3Z^f$ across the series connection. Thus we are led to Fig. 13.10. Since n, n', and g are at the same potential in the positive- and negative-sequence networks, we only show g.

    Exercise 4.   Check that I^f calculated from Fig. 13.10 equals the result in (13.15).

    Exercise 5.   Suppose that in Fig. 13.2, (a) the load is not grounded, or (b) the source is grounded. How would this affect Fig. 13.10 and the magnitude of the fault current I^f. Does it make sense physically?

    It should be noted that Fig. 13.10 describes the network for any choice of Z^f, including $Z^f = \infty$. Suppose that $Z^f = \infty$. In that case it is physically clear that there

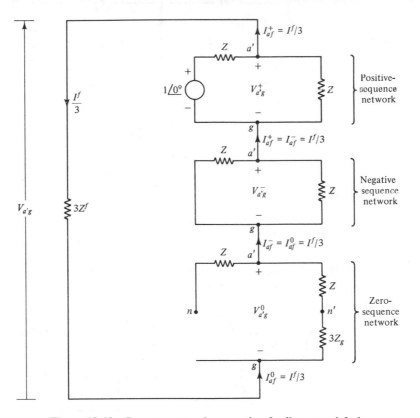

Figure 13.10   Sequence network connection for line-ground fault.

is no fault, $I^f = 0$, and $V_{a'g} = V_{a'g}^+$ is simply $V_{a'g}^{pf}$, the prefault phase voltage (line-neutral) at point a', during normal balanced operating conditions. Taking another point of view, suppose that we replace the positive-sequence network between the terminals a' and g by a Thévenin equivalent circuit. Then with $I^f = 0$ we get the (open-circuit) Thévenin equivalent voltage across the terminals a' and g. Thus $V_{a'g}^{pf}$ is also the Thévenin equivalent voltage $V_{a'g}^{oc}$ of the positive-sequence network. The Thévenin series impedance is found in the usual way by setting all independent sources equal to zero. This interpretation is frequently used as in the following example.

Example 13.3   Prefault Voltage Specified

A transmission line modeled by series reactance $X_L = j0.1$ delivers power to a wye-connected load [Fig. E13.3(a)]. Then there is an SLG fault halfway down the line. The network is balanced before the fault. The voltage sources are positive sequence. Before the fault, $V_{a'n} = V_{a'g} = 1$. Find I^f, $V_{a''g}$, and $V_{b''g}$ after the fault.

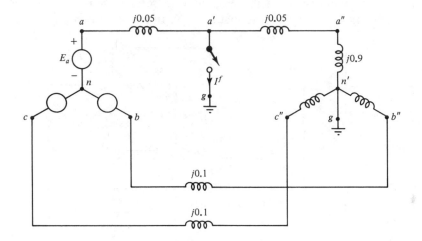

Figure E13.3(a)

Solution   We start with the series connection of sequence networks shown in Fig. E13.3(b). We are not given E_a. In fact, it is simpler to proceed without calculating it. At the terminals $a'g$ we replace the positive-sequence network by a Thévenin equivalent network with $V_{a'g}^{oc} = V_{a'g}^{pf} = 1\underline{/0°}$ in series with the Thévenin series impedance. We show the series impedance unreduced, in the original network topology. Thus we get Fig. E13.3(c). From the figure we get (after familiar simplifying network reductions)

$$\frac{I^f}{3} = \frac{1\underline{/0°}}{j0.0475 + j0.0475 + j0.95} = -j0.957$$

Thus $I^f = -j2.87$. We have found one of the required answers.

     Turning next to the calculation of $V_{a''g}$ and $V_{b''g}$, as a preliminary we calculate the following from Fig. E13.3(c)

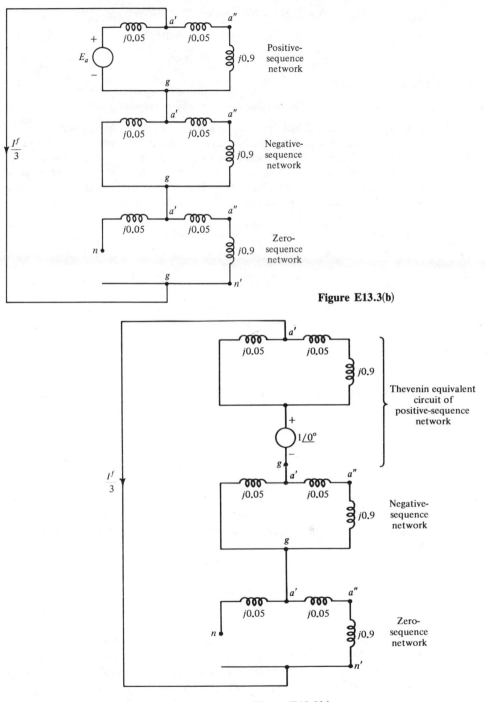

Figure E13.3(b)

Figure E13.3(c)

$$V_{a'g}^+ = 1 + j0.957 \cdot j0.0475 = 0.9545$$

$$V_{a'g}^- = j0.957 \cdot j0.0475 = -0.0455$$

$$V_{a'g}^0 = j0.957 \cdot j0.95 = -0.9091$$

As a check we note that

$$V_{a'g} = V_{a'g}^+ + V_{a'g}^- + V_{a'g}^0 = 0$$

which confirms the physical fact that there is a solid short to ground at point a'.

Next we calculate $V_{a''g}^+$, $V_{a''g}^-$, and $V_{a''g}^0$. Since a'' is a point interior to the Thévenin equivalent circuit of the positive-sequence network, some care is needed in finding $V_{a''g}^+$. We will discuss two methods. Both methods require that we go back to the original positive-sequence circuit [i.e., Fig. E13.3(b)]. Thus we can proceed as follows. Since we know $V_{a'g}^+$, $V_{a'g}^-$, and $V_{a'g}^0$, we can use the voltage-divider law to find

$$V_{a''g}^+ = \frac{0.9}{0.95} V_{a'g}^+ = 0.9043$$

$$V_{a''g}^- = \frac{0.9}{0.95} V_{a'g}^- = -0.0431$$

$$V_{a''g}^0 = \frac{0.9}{0.95} V_{a'g}^0 = -0.8612$$

We can now easily find

$$V_{a''g} = V_{a''g}^+ + V_{a''g}^- + V_{a''g}^0 = 0$$

which may be verified by checking Fig. E13.3(a). We are also asked to find $V_{b''g}$. It is almost as easy to find.

$$V_{b''g} = V_{b''g}^+ + V_{b''g}^- + V_{b''g}^0$$

$$= \alpha^2 V_{a''g}^+ + \alpha V_{a''g}^- + V_{a''g}^0$$

$$= 0.9043 \underline{/-120°} - 0.0431 \underline{/120°} - 0.8613$$

$$= 1.77 \underline{/-136.8°}$$

This example illustrates the important technique: Solve the network for the a phase symmetrical components first. Then convert to the required b or c phase symmetrical components to calculate the b or c phase quantities.

We next turn to a second method for finding $V_{a''g}^+$. In the original positive-sequence network, treat $I^f/3 = -j0.957$ as an independent source; we attempted to justify this viewpoint in Section 13.2. We can find $V_{a''g}^+$ by the method of superposition. The voltage source acting alone (with the current source set equal to zero) gives $V_{a''g}^+ = 0.9/0.95 = 0.9474$; physically, this is the prefault voltage at the point a''. The current source acting alone (with the voltage source set equal to zero) gives the voltage $V_{a''g}^+ = j0.957 \cdot j0.0475 \cdot 0.9/0.95 = -0.0431$. Thus in the actual case (with both sources present) we find that $V_{a''g}^+ = 0.9474 - 0.0431 = 0.9043$. This checks the result of the previous calculation. The rest of the calculation is the same as before. Note that we did not need to know the value of the voltage source, E_a, in doing the calculation.

We next develop the idea of using $I^f/3$ as an independent input.

13.4 FAULT CALCULATION USING Δ NETWORK

Once we have found $I^f/3$, we can find the remaining fault variables by the use of superposition as follows. Suppose that we pick any line or load current, or any phase voltage, in the faulted network. We can determine this quantity by adding the prefault value (i.e., the response to the original positive-sequence voltage sources) to the value found from the passive network (with all voltage sources set equal to zero) with the sequence fault currents $I_{af}^+ = I_{af}^- = I_{af}^0 = I^f/3$ applied as independent inputs. The passive network is called a Δ network. We note that if the prefault values are specified, the use of the Δ network has real advantages over starting from scratch. We illustrate this in the next example.

Example 13.4.   Fault Calculation Using Δ Network

The prefault network shown in Fig. E13.4(a) is symmetrical. Sources are positive sequence. Prefault variables: $I_a^{pf} = 1$, $I_{a'}^{pf} = -j1$, $I_{a''}^{pf} = -1 - j1$, and $V_{a'n}^{pf} = V_{a'g}^{pf} = 1$. Find the values of these currents, and I^f, in the faulted network. Also find I_b and $V_{b'g}$ in the faulted network.

Solution   The appropriate interconnection of networks is shown in Fig. E13.4(b). One of the desired variables, I_a, is shown on the diagram. The others may similarly be identified. We first need to calculate the fault current I^f. This can be done most simply

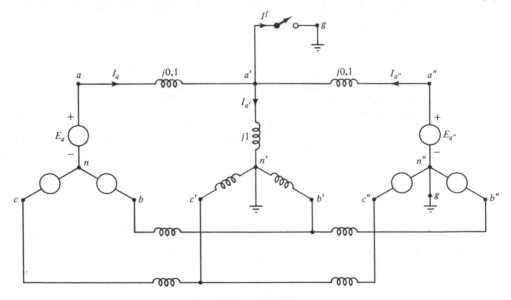

Figure E13.4(a)

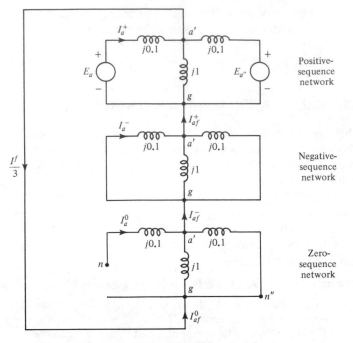

Figure E13.4(b)

by replacing the positive-sequence network by its Thévenin equivalent circuit. In finding the open-circuit voltage we do not need to calculate E_a and $E_{a''}$ since the prefault value of $V_{a'g} = V_{a'g}^{oc} = 1 \underline{/0°}$ is already given. We calculate the Thévenin impedance (three impedances in parallel) and the other parallel impedances and find

$$\frac{I^f}{3} = \frac{1 \underline{/0°}}{j0.0476 + j0.0476 + j0.0909} = -j5.372$$

We next use the method of superposition to find the desired faulted network variables. Since the prefault values (the responses to the voltage sources) are already given, we only need to consider the response of the passive network to $I^f/3 = -j5.3721$ treated as an independent source. To emphasize this we redraw the circuit as shown in Fig. E13.4(c), and label the increments (to the prefault values) with a Δ. We will also refer to this circuit as a Δ network. We also refer to the individual sequence network as Δ networks. We find the following responses to $I^f/3$.

$$\Delta I_a^0 = 0$$

$$\Delta I_a^+ = \Delta I_a^- = \frac{j0.0476}{j0.1}(-j5.372) = -j2.558$$

Thus $\Delta I_a = \Delta I_a^0 + \Delta I_a^+ + \Delta I_a^- = -j5.116$. In this calculation we used the current-divider law. Similarly, we find

$$\Delta I_{a''}^+ = \Delta I_{a''}^- = -j2.558$$

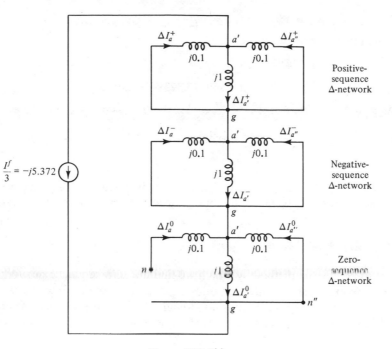

Figure E13.4(c)

$$\Delta I_{a''}^0 = \frac{j0.0909}{j0.1}(-j5.372) = -j4.884$$

Thus $\Delta I_{a''} = \Delta I_{a''}^0 + \Delta I_{a''}^+ + \Delta I_{a''}^- = -j10.0$. By the same technique, we find that $\Delta I_{a'} = j1$.

We now calculate the actual fault currents by adding the increments to the prefault values. Thus

$$I_a = I_a^{pf} + \Delta I_a = 1 - j5.116 = 5.2128\ \underline{/-78.94}$$

$$I_{a'} = I_{a'}^{pf} + \Delta I_{a'} = -j1 + j1 = 0$$

$$I_{a''} = I_{a''}^{pf} + \Delta I_{a''} = -1 - j1 - j10 = 11.0454\ \underline{/-95.19}$$

As a check we note that

$$I^f = I_a - I_{a'} + I_{a''} = -j16.116$$

which checks the value calculated earlier.

Now we calculate I_b. By the same superposition technique,

$$I_b = I_b^{pf} + \Delta I_b = 1\ \underline{/-120°} + \Delta I_b^0 + \Delta I_b^+ + \Delta I_b^-$$

We have already calculated $\Delta I_a^+ = \Delta I_a^- = -j2.558$, $\Delta I_a^0 = 0$. Thus

$$I_b = 1\underline{/-120°} - j2.558(1\ \underline{/-120°} + 1\ \underline{/120°}) = 1.76\ \underline{/106.5°}$$

Finally, we calculate $V_{b'g}$. We get

$$V_{b'g} = V_{b'g}^{pf} + \Delta V_{b'g} = 1 \underline{/-120°} + \Delta V_{b'g}^0 + \Delta V_{b'g}^+ + \Delta V_{b'g}^-$$

From the Δ network we find that

$$\Delta V_{a'g}^+ = \Delta V_{a'g}^- = j0.0476 \cdot j5.372 = -0.2557$$

$$\Delta V_{a'g}^0 = j0.0909 \cdot j5.372 = -0.4883$$

As a check we note that $V_{a'g} = V_{a'g}^{pf} + \Delta V_{a'g} = 0$. Now we return to the calculation of $V_{b'g}$.

$$V_{b'g} = 1 \underline{/-120°} - 0.4883 - 0.2557 (1 \underline{/-120°} + 1 \underline{/120°})$$

$$= 1.134 \underline{/-130°}$$

In a similar manner any quantity of interest may be calculated. The point we are trying to make is that if the prefault values are specified, it is very efficient to use the Δ network.

13.5 ZERO-SEQUENCE NETWORK

We want to review the method by which we find the zero-sequence network from the original network. First, in order for zero-sequence currents to flow, there must be a ground return. In Fig. 13.11 there is no ground return, zero-sequence currents cannot flow, and in the corresponding zero-sequence networks these networks may be represented by open circuits.

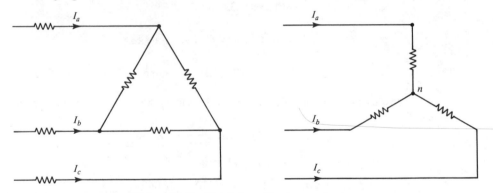

Figure 13.11 Networks without ground return.

We will only consider networks which prior to the fault were symmetrical. Thus the most general impedance element we need to consider in detail is the wye-connected set of impedances with the neutral connected to ground through an impedance Z_g. In this case we have already seen that we replace Z_g by $3Z_g$ in the (single-phase) zero-sequence diagram. The reader can check that the same is true if we are dealing with a wye-connected set of voltage sources. We next consider an example.

Example 13.5

Find the zero-sequence equivalent circuit of the three-phase network shown in Fig. E13.5(a).

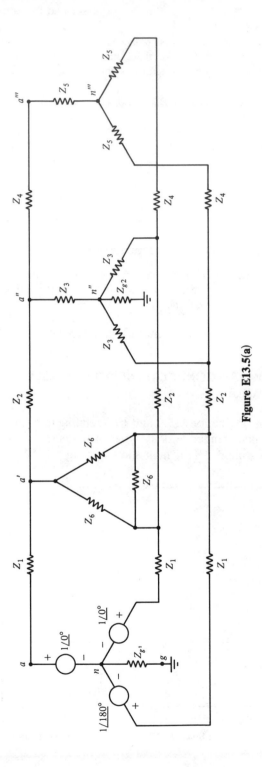

Figure E13.5(a)

Solution This source is not balanced. It has a zero-sequence component with $E_{an}^0 = E_{bn}^0 = E_{cn}^0 = \frac{1}{3}\underline{/0°}$. The delta and ungrounded wye can be replaced by open circuits. The zero-sequence circuit is shown in Fig. E13.5(b).

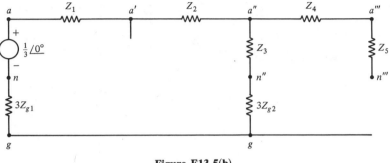

Figure E13.5(b)

Note 1: The "terminals" of this network depend on where the fault is. If the fault is at point a, we attach terminals at a and g. If the fault is at point a', a'', or a''', we attach terminals at a', a'', or a''' and g.

Note 2: If $Z_{g2} \rightarrow \infty$, we get the situation shown for Z_5.

13.6 SEQUENCE NETWORK CONNECTIONS FOR DOUBLE LINE-GROUND FAULT

The developments in Sections 13.2 and 13.3 relating to a single line-to-ground (SLG) fault may easily be extended to other types of faults. For example, consider the case of a double line-ground (DLG) fault shown in Fig. 13.12.

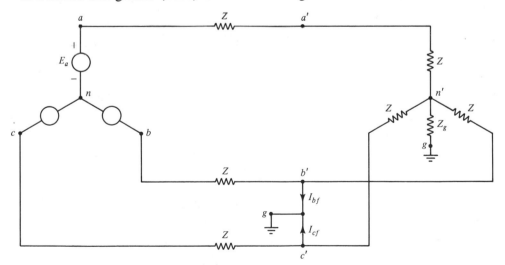

Figure 13.12   Double line-ground fault.

Assume that the voltage sources are a positive-sequence set. Suppose that we wish to find the fault currents and, say, $V_{a'g}$, V_{ag}, and V_{bg}. In solving the problem, we can carry out steps parallel to those discussed in Section 13.2. In step 1 we get two nonzero source currents. In step 2, in the stub, we have $I_{af} = 0$, $I_{bf} \neq 0$, and $I_{cf} \neq 0$ replaced by the nine symmetrical components. In step 3 we use superposition to find $V_{a'g}^+$ in terms of E_a and I_{af}^+, $V_{a'g}^-$ in terms of I_{af}^-, and $V_{a'g}^0$ in terms of I_{af}^0. (Even though $I_{af} = 0$, I_{af}^+, I_{af}^-, and $I_{af}^0 \neq 0$!) We are led to consider the same positive-sequence, negative-sequence, and zero-sequence (single phase) networks as in Section 13.2.

We can bypass the remaining two steps and proceed directly to the connection of the sequence networks. We need to look at the constraints on $V_{a'g}^+$, $V_{a'g}^-$, $V_{a'g}^0$ and on I_{af}^+, I_{af}^-, I_{af}^0 imposed by the DLG fault shown in Fig. 13.12. We note that $V_{b'g} = V_{c'g} = 0$ and $I_{af} = 0$. From (13.9), as applied to the fault voltages, we get

$$\begin{bmatrix} V_{a'g}^0 \\ V_{a'g}^+ \\ V_{a'g}^- \end{bmatrix} = \frac{1}{3} \begin{bmatrix} 1 & 1 & 1 \\ 1 & \alpha & \alpha^2 \\ 1 & \alpha^2 & \alpha \end{bmatrix} \begin{bmatrix} V_{a'g} \\ 0 \\ 0 \end{bmatrix}$$

and therefore

$$V_{a'g}^0 = V_{a'g}^+ = V_{a'g}^- \tag{13.18}$$

We also have, from (13.1) and Fig. 13.12,

$$I_{af} = I_{af}^0 + I_{af}^+ + I_{af}^- = 0 \tag{13.19}$$

Equations (13.18) and (13.19) imply that the positive-, negative-, and zero-sequence networks should be put in parallel. Thus we get Fig. 13.13.

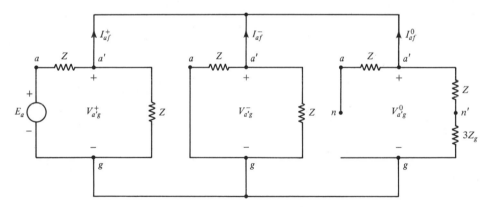

Figure 13.13 Connection of sequence networks for double line-ground fault.

Solving the circuits problem for the sequence fault currents I_{af}^+, I_{af}^-, and I_{af}^0, we can then calculate

$$I_{bf} = I_{bf}^+ + I_{bf}^- + I_{bf}^0$$

$$= \alpha^2 I_{af}^+ + \alpha I_{af}^- + I_{af}^0$$

$$I_{cf} = \alpha I_{af}^+ + \alpha^2 I_{af}^- + I_{af}^0$$

We can easily calculate

$$V_{a'g} = V_{a'g}^+ + V_{a'g}^- + V_{a'g}^0 = 3V_{a'g}^+$$

Looking at point a identified in each of the sequence networks in Fig. 13.13, we can calculate

$$V_{ag} = V_{ag}^+ + V_{ag}^- + V_{ag}^0 = E_a + 0 + V_{a'g}^0$$

and

$$V_{bg} = V_{bg}^+ + V_{bg}^- + V_{bg}^0 = E_a \underline{/-120°} + 0 + V_{a'g}^0$$

These calculations typify the calculation of any of the fault variables. *Note:* If we are not given E_a but are given the prefault voltage $V_{a'g}^{pf} = V_{a'g}^{oc}$, then we can use the Thévenin equivalent circuit for the positive-sequence network just as in Example 13.3. If we are given prefault voltages and currents, it is advantageous to use the Δ network as in Example 13.4, with the sequence fault currents I_{af}^+, I_{af}^-, and I_{af}^0 as independent inputs to calculate the voltage and current increments. *Note:* In the case of DLG fault, I_{af}^+, I_{af}^-, and I_{af}^0 are not equal.

Exercise 6.   In Fig. 13.12 consider the modification of the fault as shown in Fig. Ex6(a). Show that Fig. 13.13 changes to Fig. Ex6(b).

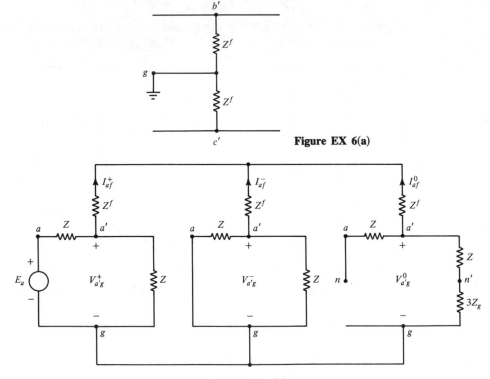

Figure EX 6(a)

Figure EX 6(b)

13.7 SEQUENCE NETWORK CONNECTIONS FOR LINE-LINE FAULT

Suppose that instead of the DLG fault we have the line-line (LL) fault shown in Fig. 13.14. Everything else is the same as in Fig. 13.12. It is clear that we will get the same sequence networks found in Section 13.6. The remaining question is how to "hook them up." We look for relations between the sequence fault voltages and currents which will give the clue. Thus from the fact that $V_{b'g} = V_{c'g}$ and using (13.9), we get

$$\begin{bmatrix} V^0_{a'g} \\ V^+_{a'g} \\ V^-_{a'g} \end{bmatrix} = \frac{1}{3} \begin{bmatrix} 1 & 1 & 1 \\ 1 & \alpha & \alpha^2 \\ 1 & \alpha^2 & \alpha \end{bmatrix} \begin{bmatrix} V_{a'g} \\ V_{b'g} \\ V_{b'g} \end{bmatrix}$$

which implies that

$$V^+_{a'g} = V^-_{a'g} \tag{13.20}$$

Also, again using (13.9),

$$\begin{bmatrix} I^0_{af} \\ I^+_{af} \\ I^-_{af} \end{bmatrix} = \frac{1}{3} \begin{bmatrix} 1 & 1 & 1 \\ 1 & \alpha & \alpha^2 \\ 1 & \alpha^2 & \alpha \end{bmatrix} \begin{bmatrix} 0 \\ I^f \\ -I^f \end{bmatrix}$$

which implies that

$$I^0_{af} = 0 \qquad I^+_{af} = -I^-_{af} \tag{13.21}$$

with $I^+_{af} = \frac{1}{3}(\alpha - \alpha^2)I^f = jI^f/\sqrt{3}$. This implies that

$$I^f = -j\sqrt{3} I^+_{af} \tag{13.22}$$

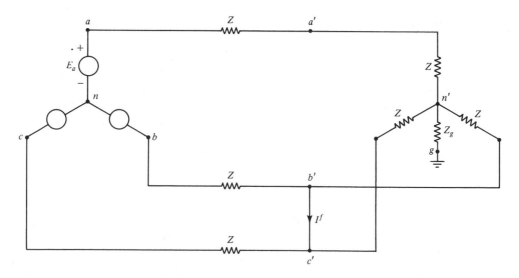

Figure 13.14 Line-line fault.

Equations (13.20) and (13.21) imply that the positive- and negative-sequence networks are in parallel. Equation (13.21) also tells us that $I_{af}^0 = 0$. In that case $V_{a'g}^0 = 0$. Thus the zero-sequence network is "dead" and has no influence on any of the fault variables. Since there is no ground connection at the fault, it is clear physically why $I_{af}^0 = 0$.

Thus we are led to consider the circuit in Fig. 13.15. We solve the circuit for any of the variables of interest, always taking into account the fact that the zero-sequence network is dead. To solve for I^f we can use (13.22).

Exercise 7. Suppose that instead of a solid connection there is an impedance Z^f between b' and c' in Fig. 13.14. Show that Fig. 13.15 will then have an impedance Z^f in the branch connecting the positive- and negative-sequence networks (a' to a').

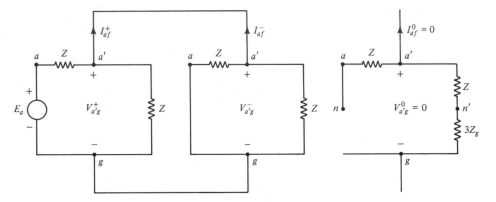

Figure 13.15   Connection of sequence networks for line-line fault.

13.8 MORE GENERAL FAULT CIRCUIT ANALYSIS

In the examples considered so far, the single phase to ground fault occurred on phase a, the line-line faults occurred between phases b and c. These choices were not arbitrary; it is simpler to deal with these "canonical" cases. Suppose then that we have a phase to ground fault on, say, phase b. The easiest way to handle this variation is to relabel the phases so that phase b is now labeled a, phase c is now labeled b, and phase a is now labeled c. A similar relabeling holds for the other possible faults to reduce them to the canonical cases. Of course, care is required so that (originally) positive-sequence sources remain positive-sequence after relabeling; this is ensured if the wires labeled abc end up being labeled either bca or cab (i.e., in the same cyclic order). With this change we can reduce to the canonical cases, and henceforth we will assume that this has been done.

Suppose now that we are given a network more general than the one we have been using as a running example. The network in general will include multiple (positive sequence) three-phase sources. Relying on basic principles, we can justify the following procedure.

General procedure

1. Find the positive-, negative-, and zero-sequence networks.
2. Attach terminal pairs at g and the point of fault.
3. Connect the networks in accordance with the type of fault (consult Figs. 13.10, 13.13, and 13.15 for some examples).
4. Using circuit analysis, find the required a phase symmetrical components; using (13.5), we find the corresponding triples of phase a, b, c variables.

A few comments on the general procedure follow.

1. Regarding item 1, we still need to introduce realistic models for generators, transformer, and transmission lines. This will be done starting in Section 13.10.
2. Regarding item 3, our catalog of faults covers some important cases but is not extensive. More complete listings, including the case of open conductors, may be found in standard references.
3. Regarding item 4, use of the Δ network as described in Example 13.4 simplifies the calculation when relevant prefault voltages and currents are already known.
4. In practice, certain simplifying assumptions are usually made. These will be discussed in Section 13.13.

13.9 POWER FROM SEQUENCE VARIABLES

Consider the complex power delivered to the network shown in Fig. 13.16. We are not restricted to balanced conditions. By the method discussed in Chapter 2, particularly Example 2.4, we can pick any node as a datum node; it is convenient to pick g as the datum node. Then, following Example 2.4,

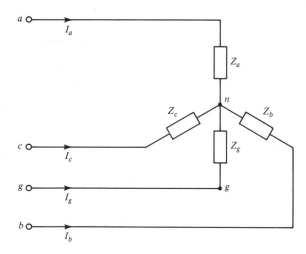

Figure 13.16 Four-wire network.

$$S_{3\phi} = V_{ag}I_a^* + V_{bg}I_b^* + V_{cg}I_c^* \qquad (13.23)$$

In the balanced case with $V_{ag} = V_{an}$, we get

$$S_{3\phi} = 3V_{an}I_a^* \qquad (13.24)$$

In the unbalanced case (13.23) may be used, or its replacement in terms of sequence variables. In considering the transformation it is convenient to use matrix notation. Thus let

$$\mathbf{I} = \begin{bmatrix} I_a \\ I_b \\ I_c \end{bmatrix} \qquad \mathbf{V} = \begin{bmatrix} V_{ag} \\ V_{bg} \\ V_{cg} \end{bmatrix}$$

then we may replace (13.23) by

$$S_{3\phi} = \mathbf{I}^*\mathbf{V} \qquad (13.25)$$

We are using $\mathbf{I}^*$ to denote the complex-conjugate transpose of $\mathbf{I}$. Using (13.7) to express $\mathbf{I}$ and $\mathbf{V}$ in terms of symmetrical components,

$$
\begin{aligned}
S_{3\phi} &= (\mathbf{AI}_s)^*\mathbf{AV}_s \\
&= \mathbf{I}_s^*\mathbf{A}^*\mathbf{AV}_s \\
&= 3\mathbf{I}_s^*\mathbf{V}_s \\
&= 3V_{ag}^0 I_a^{0*} + 3V_{ag}^+ I_a^{+*} + 3V_{ag}^- I_a^{-*} \qquad (13.26)
\end{aligned}
$$

In going from the first to the second line we have used the fact that $(\mathbf{AB})^* = \mathbf{B}^*\mathbf{A}^*$. The third line follows from the second because $\mathbf{A}^*\mathbf{A} = 3\mathbf{1}$, where $\mathbf{1}$ is the identity matrix. Equation (13.26) shows that if we have been working with the sequence variables or networks we can calculate $S_{3\phi}$ without converting back to phase variables. This can be seen in the following example.

Example 13.6   Complex Power Calculation

For the faulted system of Example 13.3, find the complex power delivered to the load.

Solution   In the solution of Example 13.3 we have already calculated $V_{a''g}^+$, $V_{a''g}^-$, and $V_{a''g}^0$, which are the sequence voltages across the load. Thus, using (13.26),

$$
S_{3\phi} = 3V_{a''g}^+\left(\frac{V_{a''g}^+}{j0.9}\right)^* + 3V_{a''g}^-\left(\frac{V_{a''g}^-}{j0.9}\right)^* + 3V_{a''g}^0\left(\frac{V_{a''g}^0}{j0.9}\right)^*
$$

$$
= \frac{3}{-j0.9}(0.9043^2 + 0.0431^2 + 0.8613^2)
$$

$$
= j5.2
$$

Exercise 8.   In Example 13.3 find the three-phase complex power delivered to the load before the fault and compare with the above. How do you interpret the result physically?

We conclude this section by noting that the impedance network in Fig. 13.16 can be replaced by a more general power network, including sinusoidal sources. Equation (13.26) holds for the more general case.

13.10 GENERATOR MODELS FOR SEQUENCE NETWORKS

For simplicity in explaining the method of symmetrical components we have used ideal voltage sources in our examples. We wish now to consider the modeling of synchronous machines (motors and generators) for use in the sequence networks; for simplicity we will refer to these as generators. We assume that the angular velocity of the generator rotor is a constant. Then the electrical equations (10.45) to (10.47) are linear and we can use superposition. Thus the method of symmetrical components based on the principle of superposition will work!

Positive-Sequence Generator Model

In accordance with the basic symmetrical components procedure, as discussed in Section 13.2, we set all negative- and zero-sequence sources to zero. We then have a model with only positive-sequence sources. We have already considered this case in Chapter 10. The steady-state model is described by (10.67). If the machine is round rotor, $X_q = X_d$ and we get the circuit model in Fig. 13.17. Usually, as an approximation, we neglect r. The model in Fig. 13.17 is also used (as an approximation), even in the case $X_q \neq X_d$. The justification is based on the following grounds. We are usually interested in the voltages and currents at or near the fault. These depend strongly on the generators nearest the fault. It turns out that for a generator near the fault, I_a^+, the positive-sequence current component lags E_a by approximately 90° (i.e., $I_a^+ \approx I_{ad}^+$ and we can ignore the quadrature component

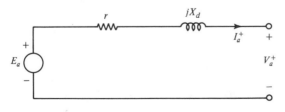

Figure 13.17   Positive-sequence circuit model (steady state).

$jX_q I_{aq}^+$); this leads to the model in Fig. 13.17. On the other hand, if a generator is far from the fault, it need not be modeled accurately because its contribution to the fault currents and voltages is small.

The model in Fig. 13.17 is appropriate for calculating steady-state behavior. We are even more interested, however, in calculating the short-circuit currents and/or voltages at the very beginning of the fault period. In Example 10.4 we calculated the currents for a symmetric short circuit and found that they were much greater than the steady-state values (in the order of 10 times greater). We would like a positive-sequence generator model which at least approximates that kind of behavior.

A model that is frequently used for this purpose is suggested by the *simplified dynamic model* in Section 10.11. Suppose that we calculate E_a' prior to the fault. With E_{fd} and P_M finite, the two differential equations (10.75) and (10.79) imply that E_a' cannot change instantaneously upon application of the fault. Then initially as an approximation we can treat E_a' as a constant in (10.73). In this case we obtain the circuit model shown in Fig. 13.18 (using the justification for neglecting the quadrature component previously discussed). This model, sometimes called the E_a', X_d' model, is used to predict the initial "transient" ac fault currents and voltages. Similarly, we can use a so-called E_a'', X_d'' model to find the "subtransient" ac currents and voltages.

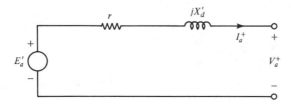

Figure 13.18　Positive-sequence circuit model (transient).

Negative-Sequence Generator Model

In this case we set all zero-sequence and positive-sequence sources to zero. Since the generator field voltage v_F is associated with the positive-sequence variables, it is set equal to zero. To find the relationship between negative-sequence terminal voltages and currents, we can conduct an experiment or test in which we apply negative-sequence currents to the generator terminals and measure the resulting voltages. As an alternative we can calculate the anticipated results by using the Park equations. The interested reader is referred to Appendix 5 for the calculation of the steady-state relations. Physically, the negative-sequence stator currents set up a flux in the air gap whose rotation is opposite to that of the rotor. In this case the damper circuits, or circuits that model the solid-iron rotor, are the dominant elements and their inclusion in the calculations lead to the result that the generator impedance $jX^- = jX_d''$.

Zero-Sequence Generator Model

Using the same approach as above, we conduct a test, or a calculation using the Park equations. The calculation is given in Appendix 5. If we consider the physical test with (identical) zero-sequence currents applied to the generator terminals, then the air-gap flux would be zero if the spatial mmf distribution for each phase were exactly sinusoidal. In this case the generator impedance would be $jX^0 = jX_l$, with X_l the leakage reactance; this result is approximately true and explains the very small values of X^0. We note, finally, that if the generator neutral is grounded through an impedance Z_g, then in the zero-sequence network we have an impedance $3Z_g$ just as described earlier for loads.

We conclude this section with some average numerical values (in per unit) shown in Table 13.1. X^+ may be taken as X_d, X_d', or X_d'', depending on the interval of interest following the imposition of the fault.

TABLE 13.1   Typical Synchronous Machine Reactances

	Turbine-generators (two-pole)		Salient-pole generators with dampers
	Conventionally cooled*	Conductor cooled	
X_d	1.20	1.80	1.25
X_q	1.16	1.75	0.70
X_d'	0.15	0.30	0.30
X_d''	0.09	0.22	0.20
X^-	0.09	0.22	0.20
X^0	0.03	0.12	0.18

*Reactances are representative of smaller air cooled and hydrogen conventionally cooled machines.

Exercise 9.   Suppose that there is an SLG fault close to the generator. Suppose that the generator neutral is solidly grounded ($Z_g = 0$) and the fault is also a solid ground. Can you justify the approximation discussed earlier—that the positive-sequence component of generator current, I_a^+, lags E_a by approximately 90°? *Hint:* Take a simple case, something similar to Fig. 13.2 (with the ideal voltage sources replaced by a realistic generator model). Assume that $|Z|$ is large compared with the generator sequence reactances.

13.11 TRANSFORMER MODELS FOR SEQUENCE NETWORKS

In Chapter 5 we discussed the per phase equivalent circuit for a three-phase transformer bank operating under balanced positive- or negative-sequence conditions.

Neglecting magnetizing inductances, we get the per phase per unit circuit model shown in Fig. 13.19 for Δ-Δ and Y-Y connections. In the case of the Δ-Y or Y-Δ connections, it was shown that the simplified circuit in Fig. 13.19 may also be used if the system is normal and only terminal behavior of the transmission links is required. The reader is reminded that we are assuming that the connection-induced phase shifts of the Δ-Y or Y-Δ connection cancel for each nonradial transmission link.

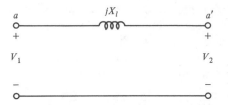

Figure 13.19 Positive- or negative-sequence simplified circuit of three-phase transformer bank.

In short-circuit studies more complete models showing connection-induced phase shifts may also be required and these are shown in Fig. 13.20.

Here we show the standard phase shift with the positive-sequence voltage on the high-voltage side advanced 30° from the voltage on the low-voltage side. Correspondingly, the negative-sequence voltage on the high-voltage side is retarded by 30°.

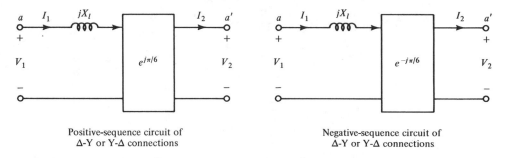

Positive-sequence circuit of
Δ-Y or Y-Δ connections

Negative-sequence circuit of
Δ-Y or Y-Δ connections

Figure 13.20 More complete circuit models.

For all but the simplest systems the representation shown in Fig. 13.20 is too cumbersome and as an alternative we propose the simpler representation shown in Fig. 13.21. The symbol $\oslash$ represents a phase increase in voltage and current in going from left to right or decrease in going from right to left. The slant of the line defines whether there is an increase or decrease in a natural way. Thus with the symbol $\obslash$ there is a decrease in phase when going from left to right. While the magnitude of the phase shift is not explicitly shown, as it is in Fig. 13.20, we will assume that it is the standard 30°.

We next turn to the zero-sequence equivalent circuits for different transformer connections. Here the situation is more variable, with very notable differences for various connections. On the assumption that magnetizing reactances can be neglected

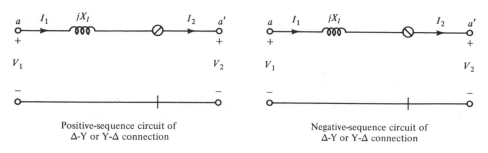

Positive-sequence circuit of
Δ-Y or Y-Δ connection

Negative-sequence circuit of
Δ-Y or Y-Δ connection

Figure 13.21   More complete circuit models.

(i.e., replaced by open circuits), we get the most common transformer connections as tabulated in Fig. 13.22.

It is easy to check the entries in the table. Apply a set of zero-sequence currents to the primary (or secondary) and calculate the resulting secondary (or primary) currents and the primary and secondary voltages. We recall that a set of zero-sequence currents are all identical sinusoids. It is helpful in understanding the circuits to notice that we get an open circuit on the primary (secondary) side if there is no ground return for primary (secondary) currents or if there is no path for the secondary (primary) zero-sequence currents. Thus in the top circuit there is a ground return for the primary zero-sequence currents, but this is not enough, because there is no path for the resulting secondary currents. In the third circuit the primary currents can flow because there is a ground return and paths for the resulting secondary circuits. On the other hand, looking into the secondary circuit we see an open circuit (no ground return).

We can verify these intuitive results by calculation. For example we consider the third circuit in more detail in the following.

Example 13.7   Zero-Sequence Equivalent Circuit

Using the single-phase transformer model in Fig. 5.4, verify the third entry in Fig. 13.22.

Solution   Using Fig. 5.4 (and leaving in the magnetizing inductances), we get Fig. E13.7 for the grounded wye-delta connection. First we might note that by symmetry the secondary voltages are all identical and to satisfy KVL they must in fact be zero. Thus we also get zero volts (a short circuit) on the primary sides of the ideal transformers. From this it follows that $V_a = jX_l I_a^0$. Alternatively, we can proceed more cautiously as follows. From the figure

$$V_a = jX_l I_a^0 + \frac{n_1}{n_2} V_{2a}$$

$$V_b = jX_l I_b^0 + \frac{n_1}{n_2} V_{2b}$$

$$V_c = jX_l I_c^0 + \frac{n_1}{n_2} V_{2c}$$

$$V_a^0 \triangleq \tfrac{1}{3}(V_a + V_b + V_c) = jX_l I_a^0$$

Symbols   Connection diagram   Zero-sequence circuit
                                                                                $Z_0 \approx Z_l$

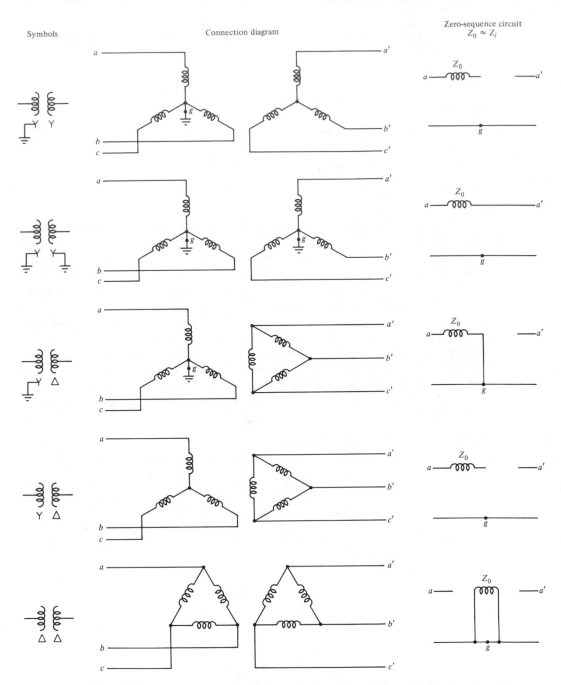

Figure 13.22 Zero-sequence equivalent circuits. (Adapted from W. D. Stevenson, Jr., *Elements of Power Systems Analysis*, 4th ed., McGraw-Hill Book Company, New York, 1982.)

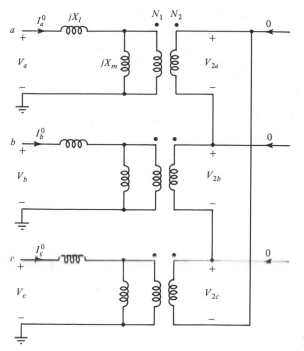

Figure E13.7

We have used the facts $I_a^0 = I_b^0 = I_c^0$, and $V_{2a} + V_{2b} + V_{2c} = 0$. Thus "looking in" from the left, the zero-sequence impedance is $Z^0 = jX_l$. On the other hand, "looking in" from the right we see an open circuit because there is no return path for the zero-sequence currents. This verifies the third entry in Fig. 13.22.

Exercise 10.   In Example 13.7, in modeling the single-phase transformers, X_m was left in; it did not matter because X_m was, effectively, shorted out. Using the more complete single-phase model (with X_m left in), derive the first zero-sequence circuit and compare with the result in the table.

13.12 TRANSMISSION-LINE MODEL FOR SEQUENCE NETWORKS

In Chapters 3 and 4 we developed a number of models suitable for use under balanced three-phase conditions. These may be used to represent the transmission line in the positive-sequence and negative-sequence networks. Usually, the simplest (short-line) model is used. The per phase transmission-line impedance is then $Z_L = R_L + jX_L$; usually, R_L is neglected.

The zero-sequence equivalent circuit is different. In the derivation of the inductances in Chapter 3, we assumed that $i_a + i_b + i_c = 0$. This condition is not satisfied if there is a nonsymmetric fault to ground, and the formula for inductance needs to be reconsidered. To give an indication of what sort of results to expect, we consider the following example.

Example 13.8   Calculation of Zero-Sequence Inductance

Consider a three-phase transmission line with zero-sequence currents in phases a, b, c. Assume that all the current is returned as ground current in an equivalent fictitious conductor. Suppose that we have the configuration shown in Fig. E13.8. Calculate the zero-sequence inductance of phase a and compare with the positive (or negative)-sequence inductance.

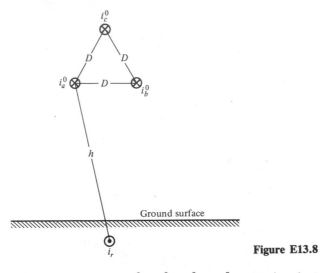

Figure E13.8

Solution   We have $i_r = i_a^0 + i_b^0 + i_c^0 = 3i_a^0$. Using (3.21) yields

$$l_a^0 = \frac{\lambda_a^0}{i_a^0} = \frac{\mu_0}{2\pi}\left(\ln\frac{1}{r'} + \ln\frac{1}{d} + \ln\frac{1}{d} - 3\ln\frac{1}{h}\right)$$

$$= \frac{\mu_0}{2\pi}\ln\frac{h^3}{r'd^2}$$

$$= \frac{\mu_0}{2\pi}\ln\frac{dh^3}{r'd^3}$$

$$= \frac{\mu_0}{2\pi}\left(\ln\frac{d}{r'} + 3\ln\frac{h}{d}\right)$$

Since the inductance for balanced (positive- or negative-sequence) operation is $l_a^+ = (\mu_0/2\pi)\ln(d/r')$, we see that with $h/d > 1$, l_a^0 is greater than l_a^+.

We have idealized matters considerably to simplify the analysis; in practice we do not have the simple case considered here. For example, there is uncertainty about the actual distribution of the ground current. In practice the zero-sequence reactance is determined by test and is found to be approximately three times the positive- or negative-sequence value.

13.13 ASSEMBLY OF SEQUENCE NETWORKS

With the generator, transformer, and transmission-line models proposed in Sections 13.10 to 13.12, we are now in a position to consider more realistic sequence networks than in the introductory examples in Sections 13.2 to 13.7.

As an example, let us consider the one-line system diagram shown in Fig. 13.23. There are two synchronous machines, two lines, and four transformers. The transformer and synchronous machine connections and grounding are shown. The points *LMNOPQRSTU* are identified.

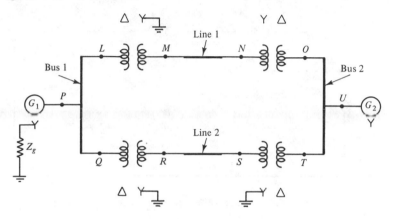

Figure 13.23   System one-line diagram.

We show the positive-sequence network in Fig. 13.24. We will always assume that the sources are balanced and positive-sequence, hence they are included in this network. In the absence of contrary information we will also assume all the Δ-Y transformer banks are connected so that the (connection-induced) positive-sequence voltage on the high-voltage side is advanced by 30° with respect to the low-voltage

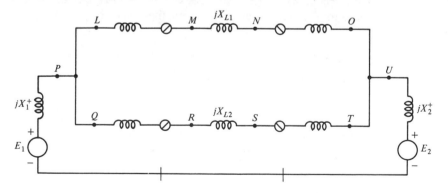

Figure 13.24   Positive-sequence network.

side. This is implied by the phase shifters above. We show the generator reactances in general terms. For steady-state calculations X_d would be used. To get results for the first few cycles, X_d' or X_d'' would be used. The unlabeled four reactances are the transformer leakage reactances of the four transformer banks.

We next consider the negative-sequence network shown in Fig. 13.25. This network differs in three ways from the positive-sequence network. There are no voltage sources; the generator reactances have negative-sequence values; the (connection-induced) phase shift from the low- to the high-voltage side of each transformer is the negative of the phase shift for the positive-sequence case.

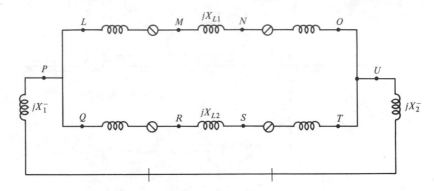

Figure 13.25   Negative-sequence network.

Finally, we find the zero-sequence network by using Fig. 13.22 to look up the two kinds of transformer connections.

We can check the validity of the network shown in Fig. 13.26 by "looking left and right" from the labeled points in Fig. 13.23 and 13.26. For example, referring to Fig. 13.23, looking to the left at point P, we have a neutral grounded through Z_g; in this case we have the zero-sequence impedance of the generator, jX_1^0, in series with $3Z_g$, and this checks with Fig. 13.26. Looking from P to the right, we see open circuits in Fig. 13.23 for the zero-sequence currents; these open circuits are also shown in Fig.

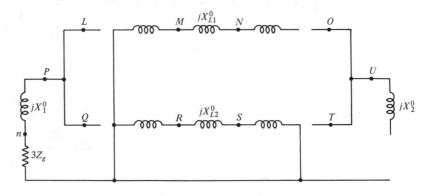

Figure 13.26   Zero-sequence network.

13.26. Looking to the left from point M in Fig. 13.23, we see the third transformer connection in Fig. 13.22, and this is shown to the left of point M in Fig. 13.26. Similarly, we can check all the connections in Fig. 13.26.

13.14 FORMULATION OF THE PROBLEM

We are now in a position to reexamine the use of symmetrical components for a realistic power system problem. At the same time we will introduce some simplifying assumptions.

As discussed in Chapter 6, a power system under normal (prefault) operating conditions is specified by complex load power at the P, Q buses, voltage at the slack bus, and active power injections and voltage magnitudes at the remaining generator buses. All the unknown (complex) bus voltages may then be found by a power flow study, and from these as many line, load, and generator currents as desired may be calculated. In principle we can also calculate the required internal (complex) generator voltages E_a, E_a', or E_a'' behind the appropriate generator impedances. For the given and calculated V_i's we can also calculate "equivalent" impedance loads to replace the specified S_{Di}; this can be done by using $Z_{Di} = V_i/I_{Di} = V_i^* V_i / V_i^* I_{Di} = |V_i|^2 / S_{Di}^*$. Thus we get one kind of model for the loads. By this means, in principle, we can obtain two things: (1) a circuit model with voltage sources, and (2) a specification of the prefault voltages and currents. Thus we are in a position to use the *general procedure* outlined in Section 13.8.

In practice this is not done; certain simplifications are made which significantly reduce the computational effort without seriously degrading accuracy. In solving practical power system fault problems, the following are observed:

1. The prefault voltage at the fault point is usually close to its nominal value (i.e., it can be taken to be $1 \underline{/0°}$).
2. Load impedances show up as "shunt" elements in the sequence networks. Generators also show up as shunt elements and usually have much lower impedances. It is also true that the load impedances are (usually) mostly resistive, whereas the generator impedances are almost purely reactive. In calculating the fault currents I^f and I_{af}^+, I_{af}^-, and I_{af}^0, we need the equivalent (parallel) impedance of each sequence network and, for the two reasons mentioned above, we find that the loads do not appreciably affect the result. Thus, at least for calculating these fault currents, we leave them out (i.e., remove the loads).
3. In solving for the other currents in the faulted system it is found that the prefault currents are much smaller than the currents of interest under faulted condition; they are also largely out of phase. Thus it is not a bad assumption to neglect the prefault currents (i.e., assume that they are zero). In this case we need only consider the responses of the Δ network to the fault currents I_{af}^+, I_{af}^-, and I_{af}^0. In this network, and for the reasons mentioned in 2, the load impedances may

(usually) be left out. We have been discussing currents; an analogous situation holds for voltages. Consistent with the neglect of prefault currents, we can neglect all prefault "series" voltage drops (due to the line currents) and set all prefault voltage magnitudes to 1.0. This is commonly called a "flat profile."

This discussion motivates the adoption of simplifying assumptions 1 and 4 below. The two additional assumptions are offered in the same spirit; they simplify the calculation without degrading the accuracy.

Assumption 1: Load impedances can be ignored.

Assumption 2: All other shunt elements in line and transformer models can be ignored.

Assumption 3: All series resistances in lines, transformers, and generators can be ignored.

Assumption 4: The prefault system is "unloaded," with zero currents, with a phase a voltage at the fault point equal to $1 \underline{/0°}$ and with a flat voltage profile.

A few additional comments follow.

1. In defining the prefault flat voltage profile, we do take into account the connection-induced phase shifts in Δ-Y transformer banks.

2. Since we assume that all prefault line currents are zero, then for the faulted system $I_{ai} = I_{ai}^{pf} + \Delta I_{ai} = \Delta I_{ai}$. In this case we can abandon the more cumbersome Δ notation. On the other hand, for voltage, we will continue to use the Δ notation, as in $V_{ai} = V_{ai}^{pf} + \Delta V_{ai}$.

3. We recall that in modeling the transformers the shunt magnetizing reactances were neglected. In modeling the transmission lines the shunt capacitances were neglected. Since these are relatively high shunt impedances, their neglect is certainly justifiable in light of the preceding discussion.

4. It is very nice that assumptions 1 and 4 are physically reasonable. It not only simplifies each calculation but greatly reduces the number of cases to be considered. We do not have to calculate for a variety of load conditions. It suffices to calculate for a variety of system configurations and fault locations and types.

5. We have ended up with a circuit model consisting only of sources, phase shifters, and pure reactances (inductances); sometimes the circuit is called a reactance diagram. The model is a boon for hand calculation since many of the calculations will be possible without using complex numbers.

We continue with an example using these simplifications and the models of power system elements discussed in Sections 13.10 to 13.12.

Example 13.9   Fault Calculation

The one-line diagram is given in Fig. E13.9 with a single line-to-ground (SLG) fault shown at the point a' at the left and of the transmission line. Assume that the prefault voltage $V_{a'n} = 1 \underline{/0°}$. Calculate I^f, I_a, I_b, I_c, $I_{a'}$, $I_{b'}$, $I_{c'}$, $V_{b'g}$, and $V_{c'g}$, using generator

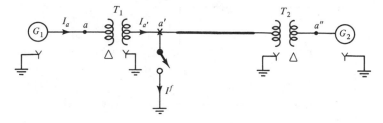

Figure E13.9(a)

transient reactances. This information is useful in designing a protective relaying scheme to protect the equipment in the event of an SLG fault at point a'.

Solution   We follow the *general procedure* described in Section 13.8, the models described in Sections 13.10 to 13.12, and the *simplifying assumptions* just described. The individual sequence networks are assembled as described in Section 13.13. In fact the one-line diagram in Fig. E13.9(a) is the same as the lower part of the one-line diagram in Fig. 13.23. Because this is a SLG fault we connect the sequence networks in series. This is shown in Fig. E13.9(b) together with the assumed values of the reactances. In the figure we also show the generator voltages found by "working back" from the given prefault voltage $V_{a'n} = V_{a'g} = 1 \underline{/0°}$.

We next find $I^f/3$ by replacing the positive-sequence network with a Thévenin equivalent. $V_{a'g}^{oc} = V_{a'g}^{pf} = 1 \underline{/0°}$. The Thévenin equivalent impedance is found by setting the voltage sources equal to zero and measuring the driving-point impedance at the terminals $a'g$. In this connection, as shown in Section 5.4, the phase shifters may be completely ignored. Using the rule for combining parallel impedances, we find the following driving-point impedances:

$$Z_{a'g}^+ = Z_{a'g}^- = j0.1714 \qquad Z_{a'g}^0 = j0.0800$$

Then

$$\frac{I^f}{3} = \frac{1 \underline{/0°}}{j0.1714 + j0.1714 + j0.0800} = -j2.365$$

Thus $I^f = -j7.095$. Next we can calculate $\Delta V_{a'g}^+$, $\Delta V_{a'g}^-$, and $\Delta V_{a'g}^0$, responses of the Δ network to $I^f/3 = -j2.365$. We find

$$\Delta V_{a'g}^+ = \Delta V_{a'g}^- = j2.365 \cdot j0.1714 = -0.4054$$

$$\Delta V_{a'g}^0 = j2.365 \cdot j0.0800 = -0.1892$$

As a check, we find that $\Delta V_{a'g} = \Delta V_{a'g}^0 + \Delta V_{a'g}^+ + \Delta V_{a'g}^- = -1.0$. Thus

$$V_{a'g} = V_{a'g}^{pf} + \Delta V_{a'g} = 0$$

which checks the physically obvious fact. Now we can calculate

$$V_{b'g} = V_{bg}^{pf} + \Delta V_{b'g}^0 + \Delta V_{b'g}^+ + \Delta V_{b'g}^-$$

$$= 1 \underline{/-120°} + \Delta V_{a'g}^0 + \alpha^2 \Delta V_{a'g}^+ + \alpha \Delta V_{a'g}^-$$

$$= 1 \underline{/-120°} - 0.1892 - 0.4054(\alpha^2 + \alpha) = 0.9113 \underline{/-108.1°}$$

$$V_{c'g} = 1 \underline{/120°} - 0.1892 - 0.4054(\alpha + \alpha^2) = 0.9113 \underline{/108.1°}$$

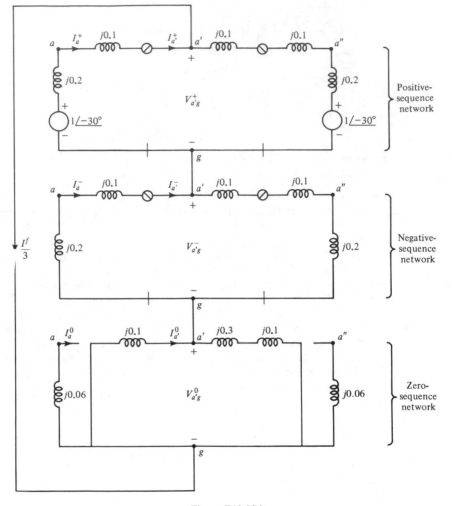

Figure E13.9(b)

Next we can find $\Delta I_{a'}^+$, $\Delta I_{a'}^-$, and $\Delta I_{a'}^0$; since the prefault currents are all assumed zero these are also the total currents I_a^+, I_a^-, and I_a^0. The easiest way to find these currents is to use the values $\Delta V_{a'g}^+$, $\Delta V_{a'g}^-$, and $\Delta V_{a'g}^0$ already calculated. For the Δ network, with all voltage sources set to zero,

$$\Delta I_{a'}^+ = \frac{-\Delta V_{a'g}^+}{j0.3} = 1.351 \underline{/-90°} = I_{a'}^+$$

$$\Delta I_{a'}^- = \frac{-\Delta V_{a'g}^-}{j0.3} = 1.351 \underline{/-90°} = I_{a'}^-$$

$$\Delta I_{a'}^0 = \frac{-\Delta V_{a'g}^0}{j0.1} = 1.893 \underline{/-90°} = I_{a'}^0$$

As an alternative we could calculate $V_{a'g}^{+} = 1 - 0.4054 = 0.5946$ and use the positive-sequence network to calculate $I_{a'}^{+}$. In either case we next find

$$I_{a'} = I_{a'}^{0} + I_{a'}^{+} + I_{a'}^{-} = 4.595 \;\underline{/-90°}$$

$$I_{b'} = I_{a'}^{0} + \alpha^2 I_{a'}^{+} + \alpha I_{a'}^{-} = 0.542 \;\underline{/-90°}$$

$$I_{c'} = I_{a'}^{0} + \alpha I_{a'}^{+} + \alpha^2 I_{a'}^{-} = 0.542 \;\underline{/-90°}$$

We note that $I_{a'} + I_{b'} + I_{c'} = 3I_{a'}^{0} = 5.679 \;\underline{/-90°}$ is the current in the grounded neutral of transformer T_1. (Certainly, this is an indicator of a fault condition and could be used as a signal to initiate protective action.)

We next consider the calculation of I_a, I_b, and I_c. For the first time we will need to consider the connection-induced phase shifts. It is important to note that the phase shifts are different for the positive- and negative-sequence components. Thus, referring to the appropriate sequence networks,

$$I_a^{+} = I_{a'}^{+} e^{-j\pi/6} = 1.351 \;\underline{/-120°}$$

$$I_a^{-} = I_{a'}^{-} e^{j\pi/6} = 1.351 \;\underline{/-60°}$$

$$I_a^{0} = 0$$

Thus

$$I_a = I_a^{0} + I_a^{+} + I_a^{-} = 2.340 \;\underline{/-90°}$$

$$I_b = I_a^{0} + \alpha^2 I_a^{+} + \alpha I_a^{-} = 2.340 \;\underline{/90°}$$

$$I_c = I_a^{0} + \alpha I_a^{+} + \alpha^2 I_a^{-} = 0$$

These are very different from the currents $I_{a'}$, $I_{b'}$, and $I_{c'}$ on the secondary side of the transformer bank! In particular, it should be noted that the corresponding currents are not related by a simple phase shift as in the balanced case; the magnitudes are also very different.

As a check we note that $I_a + I_b + I_c = 0$. This is consistent with the lack of a return path for the primary currents into the transformer bank. The reader may wish to check further the consistency of the results regarding primary and secondary currents by considering the currents in the actual Δ-Y transformer connection in more detail; the connection is shown in Fig. 5.7.

In the examples with $V_{a'n}^{pf}$ specified, the connection-induced phase shifts could be completely ignored in calculating all quantities of interest on the fault side of the transformer banks. Such variables included $V_{a'g}$, $V_{b'g}$, $I_{a'}$, $I_{b'}$, and $I_{c'}$. In fact, this is true in general for a normal system of arbitrary configuration. More precisely, we have the following rule.

Rule.   In the Δ network, for calculating all quantities of interest on the fault side of transformer banks, all connection-induced phase shifts may be completely ignored.

We conclude this section with a few additional observations and comments.

1. We have ignored the unidirectional, or dc offset, component of generator current; this component was discussed in Example 10.4 on generator short circuits. The effect of this component is to increase the maximum instantaneous current seen by bus structures, circuit breakers, and other equipment immediately following a short circuit. In designing circuit breakers to withstand the associated large mechanical forces, it is considered adequate to multiply the calculated subtransient current by a factor of 1.6. These unidirectional components decay very rapidly but may still be present in high-speed circuit breakers at the time the breaker contacts open. Again, design factors may be used to include this effect when specifying the "interrupting capacity" of a circuit breaker.

2. Another factor that tends to increase the initial short-circuit current is the presence of synchronous and induction motors. Due to inertia these machines continue to turn and, under short-circuit conditions, can act as generators, at least momentarily. It is conventional to model the larger of these machines in short-circuit studies.

13.15 MATRIX METHODS

For all but the simplest networks, computer solutions are necessary in solving for the currents and voltages of the faulted network. The principles are the same as those we have been using; we simply use matrix notation.

For simplicity we will assume that faults can only occur at buses. We will make the following assumptions initially; later, we can introduce the same simplifications made in Section 13.14.

1. The generators, transformers, and transmission lines are modeled as discussed in Sections 13.10 to 13.12.
2. Loads, if included, are represented by impedances. If, instead, the complex load powers are given, these may be replaced by $Z_{Di} = |V_i|^2/S_{Di}^*$.
3. There are n buses. The fault occurs at bus q.
4. Except for the fault the network is symmetrical.

The situation is shown in Fig. 13.27. A fault of general type is shown at bus q. The fault impedance network in general is not symmetrical. It can be picked to represent any type of fault we have considered so far. As shown in Fig. 13.27, the reference direction for fault current is away from the bus. Note that the fault bus location is arbitrary; it may be a generator bus or a load bus. We also note that generator buses may also have loads attached.

The objective is to find a desired set of currents and/or bus voltages in the faulted network. As an intermediate step we seek to find *all* the bus voltages. We find these bus voltages by superposition, treating the fault currents as independent current sources. Suppose that the vector fault current at bus q is $\mathbf{I}^f$, where the components of

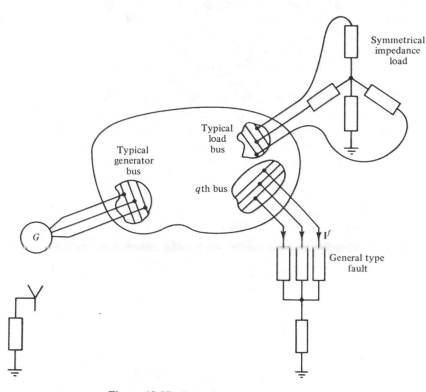

Figure 13.27 Power system representation.

$\mathbf{I}^f$ are the fault phase currents I_{af}, I_{bf}, and I_{cf}. In accordance with the method of symmetrical components, we replace $\mathbf{I}^f$ by

$$\mathbf{I}^f = \mathbf{I}_f^0 + \mathbf{I}_f^+ + \mathbf{I}_f^- \tag{13.27}$$

as in (13.3). We then use the method of superposition to find the responses of the system in Fig. 13.27 to the generator voltages and to the current sources $\mathbf{I}_f^+$, $\mathbf{I}_f^-$, and $\mathbf{I}_f^0$. In finding the response to the generator voltages and to $\mathbf{I}_f^+$ we use the (three-phase) positive-sequence network version of the system in Fig. 13.27; in finding the response to $\mathbf{I}_f^-$ we use the (three-phase) negative-sequence network, while the (three-phase) zero-sequence network is used in finding the response to $\mathbf{I}_f^0$.

We assume that we already know the prefault (positive-sequence) bus voltages; later as an approximation we can assume a flat profile. Thus we know $\mathbf{V}^{pf}$, with components V_i^{pf}, $i = 1, 2, \ldots, n$. As usual, the subscript a is understood.

In finding the responses to the sources $\mathbf{I}_f^+$, $\mathbf{I}_f^-$, and $\mathbf{I}_f^0$ we set all voltage sources to zero (i.e., we deal with the Δ network and will use the Δ notation to indicate the responses). Just as earlier, in Section 13.2, we make full use of the three-phase symmetry in the system and deal with the single-phase sequence networks. Using the

Δ notation, $\Delta\mathbf{V}^+$ is the response of the (single-phase) positive-sequence network to I_{af}^+; $\Delta\mathbf{V}^-$ is the similar response to I_{af}^-; $\Delta\mathbf{V}^0$ is the response to I_{af}^0. The vectors are n-vectors composed of the a phase components at each bus.

Relating bus voltages (i.e., line-neutral voltages) to injected currents is a problem that we considered in Section 6.1. There we used the bus admittance matrix $\mathbf{Y}_{bus}$ to relate bus voltages to (injected) bus currents.

In fact, the positive-sequence Δ network differs from the network used for power flow studies only in the addition of source and load impedances to neutral. The addition of such an impedance between the ith node and the reference or neutral node adds a corresponding admittance to the ith (diagonal) element of $\mathbf{Y}_{bus}$. The off-diagonal elements remain unchanged. Thus we can form

$$\mathbf{Y}_{bus}^+ = \mathbf{Y}_{bus} + \mathbf{Y}_G^+ + \mathbf{Y}_D \tag{13.28}$$

where $\mathbf{Y}_G^+ = \text{diag}\{Y_{Gi}^+\}$, $\mathbf{Y}_D = \text{diag}\{Y_{Di}\}$, and $Y_{Gi}^+ = (Z_{Gi}^+)^{-1}$, $Y_{Di} = (Z_{Di})^{-1}$.

Using $\mathbf{Y}_{bus}^+$ to relate $\Delta\mathbf{V}^+$ to the injected fault current at bus q, we get

$$\begin{bmatrix} 0 \\ \cdot \\ 0 \\ -I_{af}^+ \\ 0 \\ \cdot \\ 0 \end{bmatrix} = \begin{bmatrix} y_{11}^+ & \cdots & y_{1q}^+ & \cdots & y_{1n}^+ \\ \cdot & & \cdot & & \cdot \\ \cdot & & \cdot & & \cdot \\ y_{q1}^+ & \cdots & y_{qq}^+ & \cdots & y_{qn}^+ \\ \cdot & & \cdot & & \cdot \\ \cdot & & \cdot & & \cdot \\ y_{n1}^+ & \cdots & y_{nq}^+ & \cdots & y_{nn}^+ \end{bmatrix} \begin{bmatrix} \Delta V_1^+ \\ \cdot \\ \cdot \\ \Delta V_q^+ \\ \cdot \\ \cdot \\ \Delta V_n^+ \end{bmatrix} \tag{13.29}$$

where only the qth row of the injected vector current is nonzero. We note that the injected current is $-I_{af}^+$ because the reference direction for fault current in Fig. 13.27 is outward.

In fact, we need the inverse relationship since we seek the voltages. Thus defining $\mathbf{Z}_{bus}^+ = [\mathbf{Y}_{bus}^+]^{-1}$ with elements Z_{ij}^+, we have

$$\begin{bmatrix} \Delta V_1^+ \\ \cdot \\ \cdot \\ \cdot \\ \Delta V_q^+ \\ \cdot \\ \cdot \\ \Delta V_n^+ \end{bmatrix} = \begin{bmatrix} z_{11}^+ & \cdots & z_{1q}^+ & \cdots & z_{1n}^+ \\ \cdot & & \cdot & & \cdot \\ \cdot & & \cdot & & \cdot \\ \cdot & & \cdot & & \cdot \\ z_{q1}^+ & \cdots & z_{qq}^+ & \cdots & z_{qn}^+ \\ \cdot & & \cdot & & \cdot \\ \cdot & & \cdot & & \cdot \\ z_{n1}^+ & \cdots & z_{nq}^+ & \cdots & z_{nn}^+ \end{bmatrix} \begin{bmatrix} 0 \\ \cdot \\ 0 \\ -I_{af}^+ \\ 0 \\ \cdot \\ 0 \end{bmatrix} \tag{13.30}$$

We note that with only a single nonzero element in the current vector

$$\Delta V_i^+ = -z_{iq}^+ I_{af}^+ \qquad i = 1, 2, \ldots, n \tag{13.31}$$

Thus for a fault at the qth bus we need only the qth column of $\mathbf{Z}_{\text{bus}}^{+}$ to find all the positive-sequence bus voltage increments due to the current I_{af}^{+}.

To find $\Delta \mathbf{V}^{-}$, the response to I_{af}^{-}, the discussion is almost exactly the same as the above with the substitution of "negative sequence" for "positive sequence." In (13.28) negative-sequence generator impedances need to be used. Thus

$$\mathbf{Y}_{\text{bus}}^{-} = \mathbf{Y}_{\text{bus}}^{-} + \mathbf{Y}_{G}^{-} + \mathbf{Y}_{D} \tag{13.32}$$

where $\mathbf{Y}_{G}^{-} = \text{diag} \{Y_{Gi}^{-}\}$ and $Y_{Gi}^{-} = (Z_{Gi}^{-})^{-1}$. Defining $\mathbf{Z}_{\text{bus}}^{-} = [\mathbf{Y}_{\text{bus}}^{-}]^{-1}$, we find

$$
\begin{bmatrix} \Delta V_1^- \\ \cdot \\ \cdot \\ \cdot \\ \Delta V_q^- \\ \cdot \\ \cdot \\ \Delta V_n^- \end{bmatrix}
=
\begin{bmatrix}
z_{11}^- & \cdots & z_{1q}^- & \cdots & z_{1n}^- \\
\cdot & & \cdot & & \cdot \\
\cdot & & \cdot & & \cdot \\
\cdot & & \cdot & & \cdot \\
z_{q1}^- & \cdots & z_{qq}^- & \cdots & z_{qn}^- \\
\cdot & & \cdot & & \cdot \\
\cdot & & \cdot & & \cdot \\
z_{n1}^- & \cdots & z_{nq}^- & \cdots & z_{nn}^-
\end{bmatrix}
\begin{bmatrix} 0 \\ \cdot \\ \cdot \\ 0 \\ -I_{af}^- \\ 0 \\ \cdot \\ 0 \end{bmatrix}
\tag{13.33}
$$

Finally, consider $\Delta \mathbf{V}^{0}$, the response to I_{af}^{0}. Here we need to deal with the (single-phase) zero-sequence network. Given the network (as, for example, in Fig. 13.26) with buses identified on it, and taking the ground node as reference, we can form $\mathbf{Y}_{\text{bus}}^{0}$. Defining $\mathbf{Z}_{\text{bus}}^{0} = [\mathbf{Y}_{\text{bus}}^{0}]^{-1}$, we find

$$
\begin{bmatrix} \Delta V_1^0 \\ \cdot \\ \cdot \\ \cdot \\ \Delta V_q^0 \\ \cdot \\ \cdot \\ \Delta V_n^0 \end{bmatrix}
=
\begin{bmatrix}
z_{11}^0 & \cdots & z_{1q}^0 & \cdots & z_{1n}^0 \\
\cdot & & \cdot & & \cdot \\
\cdot & & \cdot & & \cdot \\
\cdot & & \cdot & & \cdot \\
z_{q1}^0 & \cdots & z_{qq}^0 & \cdots & z_{qn}^0 \\
\cdot & & \cdot & & \cdot \\
\cdot & & \cdot & & \cdot \\
z_{n1}^0 & \cdots & z_{nq}^0 & \cdots & z_{nn}^0
\end{bmatrix}
\begin{bmatrix} 0 \\ \cdot \\ \cdot \\ 0 \\ -I_{af}^0 \\ 0 \\ \cdot \\ 0 \end{bmatrix}
\tag{13.34}
$$

We consider next a useful merging of (13.30), (13.33), and (13.34) into a single matrix equation. Define

$$
\Delta \mathbf{V}_{si} =
\begin{bmatrix} \Delta V_i^0 \\ \Delta V_i^+ \\ \Delta V_i^- \end{bmatrix}
\qquad
\mathbf{I}_s^f =
\begin{bmatrix} I_{af}^0 \\ I_{af}^+ \\ I_{af}^- \end{bmatrix}
\qquad
\mathbf{Z}_{sij} =
\begin{bmatrix} z_{ij}^0 & 0 & 0 \\ 0 & z_{ij}^+ & 0 \\ 0 & 0 & z_{ij}^- \end{bmatrix}
\tag{13.35}
$$

Then it is easy to check that we can replace (13.30), (13.33), and (13.34) with the following single matrix equation:

$$
\begin{bmatrix}
\Delta \mathbf{V}_{s1} \\
\cdot \\
\cdot \\
\cdot \\
\Delta \mathbf{V}_{sq} \\
\cdot \\
\cdot \\
\cdot \\
\Delta \mathbf{V}_{sn}
\end{bmatrix}
=
\begin{bmatrix}
\mathbf{Z}_{s11} & \cdots & \mathbf{Z}_{s1q} & \cdots & \mathbf{Z}_{s1n} \\
\cdot & & \cdot & & \cdot \\
\cdot & & \cdot & & \cdot \\
\mathbf{Z}_{sq1} & \cdots & \mathbf{Z}_{sqq} & \cdots & \mathbf{Z}_{sqn} \\
\cdot & & \cdot & & \cdot \\
\cdot & & \cdot & & \cdot \\
\cdot & & \cdot & & \cdot \\
\mathbf{Z}_{sn1} & \cdots & \mathbf{Z}_{snq} & \cdots & \mathbf{Z}_{snn}
\end{bmatrix}
\begin{bmatrix}
0 \\
\cdot \\
0 \\
-\mathbf{I}_s^f \\
0 \\
\cdot \\
0
\end{bmatrix}
\tag{13.36}
$$

It should be noted that while the Z matrix is very large, $3n \times 3n$, it may be viewed simply as a repository of impedance information that will permit us to calculate any bus voltage for a fault at any bus location. In calculations we deal with the individual $\mathbf{Z}_{sij}$, which are 3×3.

Equation (13.36) gives the response of the Δ network to the sequence fault currents. If we want the total response, we need to add the prefault bus voltages. These are positive-sequence voltages; the prefault negative-sequence and zero-sequence components are zero. Thus, define the 3-vector

$$
\mathbf{V}_{si}^{pf} =
\begin{bmatrix}
0 \\
V_i^{pf} \\
0
\end{bmatrix}
\qquad i = 1, 2, \ldots, n
\tag{13.37}
$$

Then, we have, by superposition, the total (vector) sequence fault voltage at the ith bus,

$$
\mathbf{V}_{si}^f = \mathbf{V}_{si}^{pf} + \Delta \mathbf{V}_{si} \qquad i = 1, 2, \ldots, n
\tag{13.38}
$$

where the superscript f designates the faulted condition. Using (13.36) in (13.38), we have

$$
\mathbf{V}_{si}^f = \mathbf{V}_{si}^{pf} - \mathbf{Z}_{siq} \mathbf{I}_s^f, \qquad i = 1, 2, \ldots, n
\tag{13.39}
$$

From this it may be seen that once we know $\mathbf{I}_s^f$ we can find all the sequence bus voltages of the faulted network. To find $\mathbf{I}_s^f$ we concentrate on conditions at the fault location (i.e., at the qth bus). From (13.39) with $i = q$,

$$
\mathbf{V}_{sq}^f = \mathbf{V}_{sq}^{pf} - \mathbf{Z}_{sqq} \mathbf{I}_s^f
\tag{13.40}
$$

This gives one relationship between fault current and bus voltage at the fault location. An additional relationship may be found by considering the constraint imposed by the fault impedance.

Suppose that we have a general fault as described in Fig. 13.28. If $Z_a = Z_b = Z_c$, we have a symmetrical three-phase fault. If $Z_a = 0$, and $Z_b = Z_c = \infty$, we have an SLG fault. If $Z_a = Z_g = \infty$, we have a LL fault. If $Z_a = \infty$, we have a DLG fault. Thus all the cases we have considered before are included, as well as some more general cases.

The fault imposes a constraint between the bus voltages (at the qth bus) and the

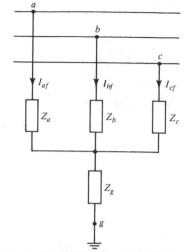

Figure 13.28 General type of fault.

fault currents. We can first state the constraint in terms of the (phase) fault currents I_{af}, I_{bf}, and I_{cf} and the corresponding (phase) voltages V^f_{ag}, V^f_{bg}, and V^f_{cg}. Defining a vector $\mathbf{V}^f_q$ with these three (phase) voltages as components, we have the following relation:

$$\mathbf{I}^f = \mathbf{Y}^f\mathbf{V}^f_q \tag{13.41}$$

assuming that the *fault admittance matrix* $\mathbf{Y}^f$ exists. Alternatively, we can consider the inverse relation

$$\mathbf{V}^f_q = \mathbf{Z}^f\mathbf{I}^f \tag{13.42}$$

with $\mathbf{Z}^f$ called the *fault impedance matrix*. We can next transform to sequence quantities and determine the relationship between $\mathbf{I}^f_s$ and $\mathbf{V}^f_{sq}$. From (13.7) applied to (13.41),

$$\mathbf{A}\mathbf{I}^f_s = \mathbf{Y}^f\mathbf{A}\mathbf{V}^f_{sq} \tag{13.43}$$

or

$$\mathbf{I}^f_s = \mathbf{A}^{-1}\mathbf{Y}^f\mathbf{A}\mathbf{V}^f_{sq}$$

$$\quad = \mathbf{Y}^f_s\mathbf{V}^f_{sq} \tag{13.44}$$

where $\mathbf{Y}^f_s \triangleq \mathbf{A}^{-1}\mathbf{Y}^f\mathbf{A}$. Thus (13.44) is a symmetrical components version of (13.41). Similarly, a symmetrical components version of (13.42) is

$$\mathbf{V}^f_{sq} = \mathbf{Z}^f_s\mathbf{I}^f_s \tag{13.45}$$

where $\mathbf{Z}^f_s = \mathbf{A}^{-1}\mathbf{Z}^f\mathbf{A}$. It is easy to show that $\mathbf{Z}^f_s$ is the inverse of $\mathbf{Y}^f_s$. We defer consideration of some examples and continue with the major development.

Equations (13.40) and (13.45) provide two equations relating $\mathbf{V}^f_{sq}$ and $\mathbf{I}^f_s$. Assuming that $\mathbf{Z}^f_s$ exists, we can proceed as follows. Using (13.45) in (13.40), we have

$$\mathbf{Z}_s^f \mathbf{I}_s^f = \mathbf{V}_{sq}^{pf} - \mathbf{Z}_{sqq} \mathbf{I}_s^f \tag{13.46}$$

Solving for $\mathbf{I}_s^f$ yields

$$\mathbf{I}_s^f = (\mathbf{Z}_s^f + \mathbf{Z}_{sqq})^{-1} \mathbf{V}_{sq}^{pf} \tag{13.47}$$

Using this result in (13.45), we find

$$\mathbf{V}_{sq}^f = \mathbf{Z}_s^f (\mathbf{Z}_s^f + \mathbf{Z}_{sqq})^{-1} \mathbf{V}_{sq}^{pf} \tag{13.48}$$

The relationships in (13.47) and (13.48) can be remembered more easily, and interpreted, using Fig. 13.29.

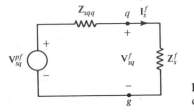

Figure 13.29 Thévenin equivalent circuit.

If $\mathbf{Z}_s^f$ does not exist, we can proceed as follows. In (13.40) use (13.44) (i.e., let $\mathbf{I}_s^f = \mathbf{Y}_s^f \mathbf{V}_{sq}^f$). Then

$$\mathbf{V}_{sq}^f = (\mathbf{1} + \mathbf{Z}_{sqq} \mathbf{Y}_s^f)^{-1} \mathbf{V}_{sq}^{pf} \tag{13.49}$$

and using (13.44) again,

$$\mathbf{I}_s^f = \mathbf{Y}_s^f (\mathbf{1} + \mathbf{Z}_{sqq} \mathbf{Y}_s^f)^{-1} \mathbf{V}_{sq}^{pf} \tag{13.50}$$

Thus we have an alternative expression for $\mathbf{I}_s^f$ which depends on $\mathbf{Y}_s^f$ rather than $\mathbf{Z}_s^f$. Either expression for $\mathbf{I}_s^f$ [i.e., either (13.47) or (13.50)] can be used to find all the (sequence) bus voltages by the use of (13.39). We note that if $\mathbf{Z}_s^f$ and $\mathbf{Y}_s^f$ both exist, then (13.47) and (13.50) are equivalent.

Exercise 11. Prove the equivalence just claimed. *Hint:* Use the matrix identity $[\mathbf{AB}]^{-1} = \mathbf{B}^{-1}\mathbf{A}^{-1}$.

While (13.47) and (13.50) give us a perfectly general and routine way to evaluate $\mathbf{I}_s^f$ for use in (13.39), we may wish to derive simpler specific formulas for particular fault types. In this connection, as an alternative to purely matrix calculations we can also derive expressions for $\mathbf{I}_s^f$ by considering the circuit implications of the physical interconnections of sequence networks. We did this earlier, in Sections 13.3, 13.6, and 13.7, for SLG, LLG, and LL faults. As an example, noting Fig. 13.10, with a (phase a) SLG fault we find the components of $\mathbf{I}_s^f$.

$$I_{af}^0 = I_{af}^+ = I_{af}^- = \frac{I^f}{3} = \frac{V_q^{pf}}{3Z^f + z_{qq}^+ + z_{qq}^- + z_{qq}^0} \tag{13.51}$$

We recognize V_q^{pf} as the Thévenin equivalent voltage, and z_{qq}^+ as the Thévenin impedance of the positive-sequence network. With $\mathbf{I}_s^f$ determined (for a SLG fault) we can proceed directly to the evaluation of $\mathbf{V}_{si}^f$ in (13.39).

Similarly, we can derive formulas for $\mathbf{I}_s^f$ corresponding to other common types of faults. For example, for a solid LL fault between phases b and c, by noting the connection between sequence networks illustrated in Fig. 13.15, we get the components of $\mathbf{I}_s^f$:

$$I_{af}^0 = 0$$

$$I_{af}^+ = \frac{V_q^{pf}}{z_{qq}^+ + z_{qq}^-}$$

$$I_{af}^- = -I_{af}^+ \tag{13.52}$$

For the case of a solid DLG fault from phase b to c to ground, illustrated by the interconnection shown in Fig. 13.13, the reader can verify that we get

$$I_{af}^+ = \frac{V_q^{pf}}{z_{qq}^+ + \dfrac{z_{qq}^- z_{qq}^0}{z_{qq}^- + z_{qq}^0}}$$

$$I_{af}^- = -\frac{z_{qq}^0}{z_{qq}^- + z_{qq}^0} I_{af}^+$$

$$I_{af}^0 = -\frac{z_{qq}^-}{z_{qq}^- + z_{qq}^0} I_{af}^+ \tag{13.53}$$

These components specify $\mathbf{I}_s^f$. In deriving (13.53) we have used the standard formula for calculating parallel impedances and also the current-divider rule. We note that (13.52) and (13.53) may easily be extended to include the effects of nonzero Z^f.

In some cases we may not know the sequence network interconnection or for some other reason may wish to use the general formulas (13.47) or (13.50) involving the matrices $\mathbf{Z}_s^f$ or $\mathbf{Y}_s^f$. The following examples illustrate the calculations thereof.

Example 13.10   Symmetrical Three-Phase Short Circuit to Ground

For the fault specified, find $\mathbf{Z}^f$ and $\mathbf{Z}_s^f$.

Solution   To represent the most general symmetric three-phase fault in Fig. 13.28, let $Z_a = Z_b = Z_c = Z^f$. In this case we get

$$V_{ag}^f = Z^f I_{af} + Z_g(I_{af} + I_{bf} + I_{cf})$$

$$V_{bg}^f = Z^f I_{bf} + Z_g(I_{af} + I_{bf} + I_{cf})$$

$$V_{cg}^f = Z^f I_{cf} + Z_g(I_{af} + I_{bf} + I_{cf})$$

These equations imply that

$$\mathbf{Z}^f = \begin{bmatrix} Z^f + Z_g & Z_g & Z_g \\ Z_g & Z^f + Z_g & Z_g \\ Z_g & Z_g & Z^f + Z_g \end{bmatrix}$$

Using $\mathbf{Z}_s^f = \mathbf{A}^{-1}\mathbf{Z}^f\mathbf{A}$, we get

$$\mathbf{Z}_s^f = \begin{bmatrix} Z^f + 3Z_g & 0 & 0 \\ 0 & Z^f & 0 \\ 0 & 0 & Z^f \end{bmatrix}$$

In the evaluation we make good use of the identity $1 + \alpha + \alpha^2 = 0$.

Note: If the fault above is not grounded (i.e., $Z_g = \infty$), then $\mathbf{Z}^f$ and $\mathbf{Z}_s^f$ are not defined. In this case it is useful to observe that $\mathbf{Y}_s^f = [\mathbf{Z}_s^f]^{-1}$ is well defined even with $Z_g = \infty$.

The fact that $\mathbf{Z}_s^f$ in Example 13.10 is diagonal has physical significance. It means that there is no coupling between the sequence networks. We can see it clearly using (13.47). Since $\mathbf{Z}_{sqq}$ is diagonal and $\mathbf{Z}_s^f$ is diagonal, the inverse will be diagonal. Then, using (13.37), only the I_{af}^+, component in $\mathbf{I}_s^f$ will be nonzero. Since the power system, aside from the fault, is symmetrical, only positive-sequence responses are possible. But we already know this result from the per phase analysis of balanced systems.

We next consider an example where there is strong coupling between sequence variables through the nonsymmetrical fault admittance.

Example 13.11   Single Line-to-Ground (SLG) Fault (Phase α)

For the fault specified, find $\mathbf{Y}^f$ and $\mathbf{Y}_s^f$.

Solution   In Fig. 13.28 let $Z_b = Z_c = \infty$, $Z_g = 0$, and $Z_a = Z^f$. We have put all the fault impedance in Z_a; since Z_g and Z_a are in series the choice is arbitrary. With $Y^f = 1/Z^f$, Fig. 13.28 gives us

$$I_{af} = Y^f V_{ag}$$

$$I_{bf} = I_{cf} = 0$$

and thus

$$\mathbf{Y}^f = Y^f \begin{bmatrix} 1 & 0 & 0 \\ 0 & 0 & 0 \\ 0 & 0 & 0 \end{bmatrix}$$

Note: $\mathbf{Y}^f$ is singular, and thus $\mathbf{Z}^f$ is not defined. We next find $\mathbf{Y}_s^f$.

$$\mathbf{Y}_s^f = \mathbf{A}^{-1}\mathbf{Y}^f\mathbf{A} = \frac{Y^f}{3}\begin{bmatrix} 1 & 1 & 1 \\ 1 & 1 & 1 \\ 1 & 1 & 1 \end{bmatrix} \tag{13.54}$$

We note that the form of (13.54) is consistent with the fact that the components of $\mathbf{I}_s^f = \mathbf{Y}_s^f\mathbf{V}_s^f$ are all equal. In fact, by using (13.54) in (13.50) we can derive (13.51) by a purely matrix calculation. The interested reader may refer to Example A6.2 in Appendix 6.

Example 13.12   Fault Current Calculation

Suppose that at the qth bus there is a phase α SLG fault. We are given the following data.

$$V_q^{pf} = 1\underline{/0^\circ} \qquad z_{qq}^+ = j0.09 \qquad z_{qq}^- = j0.06 \qquad z_{qq}^0 = j0.03 \qquad Z^f = j0.01$$

Find the fault current I^f using the formula in (13.50) (i.e., carry out the matrix operations).

Solution   Using (13.54) in (13.50) and substituting numerical values, we have

$$\mathbf{I}_s^f = -j\frac{100}{3}\begin{bmatrix} 1 & 1 & 1 \\ 1 & 1 & 1 \\ 1 & 1 & 1 \end{bmatrix}\left\{\begin{bmatrix} 1 & 0 & 0 \\ 0 & 1 & 0 \\ 0 & 0 & 1 \end{bmatrix} + \begin{bmatrix} j0.03 & 0 & 0 \\ 0 & j0.09 & 0 \\ 0 & 0 & j0.06 \end{bmatrix}\frac{-j100}{3}\begin{bmatrix} 1 & 1 & 1 \\ 1 & 1 & 1 \\ 1 & 1 & 1 \end{bmatrix}\right\}^{-1}\begin{bmatrix} 0 \\ 1 \\ 0 \end{bmatrix}$$

Reducing the contents of the braces and taking the inverse, we have

$$\begin{bmatrix} 2 & 1 & 1 \\ 3 & 4 & 3 \\ 2 & 2 & 3 \end{bmatrix}^{-1} = \begin{bmatrix} 0.857 & -0.1429 & -0.1429 \\ -0.4285 & 0.5714 & -0.4286 \\ -0.2857 & -0.2857 & 0.7143 \end{bmatrix}$$

Carrying out the remainder of the calculation, we get

$$\mathbf{I}_s^f = \begin{bmatrix} I_{af}^0 \\ I_{af}^+ \\ I_{af}^- \end{bmatrix} = -j4.76\begin{bmatrix} 1 \\ 1 \\ 1 \end{bmatrix}$$

Thus $I^f = I_{af}^0 + I_{af}^+ + I_{af}^- = -j14.28$.

Note: We get the same result, with less effort and with more insight, by using (13.51).

Once we find $\mathbf{I}_s^f$, we can use (13.39) to obtain the sequence voltages at all the buses of the faulted network. We can also use the sequence bus voltages to find any sequence line or load current in the faulted network. Finally, the corresponding phase voltages and line currents may then be found by using the symmetrical components transformation matrix as in (13.5).

13.16 CALCULATION OF Z MATRICES

Once we have determined $\mathbf{I}_s^f$, and the $\mathbf{Z}_{sij}$ matrices, it is easy to find the sequence bus voltage profile in the faulted network by using (13.39). And, to reiterate, with the sequence bus voltages known, we can find the sequence line and/or load currents (by using the corresponding line or load admittances).

In principle, to find the $\mathbf{Z}_{sij}$ matrices, or equivalently $\mathbf{Z}_{bus}^0$, $\mathbf{Z}_{bus}^+$, and $\mathbf{Z}_{bus}^-$, we invert $\mathbf{Y}_{bus}^0$, $\mathbf{Y}_{bus}^+$, and $\mathbf{Y}_{bus}^0$. We note that we can evaluate the admittance matrices by inspection (by the methods of Section 6.1) once we have the corresponding Δ networks; algorithmic assembly of these admittance matrices by computers is also very easy. Unfortunately, we cannot write the impedance matrices by inspection and in practice, (brute force) direct matrix inversion is not feasible for typical power systems because of the large number of buses.

As an alternative we will be considering a computationally attractive method for "building" or "assembling" the impedance matrices step by step. In effect we obtain (an indirect) matrix inversion of the admittance matrices.

As a preliminary we will consider the following problem. Suppose that we are given the impedance matrix $\mathbf{Z}_{bus}$ of a specified r-bus network and wish to calculate the impedance matrix of a certain modification of the originally specified network. We will call the new, modified impedance matrix $\mathbf{Z}_{bus}^n$. Note that $\mathbf{Z}_{bus}$ may stand for the positive-, negative-, or zero-sequence matrices we have been considering.

In the following we will consider three types of circuit modifications and give rules for calculating $\mathbf{Z}_{bus}^n$ in terms of $\mathbf{Z}_{bus}$. The justification of the rules may be found in Appendix 7.

Modification 1. Add a branch with impedance Z_b from a new $(r + 1)$st bus (node) to the reference node.

Rule 1. $\mathbf{Z}_{bus}^n$ is given by the following $(r + 1) \times (r + 1)$ matrix.

$$\mathbf{Z}_{bus}^n = \begin{bmatrix} \mathbf{Z}_{bus} & \mathbf{0} \\ \mathbf{0}^T & Z_b \end{bmatrix} \tag{13.55}$$

Modification 2. Add a branch Z_b from a new $(r + 1)$st node to the ith node.

Rule 2. Suppose that the ith column of $\mathbf{Z}_{bus}$ is $\mathbf{Z}_i$ and the iith element of $\mathbf{Z}_{bus}$ is z_{ii}, then

$$\mathbf{Z}_{bus}^n = \begin{bmatrix} \mathbf{Z}_{bus} & \mathbf{Z}_i \\ \mathbf{Z}_i^T & Z_b + z_{ii} \end{bmatrix} \tag{13.56}$$

Note: We emphasize, $\mathbf{Z}_i$ is simply a replication of the ith column of $\mathbf{Z}_{bus}$.

Modification 3. Add a branch Z_b between (existing) ith and jth nodes.

Rule 3. Suppose that the ith and jth columns of $\mathbf{Z}_{bus}$ are $\mathbf{Z}_i$ and $\mathbf{Z}_j$ and the iith, jjth, and ijth elements of $\mathbf{Z}_{bus}$ are z_{ii}, z_{jj}, and z_{ij}, respectively, then

$$\mathbf{Z}_{bus}^n = \mathbf{Z}_{bus} - \gamma \mathbf{bb}^T \tag{13.57}$$

where $\mathbf{b} = \mathbf{Z}_i - \mathbf{Z}_j$ and $\gamma = (Z_b + z_{ii} + z_{jj} - 2z_{ij})^{-1}$.

Note: By the rules of matrix multiplication,

$$\mathbf{bb}^T = \begin{bmatrix} b_1 \\ b_2 \\ \cdot \\ \cdot \\ \cdot \\ b_r \end{bmatrix} [b_1 b_2 \cdots b_r] = \begin{bmatrix} b_1^2 & b_1 b_2 & \cdots & b_1 b_r \\ b_2 b_1 & b_2^2 & \cdots & b_2 b_r \\ \cdot & \cdot & & \cdot \\ \cdot & \cdot & & \cdot \\ \cdot & \cdot & & \cdot \\ b_r b_1 & b_r b_2 & \cdots & b_r^2 \end{bmatrix}$$

Example 13.13

Suppose that we are given the impedance matrix $\mathbf{Z}_{bus}$ for the three-bus network shown in Fig. E13.13. Assume that

$$Z_1 = j1.0$$

$$Z_2 = j1.25$$

$$Z_3 = j0.1$$

$$Z_4 = j0.2$$

$$Z_5 = j0.1$$

$$\mathbf{Z}_{bus} = j \begin{bmatrix} 0.5699 & 0.5376 & 0.5591 \\ 0.5376 & 0.5780 & 0.5511 \\ 0.5591 & 0.5511 & 0.6231 \end{bmatrix}$$

Suppose that the network is now modified as follows: Z_5 is removed. Find $\mathbf{Z}_{bus}^n$.

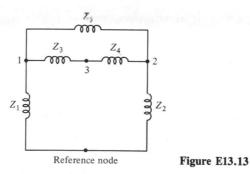

Reference node   **Figure E13.13**

Solution   The modification is of type 3. If we add an impedance $Z_b = -j0.1$ between nodes 1 and 2, we will have Z_b in parallel with Z_5 (i.e., $-j0.1$ in parallel with $j0.1$). The resulting parallel impedance is infinity (i.e., we have an open circuit). This is equivalent to removing Z_5.

Using rule 3 we first calculate

$$\gamma = (Z_b + z_{11} + z_{22} - 2z_{12})^{-1}$$

$$= [-j0.1 + j0.5699 + j0.5780 - 2(j0.5376)]^{-1}$$

$$= j36.63$$

Next we calculate

$$\mathbf{b} = \mathbf{Z}_1 - \mathbf{Z}_2 = j \begin{bmatrix} 0.0323 \\ -0.0404 \\ 0.0080 \end{bmatrix}$$

from which we find

$$\mathbf{bb}^T = j^2 \begin{bmatrix} 0.001043 & -0.001305 & 0.000258 \\ -0.001305 & 0.001632 & -0.000323 \\ 0.000258 & -0.000323 & 0.000064 \end{bmatrix}$$

Substituting these results in (13.57), we get

$$\mathbf{Z}_{bus}^n = \mathbf{Z}_{bus} - \gamma \mathbf{b}\mathbf{b}^T$$

$$\mathbf{Z}_{bus}^n = j\begin{bmatrix} 0.6081 & 0.4898 & 0.5686 \\ 0.4898 & 0.6378 & 0.5393 \\ 0.5686 & 0.5393 & 0.6254 \end{bmatrix}$$

All these operations are nicely automated using computers.

We note that to appreciate the advantages of this method of matrix modification over direct inversion of Y_{bus}^n, we should consider a matrix of high dimension.

In Example 13.13 we removed Z_5. The reader may wish to check the result, and his or her understanding of the method, by adding it back in to get the original $\mathbf{Z}_{bus}$.

We next return to the problem originally posed, that of finding the impedance matrix of a given network without (direct) matrix inversion. Following the three rules governing network modifications, we can start with a simple network and then add elements until we have the complete network. The process is called *Z building*, and the procedure is given as follows.

Z-building procedure.   Suppose that we wish to find $\mathbf{Z}_{bus}$ for a given network.

Step 0: Number nodes of the given network starting with those nodes at the ends of branches connected to the reference node.

Step 1: Start with a network composed only of all those branches connected to the reference node. Correspondingly, we find that $\mathbf{Z}_{bus}^{(0)}$ is diagonal with the impedance values of the branches on the diagonal. This result is consistent with the repeated use of rule 1.

Step 2: Add a new node to the ith node of the existing network using rule 2. Continue until all the nodes of the complete network are attached.

Step 3: Add a branch between ith and jth nodes using rule 3. Continue until all the remaining branches are connected.

We conclude with an example.

Example 13.14

Find $\mathbf{Z}_{bus}$ for the three-bus (node) network (shown in Fig. E13.14(a) using the Z-building procedure. Assume that

$$Z_1 = j1.0$$

$$Z_2 = j1.25$$

$$Z_3 = j0.1$$

$$Z_4 = j0.2$$

$$Z_5 = j0.1$$

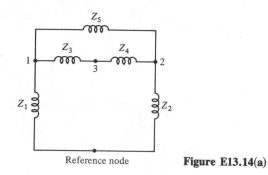

Figure E13.14(a)

Note: The network is the same as in Example 13.13. Now we will show how to find the $\mathbf{Z}_{bus}$ which was given in Example 13.13.

Solution

Step 0: The nodes connected to the reference node are already correctly numbered 1 and 2.

Step 1: We start by considering $\mathbf{Z}_{bus}^{(0)}$ for the network composed of all the branches directly connected to the reference node [Fig. E13.14(b)].

$$\mathbf{Z}_{bus}^{(0)} = j\begin{bmatrix} 1 & 0 \\ 0 & 1.25 \end{bmatrix}$$

Step 2: We next attach a new node, node 3, to node 1 through branch Z_3 using rule 2. With $\mathbf{Z}_1^T = j[1 \quad 0]$, $z_{11} = j1$, and $Z_b = Z_3 = j0.1$, we get

$$\mathbf{Z}_{bus}^{(1)} = j\begin{bmatrix} 1 & 0 & \vdots & 1 \\ 0 & 1.25 & \vdots & 0 \\ \cdots & \cdots & \vdots & \cdots \\ 1 & 0 & \vdots & 1.1 \end{bmatrix}$$

Step 3: Now all the nodes are in place. All other modifications consist of adding a branch between existing nodes. We can add Z_4 next. Z_4 connects nodes 2 and 3, so by rule 3 we calculate

$$\mathbf{b} = \mathbf{Z}_2 - \mathbf{Z}_3 = j\begin{bmatrix} 0 \\ 1.25 \\ 0 \end{bmatrix} - j\begin{bmatrix} 1.0 \\ 0 \\ 1.1 \end{bmatrix} = j\begin{bmatrix} -1.0 \\ 1.25 \\ -1.1 \end{bmatrix}$$

$$\gamma = (Z_4 + z_{22} + z_{33} - 2z_{23})^{-1} = (j0.2 + j1.25 + j1.1)^{-1} = (j2.55)^{-1}$$
$$= -j0.3922$$

$Z_1 = j1$ $Z_2 = j1.25$

Reference node   **Figure E13.14(b)**

where the impedance values are derived from $\mathbf{Z}_{bus}^{(1)}$. Then, using (13.57) and (13.58), we get

$$\mathbf{Z}_{bus}^{(2)} = \mathbf{Z}_{bus}^{(1)} + j0.3922 \begin{bmatrix} -1.0 & 1.25 & -1.1 \\ 1.25 & -1.563 & 1.375 \\ -1.1 & 1.375 & -1.21 \end{bmatrix}$$

$$= j \begin{bmatrix} 0.6078 & 0.4902 & 0.5686 \\ 0.4902 & 0.6372 & 0.5392 \\ 0.5686 & 0.5392 & 0.6255 \end{bmatrix}$$

Finally, we add Z_5 between nodes 1 and 2.

$$\mathbf{b} = \mathbf{Z}_1 - \mathbf{Z}_2 = j \begin{bmatrix} 0.6078 \\ 0.4902 \\ 0.5686 \end{bmatrix} - j \begin{bmatrix} 0.4902 \\ 0.6372 \\ 0.5392 \end{bmatrix} = j \begin{bmatrix} 0.1176 \\ -0.1471 \\ 0.0294 \end{bmatrix}$$

$$\gamma = (Z_5 + z_{11} + z_{22} - 2z_{12})^{-1}$$

$$= [j0.1 + j0.6078 + j0.6372 - 2(j0.4902)]^{-1} = -j2.742$$

where the impedance values are derived from $\mathbf{Z}_{bus}^{(2)}$. Then

$$\mathbf{Z}_{bus}^{(3)} = \mathbf{Z}_{bus}^{(2)} + j2.742 \begin{bmatrix} -0.0138 & 0.0173 & -0.0035 \\ 0.0173 & -0.0216 & 0.0043 \\ -0.0035 & 0.0043 & -0.0009 \end{bmatrix}$$

$$= j \begin{bmatrix} 0.5699 & 0.5376 & 0.5591 \\ 0.5376 & 0.5780 & 0.5511 \\ 0.5591 & 0.5511 & 0.6231 \end{bmatrix}$$

This is the desired impedance matrix $\mathbf{Z}_{bus}$.

We note again that all the operations required above are well suited for computer implementation.

13.17 SUMMARY

Faults are responsible for unbalanced conditions in power systems. Normal power flow may be interrupted, and possibly destructive currents may flow. To design and monitor a system protection scheme, it is necessary to analyze a power system operating under unbalanced conditions.

The method of symmetrical components provides such a method. Every vector of phase voltages V_a, V_b, and V_c, or line currents I_a, I_b, and I_c, is resolved into the sum of three vectors: a zero-sequence vector, a positive-sequence vector, and a negative-sequence vector. Only at the fault point is there coupling between the three different sequence systems, and this may be exploited to simplify the analysis problem. We are led to consider a single-phase version of the three-phase problem with three different sequence networks connected in various ways depending on the type of fault. This generalizes the use of the per phase analysis of balanced three-phase.

In fact, one of the networks, the positive-sequence network, is the same as used previously in per phase analysis. The negative-sequence network has the same topology but differs as follows. The positive-sequence voltage sources which model the generator internal voltages are absent and the generator reactances are negative sequence. On the other hand, the zero-sequence network looks very different because of the variety of transformer connections and grounding schemes.

The sequence networks help in understanding how the currents and voltages in the faulted network depend on the system parameters, the location of the fault, the type of fault, the type of grounding, and the type of transformer connection. This understanding is more important in design than the purely computational aspects. For computations in systems of reasonable size, computer methods would certainly be used. Once again these are based on the method of symmetrical components. We need to find the bus impedance matrices associated with the corresponding sequence networks. Given a sequence network, the Z-building procedure is a computationally attractive method for finding the corresponding bus impedance matrix.

PROBLEMS

13.1. (a) Find the (leading) symmetrical components I_a^0, I_a^+, and I_a^- for $I_a = 1$, $I_b = 10$, and $I_c = -10$.

   (b) Check by sketching I_a, I_b, and I_c as the sum of appropriate symmetrical components.

13.2. Find the symmetrical components of $E_a = 1\ \underline{/0°}$, $E_b = 1\ \underline{/-90°}$, and $E_c = 2\ \underline{/135°}$.

13.3. Refer to Fig. P13.3 and assume that

$$E_a = 1$$
$$E_b = -1$$
$$E_c = j1$$

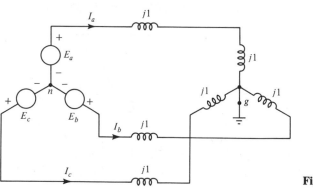

Figure P13.3

(a) Describe how you would use the *method of symmetrical components* to find I_a, I_b, I_c, and V_{ng}.

(b) Carry out the procedure.

13.4. In Fig. P13.4, the source voltages are positive-sequence sets and all impedances $= Z$. Using an appropriate interconnection of sequence networks, find I^f (in terms of Z) and $V_{a''g}$.

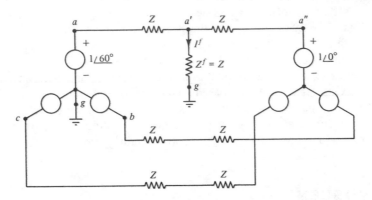

Figure P13.4

13.5. In Fig. P13.5, the source voltages are in positive-sequence sets. All impedances are $Z = j0.1$. Find I^f, I_{b1}, and I_{b2}.

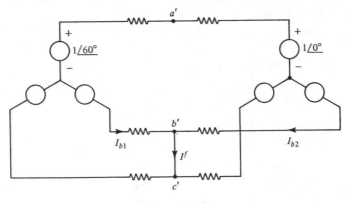

Figure P13.5

13.6. Suppose that the fault is as shown in Fig. P13.6(a). Show that the proper connection of sequence networks is as shown in Fig. P13.6(b). *Hint:* Start by calculating $V_{a'g}^+ - V_{a'g}^0$.

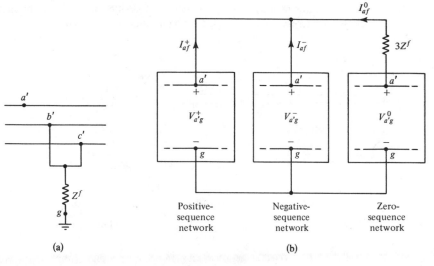

Figure P13.6

13.7. In Fig. P13.7, voltages are positive sequence, all Z are equal, and $Z^f = 0$. Of the following, which fault current magnitude is biggest?
   (a) a' to ground.
   (b) b' to c'.
   (c) b' to c' to ground. (Find I_{bf}.)
   (d) 3ϕ ground.

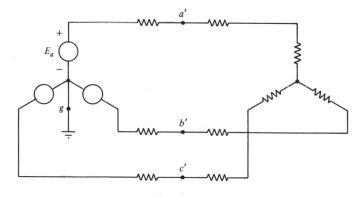

Figure P13.7

13.8. Refer to Fig. P13.8 and assume that

$$Z = j0.1 \qquad \text{load impedances}$$

$$\left.\begin{array}{l} Z^+ = j1.0 \\[4pt] Z^- = j0.1 \\[4pt] Z^0 = j0.005 \end{array}\right\} \begin{array}{l} \text{generator} \\ \text{impedances} \end{array}$$

The generator neutral is not grounded. Before the fault occurs, the generator is supplying positive-sequence voltages and currents. $V_{a'g} = 1 \underline{/0°}$. Find the fault current I^f.

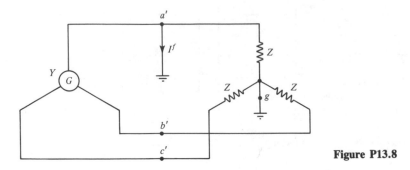

Figure P13.8

13.9. In Fig. P13.9, a *solid* phase a-to-ground fault occurs at bus 1 (i.e., the switch closes on phase a).

 (a) Show the complete sequence network connection for this fault. Neglect all resistances but show (and label) all the impedances.

 (b) Assume that $Z^+ = j1.0$, $Z^- = j0.1$, and $Z_0 = j0.005$ for the generators and that *all* the remaining impedances have the (same) value $j0.1$. Assume that the prefault voltage $V_{a'n} = V_{a'g} = 1 \underline{/0°}$ at bus 1. Find I^f.

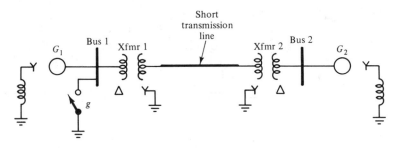

Figure P13.9

13.10. Draw the zero-sequence network for the one-line diagram shown in Fig. P13.10. Neglect transmission-line reactances, but show transformer leakage reactances and generator sequence impedances. Identify the points P, Q, and R on your diagram.

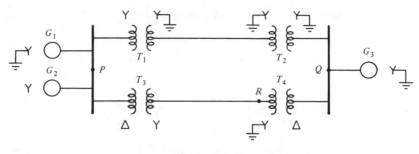

Figure P13.10

13.11. Repeat Problem 13.10 using Fig. P13.11. Show points P, Q, R, S, T, U, V, and W on your diagram.

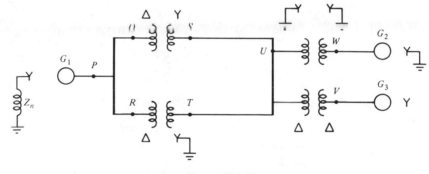

Figure P13.11

13.12. A line-line fault occurs at the point shown in Fig. P13.12. Find the steady-state fault current I^f. $Z = j0.05$ for all impedances, and all sources are positive sequence. *Note:* The fault occurs between phases a and b.

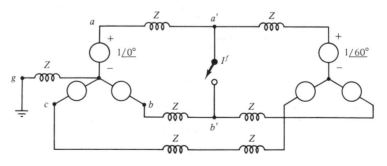

Figure P13.12

13.13. Refer to Fig. P13.13(a) and assume that

$$Z^+ = j1.0$$
$$Z^- = j0.5$$
$$Z^0 = j0.1$$
$$Z = j1.0$$
$$Z_l = j0.1$$

The generator neutral is not grounded. Before the fault $V_a = V_{an} = 1 \underline{/0°}$.
(a) Calculate I^f with the breaker open.
(b) Calculate I^f with the breaker closed.
Use Fig. P13.13(b) to model each phase of the transformer.

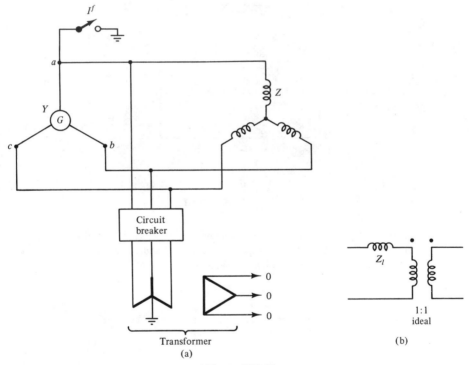

Transformer
(a)

(b)

Figure P13.13

13.14. Refer to Fig. P13.14 and assume the following:

Generators: $X^+ = X^- = 0.2, X^0 = 0.05$

Transformers: $X_l = 0.05$

Lines: $X^+ = X^- = 0.1, X^0 = 0.3$

Prefault (line-neutral) voltage $V_{an}^{pf} = V_{ag}^{pf} = 1 \underline{/0°}$ at bus 3. For a phase a-to-ground fault at bus 3, find I_{af}, I_{bf}, I_{cf}, V_{ag}, V_{bg}, and V_{cg} (all at bus 3).

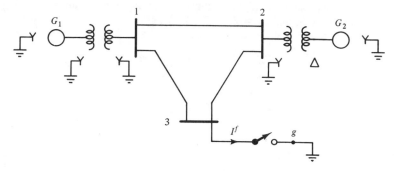

Figure P13.14

13.15. Refer to Fig. P13.15 and assume the following:

$$\text{Generator: } X^+ = X^- = 0.2, X^0 = 0.05$$

$$\text{Transformer: } X_l = 0.05$$

$$\text{Line: } X^+ = X^- = 0.2, X^0 = 3X^+ = 0.6$$

At point a at the near end of the (radial) transmission line, we measure the "impedance" of each phase of the line (i.e., $Z_a = V_{ag}/I_a$, $Z_b = V_{bg}/I_b$, $Z_c = V_{cg}/I_c$). Plot $|Z_a|$, $|Z_b|$, $|Z_c|$ versus λ for:
(a) 3ϕ fault.
(b) Single line-to-ground fault.
(c) Line-to-line fault.
(d) Double line-to-ground fault.
Note: If $|Z| = \infty$, simply note the fact.

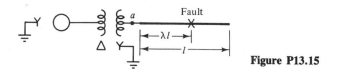

Figure P13.15

13.16. Repeat Example 13.9 if there is an SLG (phase a) fault in the middle of the transmission line. Assume that the prefault (phase a) voltage at the fault point is $1 \underline{/0°}$.

13.17. Repeat Example 13.9 if there is a DLG fault (between phases b and c) at the terminals of G_1. Assume that the prefault (phase a) voltage at the fault point is $1 \underline{/0°}$. Find the fault currents I_{bf} and I_{cf}.

13.18. Given an SLG (phase b) to ground fault with $Y^f = 1/Z^f$ given. Find (**a**) $\mathbf{Y}^f$ and (**b**) $\mathbf{Y}_s^f$.

13.19. Repeat Example 13.14 (using the Z-building procedure) assuming that $Z_1 = j2.0$, $Z_2 = j1.0$, and $Z_3 = Z_4 = Z_5 = j0.2$.

System Protection

chapter 14

14.0 INTRODUCTION

In this chapter we consider the problem of power system protection. Good design, maintenance, and proper operating procedures can reduce the probability of occurrence of faults, but cannot eliminate them. Given that faults will inevitably occur, the objective of protective system design is to minimize their impact.

Faults may have very serious consequences. At the fault point itself, there may be arcing, accompanied by high temperatures and, possibly, fire and explosion. There may be destructive mechanical forces due to very high currents. Overvoltages may stress insulation beyond the breakdown value. Even in the case of less severe faults, high currents in the faulted system may overheat equipment; sustained overheating may reduce the useful life of the equipment. Clearly, faults must be removed from the system as rapidly as possible. In carrying out this objective an important, but secondary objective is to remove no more of the system than absolutely necessary, in order to continue to supply as much of the load as possible. In this connection, we note that temporary loss of lighting or water pumping or air-conditioning load is not usually serious, but loss of service to some industrial loads can have serious consequences. Consider, for example, the problem of repair of an electric arc furnace in which the molten iron has solidified because of loss of power.

Faults are removed from a system by opening or "tripping" circuit breakers. These are the same circuit breakers used in normal system operation for connecting or disconnecting generators, lines, and loads. For emergency operation the breakers are tripped automatically when a fault condition is detected. Ideally, the operation is highly selective; only those breakers closest to the fault operate to remove or "clear" the fault. The rest of the system remains intact.

Fault conditions are detected by monitoring voltages and currents at various critical points in the system. Abnormal values individually or in combination cause relays to operate, energizing tripping circuits in the circuit breakers. A simple example is shown in Fig. 14.1.

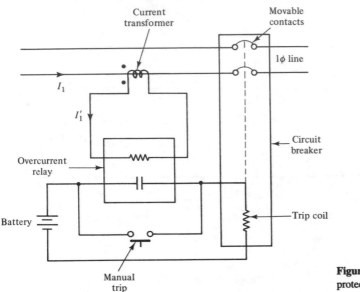

Figure 14.1   Schematic of overcurrent protection.

A description of the operation is as follows. The primary current in the current transformer is the line current I_1. The secondary current I_1' is passed through the "operating" coil of an overcurrent relay. When $|I_1'|$ exceeds a specified "pickup" value, the normally open relay contacts close. If the relay is of the "plunger" type, like that described in Appendix 2, the contacts close "instantaneously." Other types of relays have an intentional and adjustable time delay. When the relay contacts close, the trip coil circuit is energized and the circuit breaker trips open. We can think of the trip coil as a solenoid whose movable core releases a latch, permitting the stored energy in a spring or compressed air tank to force open mechanically the movable contacts of the circuit breaker. We note in passing that the arc which forms between the circuit-breaker contacts when they open is extinguished by blowing away the ionized medium between the contacts using a blast of air or a transverse magnetic field; since the ac arc current is zero twice each cycle, the arc goes out once the insulating properties of the medium are restored.

In more general cases than the one we have been describing, potential (voltage) transformers (PTs) as well as current transformers (CTs) are used. Collectively these transformers are known as *instrument transformers*. We note that these transformers are used for measurements (instrumentation) as well as in protective schemes. In either application the instrument transformers isolate the dangerous line voltages and currents (in the kilovolt and kiloampere range) from the secondary circuits. A very wide range of rated primary voltages and currents are commercially available. The secondary voltages are usually standardized to 116 V in the case of line-line voltages and to 67 V $= 116/\sqrt{3}$ V in the case of line-neutral voltages; the standard secondary current rating is 5 A.

We will be assuming that the instrument transformers are ideal so that the secondary voltages and currents are simply scaled-down versions of their primary counterparts. We will also assume, except where noted, that the measurement point for voltage or current is at the circuit-breaker location. This being the case, we will frequently describe relay action in breaker terms.

We note that it is not practical to use potential transformers at the very high line voltages frequently encountered. Instead, a portion of the line voltage is used, derived from a voltage-divider circuit composed of series capacitors. The output is then fed through a series inductance to a potential transformer, where the voltage is reduced further. The series inductance can be picked so that, together with the leakage inductance of the potential transformer, the capacitance is "tuned" out and the Thévenin equivalent impedance of the overall capacitor-coupled voltage transformer is practically zero in the sinusoidal steady state. For simplicity we will not distinguish between the low-voltage PT measuring device and this high-voltage realization since we assume ideal operation for both.

In the following exercise we pause, to ask the reader to apply a fundamental objective in providing system protection, before we go on to consider the simple case of protection of radial lines.

Exercise 1.   We are given the two alternative arrangements of circuit breakers shown in Fig. Ex1. Which breaker scheme is preferable from the point of view of minimizing the loss of load in the event of faults on buses and/or lines? *Note:* Three breakers are employed in both cases.

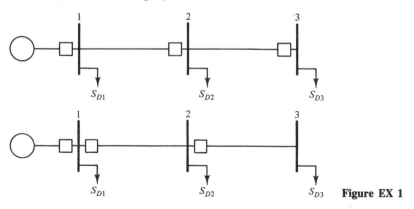

Figure EX 1

14.1 PROTECTION OF RADIAL SYSTEMS

We start by considering the simplest case, that of protecting a radial system, as shown in Fig. 14.2. Our objective is to specify a relaying scheme to satisfy the following criteria:

1. Under normal load conditions the circuit breakers B_0, B_1, and B_2 do not operate.
2. Under fault conditions only the closest breaker to the left of the fault should operate; this protects the system by clearing (and deenergizing) the fault while maintaining service to as much of the system as possible.
3. In the event that the closest breaker fails to operate, the next breaker closer to the power source should operate. This provision is called *backup protection*.

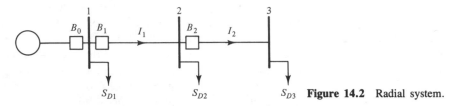

Figure 14.2   Radial system.

Suppose that there is a 3ϕ fault to the right of B_2. The current I_2, at B_2, will increase from its prefault value to a, presumably, much larger fault value. The presence of this large current can be detected by an overcurrent relay at the breaker location and used to initiate the triggering of B_2. Stated more simply and directly in breaker terms, we can trip B_2 when its fault current magnitude exceeds a specified preset pickup value of current. If we set the pickup value of current low enough to trip B_2 for a fault at bus 3, then B_2 will certainly trip for the larger current accompanying a "closer in" fault anywhere on the transmission line to the right of B_2. Thus B_2 protects the line and bus to its right, or stated differently, the transmission line and bus are in the "protection zone" of B_2.

What if B_2 fails to operate for a fault in its protection zone? We note that the fault current at B_1 is essentially the same as the fault current at B_2, since the load current (supplying S_{D2}) is small compared with the fault current. We could therefore set B_1 for the same pickup current value as B_2 but introduce a time delay so that B_2 would normally operate first; this deliberately introduced time delay is called a *coordination time*. Would this method of setting the pickup current value of B_1 interfere with its primary function in protecting against faults in its protection zone (line and bus between B_1 and B_2)? Fortunately not, since faults within its protection zone are closer to the source and produce larger fault currents. If B_1 is set to trip for remote faults (in the protection zone of B_2), it certainly can be expected to trip for faults within its own protection zone. The discussion above for the two breakers B_1 and B_2 extends pairwise to any number of breakers in a radial system. A difficulty in extending the number of breakers, however, is the possibly excessive trip time for the breakers nearest the source because of the summation of coordination time delays. Later we will introduce a relaying method without this disadvantage.

While for simplicity we have been discussing a symmetric 3ϕ fault, we certainly will want to protect our system against the more common general faults. The extension is simple enough. For example, for B_2, the pickup current for each phase should be low enough that tripping occurs for any type fault of interest in its zone of protection. The implementation will require three overcurrent relays with their contacts in parallel in the trip circuit of the 3ϕ circuit breaker. Thus a fault on any phase trips the breaker.

In implementing the protection scheme we need overcurrent relays with adjustable pickup current values and adjustable time delays. We note that the plunger-type relay described in Appendix 2 can be used as an overcurrent relay with adjustable pickup current. It can also be fitted with an oil dash pot or air bellows to provide an adjustable time delay. In practice, however, the time-delay adjustment is not accurate enough. To obtain a more accurate time delay, a relay of the induction-disk type is usually used. The basic torque generating principle is that of the common household watthour meter. In a watthour meter the angular velocity of the rotating disk is proportional to the average power supplied; in this way by counting the number of revolutions we measure the energy (i.e., the integral of power). The desired characteristic is achieved by designing the watthour meter so that the torque that drives the disk is proportional to the average power supplied:

$$T = k|V||I| \cos \phi \qquad (14.1)$$

where ϕ is the phase angle between V and I. There are also damping or braking magnets across the disk that cause an opposing torque proportional to the speed. Thus in equilibrium we get the desired proportionality between angular velocity and average power.

Although the actual construction of an induction-disk relay differs in detail, we can think of using a modified watthour meter element with the voltage and current coils connected externally to give a torque proportional to $|I|^2$ alone. If the disk is then restrained by a spiral spring, it will not turn until the current exceeds a certain pickup value. Beyond that value the higher the current magnitude, the faster the disk rotates and the less time it takes to rotate through a given angle. By having a moving contact that rotates with the disk "makeup" with a stationary contact, an "inverse" time-current characteristic may be obtained. By adjusting the position of the stationary contact, the time delay may be varied and a family of time-current characteristics is obtained.

To obtain an adjustable pickup current value, one might make the spiral spring tension adjustable. In fact, the adjustment is obtained by varying the number of turns in the current coil of the relay by means of "taps."

A typical set of overcurrent relay characteristics is shown in Fig. 14.3. The two adjustable settings are labeled "current tap setting" and "time dial setting." The current tap setting designates the relay pickup current. For example, with a tap setting of 7, the relay will pick up when the current is 7A.

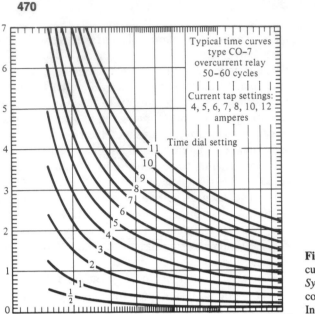

Figure 14.3 CO-7 time-delay over-current relay characteristics. (From *Power Systems Analysis* by Charles A. Gross, copyright © 1979, John Wiley & Sons, Inc. Reprinted by permission of John Wiley & Sons, Inc. Also courtesy of Westinghouse Electric Corporation.)

Note that for a given time dial setting, the relay operating time varies inversely with relay current. The curves are asymptotic to the vertical axis; thus when the relay current equals the pickup current, it takes an indefinitely long time for actual pickup. In the example to follow we will show how to compensate for this peculiar feature when choosing the value of pickup current.

Example 14.1

Consider the radial system shown in Fig. E14.1(a), which represents a portion of a larger system. Bulk power is fed from a substation represented by the source on the left. We make the following assumptions:

1. The source is an infinite bus; the voltage on the low (right) side of the transformer is 34.5 kV (line-line).

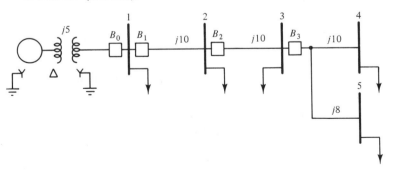

Figure E14.1(a)

2. Prefault currents are negligible compared with fault currents. The prefault voltage profile is flat.

3. The model shows actual impedances in ohms; positive, negative, and zero sequence impedances are assumed equal. Resistances are neglected.

4. The transformer impedance is $j5$ referred to the 34.5-kV side.

5. Faults may occur anywhere to the right of B_1. Faults may be 3ϕ, SLG, LL, and DLG. We assume a solid fault (i.e., $Z^f = 0$).

6. For each breaker there are three overcurrent relays (one for each phase). We wish to trip the (3ϕ) breaker if a fault condition is detected on any phase. A schematic of the setup is shown in Fig. E14.1(b). Note that the breaker trip circuit is activated if any overcurrent relay operates.

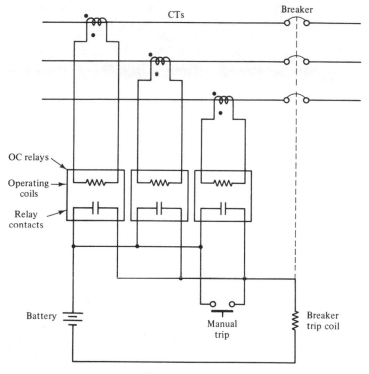

Figure E14.1(b)

The objective is to pick CTs and relay settings to protect the lines and buses 2, 3, 4, 5. We also want B_2 to back up B_3 and B_1 to back up B_2. Suppose that the coordination time delay is specified to be 0.3 sec.

Solution   With a flat prefault voltage profile the prefault Thévenin equivalent voltage at any fault point can be taken to be $34.5/\sqrt{3}$kV. Consider first setting R_3, the relay that trips B_3.

Setting of R_3. We want the relay to operate for any type (solid) fault to the right of B_3. We therefore need to know the smallest line current at location B_3 under any fault condition. Using the sequence networks and considering all the alternatives, it becomes apparent that a line-line fault at bus 4 (the electrically most remote location) gives the smallest currents. If the fault is between phases b and c, we get, by a very simple calculation, using (13.22) and the appropriate connection of networks shown in Fig. 13.16,

$$I_{34}^a = 0$$

$$I_{34}^b = -j\sqrt{3}I_{af}^+ = -j\sqrt{3}\,\frac{34{,}500/\sqrt{3}}{2 \times j35} = -492.86 \text{ A}$$

$$I_{34}^c = 492.86 \text{ A}$$

Thus with a fault between any two phases, two of the line currents will have magnitude 492.86 and we would like to trip reliably for those currents. In fact, we would like to initiate tripping action at a much lower value of current. In this way we have a safety factor with respect to reasonable modeling errors and also compensate for the peculiarity of the CO-7 characteristic mentioned earlier (i.e., wherein it would take an indefinitely long time to trip if we picked 492.86 as the tripping current).

Taking a commonly used safety factor of 3, we therefore wish to initiate tripping when

$$|I_{34}| \geq 165A \approx \frac{492.86}{3}$$

Of course, in practice we should check that the expected normal line currents are significantly less than this value; we assume this to be the case. Picking a CT ratio of 150:5,* we get the corresponding desired relay pickup value of 5.5 A. Looking at the CO-7 characteristic we cannot choose 5.5; we can choose a tap setting of 5.0 A, which will improve the safety margin; 5.0 A corresponds to a line current of 150 A. It remains to choose the time dial setting. For fastest action pick the value $\frac{1}{2}$. When $|I_{34}| = 492.86$ A, the relay current is $492.86 \times 5/150 = 16.43$, which is 3.29 times the tap setting. For the characteristic curves we trip in approximately 0.2 sec. Note that the safety factor of 3 has "biased" us nicely into the central region of the characteristic. We next consider the setting of R_2.

Setting of R_2. Since R_2 must provide backup protection for R_3, pickup is required under the same current conditions as for R_3. It is convenient therefore to pick the same CT ratio and tap setting. The time dial setting must provide for a 0.3-sec time delay for coordination with R_3; we do not want the R_2 contacts to close until B_3 has had a chance to trip into the open position. Since with a current of 492.86 A, R_3 picks up in 0.2 sec, we wish R_2 to pick up in $0.2 + 0.3 = 0.5$ sec. The characteristic curves show that a time dial setting of approximately 1 would suffice. However, looking at the curves we see that for higher currents the time delay would be too short and the coordination time delay would be less than 0.3 sec. Because the curves are closer together at the right, it is clear that to maintain the coordination time delay the time dial setting will need to be determined by considering the largest current value.

In the backup mode the largest current occurs for a 3ϕ fault just to the right of B_3 and has the value

*Standard primary ratings are 50, 100, 150, 200, 250, 300, 400, 450, 500, 600, 800, 900, 1000, and 1200 A. Secondary ratings are always 5 A.

$$I^a_{23} = I^a_{34} = \frac{34{,}500/\sqrt{3}}{j25} = -j796.74$$

Thus in R_2 and R_3 we get relay currents of $796.74 \times 5/150 = 26.56$, which are 5.31 times the tap settings. From the curves R_3 operates in 0.15 sec and we thus want R_2 to operate in $0.15 + 0.3 = 0.45$ sec. Interpolating the curves, we get a time dial setting of approximately 1.5, which is realizable since the settings are continuously adjustable.

In this manner, from considerations of backup protection, we have specified R_2, but it will certainly also pick up for all faults within its primary zone of protection. The smallest current will be for a line-line fault at bus 3 and that current will have magnitude $492.86 \times 35/25 = 690$ A. This is 4.6 times the pickup value and R_2 will trip in about 0.5 sec.

Setting of R_1. Again, we pick the settings of R_1 based on the requirement of backup for R_2. In this mode, the smallest current that R_1 will see, together with R_2, is 690 A. Again, introducing the safety factor of 3, we wish to initiate tripping for breaker currents greater than $690/3 = 230$ A. Picking at CT ratio of 200:5, for example, we get a corresponding relay pickup value of 5.75 A. Looking at the CO 7 characteristic, we can pick a tap setting of 5.0 A or 6.0 A. Suppose that we pick 6.0 A. This corresponds to a line current of $6.0 \times 200/5 = 240$ A, which is close enough to 230 A. To find the time dial setting, consider the largest short-circuit current seen by R_2 and R_1 (in its backup mode). This is a 3ϕ short just to the right of B_2. The current is

$$I^a_{12} = I^a_{23} = \frac{34{,}500/\sqrt{3}}{j15} = -j1327.91$$

Thus in R_1 and R_2 we get relay currents of $1327.91 \times 5/200 = 33.20$, which is $33.20/5 = 6.64$ times the tap setting for R_2 and is $33.20/6 = 5.53$ times the tap setting for R_1. Using the curves, we see that R_2, with a time dial setting of 1.5, operates in 0.45 sec. Considering the coordination time delay, we wish R_1 to operate in $0.4 + 0.3 = 0.7$ sec. From the curves we find that for a current 5.53 times the tap setting, the time dial setting is approximately 2.2 sec.

This completes the selection of CT ratios and relay settings of R_1, R_2, and R_3 associated with breakers B_1, B_2, and B_3, respectively.

14.2 SYSTEM WITH TWO SOURCES

Let us consider what happens if we modify the system in Fig. 14.2 by adding a source to bus 3. The advantage of the modified system compared with a radial system is improved continuity of service because the power can be supplied from either end. Additional circuit breakers are needed as shown in Fig. 14.4.

Note that a fault on bus 1, or the line between bus 1 and bus 2, need no longer interrupt service to buses 2 and 3. This, of course, presumes correct operation by the breakers.

Correct operation cannot be accomplished by the method discussed in the preceding section. To see this, suppose that there is a solid 3ϕ fault at point (x) in Fig. 14.4. We want B_{23} and B_{32} to open to clear the fault. We do not want B_{12} and/or B_{21}

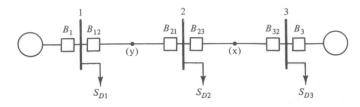

Figure 14.4 System with two sources.

to operate and interrupt service to bus 2. Suppose that we attempt to use the method of the preceding section. We would set B_{23} to trip faster than B_{21} (and B_{21} to trip faster than B_{12}), which would imply correct action for a fault at point (x). But suppose that the fault were at point (y) instead. With relays set as described above, B_{23} would trip before B_{21}, which would isolate bus 2 unnecessarily. When the fault may be fed from either left or right, there is no way to coordinate the relay time delays properly.

To remove faulted lines correctly, we require relays to respond only to faults occurring on their "forward" or "line" sides. The relays must therefore be "directional." In this case B_{21} would not operate for a fault at location (x); B_{23} and B_{12} could be coordinated so that B_{23} would operate first. On the other hand, with a fault at location (y), B_{23} would not operate and B_{21} and B_{32} could be coordinated so that B_{21} would operate first. In fact, with this directional feature, and for a system like that in Fig. 14.4 (i.e., with a string of lines and buses fed from both ends), the coordination may be successfully accomplished just as in the case of a radial system, and the two breakers closest to the fault would be expected to operate first. Note that these two breakers would not generally operate simultaneously since the short-circuit currents and coordination time delays would be different for the breakers to the left and the right of the fault point.

Note, finally, that with directional relays the buses themselves are still protected. A fault at bus 2, for example, is on the line sides of B_{12} and B_{32} and would be "seen" by these breakers.

We next consider how this directional feature may be obtained. It is easiest to explain the basic idea by considering only solid 3ϕ faults. Suppose that in Fig. 14.4 at B_{23}, using instrument transformers, we can measure the following phase a quantities: V_2, the phase-neutral voltage, and I_{23}, the line current flowing from bus 2 to bus 3.

Suppose that there is a symmetric 3ϕ fault on the line to the right of B_{23} [i.e., in the "forward" direction at some point such as (x)]. Then I_{23} will be in the form

$$I_{23} = \frac{V_2}{\lambda Z} \tag{14.2}$$

where Z is the total series impedance of the line between buses 2 and 3 and λ is the fraction of the total line between the breaker and the fault (i.e., a number between 0 and 1). $Z = |Z|\underline{/Z}$ is mostly reactive with $\underline{/Z} = \theta_Z$ in the range 80 to 88°. Thus in practice, I_{23} lags V_2 by a large angle of almost 90°.

On the other hand, suppose that there is a solid 3ϕ fault in the line to the left of B_{23} [i.e., in the "reverse" direction, say at point (y)]. In this case, neglecting the relatively small load current supplying S_{D2}, we find that

$$I_{23} = -I_{21} = -\frac{V_2}{\lambda Z'} \tag{14.3}$$

where Z' is the total series impedance of the line between buses 1 and 2, and λ has the same meaning as before. $\underline{/Z'}$ will be in the same range as $\underline{/Z}$ (i.e., in the range 80 to 88°); thus from (14.3) we see that in this case I_{23} leads V_2 by an angle between 92 and 100°. Thus the phase of I_{23} relative to that of V_2 is a distinctive feature which can be used to determine the fault direction.

The physical implementation uses an induction disk element. While differing in construction detail, we can understand the basic idea if we think of an implementation using a watthour meter element. Without loss of generality, suppose that V_2 is real (i.e., $V_2 = |V_2|$) and $I_{23} = |I_{23}| e^{j\phi_{23}}$, where $\phi_{23} = \underline{/I_{23}}$. Suppose that we pass V_2 through a phase-shifting network which retards the phase of V_2 by $\theta_z = \underline{/Z}$. In this way we obtain $|V_2| e^{-j\theta_z}$. If I_{23} and the phase shifted version of V_2 are applied to a watthour meter element, then, from (14.1), we get

$$T = k|V_2||I_{23}| \cos(\phi_{23} + \theta_z) \tag{14.4}$$

We get positive torque if $-\pi/2 < \phi_{23} + \theta_z < \pi/2$, zero torque if $\phi_{23} + \theta_z = \pm \pi/2$, and negative torque if $\pi/2 < \phi_{23} + \theta_z < 3\pi/2$. If I_{23} is the short-circuit current for a fault in the forward direction, then from (14.2), we find that $\phi_{23} = -\theta_z$. Then, from (14.4), we get the maximum possible positive torque. This result, of course, is intentional. Using the phase-shifting network, we "tune" the relay for maximum positive torque. On the other hand, if I_2 corresponds to a fault in the reverse direction, we get a strong negative torque. These results can be presented graphically as shown in Fig. 14.5.

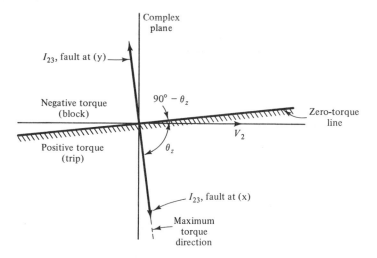

Figure 14.5 Operating characteristic of directional relay.

The zero-torque line divides the complex plane into two parts. In the lower half-plane we get positive torque; the disk will rotate so as to close contacts consistent with tripping the associated circuit breaker. In the upper half-plane we get negative torque (the watthour meter would try to turn backwards); the contacts remain open, consistent with blocking the tripping of the circuit breaker.

Also shown in Fig. 14.5 are the phasors V_2 and I_{23}, corresponding to short circuits in the forward and reverse directions. The discrimination between these two extreme cases is excellent. It is also necessary to consider cases when the fault impedance is not zero, or when the bus load currents are not completely negligible, or when the fault occurs in a more remote section of line (not the one adjacent to the observation point). More general nonsymmetric faults also need to be considered. The result of such a study is that in general, directional relays tend to operate for faults in their forward direction and block for faults in their reverse direction.

In this way at each breaker we can discriminate on the basis of the direction of the fault and trip the breaker if we get the conjunction of (1) the fault is in the forward (line) direction and (2) the fault current is greater than the pickup value.

A simplified schematic version of a physical implementation is shown in Fig. 14.6. Only phase a is shown in detail; however, the phase b and phase c relay contacts are also shown to indicate the breaker tripping logic.

It should be noted that there are a number of variations and useful refinements in the practical realizations, which have not been discussed.

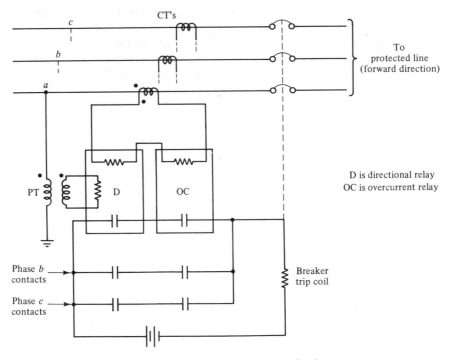

Figure 14.6 Directional relay application.

Exercise 2.   Redraw the loop system shown in Fig. Ex2 in the configuration of Fig. 14.4 (with the ideal source at both ends). Then, in the light of the previous discussion, show that the loop system can be protected by directional and time-delay overcurrent relays.

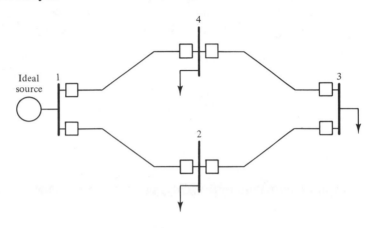

Figure EX 2

14.3 IMPEDANCE (DISTANCE) RELAYS

There are some problems with coordinating time-delay overcurrent relays even in the case of radial systems. If the string of lines and buses is too long, the time for operation of breakers close to the source becomes too large. To overcome this problem a different principle can be employed. It is true that overcurrent is a characteristic feature indicating a fault condition. An even more distinctive feature, however, is the ratio of voltage to current magnitudes. When there is a symmetric 3ϕ fault, the voltage drops and the current rises. Suppose that the voltage drops to 50% of its normal value while the current increases to 200% of its normal value. Then there is a 4:1 change in the voltage-to-current ratio compared with only a 2:1 change in the current.

A relay that operates on the basis of a voltage-to-current ratio is naturally called a *ratio* or *impedance relay*. It is also called a *distance relay;* the reason will be seen when we describe its operation.

Consider Fig. 14.4 again and suppose that there is a solid 3ϕ fault at point (x). Assume again that at B_{23} we measure the phase-neutral voltage V_2 and the current I_{23}. In this case the ratio V_2/I_{23} is the driving-point impedance of the portion of the line between B_{23} and the fault point (x):

$$\frac{V_2}{I_{23}} = \lambda Z \tag{14.5}$$

where Z is the total line impedance and λ is the fraction of the line between B_{23} and the fault. In general, V_2/I_{23} is not a driving-point impedance and it is probably less confusing to consider it simply as a measured quantity at the observation point.

The relay is designed to operate if $|V_2/I_{23}|$ is small enough. Suppose that we wish 80% of the line to be in the zone of protection; then the relay would be designed to operate if $|V_2/I_{23}| \le R_c \triangleq 0.8|Z|$. The relay would then operate for any (solid 3ϕ) fault located within the nearest 80% of the line (i.e., within the distance 80% of the length of the line). This explains the designation "distance" relay. We also speak of the "reach" of the relay; in this case the reach of the relay is 80% of the length of the line.

The region of operation of the relay is seen to be any complex value of V_2/I_{23} that lies within a radius R_c of the origin of the complex plane. In Fig. 14.7 we show several possible outcomes (i.e., values of V_2/I_{23} under different conditions); the conditions refer to the fault points labeled in Fig. 14.8. These points may be described as follows:

(a) Fault on line at 60% point (trip)

(b) Fault on line at 80% point (marginal trip)

(c) Fault on line at 100% point (block)

(d) Fault on system just beyond protected line (block)

(e) Fault on system much beyond protected line (block)

(f) Typical normal operating condition (block)

(g) Fault on line to the left of protected line (trip)

(h) More distant fault in same line as (g) (block)

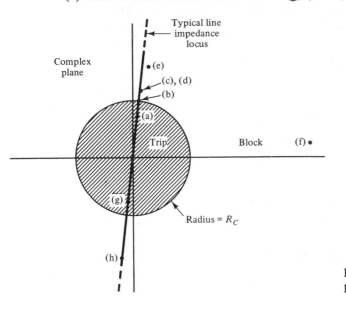

Figure 14.7 Zone of operation of impedance relay.

To avoid an undesired trip under condition (g) we can include a directional relay, just as we did previously in the case of overcurrent relays. Also, as with the overcurrent relays, we can provide backup protection. Consider a section of the trans-

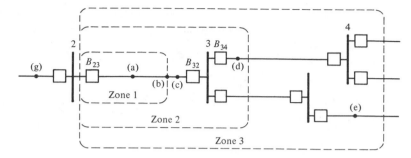

Figure 14.8 Fault points and protected zones.

mission system as shown in Fig. 14.8. We are considering only the relaying at the B_{23} location for the purpose of tripping B_{23}. We note that the other breakers (for example, B_{34}) are similarly equipped.

Assuming the use of directional relays, we need only consider the system to the right of B_{23}. At B_{23} we install three impedance relays. The first, already described, has a reach out to, say, 80% of the line and is set to trip instantaneously if $|V_2/I_{23}| \leq R_1$. This protected region is designated zone 1; it is also described as the primary, or instantaneous trip, zone. The second impedance relay has a longer reach; it includes all of zone 2 (which contains zone 1). If there is a fault at point (d), for example, then, after a time delay T_2 the breaker B_{23} will open. Presumably point (d) is in the primary protection zone of breaker B_{34} and therefore B_{34} should open instantaneously before B_{23} has opened. However in the event B_{34} does not open, B_{23} will provide backup protection. Similarly, there is the action of the third impedance relay. After an even longer time delay T_3, B_{23} provides backup protection for a zone 3 fault, for example, at point (e) just beyond the next set of "downstream" breakers.

We have been discussing only the backup protection afforded by B_{23}. There is, of course, the backup protection by other breakers afforded to B_{23}. We note that directional-impedance relaying has the flexibility to effectively protect general transmission systems, not just radial ones. Although the overall transmission system may be very complex, including many loops, the impedance relaying principle allows us to limit our attention to a much smaller domain. The system, locally, may be as simple as the one shown in Fig. 14.8.

The reader may wonder why zone 1 in Fig. 14.8 was not specified to include the entire transmission line (i.e., 100% of the line). The reason may be explained as follows. The impedance seen by the impedance relay for a fault at a point (c) just to the left of bus 3 and at a point (d) just to the right of the bus are essentially the same; the impedance of the closed breakers and the bus structure is negligible. Thus if point (c) were included in zone 1 (instantaneous trip), a fault at point (d) would also cause the instantaneous trip of B_{23}. This would be an undesirable trip since B_{34} should be allowed to clear the fault. The advantage of not tripping B_{23} when the fault is at (d) more than compensates for the disadvantage of not tripping instantaneously when the fault is at (c). Note that a fault at (c) still causes (delayed) tripping of B_{23}.

Exercise 3. Justify the use of directional-impedance relaying in the case shown in Fig. Ex3. Is the line between buses 1 and 6 in the forward direction of the directional relay at B_{12}?

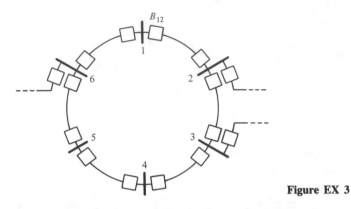

Figure EX 3

We consider next a possible physical implementation of an impedance relay with instantaneous trip. The working principle is suggested schematically in Fig. 14.9. The relay is of a type called a *balanced-beam movement*. In the absence of voltage and currents inputs, the beam is balanced. The voltage applied to the solenoid on the left causes a proportional current to flow and exerts a downward pull on the left side of the beam. The current applied to the right side solenoid exerts a downward pull on the right side. If the "voltage pull" is greater than the "current pull," the relay contacts are open. If the reverse is true, the relay contacts close. The solenoid action is described in Appendix 2, where it is seen that the average forces exerted are proportional to the squares of the phasor current magnitudes. Thus the relay tripping condition is

$$k_1 |V|^2 \le k_2 |I|^2 \tag{14.6}$$

where k_1 and k_2 depend on the design parameters discussed in Appendix 2, and in addition k_1 depends on the impedance of the voltage solenoid. As indicated in (14.6), it is convenient to assume that the relay trips when the current and voltage pulls are equal.

Equivalent to (14.6), the relay trips if

$$\left| \frac{V}{I} \right| \le \left(\frac{k_2}{k_1} \right)^{1/2} \triangleq R_c \tag{14.7}$$

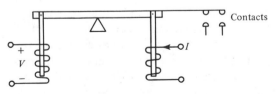

Figure 14.9 Balanced-beam movement.

By adjustment of the number of turns (by taps) and the air gap (core screw), the critical value R_c may easily be changed.

Finally, we remind the reader that, for simplicity, we have been considering only the case of symmetric 3ϕ faults. When we consider the variety of nonsymmetric faults there are additional complications. Even for a single zone application we note that we need three (1ϕ) impedance relays for each circuit breaker, one per phase. Breaker trip should occur for impedance below critical on any phase. Then we observe that the reach of each relay depends on the type of fault. For example, if we set the relays for a reach of 80% of the length of line for a symmetric 3ϕ fault, they may only protect, say, 50% of the line for an SLG fault or 10% of the line for an LL fault. On the other hand, if, for example, we set the reach for correct action for an LL fault, the relays may "overreach" for other types of faults.

The impedance relays we have been considering are called *ground relays* and work with phase-neutral voltages and line currents. They are very effective in responding to 3ϕ, SLG, and DLG faults. The relative insensitivity of ground relays in responding to LL faults has led to a modification called *phase relays*, where the same impedance relays are used, but their voltage inputs are line-line voltages (V_{ab} instead of V_{an}, for example) and the line current differences $I_a - I_b, I_b - I_c$, and $I_c - I_a$. With these voltages and current the impedance relays are quite sensitive for LL faults but relatively insensitive for SLG faults. In general, therefore, both ground and phase impedance relays need to be used.

14.4 MODIFIED-IMPEDANCE RELAYS

In the preceding section we considered the use of impedance relays in conjunction with directional relays. We turn next to a simple modification of the basic impedance relay design which gives them a directional characteristic and permits stand-alone operation.

The balanced-beam movement of Fig. 14.9 may be modified by coupling the voltage and current circuits as in Fig. 14.10. For simplicity we will neglect the (small)

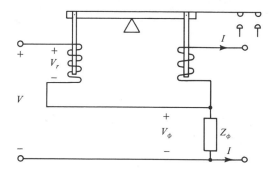

Figure 14.10 Modified-impedance relay schematic.

current in the (high-impedance) voltage solenoid in calculating the voltage across Z_ϕ. In this case the voltage across Z_ϕ is $V_\phi = Z_\phi I$ and at the voltage coil we have

$$V_r = V - Z_\phi I \tag{14.8}$$

Using (14.7), the relay will trip if

$$\left| \frac{V_r}{I} \right| = \left| \frac{V - Z_\phi I}{I} \right| = \left| \frac{V}{I} - Z_\phi \right| \le R_c \tag{14.9}$$

The region of operation is easy to show in the complex plane. In Fig. 14.11 we show Z_ϕ, V/I, and their difference.

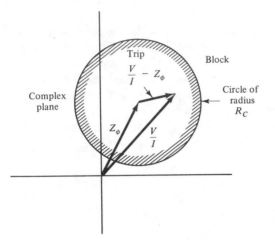

Figure   14.11 Operating   zone   of modified relay.

The condition (14.9) is satisfied, and the relay will trip, if V/I is inside the circle with center Z_ϕ and radius R_c. Compared with the operating zone of the impedance relay (see Fig. 14.7), we have the flexibility of shifting the center. We can pick Z_ϕ and R_c to make the center of the circle fall on the line impedance locus and the circle go through the origin. The resulting modified operating characteristic, also known as a *mho characteristic,* is highly directional. Just as in the case of impedance relays we can provide three-zone protection by introducing three different reaches and appropriate operating delays. In Fig. 14.12 we compare the operating zones for a three-zone impedance relay scheme operating in conjunction with a directional relay and the three-zone modified impedance relay scheme we have just described. The three zones have been drawn so that the corresponding reaches are the same. The difference, then, is in the "skirts," and here the modified impedance relay offers advantages. Loads in general have higher power factors than faults. Note then, in Fig. 14.12, the much improved selectivity of the modified impedance relay with respect to high power factor loads. In particular, compare their behavior with respect to the load impedance (f) in Fig. 14.12.

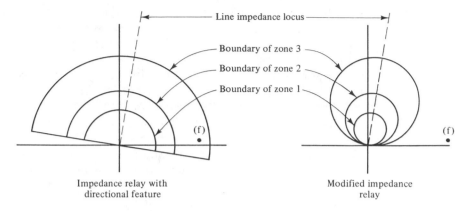

Figure 14.12 Comparison of operating zones.

14.5 DIFFERENTIAL PROTECTION OF GENERATORS

We consider next a very reliable method of protecting generators, transformer banks, buses, and transmission lines from internal faults. The basic idea as applied to generator protection may be seen in Fig. 14.13. Only the protection of phase a is shown. The scheme is repeated for the other phases. Using the dot convention, the reader may check that the reference directions for primary and secondary current in the CTs is consistent.

If there is no internal fault, $I_2 = I_1$, hence with identical current transformers, $I_2' = I_1'$; since the CTs are connected in series, the current flows from one secondary

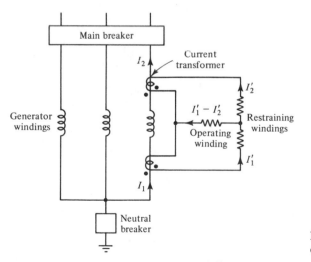

Figure 14.13 Generator protection by differential relaying.

to the other and there is no current in the operating winding of the relay. Now suppose that there is a ground in the winding, or a phase-phase short. Then $I_2 \neq I_1$, and thus $I_2' \neq I_1'$ and a differential current $I_1' - I_2'$ flows in the operating winding. It is natural to call such a relay a *differential relay*.

It would seem reasonable to detect the differential current with an overcurrent relay, but in practice this is not done. The reason is the difficulty in matching the current transformer characteristics; they obviously cannot be identical. Even very small mismatches in, say, the turns ratio might cause significant differential current under full-load conditions or in the case of high currents due to fault conditions elsewhere. On the other hand, raising the pickup value for the overcurrent relay to prevent improper operation at high generator currents would degrade the protection of the generator under light-load conditions.

A solution to the problem is achieved by using a "proportional" or "percentage" type of differential relay, where the differential current required for tripping is proportional to the generator current. A balanced-beam movement with a center-tapped "restraining" winding can be used to obtain this behavior. The general idea is shown in Fig. 14.14.

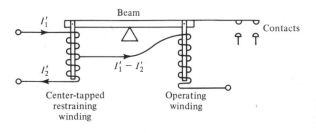

Figure 14.14 Proportional differential relay.

As in previous cases, the relay closes if the downward pull on the right side exceeds the downward pull on the left. The average downward pull on the right is proportional to the square of the differential current (i.e., to $|I_1' - I_2'|^2$). The average downward pull on the left (the restraining force) is proportional to $|I_1' + I_2'|^2$; this result is shown in Appendix 2. Thus the condition for operation can be stated as

$$|I_1' - I_2'| \geq k\frac{|I_1' + I_2'|}{2} = k|I_{\text{average}}'| \tag{14.10}$$

and has the desired proportional property. k is a constant that is adjustable by tap and air-gap (core screw) adjustments. Notice that as k is increased the relay becomes less sensitive.

Assuming that I_1' and I_2' are in phase, it is easy to find the region of operation given by (14.10). This is shown in Fig. 14.15 for a value of $k = 0.1$. We note that the characteristic can be easily modified to give an unambiguous block zone in the neighborhood of the origin. We also note that as an alternative to the balanced-beam relay we have been describing, there are induction-type relays that also operate on differential currents and have restraining windings. They have similar characteristics to Fig. 14.15.

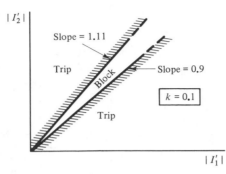

Figure 14.15 Operating region of differential relay.

We have been considering the differential protection of phase a of the generator winding; phases b and c are similarly protected. When any one of the three relays operates, the 3ϕ line (main) circuit breaker and the neutral breaker open, isolating the generator from the rest of the system. The breakers, but not the tripping circuits, are shown in Fig. 14.13. In addition, the generator field breaker (not shown) would be tripped to deenergize the stator windings. This protects the generator in the event of (internal) phase-phase shorts which would not be isolated with the opening of the main and neutral breakers. Since we need to open three different circuit breakers (main, neutral, and field), three sets of contacts must be available on each differential relay.

14.6 DIFFERENTIAL PROTECTION OF TRANSFORMERS

To protect a 3ϕ transformer bank is somewhat more complicated because we must take into account the change in current magnitude and phase in going from the low-voltage to the high-voltage side. The differences must be "canceled out" before being applied to a differential relay. Consider first the case of Y-Y or Δ-Δ connections. In this case, for practical purposes, the primary and secondary currents are in phase in the steady state (exactly so if the magnetizing inductances are infinite) and only the current ratio changes need to be canceled out. A possible realization for a Y-Y connection is shown in Fig. 14.16. Only one phase of the 3ϕ relay protection is shown; the protection for phases b and c would be the same.

Also shown in the figure are the phase a transformer bank currents I_1 and I_2 which are related by the current gain $a = 1/n$, where n is the voltage gain. If, under normal, nontrip conditions we want the CT secondary currents to be equal (i.e., $I_2' = I_1'$) while the primary CT currents are related by $I_2 = aI_1$, we need to satisfy the following condition:

$$\frac{I_2'}{I_2} = \frac{I_1'}{aI_1} \tag{14.11}$$

Since $a_2 \triangleq I_2'/I_2$ and $a_1 \triangleq I_1'/I_1$ are the current gains of CT2 and CT1, respectively, we get the following simple relation between current gains:

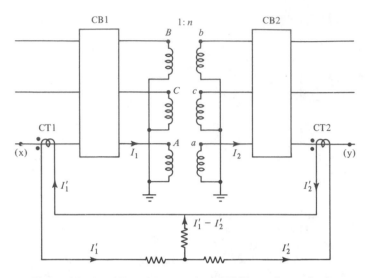

Figure 14.16 Differential protection of Y-Y transformer bank.

$$a = \frac{a_1}{a_2} \tag{14.12}$$

From (14.12) the desired CT ratings may be selected. For example, suppose that the voltage gain or step up ratio $n = 10$. Then $a = a_1/a_2 = 0.1$. We might pick the primary rating of CT1 to be 500; thus $a_1 = 5/500 = 0.01$. Picking the primary rating of CT2 to be 50, we get $a_2 = 5/50 = 0.1$. Thus $a_1/a_2 = 0.1$, as required. The selection of CT ratings is natural; we need a higher rating (10 times higher) in the high-current side of the 3ϕ transformer bank than in the low-current side.

In other cases there may be difficulty in finding standard CTs with the necessary ratios. Other methods for accomplishing the same purpose include the use of auto-transformers connected to the CTs with multiple taps which provide additional flexibility. Taps may also be provided in the differential relay itself.

There is an interpretation of the condition (14.12) implying zero differential current, which will be helpful in considering the next case: the differential protection of Δ-Y transformer banks. Consider the current gain in going from point (x) to point (y) in Fig. 14.16. There are two possible paths. If we go through the 3ϕ transformer bank, the gain is $I_2/I_1 = a$. If we go through the CTs, the gain is $(I_1'/I_1)(I_2/I_2') = a_1/a_2$. Thus to get zero differential current, (14.12) says that the current gain must be the same for the two parallel paths. It is easy to show this condition must be true even if the current gains are complex numbers. The reader may note the connection with "normal" systems. Equation (14.12) says that the CT ratios and phase shifts should be in accordance with that of a normal system.

In Fig. 14.17 we show the differential protection of a Δ-Y transformer bank. The basic idea can best be understood from a one-line diagram.

The point to note is that the secondaries of CT2 are connected in Δ. Thus when we consider the phase a (per phase) current gain a_2, we get a phase shift just as we

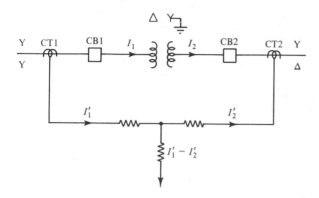

Figure 14.17 Differential protection of Δ-Y transformer bank.

do in the case of the Δ-Y power transformer bank; (14.12) requires that the phase shifts be the same from the Δ to the Y sides. By wiring the sets of Δ-Y transformers identically, we get the same (connection induced) phase shifts in going from the Δ to the Y sides. Of course, the current-gain magnitudes must also be equal, as required by (14.12).

The differential protection of transformers is beset by two problems we will touch upon briefly. The first is the problem of transformer "in-rush" current. This is a transient phenomenon in which the transformer magnetizing current may reach values several times full-load current when a transformer is first energized. Since as shown in Chapter 5, the magnetizing current flows in a shunt path, it looks exactly like a ground fault to the differential relay.

To prevent improper operation during normal switching transients, a number of methods may be used. These include desensitizing the relay [increasing the constant k in (14.11)], and using relays in which the unidirection component and the strong harmonic content of the in-rush current inhibit tripping.

The second problem concerns regulating transformers used to control complex power flows by varying voltage ratios and phase angles. These were discussed in Section 5.9. Clearly, the differential relays must not operate for the entire range of voltage ratios and phase angles associated with normal operation. One way this can be accomplished is by desensitizing the differential relay. In this way we compromise between the need to trip the relay when there is an internal fault and the need not to trip when there is not a fault.

14.7 DIFFERENTIAL PROTECTION OF BUSES AND LINES

The differential protection of buses may be illustrated by the one-line diagram shown in Fig. 14.18. Assume that all the current transformers have the same ratio and are ideal. The secondaries of CT1 and CT2 are in parallel, hence their currents add as shown in the figure. If there is no bus fault, $I_1 + I_2 = I_3$, and because all the CTs have the same ratio, $I_1' + I_2' = I_3'$. Then the differential relays do not operate. As with the previous applications, identical CTs are not required because of the restraining wind-

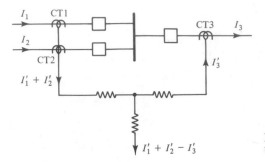

Figure 14.18 Differential protection of buses.

ings. The actual implementation would require three differential relays, one for each phase. Operation of any of the three relays would activate the trip coils of all three CBs, thereby isolating the bus.

This technique extends naturally to a larger number of lines. Some care is required in grouping the lines associated with each restraining coil in order not to lose the restraining action. In an extreme case the currents, while individually large, might sum to zero in each restraining coil. This would be the case if in Fig. 14.18, $I_3 = 0$ (line out of service?). Then the restraining action is lost even though I_1 and I_2 may individually be large. This problem, and others involving magnetic saturation of CTs, has been addressed by some alternative differential relaying schemes.

As we have seen in the last few sections, the differential relaying of generators, transformers, and buses all have their own special features but are fundamentally the same. One feature in common is the physical proximity of the current measurement points (location of CTs). This physical proximity is lost if one attempts to apply differential relay protection to transmission lines (and underground cables). Differential protection of lines has many desirable features. Simultaneous and rapid operation of circuit breakers (in approximately 1 cycle), unconstrained by problems of coordination with other relays, would be very desirable.

Differential protection of lines can be provided through the use of auxiliary communication channels which link the two ends of each line. These channels can take the form of telephone or other wires (pilot wires), microwave links, or a system known as power-line carrier. In the latter system the power lines themselves carry the signals in the form of modulated carriers in the frequency range 30 to 200 kHz (i.e., approximately the radio AM band). It is to be noted that the communication channels are also used for system communication, telemetering, remote control, and relaying schemes other than differential protection.

14.8 OVERLAPPING ZONES OF PROTECTION

In Section 14.3, in connection with transmission-line protection, we introduced the notion of the reach of a relay and its primary zone of protection. We also discussed the backup protection afforded by a relay to another relay's primary zone of protection. Thus, necessarily, there is overlap between backup protection zones and primary protection zones.

It is also very desirable to have overlaps between the *primary zones* of protection. In this way every point in the system lies within the primary protection zone of some relay. We consider next an example of such an arrangement, shown in Fig. 14.19.

The protection zones are defined by the locations of the CTs associated with the particular protected element. The location of the CTs with respect to the circuit breakers is the same as those shown in Fig. 14.6 for lines, Fig. 14.16 for transformers, and Fig. 14.18 for buses. In these cases the order of arrangement is always the same. First the element to be protected, then the circuit breaker, and then the CT. Since the circuit breakers link the different elements, this guarantees that the protection zone overlap. Note also that each circuit breaker then falls into two zones and is efficiently used. Note that a fault at point (x) in Fig. 14.19, which lies within two overlapping zones, trips all the bus and transformer breakers. A fault at point (y) trips only the bus breakers. A fault at point (z) trips the bus breakers (one of which doubles as a line breaker), and if the line is protected by differential relaying, also trips the line breaker at the other end of line 2 (not shown).

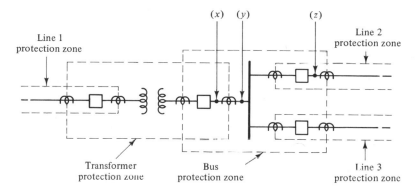

Figure 14.19 Overlapping protection zones.

14.9 SEQUENCE FILTERS

We will conclude our discussion of relay types by considering those that operate on the bases of abnormal negative-sequence and/or zero-sequence currents and voltages. Negative- and zero-sequence quantities individually and in combination are a strong distinguishing feature, or discriminating feature, associated with a nonsymmetric fault. In addition to indicating a (nonsymmetric) fault, it may be important to recognize the existence of negative- and zero-sequence currents for their own sakes. For example, consider the consequences of negative-sequence generator currents due to unbalanced loads or nonsymmetric faults. These negative-sequence currents set up a flux which rotates in the air gap at an angular velocity ω_0 (for a two-pole machine) and in a direction opposite to that of the rotor rotation. Thus the rotor experiences rapid flux variations which cause large currents to flow on the rotor surface. Even relatively small negative-sequence currents, if allowed to continue, will cause excessive heating of the generator rotor. To a lesser extent there is a similar problem with zero-sequence stator currents.

The detection of negative-sequence or zero-sequence currents and voltages is done in sequence filters. We will consider two simple examples. A zero-sequence current filter is easily made using three CTs. Suppose that the primary currents are I_a, I_b, and I_c. If we put their secondaries in parallel across a (low) resistance, we get a voltage across the resistance proportional to $I^0 = \frac{1}{3}(I_a + I_b + I_c)$. Note that the filter does not respond to positive- and negative-sequence currents. In some cases we can also read the neutral current of a grounded generator or Y-connected transformer bank which will be proportional to I^0.

To measure the negative-sequence voltage, the circuit shown in Fig. 14.20 may be used. Assume that the transformers are ideal, with the secondary of the right-hand transformer center-tapped. We find that

$$I = \frac{V_{c'a'}}{R + jX} \tag{14.13}$$

$$V = \tfrac{1}{2}V_{b'c'} + \frac{R}{R + jX} V_{c'a'} \tag{14.14}$$

Using the condition $X = \sqrt{3}\,R$, we get

$$V = \tfrac{1}{2}(V_{b'c'} + V_{c'a'}e^{-j\pi/3}) \tag{14.15}$$

Now suppose that the line voltages are a negative-sequence set. Then $V_{c'a'}^- = V_{b'c'}^- e^{j2\pi/3}$ and

$$V = \tfrac{1}{2}V_{b'c'}^-(1 + e^{j\pi/3})$$

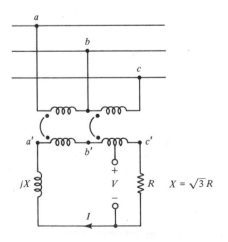

Figure 14.20 Negative-sequence voltage filter.

from which we find that

$$|V| = \frac{\sqrt{3}}{2} |V_{b'c'}^-| = \frac{3}{2} |V_{b'n'}^-| \qquad (14.16)$$

Thus $|V|$ is proportional to the negative-sequence secondary voltage. To find the corresponding primary values, we multiply by the PT ratio.

If the line voltages are a positive-sequence set, $V_{c'a'}^+ = V_{b'c'}^+ e^{-j2\pi/3}$ and

$$V = \tfrac{1}{2} V_{b'c'}^+ (1 + e^{-j\pi}) = 0 \qquad (14.17)$$

The response to zero-sequence line voltages need not be considered, since these voltages are necessarily zero. We can see this as follows. For zero-sequence voltages $V_{an}^0 = V_{bn}^0 = V_{cn}^0$. Then $V_{ab}^0 = V_{an}^0 - V_{bn}^0 = 0$. Similarly, $V_{bc}^0 = V_{ca}^0 = 0$. Thus the filter responds only to negative-sequence voltages, or to the negative-sequence component in the nonsymmetric 3ϕ voltage.

Exercise 4.   Show a circuit using PTs for obtaining a voltage proportional to the zero-sequence voltage of a 3ϕ line.

14.10 COMPUTER RELAYING

A feature of the protection system we have been discussing is that, except for the case of differential line relaying, each relay operates on the basis of local information, provided by nearby PTs and/or CTs and trips only local breakers. Excluding the exception, the protection system is therefore entirely decentralized, in the sense that no information is exchanged directly between relays at different locations, nor are tripping instructions from one relay sent to a remote breaker. Of course, in a sense, information is sent from one location to another through the intermediary of system behavior; backup protection furnishes an example. But, we repeat, among the schemes we discussed, only the differential relaying of lines uses a *direct* exchange of information or command signals.

We should note that recently there has been interest in the possible application of *computer relaying*. The idea of a completely centralized protection system (i.e., a central computer to replace all the relays) is not attractive and is not being seriously considered. Under consideration are hierarchical structures involving dedicated relaying computers at the bottom level interacting with one or more higher levels of supervisory computers. The most active proposal visualizes the one-to-one replacement of each electromagnetic relay with a dedicated "relaying computer" of similar general type. Thus there would be impedance relays, differential relays, and so on. The relaying computer would be a solid-state switching relay driven by microprocessor logic. An advantage is the high degree of flexibility, in terms of programmable tripping characteristics, compared with the hardware-limited tripping characteristics of electromechanical relays. Another advantage is the ease with which these tripping characteristics could be changed by software, and from a remote location. Thus there is the potential to "shape" the tripping characteristic in "real time" in accordance with operating conditions. For example, suppose that the system is operating under very high, but normal load conditions. The pattern of high power flows could be determined by a supervisory computer and used to change the tripping characteristics of individual (computer) relays to avoid tripping in the absence of a fault. Under light-load conditions the setting could be made more sensitive.

Another potential advantage of computer relaying is the possibility of incorporating a broader perspective into the problem of system protection. In this chapter the primary emphasis has been on protecting the equipment of the system. This is certainly the primary concern, but it must be balanced with the objective of maintaining the structural integrity of the system to provide continuity of service. For example, suppose that a line is overloaded. With system protection the sole criterion, line tripping would be indicated. However, if loss of load, or transient stability problems, follow as a consequence, perhaps the line should remain in service despite the overload.

It should be noted that system protection and the maintenance of service need not necessarily be at odds. For example, as discussed in Chapter 9, line circuit breakers have a very desirable reclosure feature which helps maintain synchronism in the event of faults of very short duration. Thus the maintenance of transient stability is enhanced with little or no degrading of system protection. In other cases, however, a role for a supervisory computer is seen, in effecting a constantly shifting balance between the possibly conflicting requirements of system protection and continuity of service.

14.11 SUMMARY

The function of power system protection is to detect and remove faults from the system as rapidly as possible while minimizing the disruption of service. To further this objective the system is designed to incorporate some desirable general features, such as zones of protection, overlapping protection zones, backup protection, and relay (time) coordination. The relays used to detect faults in their zones of protection include electromechanical devices of the plunger, induction-disk, and balanced-beam type, among others. These devices may be used to obtain relays having the following functions: instantaneous overcurrent, time-delay overcurrent, directional relaying, impedance, modified impedance, and differential relaying. Important relay characteristics are their operating characteristics (trip or block regions) and their time delays.

PROBLEMS

14.1. In Example 14.1 it is stated that the smallest fault current seen at breaker B_3 is an LL fault at bus 4. Verify the statement by checking the alternatives.

14.2. Suppose that in Example 14.1, an SLG fault occurs at the midpoint of the line between buses 3 and 4. Which breaker should operate, and how long should it take to clear the fault? If the breaker does not operate, how long does it take the backup breaker to operate?

14.3. In Example 14.1, recalculate the settings of relays R_3 and R_2 under the assumption that the zero-sequence line impedances are three times that of the positive (and negative)-sequence line impedances. *Note:* In this case the smallest fault current seen at breaker B_3 may occur for a different type of fault.

14.4. Repeat Problem 14.3, but assume that the positive (and negative) line impedances are $j2$ ohms (instead of $j10$ ohms).

14.5. Suppose that for the system and settings found in Problem 14.4, there is a 3ϕ short at the midpoint of the line between buses 2 and 3. Find the time it takes to clear the fault.

14.6. Verify that Fig. 14.15 shows the correct tripping and blocking regions [i.e., in accordance with (14.10)] for $k = 0.1$. Show the regions for $k = 0.2$.

Appendices

APPENDIX 1: RELUCTANCE

In Example 3.1 the flux in the core, ϕ, is found to be proportional to the mmf, Ni. We can write the relationship as

$$\phi = \frac{Ni}{l/\mu A} = \frac{F}{\mathcal{R}} \tag{A1.1}$$

where F is the mmf and $\mathcal{R} = l/\mu A$ is defined to be the reluctance of the magnetic circuit. There is an analogy to Ohm's law for a simple resistive circuit, where $i = v/R$. In fact, since for a conductor of uniform cross section A, length l, and conductivity σ, the resistance $R = l/\sigma A$, the analogy extends to the geometry as well.

Consider the same toroid as in Example 3.1 but with a small air gap cut in the core. If, as in Example 3.1, we now apply Ampère's circuital law to this case, we get

$$F = Ni = \int_\Gamma \mathbf{H} \cdot d\mathbf{l} = H_1 l_1 + H_2 l_2 \tag{A1.2}$$

where H_1 and H_2 are the scalar magnetic intensities and l_1 and l_2 are the path lengths in the iron core and air gap, respectively. Since the flux in the air gap and the core are the same, $\phi = B_1 A = B_2 A$; as an approximation, we are assuming no "fringing" of the flux in the air gap. We also have $B_1 = \mu_1 H_1$ and $B_2 = \mu_2 H_2$. We can now rewrite (A1.2) in terms of ϕ as follows:

$$F = Ni = \frac{\phi l_1}{\mu_1 A} + \frac{\phi l_2}{\mu_2 A} \qquad\qquad (A1.3)$$

Defining the core reluctance $\mathcal{R}_1 = l_1/\mu_1 A$ and the air-gap reluctance $\mathcal{R}_2 = l_2/\mu_2 A$, we get

$$\phi = \frac{F}{\mathcal{R}} \qquad\qquad (A1.4)$$

where $\mathcal{R} = \mathcal{R}_1 + \mathcal{R}_2$ is the total reluctance of the "series" magnetic circuit. More generally, for any series magnetic circuit (with the same ϕ in each section) we can use (A1.4) with $\mathcal{R} = \Sigma \mathcal{R}_i$ as the total reluctance and $F = \Sigma F_i$ as the total mmf for the series circuit. The analogy to a series circuit containing resistances and voltage sources is clear. It can be shown that the analogy extends in a natural way to more complicated magnetic structures, including series and parallel magnetic paths and mmfs in the different "legs."

Our understanding of magnetic circuits may be greatly enhanced by using reluctance and its resistance analogy in a qualitative way. As an example of this use of reluctance, consider the circuit shown in Fig. A1.1, in which some of the flux lines are sketched. Neglecting the "leakage flux," we have a series circuit composed of four reluctances corresponding to two iron path sections and two air gaps. Since the permeability of the iron is in the order of 1000 times that of the air, the iron reluctances may be negligible compared with the air-gap reluctances. The air-gap reluctances depend on θ. The minimum reluctance occurs when the average air-gap distance is a minimum (i.e., for $\theta = 0, \pi$); this corresponds to a maximum ϕ.

Suppose that we consider the leakage flux. Then, noting the figure, we have the reluctance from a to b (mainly the sum of air-gap reluctances) in parallel with the much greater reluctances of the relatively long leakage paths in air. As an approximation these leakage reluctances may therefore be neglected.

Finally, we note that unlike the variable reluctances of iron sections (because μ

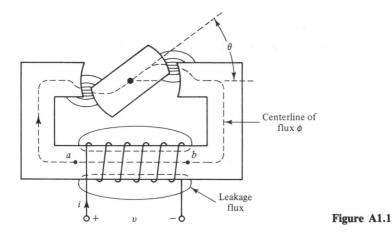

Figure **A1.1**

is not constant), those of circuits in the air are constant. In a series circuit with reluctance dominated by an air-gap term, we can therefore expect a fairly linear ϕ–F relationship, at least until the iron "saturates" and its reluctance increases significantly.

APPENDIX 2: FORCE GENERATION IN A SOLENOID

One type of electromagnet, or solenoid, is shown in Fig. A2.1. The current in the coil produces a force on the plunger, or movable core. There are applications in "stepper" motors, brakes, clutches, releases, and so on. With the addition of electrical contacts on the plunger, we get a relay or contactor whose contacts close (or open) when the current is sufficiently large. In Chapter 14 we consider some relay applications to the problem of power system protection.

We can calculate the force on the plunger by using the following modification of (10.9):

$$f_E = \tfrac{1}{2}\mathbf{i}^T \frac{d\mathbf{L}(y)}{dy}\mathbf{i} \tag{A2.1}$$

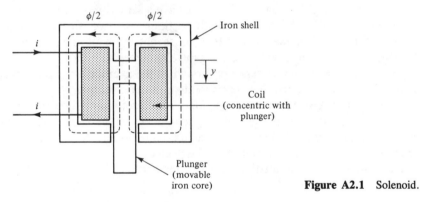

Figure A2.1   Solenoid.

In fact, in the case of Fig. A2.1, all the quantities in (A2.1) are scalars.

In calculating the inductance $L(y)$, it is convenient to use the notion of reluctance. Thus using the result in Appendix 1, we find that the reluctance of the air gap is

$$\mathscr{R} = \frac{y}{\mu_0 A} \tag{A2.2}$$

We will not calculate the reluctance of the (series) iron path; we assume that it is negligible compared with the value from (A2.2). Then, using (A2.2), we have

$$\phi = \frac{Ni}{\mathscr{R}} = \frac{\mu_0 A N i}{y} \tag{A2.3}$$

Assuming that the flux links all the turns, we have

$$L = \frac{\lambda}{i} = \frac{N\phi}{i} = \frac{\mu_0 A N^2}{y}$$ (A2.4)

Thus

$$\frac{dL}{dy} = -\frac{\mu_0 A N^2}{y^2}$$ (A2.5)

and using (A2.1), we get

$$f_E = -\frac{\mu_0 A N^2 i^2}{2y^2}$$ (A2.6)

The negative sign means that the force on the plunger is opposite to the direction of increasing y (i.e., is upward). Physically, we have a manifestation of the attraction of opposite magnetic poles.

The equation above gives the instantaneous values of f_E. The average value for a sinusoidal current is

$$f_{Eav} = -\frac{\mu_0 A N^2 |I|^2}{2y^2}$$ (A2.7)

where $|I|$ is the effective or root-mean-square (rms) value of i.

To get some idea of reasonable values of force, assume that the plunger is 1 in. in diameter, $y = 1/8$ in., $N = 1000$, $|I| = 1$ A. Then

$$f_{Eav} = -\frac{4\pi \times 10^{-7} \times 5.07 \times 10^{-4} \times 10^6 \times 1^2}{2 \times (3.175 \times 10^{-3})^2}$$

$$= -31.58 \text{ newtons} = -7.10 \text{ pounds}$$

Suppose now that the coil in Fig. A2.1 has a "tap," so that schematically we have the electrical circuit shown in Fig. A2.2. This time in using (A2.1), $\mathbf{i}$ has two components, i_1 and i_2, and $\mathbf{L}(y)$ is a 2×2 matrix. We can find $\mathbf{L}$ by using superposition as follows. If ϕ_{11} is the flux in coil 1 when only i_1 flows, then

$$\phi_{11} = \frac{N_1 i_1}{\mathcal{R}} \Rightarrow L_{11} = \frac{N_1 \phi_{11}}{i_1} = \frac{N_1^2}{\mathcal{R}}$$ (A2.8)

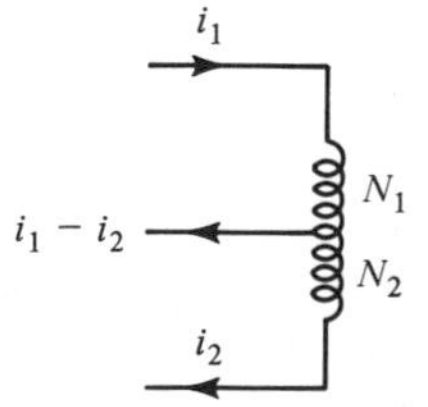

Figure A2.2   Tapped coil.

Similarly, $L_{12} = L_{21} = N_1 N_2/\mathcal{R}$, $L_{22} = N_2^2/\mathcal{R}$. Thus

$$\mathbf{L}(y) = \frac{1}{\mathcal{R}(y)} \begin{bmatrix} N_1^2 & N_1 N_2 \\ N_1 N_2 & N_2^2 \end{bmatrix} \tag{A2.9}$$

Then from (A2.2), using $\mathcal{R}(y) = y/\mu_0 A$, we get

$$f_E = -\frac{\mu_0 A}{2y^2} (N_1 i_1 + N_2 i_2)^2 \tag{A2.10}$$

Suppose that $N_1 = N_2$ (i.e., the coil in Fig. A2.2 is center tapped). Then (A2.10) reduces to

$$f_E = \frac{\mu_0 A N_1^2}{2y^2} (i_1 + i_2)^2 \tag{A2.11}$$

and for sinusoidal currents,

$$f_{Eav} = -\frac{\mu_0 A N_1^2}{2y^2} |I_1 + I_2|^2 \tag{A2.12}$$

where I_1 and I_2 are the (effective) phasor representations of i_1 and i_2, respectively.

APPENDIX 3: METHOD OF LAGRANGE MULTIPLIERS

Consider the following geometric problem: Find the rectangle of largest area inscribed in an ellipse given by

$$\frac{x^2}{a^2} + \frac{y^2}{b^2} = 1 \tag{A3.1}$$

Since the area of the rectangle is $4xy$, the problem may be restated as follows:

$$\text{maximize } f(x, y) = 4xy \tag{A3.2}$$

with (x, y) satisfying (A3.1).

In dealing with the constraint, or "side condition" (A3.1), it is convenient to introduce notation and replace it by

$$g(x, y) = \frac{x^2}{a^2} + \frac{y^2}{b^2} - 1 = 0 \tag{A3.3}$$

A geometric approach to the solution reveals a useful property of the optimal solution. Suppose that we plot the constraint $g(x, y) = 0$ in the (x, y) plane along with the plots of $f(x, y) = \alpha$ for different (constant) α. In Fig A3.1 we see that for $f(x, y) = \alpha_1$ there are two solutions that satisfy the constraint. $f(x, y) = \alpha_3$ is not feasible because it does not meet the constraint. It is clear geometrically that $f(x, y) = \alpha_2$ is the optimal or largest value of $f(x, y)$ consistent with meeting the constraint, and that (x^0, y^0) is the corresponding optimal value of (x, y).

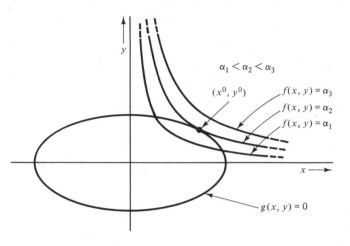

Figure A3.1 Geometric property of optimal solution.

The useful property we alluded to earlier is that at the optimal point (x^0, y^0) the curve $f(x, y) = \alpha_2$ and the ellipse $g(x, y) = 0$ are tangent. Equivalently, and more computationally convenient, at (x^0, y^0) the gradient vectors of $f(x, y)$ and $g(x, y)$ point in the same direction.

Using ∇ to indicate the gradient operator, this tangency condition can be restated as follows:

$$\nabla f = \lambda \, \nabla g \tag{A3.4}$$

where λ is a real scalar, not yet known. In terms of components, for the present problem, we get the following two scalar equations:

$$\frac{\partial f}{\partial x} = \lambda \frac{\partial g}{\partial x}$$

$$\frac{\partial f}{\partial y} = \lambda \frac{\partial g}{\partial y} \tag{A3.5}$$

Using (A3.2) and (A3.3) we calculate the following terms:

$$\frac{\partial f}{\partial x} = 4y \qquad \frac{\partial f}{\partial y} = 4x \qquad \frac{\partial g}{\partial x} = \frac{2x}{a^2} \qquad \frac{\partial g}{\partial y} = \frac{2y}{b^2}$$

Substituting in (A3.5), we get

$$4y = \frac{\lambda 2x}{a^2} \qquad 4x = \frac{\lambda 4y}{b^2} \tag{A3.6}$$

Equations (A3.6) and (A3.3) give three equations in the three unknowns x, y, and λ. Equation (A3.6) yields $\lambda = 2ab$ as the only possible solution for λ corresponding to nonzero solutions for x and y. Then substituting $y = bx/a$ in (A3.3), we get $x = a/\sqrt{2}$. Finally, $y = b/\sqrt{2}$. This defines the optimal rectangle. We note that with these values of x and y, $f(x, y) = 4xy = 2ab$. Thus the maximum area is $2ab$.

These useful notions extend to more general problems. An appropriate version for our purposes is the following. Suppose that we wish to find the critical or stationary points of a scalar "cost" function $f(x_1, x_2, \ldots, x_n)$ subject to a single scalar "equality" constraint or side condition $g(x_1, x_2, \ldots, x_n) = 0$. We form an "augmented" cost function or Lagrangian:

$$\tilde{f}(x_1, x_2, \ldots, x_n, \lambda) \triangleq f(x_1, x_2, \ldots, x_n) - \lambda g(x_1, x_2, \ldots, x_n)$$

$$(A3.7)$$

where the real scalar λ is known as a *Lagrange multiplier*. Then we set all the partial derivatives of the Lagrangian function equal to zero, exactly as we do for an unconstrained problem. We get

$$\frac{\partial \tilde{f}}{\partial x_i} = \frac{\partial f}{\partial x_i} - \lambda \frac{\partial g}{\partial x_i} = 0 \qquad i = 1, 2, \ldots, n \qquad (A3.8)$$

Equation (A3.8) is the same as (A3.4), so it leads to the same geometric interpretation in terms of tangents. In addition to (A3.8), we still need to satisfy the side condition. Thus we have $n + 1$ equations to solve for the $n + 1$ unknowns $x_1, x_2, \ldots, x_n, \lambda$. A nice refinement is to state the satisfaction of the side condition as follows: Satisfy

$$\frac{\partial \tilde{f}}{\partial \lambda} = -g(x_1, x_2, \ldots, x_n) = 0 \qquad (A3.9)$$

Thus a necessary condition for optimality is that all the partial derivatives of $\tilde{f}$ with respect to the x_i and with respect to λ must be zero.

APPENDIX 4: ROOT-LOCUS METHOD

The method is useful in characterizing the transient behavior of a closed-loop linear time-invariant system in terms of the transfer functions of the open-loop subsystems. The method is particularly useful if the effect of loop gain as a parameter is of interest. We assume that the reader is familiar with the basics of the Laplace transform, transfer functions, poles and zeros, block diagrams, and the notion of feedback.

To motivate our discussion of the root-locus method, consider the negative-feedback system shown in Fig. A4.1. Suppose that we want to know the poles of the closed-loop transfer function as a function of the loop gain K. By the standard calculation the closed-loop transfer function

$$G_c(s) \triangleq \frac{\hat{y}(s)}{\hat{y}_d(s)} = \frac{K/(s(s+2))}{1 + K/(s(s+2))} = \frac{K}{s^2 + 2s + K} \qquad (A4.1)$$

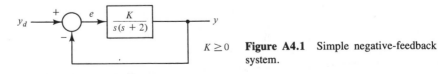

$K \geq 0$ **Figure A4.1**   Simple negative-feedback system.

We can easily find the poles of $G_c(s)$ by factoring the denominator polynomial; the pole locations will depend on K. It is instructive to plot the loci of pole locations as K varies from zero to infinity, as shown in Fig. A4.2.

This plot of closed-loop poles as K goes from zero to infinity is called a (negative-feedback) *root-locus plot*. The plot consists of branches along which the poles shift as continuous functions of K. The arrows indicate the direction of motion as K increases. The two loci originate on the open-loop poles (at 0 and -2), and as K increases the poles move along the negative real axis until they meet. Further increases in K cause the poles to become complex (in complex-conjugate pairs) with an increasing imaginary component. We can infer from the diagram that the transient behavior (e.g., step response) becomes ever more oscillatory as K increases but never reaches the point of instability.

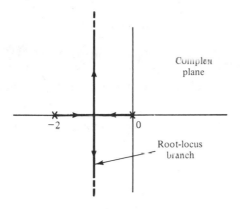

Figure A4.2   Root locus.

We next turn to the more general feedback case shown in Fig. A4.3. In terms of the open-loop transfer functions, we find the closed-loop transfer function

$$G_c(s) = \frac{\hat{y}(s)}{\hat{y}_d(s)} = \frac{KG(s)}{1 + KG(s)H(s)} \qquad (A4.2)$$

We are still interested in the root-locus plot as K varies from zero to infinity but wish to avoid the task of factoring the denominator polynomial for each K. The root-locus method offers an alternative. The basic idea is developed by making the following observations.

1. The poles of $G_c(s)$ are the zeros of $1 + KG(s)H(s)$. [For simplicity, we assume that no zero of $H(s)$ cancels a pole of $G(s)$.]
2. In this case the poles of $G_c(s)$ are the values of s such that

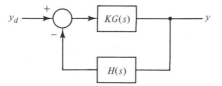

Figure A4.3   Negative-feedback configuration.

$$G(s)H(s) = -\frac{1}{K}$$

3. This condition, in turn, implies two conditions:

$$|G(s)H(s)| = \frac{1}{K} \tag{A4.3}$$

$$\underline{/G(s)H(s)} = 180° \text{ (mod } 360°) \tag{A4.4}$$

4. Suppose now that $G(s)H(s)$ may be expressed in the following pole-zero form and consider any specific value of s, such as shown in Fig. A4.4. We get

$$G(s)H(s) = \frac{(s - z_1)(s - z_2) \cdots (s - z_m)}{(s - p_1)(s - p_2) \cdots (s - p_n)} \tag{A4.5}$$

$$= \frac{l_1 e^{j\theta_1} l_2 e^{j\theta_2} \cdots l_m e^{j\theta_m}}{d_1 e^{j\phi_1} d_2 e^{j\phi_2} \cdots d_n e^{j\phi_m}} \tag{A4.6}$$

where $l_i e^{j\theta_i}$ is the polar representation of $s - z_i$ and $d_i e^{j\phi_i}$ is the polar representation of $s - p_i$. The connection between the two representations in (A4.5) and (A4.6) is shown in Fig. A4.4.

In terms of the polar representation, the magnitude condition (A4.3) may be replaced by

$$\frac{l_1 l_2 \cdots l_m}{d_1 d_2 \cdots d_n} = \frac{1}{K} \tag{A4.7}$$

while the angle condition (A4.4) translates into

$$\theta_1 + \theta_2 + \cdots + \theta_m - (\phi_1 + \phi_2 + \cdots + \phi_n) = 180° \text{ mod } (360°) \tag{A4.8}$$

For our purposes the angle condition is the more important one. We can check whether a particular test point s lies on the root locus by checking whether (A4.8) is satisfied [i.e., the sum of the angles from the open-loop zeros to the test point s, minus

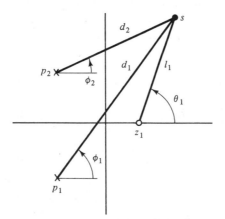

Figure A4.4   Polar form.

the sum of the angles from the open-loop poles to the point s is $180°$ mod $(360°)$]. The reader may verify that this calculation applied to the system in Fig. A4.1 yields the root locus in Fig. A4.2. If a point s lies on the root locus, the value of K that puts it there may be found from the magnitude condition (A4.7).

There is no need to resort to trial and error in seeking the values of s that satisfy (A4.8). Certain rules may be derived on the basis of (A4.7) and (A4.8) which help locate the curves at least approximately. Some of the most important rules are given below for the negative-feedback configuration of Fig. A4.3, with $G(s)H(s)$ in the form (A4.5) and $K \geq 0$. We also assume more poles than zeros (i.e., $n > m$).

Rules for plotting negative-feedback root locus

1. $K \rightarrow 0$. Each open-loop pole is the origin of a root-locus branch.
2. $K \rightarrow \infty$. Each open-loop zero is a terminating point for a particular root-locus branch. The remaining branches tend to infinity.
3. A point on the real axis lies on a branch of the (negative-feedback) root locus if and only if it lies to the left of an odd number of real-axis poles and zeros.
4. Suppose that $G(s)H(s)$ has n poles, $p_1, p_2, \ldots, p_n$, and m zeros, $z_1, z_2 \ldots, z_m$. Then there are n root-locus branches, of which m tend to the m zeros and $n - m$ tend to infinity along certain asymptotes. Concerning these asymptotes:
   a. They originate at
   $$s_0 = \frac{\sum_{i=1}^{n} p_i - \sum_{i=1}^{m} z_i}{n - m} \qquad \text{(center-of-gravity rule)}$$
   b. One asymptote is at an angle of $180°/(n - m)$.
   c. The angle between any two neighboring asymptotes is $360°/(n - m)$.
5. The angle of departure of a root-locus branch from a pole (or arrival at a zero) is found by applying the angle condition $\theta_1 + \theta_2 + \cdots + \theta_m - (\phi_1 + \phi_2 + \cdots + \phi_n) = 180°$ mod $(360°)$ to a point s arbitrarily close to the appropriate pole (or zero). All the angles but one (the angle sought) are known. Thus that angle may easily be found.

These five rules are sufficient for our purposes.

Example A4.1

Find the root locus for the system shown in Fig. EA4.1(a).

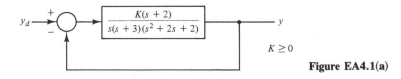

$$\frac{K(s + 2)}{s(s + 3)(s^2 + 2s + 2)}$$

$K \geq 0$

Figure EA4.1(a)

Solution   We note that we have a negative-feedback configuration with $K \geq 0$, with $G(s)H(s)$ in the form (A4.5), and with three more poles than zeros. We can then use the five rules provided.

The first step is to locate the poles and zeros in the complex plane [Fig. EA4.1(b)]. There will be four root-locus branches. One (on the real axis) tends to the zero. The remaining three locus tend to infinity along asymptotes originating at

$$s_0 = \frac{(-1 - 1 - 3) - (-2)}{3} = -1$$

and at angles of $\pm 60°$ and $180°$. In finding the angle of departure from the pole at $-1 + j1$, it helps to draw a sketch showing the angles to a test point s near the pole [Fig. EA4.1(c)]. The point s should really be very close to the pole at $-1 + j1$ but is shown more distant to make the drawing clear. The numerical values of the angles, however, are evaluated for s arbitrarily close to $-1 + j1$. Thus, applying rule 5, we have

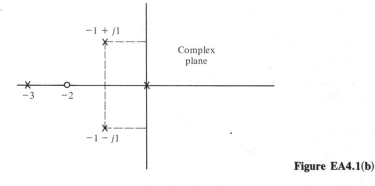

Figure EA4.1(b)

$$45° - (135° + 26.6° + 90° + \phi_4) = 180° \bmod (360°)$$

Solving for ϕ_4, we get

$$\phi_4 = -386.6° \bmod (360°)$$

$$= -26.6°$$

We now can draw the four root-locus branches shown in Fig. EA4.1(d). In this case we see that as K increases, at a certain value the system goes unstable.

So far we have been dealing with the negative-feedback case. We will have occasion to consider positive feedback as in Fig. A4.5. In this case the calculation of $G_c(s)$ yields

$$G_c(s) = \frac{\hat{y}(s)}{\hat{y}_d(s)} = \frac{KG(s)}{1 - KG(s)H(s)} \tag{A4.9}$$

Aside from this change in configuration, we will make the same assumptions as for the negative-feedback configuration [i.e., $G(s)H(s)$ in the form (A4.5) with $n > m$, and also $K \geq 0$]. In this case (A4.8) changes to

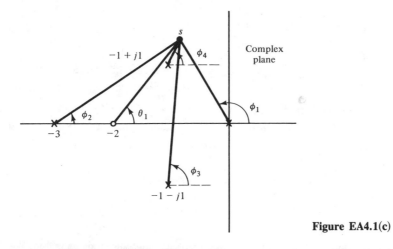

Figure EA4.1(c)

$$\theta_1 + \theta_2 + \cdots + \theta_m - (\phi_1 + \phi_2 + \cdots + \psi_n) = 0^u \bmod (360^\circ)$$

$$(A4.10)$$

but (A4.7) is unchanged. Compared with the negative-feedback rules, we have the following changes:

Rules for plotting positive-feedback root locus

1. Same as for negative feedback.
2. Same as for negative feedback.

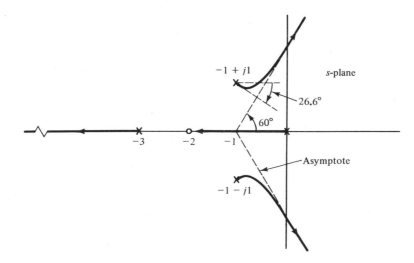

Figure EA4.1(d)

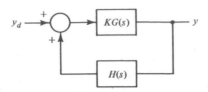

Figure A4.5 Positive-feedback
configuration.

3. A point on the real axis lies on a branch of the (positive-feedback) root locus if
   and only if it lies to the *right* of an odd number of real-axis poles and zeros.

4. (a) and (c) are the same. Replace (b) by: One asymptote is at an angle of
   $180°/(n - m) + 180°$.

5. Same   except   the   angle   condition   is $(\theta_1 + \theta_2 + \cdots + \theta_m) -$
   $(\phi_1 + \phi_2 + \cdots + \phi_n) = 0° \bmod (360°)$.

Example A4.2

Repeat Example A4.1 but with a positive-feedback configuration, as in Fig. A4.5.

Solution   Applying the rules for positive feedback, we get the root locus shown in Fig.
EA4.2.

Note 1: As K increases from zero, the system goes unstable immediately.

Note 2: Consider Example A4.1 with negative K instead of positive K. The poles of
$G_c(s)$, which are the zeros of $1 + KG(s)H(s) = 1 - |K|G(s)H(s)$, are therefore the
same as in Example A4.2. In other words, the negative-feedback configuration with
negative K gives the same root locus as the standard positive-feedback configuration with
positive K. Similarly, the positive feedback configuration with negative K gives the same
root locus as the standard negative-feedback configuration with positive K.

Note 3: Along these lines we may view Fig. EA4.2 as the smooth continuation of the

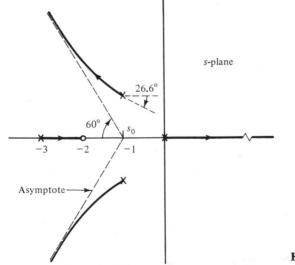

Figure EA4.2

graph in Fig. EA4.1(d) for $-\infty < K < \infty$. Only the arrows on the branches need to be reversed.

We conclude by considering the following question: How do we know if we have negative or positive feedback? If the system is in the negative-feedback configuration (like Fig. A4.3), with $G(s)H(s)$ in the pole-zero form (A4.5) and $K \geq 0$, then we have a negative-feedback system. Change only the negative-feedback configuration to a positive-feedback configuration (like Fig. A4.5); then we have positive feedback.

All the cases may be reduced to one or the other of the above. In some cases it may be helpful to make the determination after carrying out the following steps.

1. If not already single loop, reduce the feedback system to a single-loop system.
2. Ignoring the summing points, write the open-loop gain $KG(s)H(s)$ with $G(s)H(s)$ in pole-zero form (A4.5). Determine if K is positive or negative.
3. Let μ be the number of sign reversals in the loop summing points. If $(-1)^{\mu}K$ is negative, we have negative feedback. If $(-1)^{\mu}K$ is positive, we have positive feedback.

APPENDIX 5: NEGATIVE- AND ZERO-SEQUENCE IMPEDANCES OF SYNCHRONOUS MACHINES

As discussed in Section 13.10, in deriving the negative- and zero-sequence impedances, it is useful to think of tests in which negative- or zero-sequence sources are applied to the generator terminals and the steady-state response is determined. We will use the Park equations in this determination.

Negative-Sequence Test

In Fig. A5.1 we show the generator prepared for the test. The field voltage $v_F = 0$. We use the generator convention on the reference direction for stator currents. The reader may wish to review the description and notation in Section 10.2 and the transformation technique outlined in Section 10.8. We make the following assumptions regarding rotor angle θ and the currents.

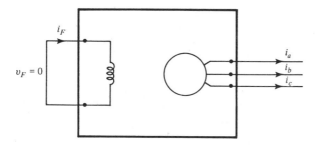

Figure A5.1 Synchronous machine under test.

1. $\theta = \omega_0 t + (\pi/2) + \delta$, δ = constant
2. i_a, i_b, and i_c are a set of negative-sequence sinusoids,

$$i_a = \sqrt{2}|I| \cos (\omega_0 t + \phi)$$

$$i_b = \sqrt{2}|I| \cos \left(\omega_0 t + \phi + \frac{2\pi}{3}\right)$$

$$i_c = \sqrt{2}|I| \cos \left(\omega_0 t + \phi - \frac{2\pi}{3}\right)$$

3. Neglect the damper currents i_D and i_Q.

The reader who did Exercise 3 in Section 8.4 knows that a negative-sequence set of sinusoidal currents generates an armature reaction flux which rotates clockwise with angular velocity ω_0. Thus since the rotor is turning counterclockwise with angular velocity ω_0, the flux linkages λ_D and λ_Q are changing rapidly (at a $2\omega_0$ rate), which implies large damper currents i_D and i_Q. Thus assumption 3 is not a good one but is made to facilitate the analysis. Later we will discuss the effect of including the neglected terms.

Applying the Park transformation (10.16), we get

$$i_0 = 0$$

$$i_d = \sqrt{3} |I| \cos \left(2\omega_0 t + \frac{\pi}{2} + \delta + \phi\right) \qquad (A5.1)$$

$$i_q = \sqrt{3} |I| \sin \left(2\omega_0 t + \frac{\pi}{2} + \delta + \phi\right)$$

To simplify the notation, let $\gamma \triangleq \pi/2 + \delta + \phi$. Using (10.31), (10.32), and (A5.1), we get

$$v_d = -ri_d - \omega_0 L_q i_q - L_d \frac{di_d}{dt} - kM_F \frac{di_F}{dt}$$

$$= -ri_d - \omega_0 L_q \sqrt{3} |I| \sin (2\omega_0 t + \gamma)$$

$$+ L_d \sqrt{3} |I| 2\omega_0 \sin (2\omega_0 t + \gamma) - kM_F \frac{di_F}{dt} \qquad (A5.2)$$

$$v_F = 0 = r_F i_F + L_F \frac{di_F}{dt} - kM_F \sqrt{3} |I| 2\omega_0 \sin (2\omega_0 t + \gamma) \qquad (A5.3)$$

$$v_q = -ri_q + \omega_0 (L_d \sqrt{3} |I| \cos (2\omega_0 t + \gamma) + kM_F i_F)$$

$$-L_q \sqrt{3} |I| 2\omega_0 \cos (2\omega_0 t + \gamma) \qquad (A5.4)$$

We may now solve (A5.3) for the steady-state i_F using phasors. We note that (A5.3) describes a series RL circuit with a voltage source $v(t) = \sqrt{2} K \sin (2\omega_0 t + \gamma)$, where $K \triangleq kM_F \sqrt{3} \, |I| \, \sqrt{2} \, \omega_0$. The circuit is shown in Fig. A5.2. The (effective) phasor representation of $v(t)$ is $V = -jKe^{j\gamma}$. Thus (noting that the frequency is $2\omega_0$) we get for the phasor representation of i_F

$$I_F = \frac{-jKe^{j\gamma}}{r_F + j2\omega_0 L_F} = -\frac{jkM_F \sqrt{3} \, |I| \, \sqrt{2} \, \omega_0 e^{j\gamma}}{r_F + j2\omega_0 L_F} \tag{A5.5}$$

Since $r_F \ll 2\omega_0 L_F$, we can neglect it. Then

$$I_F \approx -\sqrt{\frac{3}{2}} \frac{kM_F}{L_F} |I| \, e^{j\gamma}$$

and converting back to instantaneous quantities,

$$i_F = -\sqrt{3} \frac{kM_F}{L_F} |I| \cos (2\omega_0 t + \gamma) \tag{A5.6}$$

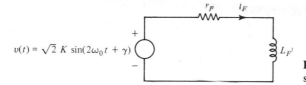

$$v(t) = \sqrt{2} \, K \sin(2\omega_0 t + \gamma)$$

Figure A5.2 Calculation of I_F in steady state.

Using (A5.6) in (A5.2) and neglecting r, we get

$$\begin{aligned}
v_d &= -\omega_0 L_q \sqrt{3} \, |I| \sin (2\omega_0 t + \gamma) \\
&\quad + 2\omega_0 L_d \sqrt{3} \, |I| \sin (2\omega_0 t + \gamma) \\
&\quad - \sqrt{3} \frac{k^2 M_F^2}{L_F} 2\omega_0 L_d \, |I| \sin (2\omega_0 t + \gamma) \\
&= -\left[\omega_0 L_q - 2\omega_0 \left(L_d - \frac{(kM_F)^2}{L_F} \right) \right] \sqrt{3} \, |I| \sin (2\omega_0 t + \gamma) \\
&= -[X_q - 2X_d'] \sqrt{3} \, |I| \sin (2\omega_0 t + \gamma) \tag{A5.7}
\end{aligned}$$

In (A5.7) we have used the definition of X_d' first seen in Example 10.4.

Next we find v_q by using (A5.6) in (A5.4) and neglecting r. We get

$$\begin{aligned}
v_q &= \omega_0 L_d \sqrt{3} \, |I| \cos (2\omega_0 t + \gamma) - \omega_0 \frac{(kM_F)^2}{L_F} \sqrt{3} \, |I| \cos (2\omega_0 t + \gamma) \\
&\quad - 2\omega_0 L_q \sqrt{3} \, |I| \cos (2\omega_0 t + \gamma) \\
&= \left[\omega_0 \left(L_d - \frac{(kM_F)^2}{L_F} \right) - 2\omega_0 L_q \right] \sqrt{3} \, |I| \cos (2\omega_0 t + \gamma) \\
&= [X_d' - 2X_q] \sqrt{3} \, |I| \cos (2\omega_0 t + \gamma) \tag{A5.8}
\end{aligned}$$

The fact that the Park variables have a frequency $2\omega_0$ is understandable in terms of the transformation to rotor-based coordinates.

We next go back to *abc* variables using the inverse Park transformation (10.17). To simplify the notation, let $A_d \triangleq -(X_q - 2X_d')\sqrt{3}\,|I|$, $A_q \triangleq (X_d' - 2X_q)\sqrt{3}\,|I|$; then $v_d = A_d \sin(2\omega_0 t + \gamma)$ and $v_q = A_q \cos(2\omega_0 t + \gamma)$. Thus

$$v_a = \sqrt{\frac{2}{3}}\,[\cos\theta \cdot A_d \sin(2\omega_0 t + \gamma) + \sin\theta \cdot A_q \cos(2\omega_0 t + \gamma)]$$

$$= \sqrt{\frac{2}{3}}\left[\frac{A_d + A_q}{2}\sin(2\omega_0 t + \gamma + \theta) + \frac{A_d - A_q}{2}\sin(2\omega_0 t + \gamma - \theta)\right] \quad (A5.9)$$

Substituting for A_d, A_q, and γ, we get

$$v_a = 3\sqrt{2}\,|I|\frac{X_q - X_d'}{2}\sin(3\omega_0 t + 2\delta + \phi)$$

$$+ \sqrt{2}\,|I|\frac{X_q + X_d'}{2}\sin(\omega_0 t + \phi) \quad (A5.10)$$

Notice that we get a $3\omega_0$ term. Neglecting this term, we find a relationship between the phasor voltage and current. Thus, using associated reference directions for impedances and noting that the current *into* the "impedance" is $-I_a$, we have

$$Z^- \triangleq \frac{V_a}{-I_a} = \frac{[(X_q + X_d')/2]|I|\,(-je^{j\phi})}{-|I|e^{j\phi}} = j\frac{X_q + X_d'}{2} \quad (A5.11)$$

The negative-sequence impedance Z^- has an interesting physical interpretation. The mmf due to the negative-sequence currents is rotating clockwise while the rotor is rotating counterclockwise. Thus the magnetic circuit is rapidly changing from one extreme in which the mmf and the direct axis of the rotor are aligned to the other extreme where they are orthogonal. If we consider the case where they are aligned, then in the steady state we might expect the phasor voltage and current to be related by the impedance jX_d'. As usual we get X_d' rather than X_d if we have the constraint $\lambda_F = $ constant. We note that $\lambda_F = $ constant is consistent with our assumption of $v_F = 0$ and $r_F = 0$.

On the other hand, when the mmf is aligned with the q axis, we expect the phasor voltage and current to be related by the impedance jX_q. Thus the actual result predicted by the analysis (that the impedance is the average of the two extremes) is a reasonable result.

We have gone into some detail on this because we now want to make the following extension. When we include the damper D, Q circuits in the analysis we get a result similar to (A5.11) but with different reactances. Instead of X_d' we get X_d'', and instead of X_q we get X_q''. X_d'' and X_q'' are subtransient reactances which depend most heavily on the rotor damper circuits. But these are quite symmetrical from d to q axis (i.e., $X_d'' \approx X_q''$). In this case in (A5.10) the $3\omega_0$ term is really negligible, and instead of (A5.11) we get

$$Z^- = j\frac{X_q'' + X_d''}{2} \approx jX_d'' \tag{A5.12}$$

Consider next v_b. Altering (A5.9) appropriately and paralleling the steps that led to (A5.10), we get

$$\begin{aligned}
v_b &= \sqrt{\frac{2}{3}}\Bigg[\cos\left(\theta - \frac{2\pi}{3}\right)\cdot A_d \sin(2\omega_0 t + \gamma) \\
&\quad + \sin\left(\theta - \frac{2\pi}{3}\right)\cdot A_q \cos(2\omega_0 t + \gamma) \\
&= 3\sqrt{2}\,|I|\left(\frac{X_q - X_d'}{2}\right)\sin\left(3\omega_0 t + 2\delta + \phi - \frac{2\pi}{3}\right) \\
&\quad + \sqrt{2}\,|I|\left(\frac{X_q + X_d'}{2}\right)\sin\left(\omega_0 t + \phi + \frac{2\pi}{3}\right)
\end{aligned} \tag{A5.13}$$

Neglecting the $3\omega_0$ term as before, we get from the ω_0 term

$$V_b = V_a e^{j2\pi/3}$$

and similarly, after algebra,

$$V_c = V_a e^{-j2\pi/3}$$

Thus the voltages form a negative-sequence set. Note that the $3\omega_0$ voltages do not form a negative-sequence set. This may be attributed to the lack of symmetry between the direct and quadrature axes magnetic circuits in the generator. Fortunately, at the subtransient level the symmetry is restored.

Zero-Sequence Test

If the generator neutral is not grounded, the zero-sequence equivalent circuit of the generator is an open circuit and need not be considered further. (We exclude from consideration the case of a short circuit at some intermediate point in the stator winding.)

Consider then the case of a generator whose neutral is grounded through an impedance Z_g. Setting all independent sources except zero-sequence sources equal to zero, we consider the following experiment. We apply zero-sequence sinusoidal currents to the generator as shown in Fig. A5.3 and measure the steady-state zero-sequence voltages. In more detail, assume that:

1. $\theta = \omega_0 t + (\pi/2) + \delta$, $\delta = $ constant
2. i_a, i_b, and i_c are zero-sequence sinusoids:

$$i_a = i_b = i_c = \sqrt{2}\,|I|\cos(\omega_0 t + \phi)$$

Applying the Park transformation, we get

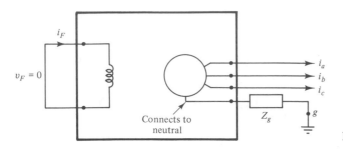

Figure A5.3   Zero-sequence test.

$$i_d = i_q \equiv 0 \tag{A5.14}$$

$$i_0 = \sqrt{\frac{2}{3}} \frac{1}{\sqrt{2}} \, 3\sqrt{2} \, |I| \cos{(\omega_0 t + \phi)} = \sqrt{6} \, |I| \cos{(\omega_0 t + \phi)} \tag{A5.15}$$

Looking at (10.31) and (10.32), we see there is no coupling between the zero sequence and direct axis or quadrature axis circuits. Thus, because $v_F = 0$, and using (A5.14), we find that in the steady state

$$v_d = v_q \equiv 0 \tag{A5.16}$$

We next find v_0 using (10.31a) and (A5.15):

$$v_0 = -ri_0 - L_0 \frac{di_0}{dt} = -\sqrt{6}r \, |I| \cos{(\omega_0 t + \phi)} + \sqrt{6}\,\omega_0 L_0 \, |I| \sin{(\omega_0 t + \phi)}$$

$$\tag{A5.17}$$

We now go back to *abc* variables. Using the inverse Park transformation (10.17) and noting (A5.16), we get

$$v_a = v_b = v_c = -\sqrt{2}r \, |I| \cos{(\omega_0 t + \phi)} + \sqrt{2}\,\omega_0 L_0 \, |I| \sin{(\omega_0 t + \phi)} \tag{A5.18}$$

Switching to phasors, we find the zero-sequence impedance Z^0:

$$Z^0 \triangleq \frac{V_a}{-I_a} = r + j\omega_0 L_0 \approx j\omega_0 L_0 = jX^0 \tag{A5.19}$$

What we have found is the zero-sequence impedance from phase to neutral. The notation v_a is a shorthand notation for $v'_{aa} = v'_{an}$. In the zero-sequence network we need a different impedance, one from phase to ground. As shown in Section 13.4, this leads to the circuit model in Fig. A5.4.

The zero-sequence reactance usually turns out to be quite small. This can be most easily understood physically by considering the air-gap fluxes set up by the three identical zero-sequence currents; we considered a similar problem in Section 8.4 except that the currents were positive sequence. For the simplest case, assume a round rotor and an air-gap mmf for each phase winding which is sinusoidally distributed in space with maximum magnitude at the center of each winding. Since the currents are in phase and the windings are displaced by 120° angles, the air-gap mmfs sum to zero. In this case the only voltage produced is in the leakage reactance, which is very small.

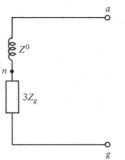

Figure A5.4 Zero-sequence model for generator.

Thus $\omega_0 L_0$ must be very small. In practice, without the idealization of total cancellation, $\omega_0 L_0$ is larger.

APPENDIX 6: INVERSION FORMULA

Suppose that we have an $n \times n$ symmetrical matrix $\mathbf{M}$ whose inverse, $\mathbf{M}^{-1}$, is known and we wish to find the inverse of $\mathbf{M} + \mu\mathbf{aa}^T$, where μ is a scalar and $\mathbf{a}$ is an n-vector. We note that by the rules of matrix multiplication

$$\mathbf{aa}^T = \begin{bmatrix} a_1 \\ a_2 \\ \vdots \\ a_n \end{bmatrix} [a_1 a_2 \cdots a_n] = \begin{bmatrix} a_1^2 & a_1 a_2 & \cdots & a_1 a_n \\ a_2 a_1 & a_2^2 & \cdots & a_2 a_n \\ \vdots & \vdots & & \vdots \\ a_n a_1 & a_n a_2 & \cdots & a_n^2 \end{bmatrix} \tag{A6.1}$$

We use the following *inversion formula:*

$$[\mathbf{M} + \mu\mathbf{aa}^T]^{-1} = \mathbf{M}^{-1} - \gamma\mathbf{bb}^T \tag{A6.2}$$

where

$$\mathbf{b} = \mathbf{M}^{-1}\mathbf{a} \tag{A6.3a}$$

$$\gamma = (\mu^{-1} + \mathbf{a}^T\mathbf{b})^{-1} \tag{A6.3b}$$

Since $\mathbf{M}^{-1}$ is assumed known, the vector $\mathbf{b}$ and the scalar γ are easily obtained. The matrix $\mathbf{bb}^T$ may easily be evaluated by noting the form of (A6.1). We note that the result in (A6.2) is a special case of a more general result commonly known as the *Householder formula.*

Proof of inversion formula.   We simply multiply the modified matrix and its inverse together and show that we get the identity matrix, $\mathbf{1}$.

$$[\mathbf{M} + \mu\mathbf{aa}^T][\mathbf{M}^{-1} - \gamma\mathbf{bb}^T] = \mathbf{MM}^{-1} - \gamma\mathbf{Mbb}^T + \mu\mathbf{aa}^T\mathbf{M}^{-1} - \mu\gamma\mathbf{aa}^T\mathbf{bb}^T$$

$$= \mathbf{1} + \text{remainder}$$

We now show the remainder $= \mathbf{0}$. To start, divide the remainder by γ. We get

$$\frac{\text{remainder}}{\gamma} = -\mathbf{Mbb}^T + \mu\mathbf{a}\frac{1}{\gamma}\mathbf{a}^T\mathbf{M}^{-1} - \mu\mathbf{aa}^T\mathbf{bb}^T$$

$$= -\mathbf{ab}^T + \mu\mathbf{a}(\mu^{-1} + \mathbf{a}^T\mathbf{b})\mathbf{b}^T - \mu\mathbf{aa}^T\mathbf{bb}^T$$

$$= \mathbf{0}$$

In line 2 above we have used $\mathbf{Mb} = \mathbf{a}$ [from (A6.3a)], $\gamma^{-1} = \mu^{-1} + \mathbf{a}^T\mathbf{b}$ [from (A6.3b)], and (A6.3a).

Example A6.1

Suppose that we are given

$$\mathbf{M} = \begin{bmatrix} 2 & 1 & 0 \\ 1 & 2 & 1 \\ 0 & 1 & 1 \end{bmatrix} \qquad \mathbf{M}^{-1} = \begin{bmatrix} 1 & -1 & 1 \\ -1 & 2 & -2 \\ 1 & -2 & 3 \end{bmatrix}$$

and wish to evaluate the inverse of $\mathbf{M}$ with m_{22} increased from 2 to 4. We then have

$$\tilde{\mathbf{M}} = \begin{bmatrix} 2 & 1 & 0 \\ 1 & 4 & 1 \\ 0 & 1 & 1 \end{bmatrix} = \mathbf{M} + 2\begin{bmatrix} 0 \\ 1 \\ 0 \end{bmatrix}[0 \quad 1 \quad 0]$$

in which we identify $\mu = 2$ and $\mathbf{a}^T = [0 \quad 1 \quad 0]$.

Solution   First find $\mathbf{b}$ and then γ.

$$\mathbf{b} = \mathbf{M}^{-1}\mathbf{a} = \begin{bmatrix} 1 & -1 & 1 \\ -1 & 2 & -2 \\ 1 & -2 & 3 \end{bmatrix}\begin{bmatrix} 0 \\ 1 \\ 0 \end{bmatrix} = \begin{bmatrix} -1 \\ 2 \\ -2 \end{bmatrix}$$

$$\gamma = (\mu^{-1} + \mathbf{a}^T\mathbf{b})^{-1} = (\tfrac{1}{2} + 2)^{-1} = 0.4$$

Substituting in the inversion formula, (A6.2), we get

$$\tilde{\mathbf{M}}^{-1} = \begin{bmatrix} 1 & -1 & 1 \\ -1 & 2 & -2 \\ 1 & -2 & 3 \end{bmatrix} - 0.4\begin{bmatrix} -1 \\ 2 \\ -2 \end{bmatrix}[-1 \quad 2 \quad -2]$$

$$= \begin{bmatrix} 0.6 & -0.2 & 0.2 \\ -0.2 & 0.4 & -0.4 \\ 0.2 & -0.4 & 1.4 \end{bmatrix}$$

The correctness of the result may be checked by direct calculation.

Example A6.2

We wish to use (13.54) in (13.50) to obtain (13.51) by matrix methods. In (13.50) we need to evaluate $(\mathbf{1} + \mathbf{Z}_{sqq}\mathbf{Y}_s^f)^{-1}$. As a preliminary we note that

$$(\mathbf{1} + \mathbf{Z}_{sqq}\mathbf{Y}_s^f)^{-1} = [\mathbf{Z}_{sqq}(\mathbf{Y}_{sqq} + \mathbf{Y}_s^f)]^{-1}$$

$$= (\mathbf{Y}_{sqq} + \mathbf{Y}_s^f)^{-1}\mathbf{Y}_{sqq} \tag{A6.4}$$

where $\mathbf{Y}_{sqq} = \mathbf{Z}_{sqq}^{-1}$, and we have used the matrix property $(\mathbf{AB})^{-1} = \mathbf{B}^{-1}\mathbf{A}^{-1}$. Since from (13.54)

$$\mathbf{Y}_s^f = \frac{Y^f}{3}\begin{bmatrix} 1 & 1 & 1 \\ 1 & 1 & 1 \\ 1 & 1 & 1 \end{bmatrix} = \frac{Y^f}{3}\begin{bmatrix} 1 \\ 1 \\ 1 \end{bmatrix}\begin{bmatrix} 1 & 1 & 1 \end{bmatrix}$$

$\mathbf{Y}_{sqq} + \mathbf{Y}_s^f$ is in the right form to use the inversion formula with $\mathbf{M} = \mathbf{Y}_{sqq}$, $\mu = Y^f/3$, and $\mathbf{a}^T = [1 \ \ 1 \ \ 1]$. We get

$$(\mathbf{Y}_{sqq} + \mathbf{Y}_s^f)^{-1} = (\mathbf{Z}_{sqq} - \gamma\mathbf{b}\mathbf{b}^T) \tag{A6.5}$$

where

$$\mathbf{b} = \mathbf{Z}_{sqq}\mathbf{a} = \begin{bmatrix} z_{qq}^0 \\ z_{qq}^+ \\ z_{qq}^- \end{bmatrix}$$

$$\gamma = (3Z^f + z_{qq}^0 + z_{qq}^+ + z_{qq}^-)^{-1}$$

and where $Z^f = 1/Y^f$. Using (A6.4) and (A6.5) in (13.50), we get

$$\mathbf{I}_a^f = \mathbf{Y}_s^f(\mathbf{Z}_{sqq} - \gamma\mathbf{b}\mathbf{b}^T)\mathbf{Y}_{sqq}\mathbf{V}_{sq}^{pf}$$

$$= \frac{Y^f}{3}\mathbf{a}\mathbf{a}^T(\mathbf{1} - \gamma\mathbf{b}\mathbf{b}^T\mathbf{Y}_{sqq})\mathbf{V}_{sq}^{pf}$$

$$= \frac{Y^f}{3}\mathbf{a}\mathbf{a}^T(\mathbf{1} - \gamma\mathbf{b}\mathbf{a}^T)\mathbf{V}_{sq}^{pf}$$

$$= \frac{Y^f}{3}\mathbf{a}\left(\mathbf{a}^t - \frac{z_{qq}^0 + z_{qq}^+ + z_{qq}^-}{3Z^f + z_{qq}^0 + z_{qq}^+ + z_{qq}^-}\mathbf{a}^T\right)\mathbf{V}_{sq}^{pf}$$

$$= \frac{V_q^{pf}}{3Z^f + z_{qq}^0 + z_{qq}^+ + z_{qq}^-}\begin{bmatrix} 1 \\ 1 \\ 1 \end{bmatrix}$$

Thus we get

$$I_{af}^0 = I_{af}^+ = I_{af}^- = \frac{V_q^{pf}}{3Z^f + z_{qq}^0 + z_{qq}^+ + z_{qq}^-}$$

which checks the result in (13.51).

APPENDIX 7: MODIFICATION OF IMPEDANCE MATRICES

Suppose that we have $\mathbf{Y}_{bus}$ for a given network. Suppose now that the value of one of the elements in the network is changed and we wish to find $\mathbf{Y}_{bus}^n$, the new admittance matrix. Consider an example. Suppose that in Fig. A7.1, Y_3 changes to $Y_3 + \Delta Y_3$. It is easy to find the new admittance matrix. Using the rules in Section 6.1, we find (by inspection)

$$\mathbf{Y}_{bus}^n = \begin{pmatrix} Y_1 + (Y_3 + \Delta Y_3) + Y_5 & -Y_5 & -(Y_4 + \Delta Y_3) \\ -Y_5 & Y_2 + Y_4 + Y_5 & -Y_4 \\ -(Y_3 + \Delta Y_3) & -Y_4 & (Y_3 + \Delta Y_3) + Y_4 \end{pmatrix}$$

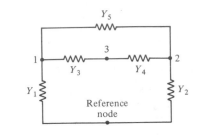

Figure A7.1 Example network.

$$= \mathbf{Y}_{bus} + \Delta Y_3 \begin{pmatrix} 1 & 0 & -1 \\ 0 & 0 & 0 \\ -1 & 0 & 1 \end{pmatrix} \qquad (A7.1)$$

where $\mathbf{Y}_{bus}$ is the original admittance matrix (with $\Delta Y_3 = 0$). Using the rules of matrix multiplication, the reader may check that (A7.1) may also be written as follows:

$$\mathbf{Y}_{bus}^n = \mathbf{Y}_{bus} + \Delta Y_3 \begin{bmatrix} 1 \\ 0 \\ -1 \end{bmatrix} [1 \quad 0 \quad -1] \qquad (A7.2)$$

Changes in other branches may be described in similar terms. We note that in the case of a branch connected to the reference node there is only a single nonzero entry in the vectors in (A7.2).

For a more general network, suppose that the admittance of the kth branch connecting nodes i and j is changed from Y_k to $Y_k + \Delta Y_k$. Then

$$\mathbf{Y}_{bus}^n = \mathbf{Y}_{bus} + \Delta Y_k \mathbf{a}_k \mathbf{a}_k^T \qquad (A7.3)$$

where $\mathbf{a}_k$ is a column vector. There are two cases to consider.

> *Case 1:* Neither node i nor j is the reference node. Then in $\mathbf{a}_k$ we can put a 1 in the ith row and a -1 in the jth row. All other entries are zero.
>
> *Case 2:* The kth branch connects the ith node to the reference node. Then we can put a 1 in the ith row of $\mathbf{a}_k$. All other entries are zero. We note in passing that in the network formulation using reference directions for branches and the reduced incidence matrix, we can take the kth column of the node-branch reduced incidence matrix as $\mathbf{a}_k$.

Now if we consider the corresponding changes in $\mathbf{Z}_{bus}$, some of this simplicity is lost. In general, a change in a branch impedance causes changes in all the elements of $\mathbf{Z}_{bus}$. Nevertheless, noting that (A7.3) is in the required form, it is still quite simple to find the new impedance matrix by application of the inversion formula derived in Appendix 6. We get

$$\mathbf{Z}_{bus}^n = [\mathbf{Y}_{bus}^n]^{-1} = [\mathbf{Y}_{bus} + \Delta Y_k \mathbf{a}_k \mathbf{a}_k^T]^{-1}$$
$$= \mathbf{Z}_{bus} - \gamma_k \mathbf{b}_k \mathbf{b}_k^T \qquad (A7.4)$$

where

$$\mathbf{b}_k = \mathbf{Z}_{bus} \mathbf{a}_k \tag{A7.5}$$

and

$$\gamma_k = \left[\frac{1}{\Delta Y_k} + \mathbf{a}_k^T \mathbf{b}_k \right]^{-1} \tag{A7.6}$$

We can evaluate $\mathbf{b}_k$ and γ_k more explicitly. Suppose that the ith and jth columns of $\mathbf{Z}_{bus}$ are $\mathbf{Z}_i$ and $\mathbf{Z}_j$, respectively, and the iith, jjth, and ijth elements are z_{ii}, z_{jj}, and z_{ij}, respectively. Noting the specification of $\mathbf{a}_k$, in place of (A7.5) and (A7.6), we get:

Case 1:

$$\mathbf{b}_k = \mathbf{Z}_i - \mathbf{Z}_j \tag{A7.7}$$

$$\gamma_k = \left(\frac{1}{\Delta Y_k} + z_{ii} + z_{jj} - 2z_{ij} \right)^{-1} \tag{A7.8}$$

Case 2:

$$\mathbf{b}_k = \mathbf{Z}_i \tag{A7.9}$$

$$\gamma_k = \left(\frac{1}{\Delta Y_k} + z_{ii} \right)^{-1} \tag{A7.10}$$

Equation (A7.4) together with (A7.7) and (A7.8) or (A7.9) and (A7.10) give a prescription for calculating $\mathbf{Z}_{bus}^{"}$ in terms of $\mathbf{Z}_{bus}$ when a branch value is changed.

Example A7.1

Suppose that for a given network we are given $\mathbf{Z}_{bus}$:

$$\mathbf{Z}_{bus} = \frac{j}{3} \begin{bmatrix} 0.200 & 0.100 & 0.100 \\ 0.100 & 0.500 & 0.200 \\ 0.100 & 0.200 & 0.200 \end{bmatrix}$$

Now the admittance of branch 2, located between nodes 1 and 2, is changed from $-j5$ to $-j15$. Find $\mathbf{Z}_{bus}^{"}$.

Solution   We have $\Delta Y_2 = -j15 - (-j5) = -j10$. Since the branch is connected between nodes 1 and 2, to find $\mathbf{b}_2$ we use (A7.7) and get

$$\mathbf{b}_2 = \mathbf{Z}_1 - \mathbf{Z}_2 = \frac{j}{3} \begin{bmatrix} 0.100 \\ -0.400 \\ -0.100 \end{bmatrix}$$

From (A7.8) we get

$$\gamma_2 = \left[\frac{1}{-j10} + \frac{j}{3}(0.2 + 0.5 - 0.2) \right]^{-1} = -j3.750$$

Using (A7.4), we get

$$\mathbf{Z}^n_{bus} = \mathbf{Z}_{bus} + j3.75 \frac{j}{3} \begin{bmatrix} 0.1 \\ -0.4 \\ -0.1 \end{bmatrix} \frac{j}{3} [0.1 \quad -0.4 \quad -0.1]$$

$$= \mathbf{Z}_{bus} - j \frac{1.25}{3} \begin{bmatrix} 0.01 & -0.04 & -0.01 \\ -0.04 & 0.16 & 0.04 \\ -0.01 & 0.04 & 0.01 \end{bmatrix}$$

$$= \frac{j}{3} \begin{bmatrix} 0.1875 & 0.150 & 0.1125 \\ 0.150 & 0.300 & 0.150 \\ 0.1125 & 0.150 & 0.1875 \end{bmatrix}$$

This is the end of the example.

We will now consider an important special case of a change in element value. Suppose that we add a branch k, between nodes i and j, where previously there was no element. Suppose that this branch has admittance Y_b and impedance $Z_b = 1/Y_b$. Then $\Delta Y_k = Y_b$ and $1/\Delta Y_k = Z_b$. In (A7.6), (A7.8), and (A7.10), therefore, we may replace $1/\Delta Y_k$ by Z_b. For example, in (A7.8), we get

$$\gamma_k = (Z_b + z_{ii} + z_{jj} - 2z_{ij})^{-1} \tag{A7.11}$$

The modification considered is an important part of the Z-building technique discussed in Section 13.16. Another important modification used in the Z-building technique consists of adding a node to the given network by attaching only one side of a branch. The other side (the dangling side) defines a new node. We wish now to consider how this modification changes $\mathbf{Z}_{bus}$. Assuming that $\mathbf{Z}_{bus}$ is an $r \times r$ matrix, let the new node be numbered $r + 1$. Assume that the branch has admittance Y_b and impedance $Z_b = 1/Y_b$.

If the new branch is connected to the reference node, we have by the rules for forming $\mathbf{Y}_{bus}$,

$$\mathbf{Y}^n_{bus} = \begin{bmatrix} \mathbf{Y}_{bus} & \mathbf{0} \\ \mathbf{0}^T & Y_b \end{bmatrix}$$

and it immediately follows that

$$\mathbf{Z}^n_{bus} = \begin{bmatrix} \mathbf{Z}_{bus} & \mathbf{0} \\ \mathbf{0}^T & Z_b \end{bmatrix} \tag{A7.12}$$

The correctness of the result can be checked by matrix multiplication.

If, instead, the new branch is connected to the ith node [i.e., we have a connection between the ith node and the $(r + 1)$st node], then

$$\mathbf{Y}^n_{bus} = \begin{bmatrix} \mathbf{Y}_{bus} & \mathbf{0} \\ \mathbf{0}^T & 0 \end{bmatrix} + Y_b \mathbf{a}_k \mathbf{a}^T_k \tag{A7.13}$$

where $\mathbf{a}_k$ has the following nonzero entries: 1 in the ith row and -1 in the $(r + 1)$st row. This looks similar to the case treated earlier, with one exception; the matrix above, involving $\mathbf{Y}_{\text{bus}}$, does not have an inverse, so we cannot use the inversion formula directly. The problem can be bypassed in typical engineering fashion by replacing the zero in the $r + 1, r + 1$ place by an admittance of ϵ (with the intention of letting $\epsilon \to 0$ later). Thus instead of (A7.13), consider

$$\mathbf{Y}_{\text{bus}}^n = \begin{bmatrix} \mathbf{Y}_{\text{bus}} & \mathbf{0} \\ \mathbf{0}^T & \epsilon \end{bmatrix} + Y_b \mathbf{a}_k \mathbf{a}_k^T \tag{A7.14}$$

with inverse now given by the inversion formula.

$$\mathbf{Z}_{\text{bus}}^n = \begin{bmatrix} \mathbf{Z}_{\text{bus}} & \mathbf{0} \\ \mathbf{0}^T & \dfrac{1}{\epsilon} \end{bmatrix} - \gamma_k \mathbf{b}_k \mathbf{b}_k^T \tag{A7.15}$$

where

$$\mathbf{b}_k = \begin{bmatrix} \mathbf{Z}_{\text{bus}} & \mathbf{0} \\ \mathbf{0}^T & \dfrac{1}{\epsilon} \end{bmatrix} \begin{bmatrix} 0 \\ \cdot \\ 1 \\ \cdot \\ 0 \\ \cdot \\ \cdot \\ 0 \\ -1 \end{bmatrix} = \begin{bmatrix} \mathbf{Z}_i \\ \dfrac{-1}{\epsilon} \end{bmatrix}$$

and

$$\gamma_k = (Z_b + \mathbf{a}_k^T \mathbf{b}_k)^{-1}$$
$$= \left(Z_b + z_{ii} + \frac{1}{\epsilon} \right)^{-1}$$

$\mathbf{Z}_i$ is the ith column of $\mathbf{Z}_{\text{bus}}$, and z_{ii} is its iith element. Substituting these results in (A7.15), we get

$$\mathbf{Z}_{\text{bus}}^n = \begin{bmatrix} \mathbf{Z}_{\text{bus}} & \mathbf{0} \\ \mathbf{0}^T & \dfrac{1}{\epsilon} \end{bmatrix} - \frac{1}{Z_b + z_{ii} + 1/\epsilon} \begin{bmatrix} \mathbf{Z}_i \mathbf{Z}_i^T & \dfrac{-\mathbf{Z}_i}{\epsilon} \\ \dfrac{-\mathbf{Z}_i^T}{\epsilon} & \dfrac{1}{\epsilon^2} \end{bmatrix}$$

$$= \begin{bmatrix} \mathbf{Z}_{\text{bus}} & \mathbf{0} \\ \mathbf{0}^T & \dfrac{1}{\epsilon} \end{bmatrix} - \frac{1}{\epsilon(Z_b + z_{ii}) + 1} \begin{bmatrix} \epsilon \mathbf{Z}_i \mathbf{Z}_i^T & -\mathbf{Z}_i \\ -\mathbf{Z}_i^T & \dfrac{1}{\epsilon} \end{bmatrix} \tag{A7.16}$$

Now as we let $\epsilon \to 0$, we get

$$\mathbf{Z}_{\text{bus}}^{n} = \begin{bmatrix} \mathbf{Z}_{\text{bus}} & \mathbf{Z}_{i} \\ \mathbf{Z}_{i}^{T} & Z_{b} + z_{ii} \end{bmatrix} \qquad (A7.17)$$

It is a simple matter to border $\mathbf{Z}_{\text{bus}}$ as shown in (A7.17).

Given a network whose impedance matrix is sought, we may start with a very simple version of the network whose impedance matrix may easily be determined (by inspection) and then, using (A7.12), (A7.17), and (A7.4), determine the new impedance matrices that accompany the adding of nodes and branches until the given network is "built." This technique, called Z building, is discussed in Section 13.16.

Selected Bibliography

ANDERSON, P. M., and FOUAD, A. A., *Power System Control and Stability*, The Iowa State University Press, Ames, Iowa, 1977.

ELGERD, O. I., *Electric Energy Systems Theory: An Introduction*, 2nd ed., McGraw-Hill Book Company, New York, 1982.

GROSS, C. A., *Power System Analysis*, John Wiley & Sons, Inc., New York, 1979.

KIMBARK, E. W., *Power System Stability: Synchronous Machines*, Dover reprint edition of the original John Wiley & Sons' book, Dover Publications, Inc., New York, 1968.

McGraw-Hill Encyclopedia of Energy, 2nd ed., McGraw-Hill Book Company, New York, 1981.

NEUENSWANDER, J. R., *Modern Power Systems*, Intext Educational Publishers, New York, 1971.

STEVENSON, W. D., JR., *Elements of Power System Analysis*, 4th ed., McGraw-Hill Book Company, New York, 1982.

WEEDY, B. M., *Electric Power Systems*, 3rd ed., John Wiley & Sons Ltd., London, 1979.

WESTINGHOUSE ELECTRIC CORPORATION, *Electric Transmission and Distribution Reference Book*, 4th ed., East Pittsburgh, Pa., 1964.

Index